USA IN SPACE

—

USA IN SPACE

Edited by
Frank N. Magill
and
Russell R. Tobias

VOLUME 2

Jo – Spa

311 – 616

Salem Press, Inc.

Pasadena, CA   Englewood Cliffs, NJ

The paper used in these volumes conforms to the American
National Standard for Permanence of Paper for Printed
Library Materials, Z39.48-1984.

Library of Congress Cataloging-in-Publication Data
USA in space/edited by Frank N. Magill and
Russell R. Tobias.
p. cm.
Includes bibliographical references and index
1. Astronautics–United States.   I. Magill, Frank
Northen, 1907-.   II. Tobias, Russell R.
TL789.8.U5U83 1996
387.8'0973--dc20   96-42044
CIP
ISBN 0-98356-924-0 (set)
ISBN 0-89356-926-7 (vol 2)

First Printing

PRINTED IN THE UNITED STATES OF AMERICA

CONTENTS

JOHNSON SPACE CENTER

Date: Beginning November 1, 1961
Type of facility: Space center

Johnson Space Center is the National Aeronautics and Space Administration's lead center for development of manned spacecraft, training of space crews, and direction of manned space missions. It is also responsible for research and development in a number of areas, including life sciences and remote sensing of Earth resources.

PRINCIPAL PERSONAGES

ROBERT R. GILRUTH,
CHRISTOPHER C. KRAFT,
GERALD GRIFFIN, and
AARON COHEN, Directors of Johnson Space Center
PAUL WEITZ, Deputy Director
GEORGE ABBEY, Flight Crew Operations Director
EUGENE F. KRANZ, Mission Operations Director
RONALD BERRY, Mission Support Director
CAROLYN HUNTOON, Space and Life Sciences
    Director
HENRY POHL, Engineering Director
ROB TILLETT, Director, White Sands Tests Facility

Summary of the Facility

Johnson Space Center was formed early in the U.S. space program to take charge of all U.S. manned spaceflights. It evolved from the Space Task Group (STG), a special detail formed by the National Aeronautics and Space Administration (NASA) on November 5, 1958, and charged with starting preliminary work on a manned spacecraft. STG was located at the Langley Research Center in Hampton, Virginia, and borrowed thirty-three of Langley's engineers. Nine days after its formation, the Space Task Group wrote "Specifications for a Manned Space Capsule," sending copies to fifty prospective bidders. On December 17, the "manned space capsule" project was named Mercury. In early 1959, STG became one of six departments in the newly formed Goddard Space Flight Center in Greenbelt, Maryland. Because of its rapid growth, though, the group never moved to Goddard. Instead, it was made an independent entity on March 1, 1961, and a search for a new home began. Encouraged by Rice University's offer of free land and by Vice President Lyndon B. Johnson's support, NASA selected Clear Lake City, a suburb of Houston, as the site of its manned spaceflight headquarters on September 9 of that year.

On November 1, the new facility was named the Manned Spacecraft Center, a name it held until February 17, 1973, when it was renamed in honor of the late President Johnson. Some STG offices moved as early as October, 1961, and were temporarily housed in quarters at nearby Ellington Air Force Base and in Houston. The move was completed after the last Mercury mission, and the center was formally opened in February, 1964.

Designed like a college campus, Johnson Space Center (JSC) has more than thirty major buildings, ranging from the office building for headquarters' personnel to the windowless Mission Control Building. The growth of the shuttle program spurred development of a number of off-site contractor support facilities to handle engineering and support work.

JSC's first director, Robert Gilruth, managed the center from its origins in the Space Task Group until January, 1972, when he was appointed to be NASA's Director of Key Personnel Development. He was succeeded by Christopher C. Kraft, a member of the original STG team and a former flight director. Kraft managed JSC until his retirement after seeing the shuttle program into its initial flights. He was replaced by Gerald Griffin, who served until after the 1986 *Challenger* accident. Griffin was succeeded by Aaron Cohen, a former manager of the shuttle orbiter project, who left in 1986 to become NASA Deputy Administrator. More recently, George W.S. Abbey became JSC Director in January 1996.

Reorganization in May, 1963, divided the center into separate operations and development branches. In November of that year, a further change gave the space center's director and his deputy joint responsibility for four assistant directors — of flight operations, administration, engineering and development, and flight crew operations — and for the Apollo and Gemini program offices. Flight operations included four major divisions: flight support, mission planning and analysis, flight control, and landing and recovery. Flight crew operations included the astronaut office, headed by a chief astronaut; an aircraft operations office; and a flight crew support division. An assistant director for life science research was added later.

The number of employees at JSC changed with the fluctuating NASA budget over the years. In 1969 the space program had become so retrenched that JSC managers could hire only one employee to replace every fourteen who left.

The Apollo-Soyuz Test Project, although criticized by some as a scientific dead end, did in fact help to retain a number of skilled flight controllers who might otherwise have left the space program in the long hiatus between Skylab and the space shuttle.

Mission Control, the best-known facility at JSC, operated as a backup during the Gemini 2 and 3 missions in 1965. Starting with the Gemini IV mission, however, Houston was the primary control center. It was manned twenty-four hours a day during manned space missions, and anyone at the center was on standby to assist with any problems that might arise during a mission. (NASA crews at JSC had worked around the clock to refine their understanding of the shuttle's strength margins when erratic winds threatened to delay the STS-26 mission in 1988.)

As the Gemini program progressed and the Apollo program began, crew members became less dependent on Mission Control for routine operations; nevertheless, they still required guidance updates and schedule revisions, since each mission was subject to minor variations. Mission Control is responsible for a mission from the time the launch vehicle "clears the tower" at the Kennedy Space Center until "wheel stop" at landing.

Mission Control's value was best shown during the Apollo 13 mission in 1971 when an oxygen tank explosion nearly killed the crew in flight. The crew, unable to diagnose the problem, relied on the alert ground control team of engineers and flight directors who quickly devised emergency life-support procedures and altered the flight plan to return the crew safely.

Although the public most often sees the "trenches" in the flight control room and hears the voice of the capcom, or Capsule Communicator, Mission Control actually comprises a number of support rooms and staff members. There are two complete facilities in the Mission Control Center. The one on the second floor is used for civilian shuttle missions; the one on the third floor is dedicated to military missions. The entire building, like other buildings at JSC, has a special security card-key system — which did not exist during the comparatively open Apollo era — and the third-floor facilities are enclosed in an electronic screen to prevent eavesdropping on discussions during a flight.

The flight control room encloses three rows of computer terminals where more than twenty flight controllers can be seated. All three rows face a wall-sized map of the world that indicates the spacecraft's position and orbital path and the locations of tracking stations. On either side of the map are technical status displays and, on the right, a television image transmitted from the spacecraft.

There are ten key positions in Mission Control. The Flight Director has overall responsibility for the conduct of the mission and final authority for major decisions. The Space Communicator — or capcom, from Mercury's "Capsule Communicator" — is an astronaut. The capcom acts as the liaison between ground and flight crews. The Flight Dynamics Officer is responsible for spacecraft maneuvers, the Guidance Control Officer oversees the on-board navigation and guidance software, and the Flight Surgeon is in charge of crew health and life-support monitoring. There are five Systems Engineers: The Data Processing Systems Engineer monitors the on-board computers, the malfunction displays, and the caution and warning system; the Booster Systems Engineer covers the solid-fueled rocket booster, main engine, and external tank during countdown and ascent; the Propulsion Systems Engineer is in charge of the reaction control and orbital maneuvering systems; the Guidance, Navigation, and Control Systems Engineer monitors those systems, along with abort conditions; and the Electrical, Environmental, and Consumables Systems Engineer is responsible for the power supply and distribution, environmental control, and life-support systems. Personnel in a number of "back rooms"— the multipurpose support room, the payload operations control center, and several mission control support areas — provide expertise for various systems and activities.

Each manned space program has required the construction of simulators in which the astronauts may train for their missions. Relatively crude simulators for Mercury and Gemini were located at other NASA centers and contractor sites as well as at JSC. More sophisticated simulators were assembled for the Apollo program. The interiors were virtually identical to the flight vehicles, but the exteriors were covered by a complex array of boxes enclosing sophisticated optical instruments that could project the same views that the crew would see in space. The lunar module simulator also had a large model of the lunar surface on which the crew would "land," using the image from a tiny television camera maneuvered by computer.

Most of the Apollo and Skylab simulators were dismantled or turned into museum pieces and replaced with new, computerized simulators for the shuttle era. It is still not possible to build a single simulator that can serve all crews in all ways, so several have been built for specific purposes. The most important are the Shuttle Mission Simulators (SMS's). The moving-base SMS takes the crew through the full range of motions experienced during a shuttle launch and landing, and the fixed-base SMS has upper and lower decks for simulating orbital activities and experiments. Each SMS is attached to a complex array of computers to link it to Mission Control and to simulate virtually every aspect of spaceflight, from instrument displays and outside scenery to the fraction-of-a-second delay in voice signals traveling through the NASA network.

The crews also train on single-system trainers, which mimic the functions of limited sets of instruments and controls on the flight deck; 1-g mock-ups, which have the shape and feel of the shuttle cabin for training in ingress and egress, housekeeping, and maintenance, but are otherwise nonfunctional; the manipulator development center, which duplicates the aft flight deck, payload bay, and robot arm for satellite handling tests with helium-filled balloon mock-ups; the shuttle engineering simulator, in which detailed procedures can be worked out before crew tests are conducted on the SMS; and the shuttle avionics integration laboratory, which mimics the functions of virtually every piece of electronic gear in the shuttle and tests flight software.

The simulators have served as more than crew training aids. They have also served as development tools for new experiments and procedures and as trouble-shooting facilities, since engineers cannot examine hardware while a spacecraft is in orbit. The astronauts, whose time is limited, do not take part until a procedure is refined and needs their comments.

Pilot astronauts maintain their flying proficiency by flying two-seat T-38 Talon jet trainers operated from Ellington Air Force Base, a few miles from JSC. In addition, two special simulators are operated from Ellington: a shuttle trainer aircraft and a KC-135 cargo jet. The shuttle trainer is a Gulfstream 2 corporate jet refitted with controls, computers, and actuators that make it handle as the shuttle orbiter does during final approach and landing. For familiarization with weightlessness, the astronauts ride aboard the KC-135, which flies roller-coaster-like parabolas to provide about thirty seconds of free-fall at each peak. Scientists and engineers also use the jet to test new designs for equipment that might be affected by weightlessness.

JSC has a number of other simulation facilities. The Space Environment Simulation Laboratory has two large chambers (36.6 meters high by 19.8 meters wide, and 13.1 meters high by 10.7 meters wide) in which the Apollo spacecraft were subjected to a spacelike vacuum and to temperatures ranging from 127 to −173 degrees Celsius. The Vibration and Acoustic Test Facility subjects hardware to the noise and vibration predicted during launch. The White Sands Test Facility, in New Mexico, tests the space shuttle's thruster modules.

A flight acceleration building once housed a centrifuge on which riders could experience a force twenty times that of gravity, for sustained periods, or thirty times, momentarily. It was dismantled in the late 1970's, and its building was renovated to house the Weightless Environment Training Facility (WETF), a water tank large enough (10 by 23.7 meters in area, 7.6 meters deep) to house a mock-up of the entire space shuttle payload bay. WETF is used to develop crew procedures for various contingencies, such as manually closing the payload bay doors in an emergency, freeing stuck solar arrays on a satellite, and repairing and maintaining orbiting satellites.

JSC also has a number of research offices. The most famous is the Lunar Sample Building, which houses the Lunar Receiving Laboratory, built to protect the Apollo lunar samples from contamination. When the laboratory was conceived, mission scientists' first and greatest concern was that lunar samples might harbor life forms hazardous to life on Earth. As a result, the Lunar Receiving Laboratory included a crew quarantine facility in which the early Apollo lunar crews

were held for two to three weeks after returning to Earth. Crew quarantine was discontinued after Apollo 12, and tests have firmly established that the Moon has no indigenous life forms; nevertheless, samples of space materials are still quarantined so that no terrestrial compounds can contaminate them. Analyses normally are carried out in isolation boxes made of stainless steel and glass and with virtually no organic materials present. To support the work at JSC, the Lunar and Planetary Institute was established at a small ranch adjoining the center.

JSC and Marshall Space Flight Center helped to define the space shuttle program in the late 1960's and early 1970's. Each center jockeyed for the lead role in the program, JSC claiming that the shuttle was a spacecraft and Marshall claiming that it was a launch vehicle. In the end the program was neatly divided between the two: JSC was named the lead center for integration and for development of the orbiter, and Marshall was given responsibility for the propulsion systems.

JSC has always had a strong interest in advanced programs, especially space stations. Between 1962 and 1968, by one accounting, the center conducted some twenty space station studies at a total cost of $6.3 million. In the late 1960's the studies took two major forms: the Apollo Applications Program, which became Skylab and was ultimately managed by Marshall Space Flight Center, and a long-term, reusable space station, which would have been launched as a single unit by the Saturn 5 rocket. When NASA stopped using the Saturns and began emphasizing the shuttle program in the early 1970's, JSC initiated studies on a number of proposed stations that would be carried up as modules by the shuttle. In 1980 these projects converged as the Space Operations Center (SOC) study. SOC was to be a facility providing maintenance and other support to satellites in nearby orbits; it was designed to expand and encompass manned research. JSC currently operates the Space Station Control Center, handles crew integration and training, and oversees systems operations for U.S. involvement in the International Space Station.

Context

The center has retained much of the organizational structure set up in the 1960's, with program offices added to handle the Space Transportation System, the shuttle orbiter, and the space station programs. Major changes were made in the hierarchy after the 1986 *Challenger* accident — several top-level managers now are attached to NASA headquarters, even though they work at JSC — and a special safety, reliability, and quality assurance office was created to provide an objective review of areas outside the control of various program offices. Other major offices include flight crew operations, mission operations (which includes Mission Control), mission support, space and life sciences, and engineering.

Following President Ronald Reagan's endorsement in January, 1984, four work packages were divided among the various NASA centers. JSC was allocated WP-2, the second-largest package. It included the outfitting of the U.S. Space

Station's crew habitation module; the building of various crew-related systems, guidance systems, and computers; the development of the station's primary structure; and the integration and assembly of the station as a whole. In 1988, advanced studies under way at JSC included the phase 2 expansion of the space station, and designers were studying the feasibility of large support beams for experiments above and below the modules.

The astronaut office supports a number of NASA programs by providing crew members who can offer users opinions about the experiments and systems that ultimately will be deployed in space. Astronauts, new and veteran, often support crew station and other program reviews. Furthermore, the astronaut office has been given a greater role in space shuttle planning and operations decisions since the *Challenger* accident. Several astronauts have taken part in analyses of failures or potential failures, and most are active in the Manned Spaceflight Awareness program, which assists NASA's quality control efforts.

The National Space Transportation System Office has a major branch and a deputy director at JSC. The office oversees all shuttle mission planning and reports to NASA headquarters in Washington, D.C., rather than to the field center.

JSC is the lead center for a number of space shuttle payloads, including those dealing with Earth observations and life sciences. The center maintains an active Earth resources office that has supported the Earth Resources Laboratory at the Stennis Space Center. Life sciences activities cover work ranging from small, occupational health experiments, designed to develop a larger data pool on spacesickness, to the dedicated SLS 1 and 2 Space Life Science missions. JSC has a large cadre of flight medicine specialists and conducts experiments using flight crews and civilian control subjects.

The Lunar Receiving Laboratory's role has expanded to include analyses of comet dust samples collected by high-flying aircraft and of meteorites, especially samples from Antarctica that are believed to be Martian rocks cast across space by massive impacts. JSC also supports NASA studies of modules and landers for manned missions to the Moon and Mars.

Bibliography

Baker, David. *The Rocket.* New York: Crown Publishers, 1978. A thorough, well-researched history of the development of rocketry and of specific launch vehicles. Although the focus is on rocket hardware, there are extensive descriptions of launch preparations and of many of the early failures and their causes.

Bilstein, Roger E. *Stages to Saturn: A Technological History of the Apollo/Saturn Launch Vehicles.* NASA SP-4206. Washington, D.C.: Scientific and Technical Information Branch, National Aeronautics and Space Administration, 1980. This history focuses on the development of the Saturn boosters and includes details on support activities at JSC.

Cooper, Henry S. F., Jr. *Before Lift-Off: The Making of a Space Shuttle Crew.* Baltimore: The Johns Hopkins University Press, 1987. A highly detailed and engrossing account of several weeks in the shuttle crew's training for the October, 1984, *Challenger* mission. The book portrays the ordinary and extraordinary events and frustrations that take place as a crew prepares for and conducts a mission and explains how those events shape mission plans. Written for a general audience.

Grimwood, James M. *Project Mercury: A Chronology.* Washington, D.C.: Government Printing Office, 1963. A detailed chronology of America's first manned space missions, from the space program's post-war beginnings through the start of the Gemini program. Although not written as a narrative, the book provides accounts of launch preparation activities.

Lay, Beirne, Jr. *Earthbound Astronauts: The Builders of Apollo Saturn.* New York: Prentice-Hall, 1971. Discusses the development of the Saturn 5 launch vehicle from the viewpoint of several engineers and managers who worked on the Saturn program. Includes many details of mission operations.

Levine, Arnold S. *Managing NASA in the Apollo Era.* NASA SP-4102. Washington, D.C.: Government Printing Office, 1982. A highly detailed history of the difficulties of and lessons learned from managing a large, growing agency as it developed a complex program to place Americans on the Moon.

Lewis, Richard S. *The Voyages of Apollo: The Exploration of the Moon.* New York: Quadrangle, 1974. Covers the development of the Apollo spacecraft and the conduct of the missions. Includes details on mission control operations, especially during the crises of the Apollo 13 flight. Suitable for general audiences.

National Aeronautics and Space Administration. *National Space Transportation System Reference.* Washington, D.C.: Author, 1988. A technical manual written to provide the media with as many details as possible about space shuttle hardware, operations, and missions. Includes descriptions of the shuttle facilities at JSC.

Pellegrino, Charles R., and Joshua Stoff. *Chariots for Apollo: The Making of the Lunar Module.* New York: Atheneum Publishers, 1985. A well-written, enjoyable history of the development of the Lunar Module. Although the book focuses on the difficulties of building the spacecraft, much attention is given to the work involved in making the module a suitable payload and in launching it into space.

Wells, H. T., et al. *Origins of NASA Names.* NASA SP-4402. Washington, D.C.: National Aeronautics and Space Administration, 1976. A reference that lists the origins of the names of NASA facilities, vehicles, and programs.

Dave Dooling

LANDSAT 1, 2 AND 3

Date: July 23, 1972, to July 16, 1982
Type of satellite: Earth observation

Landsats 1, 2, and 3 were satellites that collected data about Earth: its agriculture, forests, flatlands, minerals, waters, and environment. These were the first Earth science satellites. They helped produce the best available maps and provided important agricultural information.

PRINCIPAL PERSONAGES

WILLIAM T. PECORA, the originator of the Landsat concept at the U.S. Geological Survey
ARCHIBALD B. PARK, the originator of the Landsat concept at the U.S. Department of Agriculture
WILLIAM NORDBERG, Director of Applications at Goddard Space Flight Center
PAUL LOWMAN, a NASA geologist

Summary of the Satellites

In 1970, the National Aeronautics and Space Administration (NASA) developed a series of satellites specifically dedicated to obtaining useful and practical information about the surface of Earth. William T. Pecora of the U.S. Geological Survey originated the idea for such an Earth observation satellite in the late 1960's. In the Department of Agriculture, Archibald Park was interested in large-scale surveys of crops and forests. Earlier in the decade, weather satellites had taken pictures of Earth, as had the Gemini astronauts. These images had shown survey scientists the orbital satellite's potential to provide maps and other data about Earth.

The Earth Resources Observation Satellite (EROS) program took form early in 1966; immediately, NASA stepped in and demanded control. In time, a cooperative effort between NASA and the Department of the Interior was effected. The satellite, originally called the Earth Resources Technology Satellite (ERTS), quickly had its name changed; Landsat, short for land satellite, provided a catchy title for promotion to members of Congress, the executive branch of the federal government, and the general public.

On July 23, 1972, Landsat 1 was launched, followed by Landsat 2 in 1975 and Landsat 3 in 1978. They were sent up on Delta rockets. The Landsat 1 spacecraft was manufactured at the General Electric Company's Spacecraft Operations facility at Valley Forge, Pennsylvania. This spacecraft, an outgrowth of the Nimbus series of meteorological satellites, was designed to carry two remote sensing systems and a system to collect data from sensors located in remote places on Earth. Landsat 3 was launched along with the Lewis Research Center Plasma Interaction Experiment and the Orbiting Satellite Carrying Amateur Radio (OSCAR) communications relay satellite.

Landsats 1, 2, and 3 were all placed in near-polar, Sun-synchronous, near-circular orbits around the poles at an altitude of approximately 920 kilometers. Landsat's orbit is 99 degrees, which means that it is only one degree off from a North-Pole-to-South-Pole orbit. A Sun-synchronous orbit means that the satellite orbits about Earth at the same rate that Earth moves around the Sun. This feature enables the spacecraft to cross the equator at the same local time every day — in this case, at about 9:30 A.M. — on the daylight side of Earth. Indeed, it crosses the equator at exactly the same local time on every pass. One benefit of a crossing at 9:30 A.M. is the presence of relatively long shadows on the underlying ground, which help show up the topographical features more clearly. By passing over at exactly the same time every day, the sensors can photograph the scenes under conditions of light and shade that are relatively constant. There are differences caused by varying amounts of cloud cover and the changing seasons of the year, but these variations are something over which scientists have no control. The orbit simply minimizes the variations.

In this orbit at this altitude, each satellite circled the globe fourteen times a day and thus took approximately 103 minutes to orbit Earth. That is, the satellite circled the globe slightly off center from a precisely north-south orbit. It is easiest to think of Earth as an enormous orange with fourteen sections. The three Landsats passed over each section twice a day, once in the daylight and once at night.

Programmed ground commands were sent from Goddard Space Flight Center, located in Greenbelt, Maryland. The orbit and field of view were such that any given place on Earth's surface is observed every eighteen days by the same satellite at the same local time. With the launch of Landsat 2, the two satellites were synchronized in orbit so that the coverage of any spot on the globe could be obtained every nine days. At sites closer to the North and South poles, the coverage was even more frequent.

For all of their tasks, the Landsat satellites required electrical power, which they obtained from the Sun. Banks of silicon cells converted the sunlight to electricity. With their Sun-synchronous orbit, the Landsat satellites spent half their time in the dark; thus, some means of storing power was required. A series of batteries stored the power collected while the satellite traveled in the sunlight, releasing it when the satellite traveled in the dark.

The first three Landsat spacecraft weighed about 950 kilograms each, were approximately 3.3 meters tall, and were stabilized by gas jets so that if the orbit varied it could be repositioned. This was a critical feature, because it guaranteed that the images recorded would be consistent over time.

The three Landsat spacecraft did not carry ordinary still cameras. Instead, each satellite was equipped with a Multispectral Scanner Subsystem (MSS) as its principal sensing instrument, three special television cameras that were known as the Return Beam Vidicon (RBV) system, and other data collection systems that could obtain information from remote, surface-based, automatic platforms around the world. These platforms were able to monitor local conditions and relay the data to central ground stations whenever the Landsat could simultaneously view the platform and a ground station.

On Landsat 1, the RBV system failed shortly after launch; thus, the key recording instrument became the MSS. This instrument had been tested on high-altitude aircraft, but it was first tested in space on Landsat 1. A scanner mirror rocked from side to side some thirteen times per second. By deflecting light from the ground into the detectors, the mirror scanned the scene below in a series of parallel swatches, rather like someone walking along while sweeping with a broom. The sensors were mounted on the bottom of the spacecraft. The area in each image measured approximately 185 square kilometers.

The MSS was the primary information gatherer. The advantages of this type of sensor over a traditional camera are that the information is sent back rapidly by radio waves to ground stations on Earth instead of being recorded on the camera film, which would then have to be physically transported back to Earth in a costly and complicated process. Moreover, scanners could capture images at wavelengths — including the important infrared region of the electromagnetic spectrum — that are invisible both to the human eye and to conventional cameras. Other variables that the use of the MSS avoided included the use of different film-developing practices of various laboratories and the subjective eyes of the technicians who interpret the photographic information.

On Landsat 3, the video equipment was greatly improved; as a consequence, so was the quality of the image resolution. The video system for Landsats 1 and 2 had consisted of three television-type cameras, each covering a different spatial region. On Landsat 3, the three cameras were replaced by two improved cameras giving side-by-side viewing at twice the spatial resolution.

These MSS data, and later video data, were principally in digital form; thus the information could be rapidly processed with computers, which could analyze the information so that it could be used quickly and precisely.

NASA operated three ground stations to receive data from Landsats 1, 2, and 3 and control their operation. These stations were located in Greenbelt, Maryland, Goldstone, California, and Fairbanks, Alaska. Any one of these three stations could detect the Landsat satellites at any time they were over the continental United States. Additional coverage was obtained from ground stations in Italy and Sweden operated by the European Space Agency and from stations in Canada, Brazil, Argentina, South Africa, India, Australia, Japan, Thailand, Indonesia, and mainland China. In addition to real-time coverage permitted by these stations, tape recorders on the satellites allowed the recording of data from areas around the world that were outside the range of the ground stations.

In the early days of the Landsat program, all the data were processed at the Goddard Space Flight Center. After 1979, Goddard and the EROS Data Center in Sioux Falls, South Dakota, shared the processing of information. The whole process took about two weeks. The information gathered by the Landsat satellites was placed in the public domain as rapidly as possible and was stored in two national data bases, one operated by the U.S. Department of the Interior in Sioux Fails, South Dakota, and the other by the Department of Agriculture in Salt Lake City, Utah.

Landsat 2 was identical in payload to Landsat 1, and despite a design lifetime of one year, it continued to function satisfactorily for five years. After that time, failure of its primary flight control mechanism became evident, and it was difficult to keep the spacecraft pointed at Earth.

When Landsat 3 was launched in March, 1978, Landsat 1 had been out of service for two months, but Landsat 2 was still operating; it was then arranged so that with overlap the repeat time for any section of the globe was again nine, not eighteen, days. This proved of great value in the surveillance of dynamic events. Landsat 3 carried an MSS with four bands identical to those of Landsat 1 and Landsat 2. It also carried a thermal infrared scanner, but this failed shortly after launch. The video cameras of Landsat 3 gave twice the ground resolution by doubling the focal length of the lens system.

Knowledge Gained

The Landsat system was designed to gather information for the better use of Earth resources. Crop information could be gained for better use of land for crop production. Landsats 1, 2, and 3 made it possible to discriminate the patterns of crops, lumber, and vegetation around the world. They could measure crop acreage by species. Precise estimates of the amount and type of lumber resources in the world could be obtained, and the strength and stress on vegetation could be determined. Soil conditions could be monitored as well as the extent of fire damage.

Moreover, for the first time, an exact determination of the water boundaries and surface water area and volume around the globe was possible. Plans could be formulated to minimize damage from flooding. Scientists were also able to survey snow areas of mountains and glacial features. The depths of the oceans, seas, lakes, and other bodies of water could be calculated and used to plan the better use of water resources. On June 26, 1978, a new experimental satellite, Seasat, was launched; it was designed to test the methods most suitable for research about oceans and seas. After only four months of operation, however, the satellite failed, and Landsat continued with this important function.

Mineral and petroleum exploration was also made easier with precise geological data. Maps of rocks and rock types were made for all parts of the globe. More was learned about rocks and soils, volcanoes, changing landforms, and precise land formations. Surface mining could be monitored and land more productively reclaimed. The science of mapping Earth was revolutionized. Urban and rural demarcations were made, and help became available for regional planners. Transportation networks were mapped as well as land and water boundaries; imprecise maps were updated. Scientists could detect living ocean forms and their patterns and movements. Precious data could be gathered on changing conditions in shorelines, on shoals and shallow areas, and on wave and ice patterns. Scientists were able to gather more precise data on air and water pollution, its sources and effects. They could determine the scope and effects of natural disasters and monitor the environmental effects of defoliation.

Context

Landsat 1 was the first of the ERTS series, the first satellite series in which satellites were used to explore and better understand the planet rather than outer space. A single image from space could encompass large-scale geological features that otherwise would take days or weeks to cover, even with aerial mapping. Better still, such photographs could show features so extensive that they would never have been noticed from the ground or would have been lost in the patchwork of aerial mosaics.

William Nordberg, Director of Applications at Goddard Space Flight Center, summarized the results when he said, "Within a few weeks after its [Landsat-1's] launch, we saw that the variety of uses to which Landsat could be used exceeded our expectations." Within a week of the launch of Landsat 1, NASA geologist Paul Lowman was able to make a new geological map of California's coastal ranges near Monterey Bay. In studying this new map, Lowman was able to identify more than thirty previously unknown features, including some geological faults.

This family of satellites has proven an invaluable component of a new approach to locating, monitoring, managing, and understanding many of the natural resources of the planet Earth. For those scientists studying Earth, Landsats 1, 2, and 3 have been among the most useful and productive of satellites ever launched by NASA. Before Landsat began its systematic sensing of Earth's changing features, cost-effective, broad-scale land monitoring was nearly impossible. Producing comparable maps by conventional methods was costly, and the time required was such that when finally produced, the maps were already out of date.

In the areas of agriculture, oceanography, geology, and environmental studies, scientists were able to gather significant and important data from Landsats 1, 2, and 3 — information that could help both in fundamental research and in governmental decision-making.

Consider but a few examples of the ways in which the Landsat system has benefited mankind. For agriculture, even in the most technologically advanced countries, up-to-date assessments of total acreage of different crops were incomplete before Landsat. After its launch, forests could be managed for both fire control and insect infestation. Wildlife habitats could be monitored and better conserved. More precise, inexpensive maps became available. In terms of land use, vast areas of Africa, Asia, and South America were poorly mapped or even incorrectly mapped before Landsat. Mountain ranges, deserts, vegetable cover, and land use could now be known for all parts of the planet. The Landsat information has proved so reliable that private industry now makes regular use of it. Landsat can map the damage from forest fires more accurately, so that planting can begin again. The same is the case for drought-stricken areas.

Since the mid-1970's, Landsat has marketed several million dollars' worth of products each year to private industry, state governments, and the U.S. Departments of the Interior, Agriculture, and Commerce. With sophisticated computers, the science of knowing Earth, its resources, and its possibilities was greatly advanced.

Bibliography

Harper, Dorothy. *Eye in the Sky*. 2d ed. Montreal: Multiscience Publications, 1983. This clear explanation of the use of satellites includes a survey of the Landsat system. For those with little or no knowledge of space science.

Short, Nicholas M. *The Landsat Tutorial Workbook: Basics of Satellite Remote Sensing*. NASA RP-1078. Washington, D.C.: Government Printing Office, 1982. A basic guide to the uses of Landsats 1, 2, and 3, this publication is aimed at the user of the data of the Landsat system. Contains numerous charts and diagrams and provides references to numerous publications. Suitable for college-level audiences.

Short, Nicholas M., Paul D. Lowman, Jr., Stanley C. Freden, and William A. Finch. *Mission to Earth: Landsat Views the World*. NASA SP-360. Washington, D.C.: Government Printing Office, 1976. This picture book describes the Landsat program and presents a multitude of maps of Earth made from information gleaned by Landsat 1. It focuses on the wonders of the maps created.

U.S. Congress. Senate. Committee on Aeronautical and Space Sciences. *An Analysis of the Future Landsat Effort*. 94th Cong., 2d sess., 1976. Committee Print. This comprehensive report adds much to the knowledge of the Landsat program, its uses and its shortcomings.

Waldrop, M. Mitchell. "Imaging the Earth: The Troubled First Decade of Landsat." *Science* 215 (March 26, 1982): 1600-1603. A comprehensive evaluation of the Landsat program for an important journal. There have been intra-agency political squabbles about the Landsat program, yet that has not stopped it from being a major success.

Williams, Richard S., and William D. Carter, eds. *ERTS-1: A New Window on Our Planet*. U.S. Geological Survey Professional Paper 929. Washington, D.C.: Government Printing Office, 1976. A short, comprehensive guide the Earth Resources Technology Satellite, which later became known as Landsat 1. This is the pioneering work that laid out the uses of the satellite to map Earth and help plan its management. It is the professional predecessor to *Mission to Earth: Landsat Views the World* (see entry above).

Douglas Gomery

LANDSAT 4 AND 5

Date: Beginning July 16, 1982
Type of satellite: Earth observation

Landsats 4 and 5 are satellites that collect data about the Earth: its agriculture, forests, flatlands, minerals, waters, and environment. These satellites are a continuation of the Landsat 1, 2, and 3 satellites, which were the first satellites devoted exclusively to Earth resources. The Landsat satellites produce the best maps available and aid farmers around the world to produce more and better crops.

PRINCIPAL PERSONAGES
   WILLIAM T. PECORA, originator of the program at
      United States Geological Survey
   ARCHIBALD B. PARK, originator of the program at the
      Department of Agriculture
   WILLIAM NORDBERG, Director of Applications at
      NASA's Goddard Space Flight Center
   PAUL LOWMAN, NASA geologist

Summary of the Satellites

In 1970, the National Aeronautics and Space Administration (NASA) proceeded to design and develop a series of satellites specifically dedicated to obtaining useful information about the surface of Earth, which could then be used on a routine and repetitive basis. William T. Pecora, of the United States Geological Survey, originated the idea in the late 1960's. In the Department of Agriculture, Archibald B. Park believed that large-scale surveys of crops and forests could be conducted with the aid of satellites. As a result, the concept for the Earth Resources Observation Satellite (EROS) took form early in 1966.

This type of satellite, originally designated the Earth Resources Technology Satellite (ERTS), was subsequently renamed Landsat to enhance recognition among members of the United States Congress and influential employees of the executive branch of the government. Landsat 1 was the first in the ERTS series in which satellites were used to explore Earth, rather than outer space.

On July 23, 1972, Landsat 1 was launched, followed by Landsat 2 in 1975, and Landsat 3 in 1978; they were all launched on Delta rockets. The Landsat spacecraft, manufactured at General Electric's Space Sciences Facility at Valley Forge, Pennsylvania, was an outgrowth of the Nimbus weather series of meteorological satellites and was designed to carry two remote sensing systems as well as a system to collect data

from sensors located in remote places on Earth.

The Landsat satellites were placed in near-polar, Sun-synchronous circular orbits at an altitude of approximately 920 kilometers. Landsat's orbit is 99 degrees, which means it is one degree off from a purely North-Pole-to-South-Pole orbit. A Sun-synchronous orbit means that the spacecraft crosses the equator at the same local time (between 9:30 and 10:00 A.M.) each day, regardless of its locale.

At 9:30 A.M., the relatively long shadows on the underlying ground help to enhance the topographical features more clearly. By passing over at exactly the same time every day, the satellite's sensors are assured of photographing day-to-day scenes under similar conditions of light and shade. In this orbit at this altitude, each Landsat satellite circled the globe fourteen times a day. It is easiest to think of Earth as an enormous orange with fourteen sections. The Landsat satellites passed over each section twice a day, once in the daylight and once at night. Programmed ground commands were sent from NASA's Goddard Space Flight Center.

NASA operates three ground stations— in Greenbelt, Maryland; Goldstone, California; and Fairbanks, Alaska— to receive data from the Landsat system and to control the satellites' operations. At least any one of these three stations can "see" the Landsat at any time it is over the continental United States or in early coastal areas. Additional coverage is obtained from ground stations operated by the European Space Agency and stations in Canada, Brazil, Argentina, South Africa, India, Australia, Japan, Thailand, Indonesia, and mainland China, operated by their respective governments.

The information gathered by the Landsat satellites is placed in the public domain as rapidly as possible and in two United States data bases operated by the Department of the Interior (EROS Data Center, Sioux Falls, South Dakota) and the Department of Agriculture in Salt Lake City, Utah.

Before launch, the fourth Landsat was referred to as Landsat-D. When it successfully reached orbital altitude, the designation was changed to the appropriate Arabic number, in this case Landsat 4. Landsat 4 uses a multimission spacecraft, an improvement over the original Nimbus series used for Landsats 1, 2, and 3. With an improved launch vehicle, the 3920 Delta, Landsat 4 was increased in size to some 2,200 kilograms and was launched in July, 1982, from the Western Test Range at Vandenberg Air Force Base in California.

Physically, Landsat 4 consists of two major sections. On board the spacecraft a computer controls power and altitude. This computer also sends commands to the propulsion module that provides the capability to adjust the orbit. The forward end of the spacecraft contains the thematic mapper and multispectral scanner. The antenna mast is 13 feet high.

The Landsat 4 orbit differs from those of Landsats 1, 2, and 3. Most significant is the nominal altitude of 705 kilometers as opposed to the 900 kilometers of the first three Landsats. Additionally, Landsat 4 circles the globe in sixteen instead of eighteen days.

Landsat 4 carries a thematic mapper, which images Earth to detect geological features of interest in mineral exploration. Together with the Multispectral Scanner Subsystem (MSS), the thematic mapper has also provided information that is useful for agriculture. The instrument was designed to be most relevant to agriculture experiments that would observe vegetation cover and measure crop acreages. The thematic mapper has better resolution than the MSS and was adopted to estimate crop acreages better in regions with small fields such as China, India, the eastern United States, and Europe. The sensor sends data directly to Earth, in this case to White Sands, New Mexico.

The speed of data collection was improved for Landsat 4. The early Landsat 1, 2, and 3 programs were primarily for researchers, and the speed of data transmission was not considered necessary. As familiarity and experience with the use of the first three Landsats increased, they were gradually adapted to digital data. Landsat 4 had it from the beginning.

The utility of Landsat was, however, somewhat compromised because of a data-relay satellite that malfunctioned at launch. Later engineering problems on board the spacecraft further reduced its performance.

Landsat 5, which was launched early in 1984 to complete the mission of Landsat 4, was the same size as its predecessor and fulfilled the same functions. Starting in January, 1983, the National Oceanic and Atmospheric Administration (NOAA), rather than NASA, assumed responsibility for operating the Landsat series of satellites. NOAA also assumed responsibility for producing and distributing data and data products, except from the thematic mapper. Landsat 5 synchronizes with Landsat 4 to provide an eight-day, full-Earth-coverage orbit. The orbit of Landsat 5 was targeted low intentionally to ensure that no orbit-lowering maneuvers would be required. Between March 7 and April 4, 1984, a series of eight orbit-raising maneuvers were performed to correct the axis so that Landsat 4 and 5 would be coordinated to cover the entire globe in eight days. Landsat 5, like Landsat 4, has a Sun-synchronous orbit and a 16-day tracking cycle around the world. The nominal orbits of the two are identical except for their phasing. The 8-day complete coverage of Earth is achieved when Landsats 4 and 5 are on opposite sides of the globe. This phasing also minimizes any interference between the two satellites. While still functional, routine collection of data

was terminated in late 1992.

Knowledge Gained

Landsats 4 and 5 provide more precise data than were gathered by Landsats 1, 2, and 3. The principal uses of the information continually gathered by Landsats 4 and 5 can be grouped into the following broad areas.

With respect to agriculture, forestry, and range land, information can be obtained for better use of land for crop production. Landsat makes it possible to discriminate the patterns of crops, lumber, and vegetation around the world, and it can measure crop acreage by species. The world's lumber resources and their strength and stress on vegetation can be determined, and soil conditions can be monitored as can the extent of fire damage.

For the first time, an exact determination of the water boundaries and surface water area and volume around the globe is possible. Information can be gathered about floods and flooding, mountains and glacial features, and depths of oceans and other bodies of water. An inventory of lakes is also kept. Mineral and petroleum exploration are made easier with precise geological data and maps of rocks and rock types. Indeed, the knowledge of geology has greatly expanded with more information about rocks and soils, volcanoes, changing land forms, and precise land formations. Surface mining can also be monitored and land more productively reclaimed, as the science of the mapping of Earth continues to be revolutionized. Urban and rural demarcations are corrected, and help is possible for regional planners. Transportation networks are also mapped.

Regarding oceanography and marine resources, floods can be monitored and measured, and damage can be delineated and repaired. In desert regions, possible water sources can be more easily identified, and water quality can be monitored. It is now possible to detect the pattern and movement of ocean life. Shoreline changes can be mapped, along with shoals and shallow areas and ice patterns for better shipping and wave patterns.

More precise data on air and water pollution, its sources and effects, are now available, and it is possible to determine the scope and effects of natural disasters. The satellites also monitor the environmental effects of human activity.

Context

Before Landsat began systematic sensing of Earth's changing features, broadscale land monitoring was severely limited. Producing comparable maps by conventional methods was costly, and the time required was such that when they were finally produced, they were already outdated.

NASA geologist Paul Lowman used the results from Landsat to make a new geological map of California's coastal ranges near Monterey Bay. Landsat had shown him more than thirty previously unknown linear features, including some geological faults. Landsats 4 and 5 continue to demonstrate a new approach to locating, monitoring, managing, and understanding many of the natural and non-natural resources on Earth. For those scientists studying Earth, the Landsat series has been among the most useful and productive satellite programs ever launched by NASA.

In the field of agriculture, even in the most technologically advanced countries, up-to-date, complete assessments of total acreage of different crops was incomplete before the arrival of the Landsat system. Forests can now be managed for fire and insect infestation, wildlife habitats can be described, and vegetation and land use can now be known for all parts of the planet. So reliable is the information, that private industry has begun to make regular use of it, accounting for one-third of the sales of Landsat images.

The use of sophisticated computers has furthered immeasurably the science of knowing Earth's resources and possibilities. Companies that explore for minerals and petroleum have been quick to realize the efficiency of the Landsat system. Landsat 4 and 5 were instrumental in gathering information about the nuclear accident at Chernobyl in the Soviet Union on April 26, 1986.

Bibliography

Baker, John. *Landsat-4 Science Investigations Summary.* NASA Conference Publication 2326. Washington, D.C.: Government Printing Office, 1984. This volume provides a summary of the success of Landsat 4 in terms of its ability to map and survey Earth's resources. Some of the material is intended for geologists, but the volume does provide basic information and descriptions for the layperson.

Harper, Dorothy, ed. *Eye in the Sky: Introduction to Remote Sensing.* 2d ed. Montreal: Multiscience Publications, 1983. This clear explanation of the use of satellites contains a survey of the Landsat system. It provides its discussion in clear language aimed for the lay reader.

Richter, Rudolf, Frank Lehmann, Rupert Haydn, and Peter Volk. "Analysis of Landsat TM Images of Chernobyl." *International Journal of Remote Sensing* 7 (December, 1986). Landsat 5 provided the images that proved useful in analyzing the results of the Chernobyl nuclear plant disaster. This work is a technical article but contains fascinating information and photographs. This article provides an important example of the success of the Landsat system.

Salomonson, V. V., and R. Kottler. *An Overview of Landsat 4: Status and Results.* Greenbelt, Maryland: Goddard Space Flight Center, 1983. This work analyzes how well the thematic mapper is working on Landsat 4. Several helpful diagrams are included, and the details of this measuring and data gathering device are explained to the reader.

Salomonson, V. V., and Harry Mannheimer. *An Overview of the Evolution of Landsat 4.* Proceedings of the Eighth Pecora Symposium. Ann Arbor, Mich.: October, 1983. This work is a fine summary of the history of the development of the Landsat 4 program. It contains several useful diagrams, and the improvements over the Landsat 1, 2, and 3 satellites are described in some detail.

Short, Nicholas M., et al. *Mission to Earth: Landsat Views the World.* NASA SP-360. Washington, D.C.: Government Printing Office, 1976. This picture book does not touch on the Landsat 4 and 5 programs specifically, but it does provide the most useful background to the goals and missions of the Landsat program. It describes the program and presents a multitude of maps of Earth made from Landsat 1.

Douglas Gomery

LANGLEY RESEARCH CENTER

Date: Beginning in 1917, as part of NASA, October 1, 1958
Type of organization: Space center

The NASA Langley Research Center, located near Hampton, Virginia, is the United States' oldest government-run aerodynamic research and testing facility. Built first as a home for civilian airplane development during World War I, Langley conducts and manages a variety of programs on advanced aerodynamics and the future of manned and unmanned space travel.

PRINCIPLE PERSONAGES
    RICHARD W. BARNWELL, Chief Scientist, NASA
        Langley Research Center
    RICHARD H. PETERSEN, Director, NASA Langley
        Research Center
    WILLIAM D. MACE, Director of Electronic Research,
        NASA Langley Research Center
    LEIGH M. GRIFFITH, Engineer in Charge, NACA
        Langley Research Center, 1923–1925
    HENRY J. E. REID, Engineer in Charge,
        NACA/NASA Langley Research Center,
        1925–1960

Summary of the Organization

Located in the Virginia Tidewater area on the Chesapeake Bay near the city of Hampton, the National Aeronautics and Space Administration's (NASA) Langley Research Center is the nation's oldest and most comprehensive aeronautics research and testing center. Established in 1917 to test civilian aircraft during World War I, Langley quickly became the central experimental and testing facility for state-of-the-art air and spacecraft technology in the United States. Today, Langley Research Center provides vital research and development information on proposed air and spacecraft, equipment, and software systems to civilian government agencies, the military, and private industry. The center also manages the Scout launch vehicle, as well as numerous space research projects such as the two Viking Mars landing missions, the Echo communications satellites, and the unmanned Lunar Orbiter from the 1960's. Langley is one of eleven such major facilities operated by NASA out of its headquarters in Washington, D.C.

Langley Research Center was originally created by the National Advisory Committee for Aeronautics (NACA) in 1917. NACA was the first U.S. federal agency dedicated to advancing the principles and practice of powered flight.

Langley, NACA's first facility, was named for American aviation pioneer Samuel P. Langley, and run by the first Engineer in Charge, Leigh M. Griffith. The first testing equipment constructed at Langley, in 1920, was a wind tunnel capable of generating wind speeds of 130 miles per hour. Over the years, Langley's contributions to aeronautic technology grew as the center's resources and responsibilities increased. Virtually all the major advancements in airplane design made since the 1920's in the United States have undergone testing at Langley Research Center prior to production by private industry or the government. When NASA was created in 1958 to oversee the United States' fledgling space program, Langley was one of the first NACA centers incorporated into the new agency. Scientists and engineers who worked at Langley were among the first in the country to look into possible spacecraft design options.

Langley personnel were also part of the team that originally conceived the United States' first program for putting men into space, the effort later named Project Mercury. The project was managed by the NASA Langley Research Center, which also served as one of the principal training facilities for the first group of seven astronauts selected in 1959. Robert Gilruth, a Langley scientist and administrator, headed the team of NACA/NASA experts who created Project Mercury. Because of Langley's wealth of expertise and technical resources, the facility was given managerial control over the project and served as a key training facility for the original seven astronauts. Langley human and organizational resources were later spun off to help create the Johnson Space Center in Houston, the home of the United States' manned space efforts ever since.

In 1965, when Project Gemini, the United States' two-man spacecraft program, was put into use, NASA used Langley personnel to open what would later be called the Johnson Space Center in Houston, Texas, and transferred responsibility for manned flights from Langley. During and after this period, Langley also managed the Viking Mars orbiter and lander program, the Lunar Orbiter program, NASA's Echo communications satellites, and experiments for various Explorer unmanned spacecraft as well as providing research and design support for other NASA manned and unmanned spaceflight projects.

Along with other NASA facilities, such as the Ames

Research Laboratory in California and the Goddard Space Flight Center in Maryland, Langley has provided key support in the development and operation of the Space Transportation System, or space shuttle. Langley scientists tested and helped to perfect the shuttle's design and to invent the heatshielding materials used to protect the shuttle from the intense heat of reentry into Earth's atmosphere. Langley has managed or assisted in the development of several space shuttle payloads: The Langley-developed Long-Duration Exposure Facility, for example, was an unmanned payload delivered into orbit by the space shuttle in 1984 to perform experiments on the long-term effects of exposure to rigors of space on man-made objects. Today, Langley scientists continue to participate in the development of second- and third-generation space shuttles.

In the late 1980's, more than 4,100 civil service and private sector contract personnel worked at Langley's nineteen major testing facilities. The center operates twenty-four of the world's most advanced conventional, supersonic, transonic, and hypersonic wind tunnels for testing air and spacecraft design performance characteristics, as well as eight structural laboratories (including a massive aircraft crash test complex), seven engineering and flight simulators, eight facilities for fabricating air and spacecraft designs, seventeen research and support aircraft, a technical reference library of nearly three million titles, and a state-of-the-art supercomputer complex for design and testing support. Langley personnel also publish more than one thousand articles for technical and professional journals, research and project reports, and papers every year.

Organizationally, the NASA Langley Research Center is divided into six research directorates and a congressionally mandated Technology Utilization Program. These directorates (Aeronautics, Electronics, Flight Systems, Space, Structures and Systems Engineering, and Operations) oversee research and testing in such areas as crew emergency rescue systems for the manned space station, the development of robot arms, wind shear modeling for aircraft, advanced avionics (aircraft electronic systems), high-temperature superconducting materials, and flight simulator designs, to name but a few. The Technology Utilization Program conducts studies to find ways to put NASA-developed technologies into everyday use. Programs in this area at Langley have included stress testing on railroad car wheels, developing improved kidney dialysis machines, and other medical applications.

Context

There are two major contexts in which to view the Langley Research Center: first, as an operational research and testing center in the NASA network of space facilities around the nation; the second context in which Langley should be studied is a historic milestone in the nation's move into the skies and heavens above the surface of the planet.

The NASA Langley Research Center is one of eleven major research and operational facilities run by the space agency through its Washington, D.C., headquarters. The others are Ames Research Center at Moffett Field, California; Dryden Flight Research Facility at Edwards Air Force Base, California; Goddard Space Flight Center in Greenbelt, Maryland; the Jet Propulsion Laboratory in Pasadena, California; the Lyndon B. Johnson Space Center in Houston, Texas; John F. Kennedy Space Center at Cape Canaveral, Florida; Lewis Research Center in Cleveland, Ohio; George C. Marshall Space Flight Center in Huntsville, Alabama; the National Space Technology Laboratories in NSTL, Mississippi; and Wallops Flight Center in Wallops Island, Virginia.

These centers give NASA a multifaceted approach to the science and business of space travel. Langley, like the other NASA facilities, plays a valuable and vital part in fulfilling the space agency's mandate to keep the United States moving forward in the progression into space. Through its advanced research and testing programs, Langley has played an important role in helping the United States to build a viable and competitive aircraft industry. This development, through commercial airlines alone, has revolutionized the culture of the twentieth century. The center's design and testing services to the nation's military establishment have reshaped the posture of the United States on the world stage. The center has also contributed significantly to the development of the nation's intercontinental ballistic missile system and other launch vehicles, including the Scout rocket.

Through its aeronautical research, which comprises about 75 percent of the center's activities, Langley also interacts with public and private sector agencies and companies to advance aviation technology. Langley scientists and engineers work to make air travel safer and more efficient through the development of better electronics and aircraft designs, which are then put into production by private companies.

Beyond this role, Langley's creation in 1917 meant that the United States could compete with the United Kingdom, France, Russia, Germany, and other countries in the development of new types of aircraft and new applications of existing aviation technology in the critical first years after the Wright Brothers' first flight in 1903. Langley, in cooperation with the branches of the military, also helped prove the viability of airplanes as weapons and tools in the nation's arsenal, thereby helping prepare the nation for pivotal air confrontations — and dramatic technological advances that resulted — in World Wars I, II, and other conflicts.

Bibliography

Ezell, Edward Clinton, and Linda Neuman Ezell. *On Mars: Exploration of the Red Planet, 1958–1978*. NASA SP-4212. Washington, D.C.: Government Printing Office, 1984. This somewhat technical treatise on the nation's efforts to reach Mars is valuable as a definitive look at an important space project. Langley made a significant contribution to this effort.

Hansen, James R. *Engineer in Charge: A History of the Langley Aeronautical Laboratory, 1917–1958*. Washington, D.C.: National Aeronautics and Space Administration, 1987. This is an interesting, well-written examination of the early history of Langley Research Center. Of particular value for its insight into the origins of the center.

Langley Research Center. *Public Information Kit*. Washington, D.C.: National Aeronautics and Space Administration, 1988. Numerous brochures, pictures, and other printed materials are available free of charge from NASA on the Langley Research Center and its various activities. Many of these materials are designed for classroom use in all primary and secondary schools.

————. *Research and Technology, 1987: Annual Report of the Langley Research Center*. NASA technical memorandum TM-4021. Washington, D.C.: National Aeronautics and Space Administration, 1988. This technical report documents current and ongoing research projects conducted at the Langley Research Center. In addition to some detailed scientific information, each section offers a layperson's guide to the importance of the different projects.

McAleer, Neil. *The Omni Space Almanac: A Complete Guide to the Space Age*. New York: Pharos Books, 1987. This volume is a compendium of information about the major developments of the space age, with emphasis on the later years and their import for the future. Valuable as a reference to programs such as Viking, Echo, Explorer and other Langley-managed programs.

Eric Christensen

LAUNCH VEHICLES

Date: Beginning January 31, 1958
Type of technology: Launch vehicles, expendable

The power of launch vehicles and booster rockets is crucial to the establishment and maintenance of a reliable space program. Without powerful rocketry, it would be impossible to leave Earth's atmosphere.

PRINCIPAL PERSONAGES

ROBERT GODDARD, the father of American rocketry
WERNHER VON BRAUN, the inventor of the V-2 rocket
and the early force behind NASA's rocket booster
development program
JOHN P. HAGEN, Director of Project Vanguard
JOHN B. MEDARIS, Director of Army Ballistic Missile
Development

Summary of the Technology

The Vanguard was designed to be the first American satellite launching vehicle. It was developed, beginning in 1955, by the Naval Research Laboratory in Washington, D.C. The Vanguard rocket was 22 meters long and 114 centimeters at its widest point. It was both cylindrical and finless. Its gross weight at launch was 10,251 kilograms. The Vanguard was designed to place a 9.75-kilogram satellite into orbit.

The first stage of the vehicle was liquid fueled and generated a lift-off thrust of some 120,000 newtons. The General Electric Company built Vanguard's first-stage engine to lift the vehicle to a point about 58 kilometers above Earth's surface and to attain a velocity of close to 6,000 kilometers per hour.

The second stage, which was also liquid fueled, contained the guidance system for the entire vehicle. The engine for this stage was built by Aerojet-General Corporation and generated 33,360 newtons of thrust at its operating altitude.

The third stage of the vehicle was propelled by solid fuels and developed about 10,230 newtons of thrust at its operating altitude. The third stage had no guidance system. Stable flight was achieved by rapidly spinning the stage by means of a mechanism that was located in the second stage. After separation of the stages, the third stage was ignited. This firing of the third stage took place at orbital heights of 450 kilometers or greater. The third stage then accelerated the satellite, which was riding on its nose. After orbital velocity was achieved, the satellite was separated from the third stage by a spring mechanism activated by a mechanical timer.

The Jupiter C was based on the Redstone medium range ballistic missile. The Redstone, which was often called an offspring of the German V-2 rocket, was developed by the U.S. Army in the early 1950's. The missile was 21 meters long and 178 centimeters wide. It was stabilized by four fins at the base of the vehicle. This Chrysler-built, single-stage missile was propelled by a rocket engine using liquid oxygen as the oxidizer and an alcohol-water mixture as the fuel. The engine developed 333,600 newtons of thrust and had a burning time of 121 seconds. The weight of the missile at launch was 27,670 kilograms.

The Jupiter C consisted of a high-performance version of the Redstone (in addition to enlarged fuel tanks, a change in fuel significantly increased the lift-off thrust) as the first stage and two clusters of solid-propellant rocket motors as the second and third stages. These solid-fueled rockets had been developed by the Jet Propulsion Laboratory in Pasadena, California. Eleven of these rockets were clustered in a ring to form the second stage; three rockets were then clustered together and fitted inside the second stage ring.

The upper stage of the Jupiter C sat in a bucketlike container atop the Redstone first stage. The bucket was rotated at high speeds to stabilize the upper stages (much like a rifle bullet is stabilized in flight).

The Jupiter C was converted into a satellite launching vehicle (Juno 1) simply by adding an instrument package and an additional solid-fueled rocket as a fourth stage.

While the Juno 1, or Jupiter C, used the Redstone as its first stage, the Juno 2 used the more powerful Jupiter intermediate range ballistic missile as its first stage. The three remaining solid-fueled stages were the same for both the Juno 1 and Juno 2 vehicles. The Juno 2, however, carried a shroud that covered the upper stages to prevent aerodynamic heating during the ascent.

The Thor intermediate range ballistic missile was developed during the 1950's as the Air Force's equivalent of the Army's Jupiter missile. The Thor is a single-stage missile with a lift-off thrust of 765,056 newtons. It is propelled by a single Rocketdyne-built main engine and two small vernier engines for stabilization. The main frame of the Thor, which was built by Douglas Aircraft, was about 17 meters long. The missile weighed 44,900 kilograms fully fueled.

The Thor was used as the first stage for several different

Launch of CRRES aboard Atlas/Centaur-69 (top left); launch of Voyager I (top right); lift off of first flight of Atlantis and the STS 51-J mission (center); the Delta rocket (bottom left); Scout launch vehicle lift off. (NASA)

missile systems that were used in the U.S. space program, including the Thor-Able, Thor-Agena, Thrust-Augmented Thor, Thor-Delta, and Thrust-Augmented Delta.

The first of these configurations, the Thor-Able, was flown successfully in July of 1958. The Able element of this system consisted of the second and third stages of the Vanguard missile. With these added stages, the Thor became a viable launch vehicle for placing Air Force payloads into space.

Early in 1960, the National Aeronautics and Space Administration (NASA) decided to use the Agena vehicle, which had been developed by the Air Force, as a second stage for the Thor. This combination, the Thor-Agena, stood 23 meters tall (without the spacecraft) and measured about 2 meters at its widest point. Like the Thor, the Agena was a liquid-propelled missile. The two combined provided a total thrust of 827,328 newtons. The Agena was able to be restarted after it had been deactivated in space. This feature permitted great precision in the selection of an orbit. The Agena as modified by NASA was designated Agena B. The Thor-Agena B was capable of sending a 726-kilogram payload into an orbit 480 kilometers high or a 272-kilogram payload into an orbit 1,931 kilometers high.

The Thrust-Augmented Thor (TAT) consisted of the Thor missile with three solid-propellant motors strapped onto its base. With these three motors, the Thor's lift-off thrust was increased to about 1.5 million newtons.

The Thor-Delta was built with the Thor as its first stage, a modified and improved second stage from the Vanguard and Thor-Able designs, and a spin-stabilized, solid-propellant third stage known as Altair. The entire vehicle stood 27 meters high and was capable of launching a 363-kilogram payload into an orbit 483 kilometers high. The gross weight of the vehicle was 50,803 kilograms, and it had a lift-off thrust of 756,160 newtons.

The Thrust-Augmented Delta (TAD) consisted of the Thor-Delta configuration with strapped-on solid boosters. This spacecraft has been continuously upgraded over the years and is today known as the Delta. The liquid-fueled first stage of the 35-meter-tall rocket is augmented by nine solid-propellant motors, six of which ignite at lift-off and three of which ignite after the first six are exhausted, about 58 seconds into the flight. The Delta generates some 3 million newtons of thrust at lift-off. It has a liquid-fueled second stage and a solid-propellant third stage. The Delta's third stage has occasionally been replaced by a Payload Assist Module (PAM). This stage boosts the spacecraft from a low Earth orbit into a higher one. With the PAM and a modification of the second stage, the Delta can lift 1,270 kilograms into orbit.

In the early 1950's the Air Force began work on its intercontinental ballistic missile programs. The first intercontinental ballistic missile to be developed was the Atlas. The Atlas, built by Convair, was considered to be a stage-and-a-half vehicle. It had two side-mounted liquid-propellant rocket boosters and a liquid sustainer engine. Two small vernier

rockets were located at the base of the Atlas on sides opposite the boosters. When the Atlas was launched, all five of the engines would be running. A total of more than 1 million newtons of thrust was produced at lift-off. The vehicle stood 21 meters high, measured about 2 meters in diameter, and weighed 113,400 kilograms at launch.

After it became operational in the late 1950's, the Atlas was used as a first stage for various spacecraft. The Able and Agena configurations, which had been used with the Thor, were now mated to the more powerful Atlas.

In 1966, the Centaur, the United States' first high-energy, liquid-hydrogen, liquid-oxygen launch vehicle stage, became operational. This vehicle was combined with the Atlas. The Atlas-Centaur stood 42 meters tall. Its first stage developed close to 2 million newtons of thrust, and its Centaur stage developed 146,784 newtons of thrust in a vacuum.

The Titan, which was developed by the Martin Company in the late 1950's, was somewhat more sophisticated than the Atlas. Like the Atlas, the Titan was an intercontinental ballistic missile with a designed range of more than 8,000 kilometers. Unlike the Atlas, it was a two-stage vehicle. The first stage produced some 1 million newtons of lift-off thrust, and the second stage produced 266,880 newtons of thrust. Both stages were liquid fueled. The Titan stood 27 meters tall and was 3 meters in diameter.

Because of military considerations, the Titan was modified and became known as the Titan 2. About the only characteristic the Titan 2 shared with the Titan was its diameter. The first stage of the Titan 2 generated close to 2 million newtons of thrust, up from the 1 million newtons of thrust generated by the Titan. Its second stage had a significant increase in thrust. The Titan 2 stood 31 meters tall and had a weight of 149,688 kilograms at launch.

After various redundant components (multiple devices capable of performing the same function) had been added to ensure the workability of backup systems, the Titan 2 became man-rated and joined the Gemini manned spaceflight program in 1965.

The Titan 2 was eventually modified by the addition of two massive solid-fueled boosters and mated to the Centaur upper stage. This configuration became known as the Titan 3E/Centaur. It was first launched in 1974 and gave the United States an extremely powerful and versatile rocket for launching large spacecraft on planetary missions.

The Scout launch vehicle, which became operational in 1960, has undergone several modifications since that time. The Scout is a solid-propellant, four-stage vehicle which stands 23 meters tall. It weighs 21,147 kilograms and has a lift-off thrust of 588,203 newtons. The Scout was originally designed to place small payloads into orbit, but its uprated third stage made it possible to orbit payloads of more than 200 kilograms.

The development of the Saturn series of rockets began in 1958. The first stage of the Saturn 1 was a cluster of eight liquid-fueled engines of the type used in the Jupiter program. Each engine was capable of generating about 800,000 newtons of thrust. The second stage had six liquid-oxygen, liquid-hydrogen engines, each rated at about 65,000 newtons of thrust. The Saturn 1 stood 38 meters tall and had a base diameter of 6.58 meters. The vehicle was capable of placing a 10,000-kilogram spacecraft into Earth orbit.

The Saturn 5 was the largest, most powerful rocket ever built. This three-stage vehicle stood 111 meters tall and weighed more than 2 million kilograms when totally fueled. Its first and second stages were each powered by five liquid-fueled engines, and its third stage was powered by a single engine. The Saturn 5 was powerful enough to launch 109,000 kilograms into Earth orbit, 41,000 kilograms on a lunar mission, and 32,000 kilograms on a planetary mission.

The space shuttle consists of a delta-winged (a delta wing is a triangular wing with a tapered leading edge and a straight trailing edge) space glider called an orbiter; two solid-propellant rocket boosters, which are also reusable; and an expendable external fuel tank containing liquid propellants for the orbiter's three main engines.

The assembled space shuttle is 56 meters long and has a wingspan of 24 meters. The shuttle weighs more than 2 million kilograms at lift-off. At ignition, the orbiter's three main engines and the two solid-propellant rocket boosters burn simultaneously, generating more than 28 million newtons of thrust.

At an altitude of about 48 kilometers, the used solid rockets are parachuted into the ocean, where they are recovered by waiting ships. The orbiter and the external tank continue toward Earth orbit. After the orbiter's engines cease to operate, the fuel tank is jettisoned into the ocean. By the use of maneuvering engines, the orbiter is guided into Earth orbit for the duration of the mission (usually two to seven days). When the mission is completed, the orbiter reenters the atmosphere and returns to Earth, gliding to a landing.

Knowledge Gained

The Jupiter C, which was used to orbit the first American satellite, was developed and first used to test nose cones for the Army's Jupiter intermediate range ballistic missile. One of the major problems in the development of ballistic missiles was the aerodynamic heating of the warhead during reentry into Earth's atmosphere. With the development of the Jupiter C, these problems were solved. This knowledge was later applied to the design of manned vehicles.

Much knowledge was gained during the early space program on the development of multistaged rockets. Since such vehicles were essential for the exploration of space, their early development was critical. Prior to the Vanguard and Jupiter C rockets, the only real experience American missile designers had had with multi-stage rockets was a configuration known as the V-2/Wac Corporal. This vehicle was developed in the late 1940's and early 1950's by combining captured German

V-2 rockets with the Army's Wac Corporal artillery missile.

In spite of some spectacular successes, early progress was slow, particularly in the Vanguard program. Typical problems encountered with multistaged rockets were premature shutdown of stages, the upper stage failing to fire, and stages firing in unintended directions.

Progress was made, however, and by the early 1960's the Redstone and Atlas missiles had been equipped with backup systems and were considered reliable enough to be man-rated. These two missiles then became the boosters for the United States' first manned spaceflight program, Project Mercury.

Although the Jupiter missile never carried men into space, it did carry two chimpanzees, Able and Baker, on a suborbital flight. Their successful recovery demonstrated that living creatures could survive the heat of reentry.

In 1965, the first multistaged rocket in NASA's arsenal became man-rated. The Titan 2 then joined the Gemini manned spaceflight program. It is important to note that during the Gemini program the technique of rendezvous and docking was mastered. Without the ability to rendezvous and dock two spacecraft while in orbit, the lunar missions could never have taken place.

The Saturn project began in 1958 with the long-range goal of producing a vehicle with the capability of orbiting very large payloads. It was decided that the Saturn 1 should be built from existing, proved hardware. Thus, the first stage of the Saturn 1 used a grouping, or cluster, of eight liquid-fueled engines of the type used in the Jupiter program. This configuration proved very reliable.

The space shuttle employed a new technology for protection from reentry heat. Previous spacecraft had been coated with layers of material on the underside (the part of the spacecraft that sustains the greatest heat upon reentry). As the layers were heated, the outer layer would ablate, or fly into space, dissipating the heat. The next layers would do the same. Since the space shuttle was designed to be a reusable vehicle, a more efficient method of protecting the spacecraft from the heat of reentry was needed. The solution to this problem was the use of some thirty-four thousand heat-resistant tiles on the underside of the orbiter. These tiles, which conduct almost no heat, were made from fibers of nearly pure silica.

Context

On July 29, 1955, President Dwight D. Eisenhower announced that the United States would launch an Earth satellite during the International Geophysical Year (the eighteen-month period between July 1, 1957, and December 31, 1958). This was the genesis of the Naval Research Laboratory's Project Vanguard. Unfortunately, the Vanguard proved to be unreliable.

The development of the Jupiter C, or Juno 1 (as it was later designated), marked the U.S. Army's entry in the space race. The Army had made a case for the use of its vehicle in the mid-1950's, but the Eisenhower Administration had instead favored the Vanguard program. While the Vanguard program was lagging, the Soviets shocked the world with the launching of Sputnik 1 on October 4, 1957. Shortly thereafter, the U.S. Secretary of Defense gave the Army permission to make launch preparations for its Jupiter C. Only three months later, Explorer 1 was fired into orbit. The first successful American satellite, Explorer 1 helped to regain some of the prestige lost because of the Sputniks and early Vanguard failures.

Shortly after its formation in October of 1958, NASA announced plans for a manned Earth satellite program. Since the Redstone had proved to be so reliable in past launches, it was selected as the vehicle that would carry the first Americans into space.

The Army's Jupiter missile did not become part of the manned spaceflight program but did play an important role in the space effort. In 1958, the upper stages of the Jupiter C missile were added to the Jupiter to form the Juno 2. On March 3, 1959, a Juno 2 vehicle sent a conical-shaped payload named Pioneer 4 past the Moon and into a solar orbit.

The Air Force's Thor intermediate range ballistic missile became the workhorse of the 1960's and the 1970's, as it was combined with various upper stages. Two of the most notable were the Thor-Agena and Thor-Delta configurations. The Thor-Agena was used successfully in the launching of meteorological, communications, and scientific satellites, including the Orbiting Geophysical Observatories and the Echo 2 communications satellite. It was also used in the launching of various military payloads for the Air Force.

First launched by NASA in May of 1960, the Thor-Delta became a reliable vehicle for a wide range of satellite missions. It launched the first orbiting solar observatory, and satellites in the TIROS, Echo, Telstar, Relay, and Syncom programs.

The intercontinental ballistic missiles Atlas and Titan were used extensively both in the manned and in the unmanned space progams. The Atlas served in Project Mercury, sending John Glenn and other astronauts into orbit. In addition, the Atlas was united with various upper stages, such as the Agena and Centaur, for unmanned satellite missions and lunar probes. These included the Applications Technology Satellites and the Ranger and Lunar orbiter projects. The Atlas was also used as a booster for probes to Venus and Mars as part of the Mariner program.

The Titan was used to launch ten manned Gemini spacecraft into Earth orbit during the years 1965 and 1966. It has since been combined with strap-on boosters to form the Titan 3E/Centaur. This vehicle successfully launched two Viking Mars landers, two Voyager spacecraft to the Jovian planets, and two Helios spacecraft toward the Sun.

The Saturn launch vehicle was used in the Apollo manned lunar landing program. After the completion of the Apollo program, the Saturn 1B was used to launch three manned missions to the Skylab space station in 1973. In 1975, it launched the American crew for the Apollo-Soyuz Test Project, the joint United States/Soviet Union orbital docking mission. The massive Saturn 5 was used to launch the space station Skylab in May of 1973.

The space shuttle, which was first flown in April, 1981, was designed to carry large, heavy payloads into Earth orbit. The shuttle was also designed to serve as a satellite checkout and repair vehicle. (In fact, the most famous success of the space shuttle was the Solar Maximum Mission satellite repair.) In addition, the shuttle has been used for meteorological, oceanographic, and cartographic study — further establishing the value of research performed beyond Earth's atmosphere.

In the wake of the Challenger accident, NASA decided to reinstate expendable launch vehicles. Through the middle of the 1990's, NASA had ten launch vehicles (or variations) in addition to the space shuttle. Pegasus is the first successful space vehicle to be launched from the air. The winged vehicle and its payload are carried to an altitude of about 12 kilometers by a Lockheed L-1011 wide-body aircraft. It can carry a 300 kilogram payload to low Earth orbit. A longer version, Pegasus XL, can loft a 450-kilogram satellite to orbit. A ground-launched variation of Pegasus is Taurus, which utilizes a wingless Pegasus atop a large solid-fuel rocket.

The Lockheed Martin Launch Vehicle, LMLV 1, uses two solid-fuel stages, while the LMLV 2 uses three stages and the LMLV 3 adds strap-on boosters to the stack. The Titan II is still in operation, launched with and without strap-on solid rocket boosters. The Titan IV variation is the primary launch vehicle for heavy Air Force payloads. The Delta has been upgraded through sixteen variations and has used as many as nine strap-on solid boosters.

NASA's oldest family of launch vehicles, Atlas, has four current models. The Atlas 1, Atlas II, and Atlas IIA retain the vehicle's original one-and-a-half stage design and carry a liquid fuel upper stage. The Atlas IIAS adds four solid-fuel strap-on boosters for carrying 18,000 kilogram payloads to low Earth orbit and 4,500 kilogram payloads to geosynchronous orbit.

Bibliography

Baker, David. *The Rocket: The History and Development of Rocket and Missile Technology*. New York: Crown Publishers, 1978. A well-illustrated, highly detailed volume recounting the history of rocketry, from the invention of gunpowder to the landing of a man on the Moon. Suitable for general readers.

Braun, Wernher von, et al. *History of Rocketry and Space Travel*. New York: Thomas Y. Crowell, 1975. A history of rocketry and spaceflight from the ancient Chinese rockets to early Apollo missions. Well illustrated.

Emme, Eugene M., ed. *The History of Rocket Technology: Essays on Research, Development, and Utility*. Detroit: Wayne State University Press, 1964. A collection of fourteen papers written by scientists and historians covering the development of rocketry from Robert Goddard's first liquid-fueled rocket through Project Mercury. Suitable for general readers.

Green, Constance M., and Milton Lomask. *Vanguard: A History*. Washington, D.C.: Government Printing Office, 1970. Part of the NASA History Series, this work traces the evolution of the Vanguard program from its genesis to the orbiting of Vanguard 1.

Haley, Andrew G. *Rocketry and Space Exploration*. New York: Van Nostrand Reinhold Co., 1958. A general history of rocketry from the ancient Chinese to the early years of the space race with the Sputniks and the Explorer satellite. Suitable for general readers.

Holder, William G. *Saturn V: The Moon Rocket*. Edited by Glenn Holder. New York: Julian Messner, 1969. A brief history of the development of the rocket from the ancient Chinese to the American manned space program. Includes an excellent description of the Saturn 5 rocket and its launching.

Ley, Willy. *Rockets, Missiles, and Men in Space*. Rev. ed. New York: Viking Press, 1968. A very detailed work starting with the ideas of ancient astronomers such as Galileo and Johannes Kepler and building up to manned spaceflight. The text is suitable for the general reader. The extensive appendices are more technical.

National Aeronautics and Space Administration. *Countdown! NASA Launch Vehicles and Facilities*. Washington, D.C.: Author, 1978. A collection of short articles on various NASA launch vehicles, both active and inactive, and a description of NASA facilities.

————. *NASA, 1958 – 1983: Remembered Images*. NASA EP-200. Washington, D.C.: Government Printing Office, 1983. A well-illustrated booklet describing the first twenty-five years of NASA achievements. Included are tables of launch vehicles and brief summaries of missions.

David W. Maguire

LEWIS RESEARCH CENTER

Date: Beginning October 1, 1958
Type of facility: Research center

Lewis Research Center performs basic and applied research to develop technology in aircraft propulsion, space propulsion, space power, microgravity science, and satellite communications. The center manages projects that validate new technology and produce new flight systems.

PRINCIPAL PERSONAGES

GEORGE W. LEWIS, first Director of Research for the National Advisory Committee for Aeronautics (NACA)

HUGH L. DRYDEN, a major contributor to the transformation of NACA to the National Aeronautics and Space Administration (NASA) and to the conduct of propulsion work at Lewis

FREDERICK C. CRAWFORD, a pioneer aircraft industrialist in Cleveland who led the effort to locate Lewis Research Center in that city

EDWARD R. SHARP, the first Director of Lewis

ABE SILVERSTEIN, the second Director of Lewis and a key figure in the reorganization of NACA into NASA

BRUCE T. LUNDIN, the third Director of Lewis and the principal architect of the thinking that led to the final structuring of NASA

JOHN F. MCCARTHY, JR., the fourth Director of Lewis, who solidified the center's position as the leading U.S. center for satellite communications

ANDREW J. STOFAN, the fifth Director of Lewis, who was responsible for securing for Lewis the assignment to develop the space station power system

JOHN M. KLINEBERG, the sixth Director of Lewis, who enhanced the center's technical excellence

DONALD J. CAMPBELL, Director of Lewis as of 1996.

Summary of the Facility

Lewis Research Center first was called the Aircraft Engine Research Laboratory of the National Advisory Committee for Aeronautics (NACA). It was renamed the Lewis Flight Propulsion Laboratory in 1948 and the Lewis Research Center in 1958, when NACA became the National Aeronautics and Space Administration (NASA).

The first buildings at Lewis were erected on an 80-hectare plot of land next to the Cleveland airport on the extreme west side of the city. Subsequently, another 60 hectares next to the airport were annexed, and 3,200 hectares of land near Sandusky, Ohio, about 80 kilometers west of Cleveland, were acquired to become Lewis' Plum Brook Station.

When it opened in the early 1940's, Lewis Research Center had the urgent but narrow task of improving the performance of piston aircraft engines. By 1950, however, the focus of its aeronautical work had shifted to gas turbine (jet) engines for aircraft propulsion systems, and the center had begun to expand into rocket propulsion systems for space exploration.

The early work on improving the performance of gas turbine engines concentrated on the basic objective of developing the fundamental operating technology. The next step, in the 1960's, was research geared to developing propulsion systems that would allow aircraft to go higher, farther, and faster and to increase reliability and maintainability. In the 1970's, however, the focus changed again. Work was begun to put propellers back on commercial transport aircraft. New life was breathed into research on a propeller system that would give the performance of a jet engine in terms of speed and altitude but with fuel savings of 15 to 30 percent over the most efficient jet engines. The new systems, however, would use modern gas turbine engines instead of piston engines, not to provide thrust but to turn the propellers.

During the 1970's and 1980's, Lewis managed the Advanced Turboprop (ATP) program, which involved both single rotation and dual counterrotation propeller systems. Although the technology of each propeller system was distinct, both research efforts were aimed at the "repropellerization" of commercial transport aircraft. The two programs reached their goals in 1987, when a series of flight tests verified the readiness of advanced turboprop technology for use in commercial transport.

One of the oldest areas of aeronautical research at Lewis deals with the causes and prevention of ice formation on aircraft during flight. The icing tunnel at Lewis, built in response to a request from the United States Army Air Corps in 1944, incorporates unique features that have allowed scientists to study icing problems using full-scale aircraft components. The oldest active icing tunnel in the world, it was declared an international historic mechanical engineering landmark in

1987 by the American Society of Mechanical Engineers.

Lewis engineers worked on liquid hydrogen as a fuel during the NACA era and developed confidence in handling it. Nevertheless, the feasibility of a liquid hydrogen-oxygen upper-stage rocket had never been demonstrated. In 1951, Lewis received its first formal appropriation for rocket research, though the number of people assigned to the rocket section was still small. Lewis' work in high-energy chemical rockets led to the development of the Centaur upper-stage vehicle. In 1962, Lewis took on the job of managing the Centaur, which is powered by two liquid hydrogen-oxygen engines with 66,720 newtons of thrust each. Centaur was the nation's first sizable space vehicle, able to shut down and restart engines in order to change direction and velocity in space. With both Atlas and Titan rockets as launch vehicles, the Centaur upper stage has been a major factor in the American thrust into space, for it has propelled Surveyor spacecraft to the Moon, scientific satellites to the outer planets, and communications satellites into orbit around Earth. In all, the Lewis-managed Centaur has made more than one hundred successful flights.

The establishing in 1960 of NASA's joint program with the Atomic Energy Commission to develop a nuclear rocket grew out of Lewis' commitment to farsighted research — in this case, investigating nuclear power as a means of aircraft propulsion. Lewis acquired a cyclotron in 1949, and in 1956 the Atomic Energy Commission approved plans for Lewis' nuclear reactor at Plum Brook.

Although nuclear propulsion for aircraft was approached by many with high hopes, problems involving the weight of the necessary shielding and heightened environmental concerns led to the ultimate cancellation of the program in 1961. Nevertheless, the experience gained in nuclear aircraft technology became the foundation for the work on nuclear rockets that continued through the Apollo decade of the 1960's.

Interest in electric rocket propulsion was stimulated originally by an idea dating back to the 1920's — that for space travel, rocket thrust could be produced by the flow of electrically charged particles. The chemical rocket was limited by the enormous amounts of propellant that would be required for flight to distant planets. The advantage of using electric rockets is that although the amount of thrust is small relative to chemical rockets, with a power source in space, electric rockets can produce thrust over longer periods of time. Thus, electric rockets might be useful for long-distance travel between planets. Yet what would their power source be? Their interest stimulated by work described in early papers by European scientists, Lewis engineers began a program to investigate the use of a nuclear reactor to power an electric rocket. Electric propulsion remains a major area of research at Lewis as the United States explores deep space.

Beginning in the early 1970's, work done at Lewis in a joint project with the government of Canada led to the development of the Communications Technology Satellite (CTS). Launched in 1976, CTS was the first communications satellite to incorporate a high-efficiency, high-power transmitting tube invented at Lewis. The tube made it possible for CTS to operate at power levels ten to twenty times higher than those of previous satellites. It also made possible operation in a new frequency band, the Ku-band. In turn, the improved broadcast capability required much smaller and less expensive ground receiving equipment; it also made possible transmission to remote areas in which terrestrial communications were not highly developed.

Lewis researchers were responsible not only for the development of the high-power transmitter tube for the CTS but also for environmental testing and for the launch vehicle and associated launch services. The spacecraft itself was developed by the Communications Research Center in Canada. The CTS transmitter was turned off in October, 1979, after more than three years of successful experimentation. In 1987, NASA received an Emmy from the National Academy of Television Arts and Sciences for developing technology that had improved television broadcasting throughout the world.

In 1984, Lewis entered the mainstream of the U.S. space program and won the contract to develop the power system for the space station. This power system serves as a miniature electric utility for a small community of people living and working in space. The total system generates, stores, conditions, and distributes electric power to support human life, operate research equipment, and transmit data.

Initially, electrical power for the space station was to be provided by solar cell arrays, which would convert the light of the Sun directly into electricity. A solar dynamic system, which would use the heat of the Sun to power a heat engine to turn a generator, was being developed for eventual use in the space station.

Lewis' work in microgravity experiments began in the early 1950's in a small drop tower that could produce reduced gravity conditions for experimental packages during a 30-meter free-fall lasting approximately 2.2 seconds. In 1966, a second, larger drop tower, some 150 meters deep, was built. It could produce conditions of one ten-thousandth of Earth gravity for more than five seconds. In 1979, Lewis acquired a Model 25 Learjet, which, when flown in a parabolic trajectory, can provide reduced gravity conditions for 18 to 20 seconds.

The Microgravity Materials Science Laboratory at Lewis opened in 1985. This laboratory is equipped with functional duplicates of the flight hardware used on the space shuttle and the space station. Its purpose is to stimulate the development of experiments in microgravity by offering U.S. scientists and engineers a low-cost, low-risk way to test new ideas for reduced-gravity materials science research. Access to such a facility gives U.S. companies a competitive edge in developing better products through microgravity research. The Micromaterials Science Laboratory has three major sections, for work in metals, alloys, and electronic crystals; glasses and ceramics; and polymers.

In the mid-1970's, Lewis began to conduct research projects on terrestrial energy technology at the request of and with funding from the Department of Energy. As in the case of the advanced turboprop work, this research was driven by a sharp rise in the price of petroleum. There were also environmental considerations. The work included automotive (electric, gas turbine, and Stirling) programs; the development of technology for wind turbine systems (two-bladed windmills) to generate electricity; the development of fuel cell technology for on-site generation; and photo-voltaic (solar cell) programs.

Despite the fact that automotive technology was developed to the point that several experimental vehicles were built at Lewis, neither electric nor gas turbine power for automotive applications has become practical. The story of the development of Stirling technology, however, is different. A Stirling engine has pistons which are inside sealed cylinders and are moved by a working gas that expands and contracts with the addition and removal of heat. In automotive applications, the heat comes from the burning of fossil fuels. The experimental vehicles can run on anything from ordinary gasoline to aftershave lotion. Stirling technology also may have application in the power system for the space station. For space uses, the source of the heat could be either solar or nuclear.

In 1987, a wind turbine with a 3.2-megawatt power rating went on line in the utility grid of the Hawaiian Electric Company. That event was the culmination of fifteen years of intensive work by Lewis engineers to harness a renewable energy source, the wind, to generate electric power. The machine has a rotor that spans more than 97 meters tip to tip and can generate more than 13 million kilowatt-hours of electrical energy annually, as much as would be used by 1,300 typical single-family homes.

A fuel cell is a device that generates electricity on site for such commercial and residential applications as apartment buildings, restaurants, retail stores, and nursing homes. In operation, natural gas or some other hydrocarbon is converted into a hydrogen-rich fuel in a fuel processor and fed to one electrode, while oxygen (in this case, from the air) is fed to another electrode. Electricity is produced by a process similar to that by which it is produced in an automobile storage battery. Since there is no combustion, no pollution-control devices are required. Furthermore, the operation of a fuel cell is virtually noiseless.

Lewis qualified itself for its space station assignment over a period of two decades with basic research in silicon and gallium arsenide solar cell technology. Indium phosphide as a solar cell material also is being investigated. In the 1970's, research moved from the laboratory into the field, in a pilot program that put solar-powered refrigerators in developing countries around the world. The purpose of the program was to demonstrate on-site use of solar cells, in this case to provide the electrical power needed to refrigerate vaccines.

In 1986, Lewis was assigned to play a major role in the development of the propulsion system for the National Aerospace Plane, a flight vehicle that would take off from a conventional runway, fly to orbit, maneuver in the upper atmosphere, and return to Earth to land at an airport anywhere in the world. The plane also could fly at hypersonic speeds (6,436 to 12,872 kilometers per hour) in the upper atmosphere. Other participants in the program are NASA's Langley Research Center, the Defense Advanced Research Projects Agency, the Air Force, the Navy, and the Strategic Defense Initiative Organization.

Context

For the most part, the political climate in Washington, D.C., in the fall of 1957 was very negative toward space, and any enthusiasm for space research was viewed suspiciously. Against that backdrop of opinion in the capital and throughout the government, Lewis was preparing for its triennial inspection in mid-October by a group of officials and experts that included NACA headquarters personnel, members of Congress, and executives of the aeronautics industry. How should the center's space-related propulsion work be displayed? Causing even more apprehension was the question of how it would be viewed. The Soviets solved both problems.

On October 4, just days before the inspection, the Soviet Union launched Sputnik, changing forever the nation's and the world's outlook on space exploration. When the visitors arrived, Lewis engineers proudly unveiled their work in space-flight propellant systems and high-energy rocket propellants. The vision and eagerness of Lewis scientists to venture into the unknown had been more than vindicated.

When NASA was created by Congress in 1958, much more was involved than a mere name change and the substitution of one letter in the acronym for another. The fundamental character of the agency was to undergo a major shift, and the technology it would have to develop would take it down untrodden paths of research. Managing a different type of organization that now would be responsible for missions and operations as well as for its traditional research role would be slightly easier, however, because of the head start in technological development that Lewis had provided.

During the 1950's, Lewis pioneered many important innovations in gas turbine propulsion systems. That work improved the reliability of the gas turbine engine, and by the late 1960's commercial jet passenger service had become a commonplace means of mass transportation over long distances. The piston engine for commercial transport aircraft was gone forever. Propeller technology, however, would return.

Lewis had been involved in advanced propeller research from its very beginning in the 1940's. Early studies indicated the potential for good performance and fuel economy for advanced propeller designs, but with the advent of the jet engine and an abundance of low-cost fuel, propeller work was shelved in the 1950's in response to the public preference for jet-powered aircraft, which could provide fast, comfortable transportation at altitudes well above normal air turbulence.

In the 1970's, however, sharply higher fuel prices and public concern for quality of life and energy use led to research that had environmental and societal goals, and Lewis began to develop technology that would produce aircraft propulsion systems that were cleaner, quieter, and more fuel efficient. The concern for energy conservation, which became a national driving force during the 1970's, led to one of the most unusual research projects ever undertaken at Lewis: work to put propellers back on commercial transport aircraft.

In 1988, the National Aeronautic Association awarded its prestigious Collier Trophy to Lewis and the NASA/Industry Advanced Turboprop Teams — which included Hamilton-Standard Company, General Electric Company, Lockheed Missiles and Space Company, Boeing Aerospace Company, McDonnell Douglas Astronautics Company, Allison Company, and Pratt and Whitney Aircraft Division — for success in the development of advanced turboprop technology for commercial transport use. Although NACA had received the award several times and NASA has received it for astronautic achievements, this was the first time that NASA had received the award for its work in aeronautics.

Bibliography

Anderson, Frank W., Jr. *Orders of Magnitude: A History of NACA and NASA, 1915–1980.* 2d ed. NASA SP-4403. Washington, D.C.: Government Printing Office, 1981. A revised and enlarged version of a work prepared originally for the United States' bicentennial celebration. A readable and well-illustrated volume. Covers the period from just before World War I to just before the launch of the first space shuttle in 1981. Recounts NACA's aeronautical achievements and the Mercury, Gemini, and Apollo projects that put Americans on the Moon and established routine access to space as a familiar concept in the American mind.

Graham, Robert W. *Four Giants of the Lewis Research Center.* NASA TM-83642. Cleveland, Ohio: Lewis Research Center, 1984. George W. Lewis, Hugh L. Dryden, Edward R. Sharp, and Abe Silverstein are singled out for their contributions to Lewis and to the development of aerospace technology.

National Aeronautics and Space Administration. *NASA, the First Twenty-five Years: 1958–1983.* NASA EP-182. Washington, D.C.: Government Printing Office, 1983. Written in a clear, easy-to-read style, this book is meant to be a reference for teachers. It includes a history of NASA and an overview of NASA's work in various fields, including the aeronautics and energy research conducted at Lewis Research Center. Also provides descriptions of the various NASA facilities. Illustrated.

Roland, Alex. *Model Research.* NASA SP-4103. Washington, D.C.: Government Printing Office, 1985. A two-volume historical work that examines NACA as an institution. Attempts to explain how and why NACA functioned and to evaluate it as a research organization.

Sloop, John L. *Liquid Hydrogen as a Propulsion Fuel, 1945–1959.* NASA SP-4404. Washington, D.C.: Government Printing Office, 1978. A misleadingly titled book that gives, in addition to what the reader expects to find, a most lively account of some of the leading figures in the history of Lewis, their contributions, and their personalities. Much of the reading is light and enjoyable.

John M. Shaw

THE LUNAR ORBITER

Date: August 10, 1966, to January 31, 1968
Type of program: Unmanned lunar probes

The Lunar Orbiter program was one of three unmanned spacecraft programs designed to help scientists select safe landing sites on the Moon for the Apollo program. In one year, the orbiters provided more new information about the Moon than had been gathered in hundreds of years of observation from Earth.

PRINCIPAL PERSONAGES
> HOMER E. NEWELL, NASA Associate Administrator
> HAROLD MASURSKY, a scientist at the Jet Propulsion Laboratory, California Institute of Technology
> LEE R. SCHERER, Director of Apollo Lunar Exploration, NASA
> CLIFFORD H. NELSON, Assistant Director of Langley Research Center
> ROBERT J. HELBERG, Manager of the Lunar Orbiter program, the Boeing Company

Summary of the Program

The 385-kilogram Lunar Orbiter spacecraft was a spaceborne photographic laboratory. Consistent with the program's primary mission, the major instrumentation aboard the Lunar Orbiters was the photographic subassembly made up of a camera with two lenses — a field lens and a telephoto lens for extreme close-up views — housed in a pressurized and temperature-controlled container. The two lenses operated simultaneously. The camera frame, film supply, and other mechanisms were the same for the two lenses. Exposed film was developed in flight by a damp, monochemical process into a high-quality photographic negative, which was then dried by a miniature heater. A tiny beam of light about as thick as a human hair then scanned the completed negative and was converted into electrical signals for radio transmission back to Earth.

As the target area moved into the sunlit zone, the Lunar Orbiter changed altitude, moving from 200 kilometers to about 46 kilometers (Orbiter 1 went within 39.7 kilometers of the surface), in order to get closer to the Moon's landscape. Each exposure of the dual-lens camera simultaneously exposed one high-resolution and one medium-resolution frame. An overlap of 50 percent or more of the medium-resolution photographs permitted stereo viewing, which discloses surface details not readily apparent from single images. The photographic system allowed versatility in selecting the site to be photographed. The exposed film was developed and stored on a take-up reel for later transmission to ground antennae.

On the ground, a series of antennae picked up the transmitted radio signals and displayed the data line-by-line on a kinescope, which looks like a television tube. Cameras on the ground were used to photograph the image on the tube in 35-millimeter film strips from which 23-by-36-centimeter photographs were made. Data recorded on film aboard the Lunar Orbiter also provided the time, spacecraft altitude, location, lighting conditions, and other information needed for analysis by scientists. The first two Lunar Orbiters, for example, covered more than 48,270 square kilometers of landing sites. Orbiter 3 rephotographed the same areas in a confirmation mission.

Without the solar panels and antennae, the Lunar Orbiter spacecraft was about 1.5 meters in diameter and 1.6 meters high. At the time of launch, the solar panels were folded up under the spacecraft base, and the antennae were fixed against the side of the structure. A payload shroud covered the entire spacecraft during the launch phase to protect the structure from atmospheric conditions that might have caused damage.

The structure of the spacecraft consisted essentially of a main equipment mounting deck and an upper module supported by trusses and an arch. The module supported the control engine and propellant tanks as well as the high-pressure nitrogen tank, which provided pressurization for the engine feed system and attitude control thrusters (a spacecraft's attitude is its pointing position during flight in reference to some fixed object in space; if the craft begins to "fall over," attitude control thrusters automatically spew nitrogen gas through small jets and force the craft to return to its original position).

The Lunar Orbiter left Launchpad 13 at Cape Canaveral, Florida, at approximately three-month intervals throughout the overall project. The launch vehicle for Lunar Orbiter was a combination of two rockets — a 27.4-meter-tall, two-stage Atlas-Agena D rocket. The Atlas rocket sent the payload to about 136 kilometers and then shut down and separated from the group, leaving the Agena to continue boosting the Lunar Orbiter on a trajectory to the Moon. The protective payload shroud also was ejected at that time. The Agena booster then ignited and continued pushing its payload forward. Once the

Lunar Orbiter 1. (NASA)

Agena established a trajectory to the Moon, it, too, shut down and separated, leaving only the Lunar Orbiter flying through interplanetary space. The work of the boosters was finished; the work of the Lunar Orbiter now began.

In translunar flight, the Lunar Orbiter solar panels opened out and the two antennae opened. Control of the mission was then transferred from the Florida launch site at Cape Canaveral to the Space Flight Operations Facility at the Jet Propulsion Laboratory (JPL), in Pasadena, California, which would control and command the Lunar Orbiter for the duration of the mission. The spacecraft then turned so that the solar panels faced the Sun and the Sun-sensor aboard the craft could identify the Sun as its reference point in space. In addition, the Lunar Orbiter had a star-tracker sensor to fix on the star Canopus as another reference point. The journey to the Moon took about three days.

As the Lunar Orbiter reached the vicinity of its target, its speed decreased, allowing the Moon's gravitational field to capture the craft and place it in an elliptical orbit. Constant trajectory corrections were automatically made by the spacecraft. It took about three orbits to establish a correct orbital path. A single orbital path of a Lunar Orbiter took approximately three and a half hours.

All five of the Lunar Orbiters carried instrumentation for additional scientific investigation of the Moon. Information was transmitted about the density of micrometeoroids in the vicinity of the lunar environment, the frequency and severity of solar flares, and the extent of magnetically trapped particles. Specific tracking techniques were designed to refine data about the Moon's gravity and shape.

The basic design of the Lunar Orbiter craft was so responsive and so versatile that it was never changed during the five flights. At the end of each separate mission, the spacecraft was intentionally crashed into the lunar surface to prevent the complications of possible overlapping radio signals from subsequent spacecraft operating near or on the lunar surface. Lunar Orbiters 1, 2, and 3 were downed on the Moon's far side.

Orbiter 1 surveyed sixteen potential Apollo landing sites; Orbiter 2 examined thirteen secondary landing sites and captured the crash site of Ranger 8. Orbiter 3 surveyed forty-one Apollo landing sites and captured Surveyor 1 sitting at its landing site. Orbiter 5 examined five Apollo sites.

Some mechanical anomalies occurred during the several flights, most relating to the camera subsystem, but only slight modifications to the original design of the spacecraft were necessary. On Orbiter 1, problems with the camera caused the high-resolution film to blur, making 25 percent of its mission unusable. The film advance mechanism on Lunar Orbiter 3 broke after achieving only 182 frames, or about 72 percent of its work. Because of the intense heat of the Sun directly on the flat panel protecting the main instruments, several hundred small mirrors, each measuring about two and one-half centimeters square, were added to the decks of Orbiters 4 and 5 to reflect sunbeams and keep the temperature within tolerable limits inside the spacecraft's thermal blanket.

The five orbiters received and executed about 25,000 commands. They performed more than 2,000 maneuvers without a failure. In essence, the Lunar Orbiter spacecraft design was an excellent one for the job. By January 31, 1968, the last orbiter was allowed to impact with the Moon's surface.

Knowledge Gained

The five Lunar Orbiter missions took place from August, 1966, to January, 1968. During this time, all five spacecraft recorded imagery of the Moon's surface, although nearly all the mission's photographic requirements were satisfied by Lunar Orbiters 1, 2, and 3. As a result, Lunar Orbiters 4 and 5 were assigned chiefly scientific activities, although a few photographic images were recorded on each flight.

The Lunar Orbiter missions increased knowledge of the Moon tenfold and represented a quantum leap in understanding the Moon as a body. The urgency of the missions to enable scientists to find suitable landing sites for Apollo astronauts was equaled by the need to determine the nature of the environment in order to provide protection for astronauts exposed to the lunar condition.

In the 800 days of the five missions, the Lunar Orbiters recorded only eighteen meteoroid hits, in a random distribution pattern, putting to rest fears that moon-walking astronauts would suffer from micrometeoroid bombardment. Radiation levels recorded near the Moon were generally low. Occasional solar flares registered relatively high levels on the radiation-detection instruments.

As a result of combined Lunar Orbiter and Surveyor photographs, eight landing sites were chosen from an original list of about thirty sites along the Moon's equator. The list of eight sites was then reduced to three or four areas in three lunar sea regions — the Sea of Tranquillity, Central Bay, and Ocean of Storms. Orbiter imagery of potential landing sites was sufficiently detailed that features on the surface measuring as small as 1 meter across in the telephoto images and 8 meters in the wide-angle photographs could be seen. In comparison, the smallest objects on the Moon that can be seen through telescopes on Earth are about eight-tenths of a kilometer across.

For the first time, the Moon's hidden side was almost completely mapped. While circling the Moon on August 8, 1967, Lunar Orbiter 5 recorded the first image of Earth from the Moon, capturing a nearly full Earth view. The missions provided the first definitive information about the Moon's gravitational field, and Lunar Orbiter 1 captured the image of the area where the earlier Surveyor 1 spacecraft had soft-landed. Finally, the mission provided the first vertical views of the Moon's poles, as well as the entire areas of the eastern and western limbs (which are only partially observable from Earth).

Maps produced from contributions of the Lunar Orbiter missions are infinitely more detailed and point-accurate than were any previous lunar maps. Stereo views of lunar surface features enabled scientists to calculate near-precise dimensions of craters, crater walls, mountains, valleys and rilles, and scarps, and to speculate on the origins of a number of special features, such as the meandering Prinz Valleys near the Harbinger Mountain. The detailed maps and individual stereo views of various regions significantly increased scientific knowledge about the possible origins of densely cratered areas, mountain ranges, and possible volcanic sites. An oblique view of Crater Copernicus, imaged by Orbiter 2 from an altitude of 45 kilometers, was hailed by scientists as "the picture of the century."

Context

The Lunar Orbiter missions were sandwiched in with three successful Ranger missions and six Surveyor soft-landing missions to the once-forbidden and hostile world of the Moon. An immense body of information was needed about the physical nature of the Moon before a successful manned expedition to the stark lunar surface could be mounted. Before then, the unanswered questions about the Moon and about human survival there were nearly endless and included everything from the possible danger of unusually high dosages of radiation (the Moon has no atmosphere to protect it from radiation) that might kill even a thick-suited astronaut to such mundane puzzles as the distance a man could walk in the deep lunar dust. In 1966, NASA was trying to design a lunar exploration schedule and had to have answers. The Lunar Orbiter provided some of those answers.

If Americans were to reach the Moon before the Russians, if astronauts were to survive the ordeal of the Moon mission, if the United States were to regain its lost international prestige, then it was necessary to define the Moon in understandable terms. The primary task facing NASA was to obtain detailed photographs of selected areas near the Moon's equator to help space scientists choose the safest sites for landings by Apollo astronauts. The Apollo zone of interest lay along the Moon's equator in an area covering east and west longitude 45 degrees and 5 degrees north and south latitude on the Moon's front face. Each of the landing sites was an oval measuring about 4.8 by 8 kilometers, with an approach path 48 kilometers long for the lunar module. At the 48-kilometer point, the module would be at an altitude of 7,620 meters. The terrain under the approach path had to be familiar to the astronauts. The actual touchdown point for the module would have to be at least 7.6 meters square, free of craters, and sloping not more than seven degrees.

Although the primary objective of the Lunar Orbiter program was to search for and identify safe landing sites, the overall result of the five missions was something close to miraculous, for what scientists discovered in the process was an entirely new concept of the Moon. The extremely detailed images of lunar topography showed a lunar surface that humans had never seen before and allowed the study of lunar terrain features that, up to the time of the Orbiter missions, had not been possible using only Earth-based telescopic instrumentation. In addition to their photographic and measuring functions, the Orbiters served as tracking targets for NASA's Manned Space Flight Network. They provided training for the tracking station crews and were used to verify computer programs and orbit determinations for the Apollo flights.

Americans were latecomers to the task of photographing the Moon at close range. At the same time that the Ranger-Surveyor-Orbiter triad was at work, the Soviets had already hard-landed Luna 2 near Crater Archimedes in September, 1959. In October of the same year, Luna 3 photographed the Moon's far side and sent the images streaming back to Earth. Then, in January of 1966, six months before the Lunar Orbiter 1 mission, the Soviets were the first to soft-land a photographic spacecraft — Luna 9 — onto the Moon in the Ocean of Storms. While the television pictures of the surrounding surface were not up to the quality of American photography, the landing proved that the lunar soil was shallow enough to support a standing human.

The Lunar Orbiter photographs of the Moon stand as the definitive source of information about the Moon. In fact, enough photographic product was generated by the five missions to keep researchers occupied for at least half a century and, together with photographs taken by the six teams of astronauts that walked on the lunar surface as well as numerous images taken from the orbiting command module, opened a new era in the scientific study of the Moon.

Bibliography

Cortright, Edgar M., ed. *Apollo Expeditions to the Moon*. NASA SP-350. Washington, D.C.: Government Printing Office, 1975. The best authoritative reference work on the U.S. assault on the Moon, this book includes an excellent but brief review of the Lunar Orbiter program. Included in the review are diagrams and photographs of the spacecraft, mission sequences, and lunar photographing sequences. The volume's most important value is its overall view of the Apollo program, with the entire American space program as its background.

Gatland, Kenneth. *The Illustrated Encyclopedia of Space Technology: A Comprehensive History of Space Exploration*. New York: Crown Publishers, 1981. The entire history of spaceflight is compiled in this comprehensive encyclopedia, including a brief description of the Lunar Orbiter program. This reliable reference includes illustrations and color photographs.

Koppes, Clayton, R. *JPL and the American Space Program: A History of the Jet Propulsion Laboratory*. New Haven, Conn.: Yale University Press, 1982. The history of the JPL facility is told in this definitive work. Virtually the story of the American planetary exploration program in its golden age, the volume includes a useful section on the lunar-exploring craft that preceded the Apollo program.

Kosofsky, L. J., and Farouk El-Baz. *The Moon as Viewed by Lunar Orbiter*. NASA SP-200. Washington, D.C.: Government Printing Office, 1970. This book contains a compilation of the photographs from the five Lunar Orbiter missions; they were chosen primarily to illustrate the heterogeneous nature of the lunar surface. Included in the introduction are brief descriptions of the Orbiter project, its spacecraft and camera systems, and the lunar surface coverage of the five missions. An appendix includes four stereoscopic views and a collapsible set of stereo eyeglasses is pasted into the book for the convenience of the reader.

Man, John, ed. *The Encyclopedia of Space Travel and Astronomy*. London: Octopus Books, 1979. This nicely illustrated compendium tends to put the global space community into perspective by discussing the technology of all nations in concurrent situations. The Lunar Orbiter, for example, is explained against the backdrop of the Soviet drive for the Moon.

Shelton, William R. *Man's Conquest of Space*. Washington, D.C.: National Geographic Society, 1968. This volume was written and published just before the era of the Apollo Moon landings and contains photographs and some data not readily available. As the definitive work on the human reach into space in 1968, it provides an overview of all the missions and spacecraft.

Weaver, Kenneth F. "The Moon: Man's First Goal in Space." *National Geographic* 135 (February, 1969): 207 – 245. Of all the excellent National Geographic magazine articles about the Moon, this one is perhaps the most meaningful because it tells the full story of the people and the processes leading to the unlocking of some of the lunar secrets. Copiously illustrated in color and black and white, it is an excellent starting point for anyone seriously interested in knowing more about the Moon and the technology that led to the Apollo astronaut landings.

Thomas W. Becker

MAGELLAN: VENUS

Date: May 4, 1989, to October 12, 1994
Type of mission: Unmanned Venus probe

Magellan's primary objective was to map the surface of Venus using powerful radar imaging instruments. The probe managed to produce a detailed map of 99 percent of the planet's surface, and also made a gravity map of Venus showing the density distribution of materials beneath the surface. The probe also contributed important information about the Venusian atmosphere.

PRINCIPAL PERSONAGES

DOUGLAS G. GRIFFITH, final Magellan
Project Manager
TONY SPEAR, first Magellan Project Manager
R. STEPHEN SAUNDERS, Magellan Project Scientist
THOMAS W. THOMPSON, Magellan Science Manager
ALLAN G. CONRAD, Mission Operations
Systems Manager
EDWARD E. KELLUM, Radar System Manager
GARY L. PARKER, Spacecraft Manager for Venus Radar
Mapping Project (VRM), which evolved into the
Magellan project
GORDON H. PETTENGILL, Principal Investigator,
Radar Investigation Group
GEORGES BAIMINO, Principal Investigator, Gravity
Investigation Group

Summary of the Mission

The Magellan mission was designed to produce an extremely detailed radar map of the surface features of the planet Venus, with a secondary goal of making a gravity map of the planet's sub-surface features. The project was named after the Portuguese explorer Ferdinand Magellan (1480-1521), who led the first successful attempt to circumnavigate the world by sea. Although Magellan did not survive the journey, his legacy profoundly changed mankind's understanding of our planet. The spacecraft that bore his name had the same effect on our understanding of the planet Venus.

Radar imaging (using high frequency radio waves as "light") is required to make images of Venus's surface because the entire planet is perpetually enshrouded in a thick sulfuric acid cloud blanket. Unlike "normal" light waves, radar is capable of penetrating these thick clouds, reflecting off surface features back to a space probe or a ground-based radar station. The first radar images of Venus were produced from Earth-bound radar facilities at Goldstone, California, and Arecibo, Puerto Rico.

The Magellan spacecraft was the culmination of a National Aeronautics and Space Administration (NASA) project begun in 1980 called the Venus Orbiting Imaging Radar Project. This project was canceled in the fall of 1981, to be resurrected later as the Venus Radar Mapper Project (VRM) with NASA's Jet Propulsion Laboratory (JPL) in Pasadena, California, given primary mission and systems development responsibilities. Tony Spear was made overall Project Manager, and Allan Conrad was assigned the task of Mission Operations Systems Manager, the office responsible for designing the actual hardware employed in the final spacecraft. VRM later evolved into Magellan, which was eventually headed by a new project manager, Douglas Griffith. Stephen Saunders was made Project Scientist, the person responsible for organizing and coordinating the gathering and analyzing of scientific data produced by the mission.

Martin-Marietta Corporation of Denver, Colorado, was contracted to build the actual spacecraft, while the construction of the critical radar sensor was assigned to Hughes Aircraft Company in El Segundo, California. Their task was complicated by budget constraints that demanded that the final craft be as cost-effective as possible. Early complicated designs produced during the VRM project were rejected as too costly. Spare parts from other missions were used in many critical hardware components, and existing hardware designs were employed where possible instead of creating new configurations. Magellan's large high-gain (high power) dish antenna, for example, was a spare left over from the Voyager project. Final costs for Magellan amounted to about 450 million dollars.

The final spacecraft was 6.4 meters tall, 3.7 meters across (high-gain antenna diameter), and weighed 3,460 kilograms. The aft section contained a network of support struts and braces supporting the four sets of attitude control rocket engines used to alter the spacecraft's orientation in space. The attitude control system was attached to the main body of the craft consisting aft of a ten-sided instrument package belt, topped off by the boxlike forward equipment module, itself capped by the large (nearly 4 meters diameter) high-gain dish antenna. Attached to one side of the large dish antenna was a horn-shaped antenna designed for surface altitude determina-

Model depicting Magellan at Venus (left); STS-30 Magellan spacecraft and IUS deployment from Atlantis' payload bay. (NASA)

tions. Projecting from opposing sides of the forward equipment module were two telescoping booms that served to deploy the nearly square solar panels away from the craft after launch and to orient them relative to the sun for maximum efficiency. The panels were capable of producing the 1,200 watts of electrical power required to operate all on-board equipment. The craft also had a round aperture on the forward equipment module, the star tracker, that was used to orient the spacecraft relative to the star field.

Magellan was originally scheduled to be launched in November, 1988, from a space shuttle, using a Centaur rocket booster to achieve initial orbital characteristics. However, as a result of the *Challenger* shuttle accident in 1986, the launch date was moved back to September, 1989. Then the Shuttle/Centaur program was canceled, necessitating a re-evaluation of the booster vehicle needed to send Magellan off to Venus. After considering three alternative launch date and booster configurations, NASA decided on a booster called Inertial Upper Stage (IUS) and a spring, 1989, launch opportunity. The precise window of opportunity for the launch was placed between April 27 and May 17, 1989.

Magellan was eventually launched from Kennedy Space Center aboard the Space Shuttle *Atlantis* on May 4, 1989. After a flawless launch from the shuttle and IUS booster burn, Magellan spent the first part of its fifteen-month jour-

ney, establishing a spiraling orbit around the Sun. It circled the Sun one and one-half times before finally achieving an elliptical, nearly polar orbit (north-south) around Venus on August 10, 1990. This orbit carried the craft about 200 kilometers from Venus's surface at its closest approach (periapsis), and about 8,500 kilometers from the surface at its farthest point (apoapsis) from the surface on the planet's opposite side. One complete orbit took about three and a quarter hours. Radar mapping was done each time the probe dipped down to its periapsis, producing high-resolution images of the surface of Venus and transmitting them back to Earth. Magellan incorporated a sophisticated Synthetic Aperture Radar (SAR) located in the large Voyager-type dish antenna that revealed objects as small as 120 meters across, one tenth the size previously detectable.

To fulfill its mapping mission Magellan would transmit a radar signal with its SAR on close approach, illuminating a 16 to 28 kilometer wide, 16,200 kilometer long swath across the planet's surface. The reflected signal was then stored on tape and periodically played back to Earth through the high-gain antenna located within the large dish mounted on the forward end of the craft. Data from Magellan was beamed back to JPL's Deep Space Network with receiving stations in Spain, Australia, and the Mojave Desert.

However, all did not go according to plan, particularly

during the initial phases of the radar mapping mission. Twice before mapping began, ground controllers at JPL thought they had lost the craft when the main computer malfunctioned and the high-gain antenna refused to point toward Earth. After thirteen hours of concentrated effort, the JPL team managed to communicate with Magellan through a smaller antenna and the mission was saved. Three more times over the next several months, however, Magellan went silent (known as loss of signal events) but was revived each time by the resourceful improvisation of the JPL mission control scientists. In September, after mapping had begun, one of Magellan's two tape recorders failed. Luckily, the remaining recorder performed flawlessly for the rest of the mission. The loss of signal events were eventually traced to a minor "bug" in the flight software. With this bug corrected, Magellan never again refused to transmit its valuable data.

For planning purposes, the Magellan mission was divided into mission cycles, one cycle being equal to the very slow rotational period of Venus on its axis, 243 Earth days. At the end of each mission cycle the JPL mission controllers assessed mapping progress by contrast to planned objectives and time tables. Magellan performed far more efficiently than anticipated. For example, at the end of the first mission cycle cumulative planetary coverage was 84 percent; at the end of the second cycle coverage had jumped to 98 percent. By the end of the third cycle 99 percent of the planet had been covered, an area three times as large as Earth's combined continental land masses. Originally, Magellan was expected to cover 70 percent of Venus's surface at most.

On September 15, 1992, Magellan initiated a fourth 243-day mission cycle, this time to perform a global gravity survey. Gravity data helps scientists to determine the thickness of sub-surface layers on a planet, and to determine relative densities of materials in the planet's interior. For example, low density areas may suggest hot rising magma— or a deep valley. By the end of mission cycle four, Magellan had generated a gravity map of most of Venus, but the most reliable data was collected when the craft was closest to the planet's surface, within 30 or 40 degrees latitude of the equator. Mission scientists, delighted at having a still functioning spacecraft after four cycles, decided to dedicate a fifth cycle to recovering high quality gravity data from the polar regions as well.

To provide better gravity data from Venus's polar regions, JPL mission specialists had to devise a way to alter Magellan's orbit so that it would swoop in closer to the polar regions than its elliptical orbit would allow. This was accomplished in a particularly imaginative fashion, using a technique first demonstrated in Arthur C. Clarke's science fiction novel and film, *2010*. This technique, called aerobraking, involves using frictional drag with the atmosphere to slow the momentum of a spacecraft so that it assumes a nearly circular orbit. In May, 1993, Magellan became the first non-fictional spacecraft to use aerobraking to alter its orbital characteristics.

Mission cycle six, Magellan's last, was used to obtain important information about Venus's atmosphere. With its attitude control fuel nearly exhausted, its useful life as a planetary space probe was nearing its end. However, the low circular orbit acquired during cycle five left the craft flying within the planet's atmosphere, allowing measurements of atmospheric density and other parameters. To measure density, Magellan was oriented so that its extended solar panels acted like the blades of a windmill. The force (expended by attitude control rockets) required to keep the craft from rotating provided a measure of the density of atoms and molecules in the atmosphere.

On October 12, 1994, Magellan's radio transmissions to Earth fell silent. Shortly thereafter, the spacecraft, which had orbited Venus for more than four years, was destroyed as it plunged through the atmosphere toward the surface.

Knowledge Gained

The Magellan mission to Venus provided mankind with the most detailed surface map of a planet in the entire solar system. The Venus revealed by Magellan shows a tortured surface, 85 percent of which is covered by volcanic lava flows that form the planet's vast plains. Much of the rest of the surface consists of mountainous highlands crisscrossed by fold belts and rift valleys in complicated patterns. Magellan also revealed for the first time the full extent of impact cratering on the surface. Impact craters are fairly evenly distributed over the entire planet and are much more abundant than on Earth.

Magellan also revealed the planet-wide extent of cratering on Venus from rocky meteoroids originating elsewhere in the solar system. It showed that impact craters less that about 30 kilometers in diameter are rare on Venus, the result of diminished momentum of small bodies encountering the thick carbon dioxide atmosphere. Using crater densities recorded by Magellan (number of craters in a given area) and comparing them to densities on the Earth and Moon, the age of the present crust of Venus was calculated at about 500 million years. Another important Magellan accomplishment was the construction of a high-resolution, planet-wide gravity map of Venus. In 1978 the Pioneer orbiter made a gravity map of Venus, but the Magellan map shows superior surface coverage and resolution of fine details. A comparison of the radar-produced topographic map with the gravity map shows an extremely close relationship between topography and gravity; high topographic features (domes, volcanic peaks) show high gravity, whereas low topographic areas (valleys) are associated with low gravity. This close topographic/gravity association is generally not observed on Earth.

Finally, Magellan contributed important information about Venus's atmosphere. Windmill experiments showed expected atmospheric densities at altitudes from 180 to 172 kilometers above the surface where atomic oxygen dominates the atmosphere, but at 160 to 150 kilometers above Venus atmospheric density was about one-half previously estimated values.

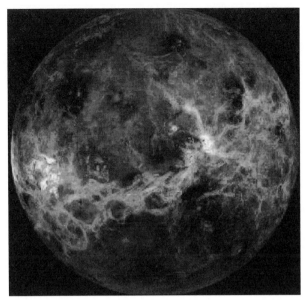

Magellan synthetic aperture radar mosaics are mapped onto a computer-simulated globe to create this image of Venus. (NASA)

Context

Magellan was the twenty-second space probe to be sent to Venus, making Venus the target of more probes than any other planet. The first interplanetary probe, Mariner 2 (Mariner 1 was destroyed on launch), was launched toward Venus in 1962 and made infrared temperature determinations as it flew by the planet. In 1967 the United States launched Mariner 5, and the Soviet Union launched Venera 4, the first spacecraft to drop a probe into the Venusian atmosphere. They confirmed the high temperature readings (minus 30 degrees centigrade at the cloud tops; 477 degrees at the surface) of Mariner 2 and established the atmospheric composition as carbon dioxide–rich and extremely thick (914,000 kilograms per square meter surface pressure; ninety times that of Earth).

Other notable U.S. probes included Mariner 10, which produced the first close-up, high-resolution pictures of the cloud formations on Venus in 1972. Pioneer Venus 1 and 2, launched about three months apart in 1978, released four "hard-landers" (Pioneer 2; probes designed to fall through the atmosphere and crash on the surface), and made the first gravity map of Venus (Pioneer 1 orbiter). Pioneer also pro-

duced radar-altimetry maps that allowed the first delineation of Venus's global topographic provinces. A few months before Magellan reached Venus in 1990, the Galileo spacecraft transmitted detailed images of cloud patterns to Earth as it flew by Venus on its way to a December, 1995, rendezvous with Jupiter.

The Soviet Union launched the most spacecraft toward Venus, fifteen in all. Beginning with Venera 5, these craft included orbiters, atmospheric probes, and soft landers capable of operating on the surface. Notable among the soft landers were Veneras 9, 10, 13, and 14, all of which carried cameras that produced the first images of the surface features. The Venera 15 and 16 orbiters were the first probes to produce radar images of Venus. The last Soviet probes were Vega 1 and 2, which launched balloons to assess wind patterns in the atmosphere, and landers to assess surface conditions.

The scientific legacy of Magellan is immense. In addition to discovering many new surface features on Venus, Magellan revolutionized our understanding of how the planet has evolved over time. Correlations of topographic features with gravity data show that, unlike Earth, Venus does not have lithospheric plates, thick slabs of rigid crust plus mantle material that slide horizontally over a weak, partially molten layer in the mantle. The sliding motion of these plates on Earth (a process called plate tectonics) accounts for most of the Earth's mountain belts, volcanic chains, ocean basins, and earthquake activity. On Venus the high surface temperatures produce ductile behavior (easily deformable) at very shallow depths in the crust, precluding the formation of strong rigid plates.

In terms of technological advances, Magellan demonstrated the feasibility of aerobraking for the first time, a technique that will be used to control the orbit of the planned Mars Global Surveyor scheduled to be launched in 1996. Magellan also demonstrated the feasibility and economy of building a very successful spacecraft out of spare parts from other missions, although this was not without its problems. For example, the Voyager-type antenna dish was not an ideal configuration for radar work and had to be jury-rigged to adjust itself three thousand times during each mapping pass. Nevertheless, the spacecraft broke records for the amount of data received, exceeding all previous U.S. missions combined. To illustrate the amount of data Magellan collected, a magnetic data tape from just one mapping cycle would stretch half way around Earth.

Bibliography

Beatty, J. Kelley, and Andrew Chaikin, ed. *The New Solar System.* 3d ed. Cambridge, Mass.: Sky Publishing Corp., 1990. This is a comprehensive, engaging account of all solar system objects from planets to comets, asteroids, and meteorites. Richly illustrated with color and monochrome photographs, drawings, graphs, and charts, this is the perfect book to start anyone on an exploration of the solar system. Information on Venus is from pre-Magellan probes but is nevertheless very timely. Has an appendix with planetary characteristics, glossary, and suggestions for further reading. Comprehensive planetary maps are also included. Suitable for general audiences.

Chapman, Clark R. *Planets of Rock and Ice: From Mercury to the Moons of Saturn.* New York: Charles Scribner's Sons, 1982. This book, by one of the premier planetary scientists in the world, is a serious but sometimes whimsical overview of the solid (as opposed to gaseous) planets and planetoids of the solar system. This is a particularly good book for the study of comparative planetology, showing how the planets evolved and what their characteristics tell us about the origin of the solar system. Includes a center section with monochrome photographs. Suitable for general audiences.

Hunten, D. M., et al., eds. *Venus.* Tucson: University of Arizona Press, 1983. This collection of essays by Soviet and U.S. authors is a comprehensive compendium of data on all missions to Venus up to that time. Has many maps, charts, and photographs illustrating important aspects of the geology and geophysical aspects of Venus. A history of Venus studies is also included. Most articles contain technically challenging material. This book is aimed at professional specialists, but should be accessible for college-level audiences.

NASA/JPL. *Magellan: Revealing the Face of Venus.* JPL 400-494 3/93. Washington, D.C.: U. S. Government Printing Office, 1993. This is a twenty-five page booklet produced by NASA to publicize the Magellan mission. It contains full color diagrams and pictures of the spacecraft itself, colorized radar images including a global view of Venus, altimetry map showing principle topographic regions of Venus, and monochrome images illustrating important geological surface features. The text is packed with interesting facts including a concise rundown on mission history, spacecraft dimensions and specifications, descriptions of mission objectives and results, facts and figures about the planet Venus, including the procedure for naming newly discovered features on its surface. It also contains a complete list of the principal scientific investigators on the Magellan project. Designed for the general public (write to: Superintendent of Documents, U.S. Government Printing Office., Wash., D.C. 20402).

Saunders, R. Stephen. "The Surface of Venus: Razor-sharp Images of the Earth's Near Twin Reveal a Mix of Familiar and Perplexing Geologic Features." *Scientific American* 263 (1990). This article by the Magellan Project Scientist at JPL presents a wide array of excellent radar images of surface features on Venus. Images include impact craters, mountains, faulted plains, and volcanoes, all shown in exquisite detail. Also shows a picture of the Magellan probe and a drawing illustrating how it obtained its images. The text gives a synopsis of the Magellan mission and explanations of the illustrated surface features. This article is easily accessible in the serials section of most libraries. It is intended for general audiences.

John L. Berkley

The Manned Maneuvering Unit

Date: Beginning June 3, 1965
Type of technology: Humans in space

The Manned Maneuvering Unit (MMU) is a device, strapped onto an astronaut's spacesuit, that allows independent movement for work in free space. The modern MMU has been modified many times, evolving from a handheld device in the Gemini program to the space shuttle MMU, which allows an astronaut to maneuver up to 100 meters from the spacecraft.

PRINCIPAL PERSONAGES

EDWARD H. WHITE, the NASA astronaut and U.S. Air Force captain who used the first-generation MMU, called the Handheld Maneuvering Unit, for the first American spacewalk

DANIEL MCKEE, director of development for the second-generation MMU, called the Astronaut Maneuvering Unit, the first backpack-mounted maneuvering unit

EDWARD G. GIVENS, a NASA astronaut and U.S. Air Force major, also responsible for developing the Astronaut Maneuvering Unit

BRUCE MCCANDLESS, the NASA astronaut who first used the MMU in spaceflight

Summary of the Technology

The Manned Maneuvering Unit is an advanced space maneuvering device that allows astronauts to engage in Extravehicular Activity (EVA), or working in free space apart from the space vehicle. Without this kind of device, an astronaut would be virtually helpless in the weightlessness of space.

The MMU is the third generation in a succession of devices designed to help American astronauts perform EVAs. The first was a handheld propulsion device used by Gemini astronaut Edward H. White on the first American spacewalk, June 3, 1965. This instrument was called the Handheld Maneuvering Unit (HHMU). It resembled a gun with antennae and was attached to its supply of compressed gas by lines running to the Gemini capsule. The astronaut needed to point the HHMU opposite the direction in which he wanted to move and pull the trigger; a burst of compressed oxygen gas would be released from the gun's antennae (which were actually thrusters), propelling the astronaut through space. The HHMU was used with limited success; its design was too simplistic, and its gas supply was insufficient to provide much control.

The National Aeronautics and Space Administration (NASA), with help from the Air Force, had plans for an improved device even as the HHMU flew on Gemini 4. The second-generation device was to consist of the same HHMU, but with modifications. Instead of being attached to the space vehicle by gas lines, the HHMU was to be attached to a supply of Freon 14 contained in a backpack called the Extravehicular Support Package (ESP). The resulting system became known as the Astronaut Maneuvering Unit (AMU). Later versions of the AMU would include twelve thrusters on the ESP backpack itself.

The AMU was designed to be worn by an astronaut during EVA outside the cramped Gemini. Designing the AMU was an Air Force project headed by Daniel McKee and Edward G. Givens. The AMU's first use was planned for Gemini VIII, in March, 1966. Unfortunately, that mission was cut short before the planned EVA when a thruster malfunctioned, a failure which nearly claimed the lives of astronauts Neil Armstrong and David R. Scott.

The AMU was flown again on Gemini IX-A in June, 1966. Astronaut Eugene A. Cernan began his EVA successfully enough, but because of excessive fogging of his faceplate and an inability to don the backpack in the weightless environment, plans for the AMU were aborted late in the EVA.

The ESP backpack was removed from the system for Gemini X (July, 1966), and the HHMU (with nitrogen gas) was used for the flight. The HHMU would be used again on Gemini XI (September, 1966). Finally, the AMU was shelved altogether in late 1966, its technical problems too complex to address during the busy pre-Apollo days.

The AMU eventually reemerged to fly in space again: inside the voluminous Skylab space station on the second Skylab mission (May, 1973). Yet the redesigned Skylab AMU had undergone significant changes under a Martin Marietta Corporation contract and hardly resembled the original AMU. It was called the M509 experimental unit. M509 was a full generation ahead of the AMU. It closely resembled and would evolve into the Manned Maneuvering Unit, to be used on the space shuttle.

The Martin Marietta Corporation retained the AMU contract during the long hiatus between the last flight of

Skylab (November, 1973) and the first space shuttle flight (April, 1981). During the intervening years, the AMU was renamed the MMU but retained many of the outward characteristics of the Skylab AMU.

The MMU is a back-mounted maneuvering unit stored in the forward end of the space shuttle orbiter payload bay. It has its own electrical power, propulsion gases (compressed gaseous nitrogen), and controls. Like the Skylab AMU, it resembles a chair with armrests supporting control devices.

The MMU is housed in the shuttle's payload bay in its flight support station, which consists of a mounting bracket and interfaces to the orbiter to supply propellant gases and power. The flight support station also provides the restraints necessary for astronauts to don and doff the MMU efficiently in weightlessness. This process takes the astronauts about fifteen to twenty minutes to accomplish. In an emergency, however, the astronaut can detach himself from the MMU in less than thirty seconds.

The MMU propulsion system consists of twenty-four thrusters, commanded by the astronaut through controls located at his fingertips: the rotational hand controller (right hand) and the translational hand controller (left hand). The MMU can be controlled either automatically or manually.

The MMU is capable of acceleration values of 0.09 meter per second translational (moving in a straight line direction) and 10.0 degrees per second rotational (moving as if on an axis), with six degrees of freedom (meaning, it can move six degrees in any direction in free space). It is designed to allow an astronaut to work at distances of up to 100 meters from the spacecraft.

The MMU's 16.8-volt batteries, which power electronic systems and lights, can be replaced in five minutes in orbit. Its gaseous propellants are stored in two 27-liter tanks located within the MMU. Each of the twenty-four thrusters can generate a thrust of 7.5 newtons for a minimum duration of 0.009 second. Drawing from supplies in the orbiter, the gas tanks can be recharged in orbit in fewer than twenty minutes. The MMU has thirty-two separate heaters located at specific points within its structure. In addition, special paint and reflective materials help control the temperature of the unit.

Each MMU has a twelve-year operating storage life. It is designed to fly in space with an active, operational life of about four hundred hours. It can be stored in the shuttle's payload bay, in space, for fifteen hundred hours. The MMU is designed in such a way that it will not move far from the orbiter if it malfunctions. Fail-safe mechanisms include those to avoid damaging collisions with the orbiter.

The first use of the MMU was made on STS 41-B on February 7, 1984. During that flight, astronaut Bruce McCandless donned the MMU, and — for the first time in history — an astronaut floated free from a manned space vehicle, becoming an independent satellite. McCandless and his MMU maneuvered as far away as 98 meters from the orbiter, and both returned safely.

The second use of the MMU occurred during STS 51-A (November, 1984) in the spectacular rescue of a disabled Westar 6 satellite. During this flight, astronaut Dale A. Gardner flew his MMU to the spinning satellite some tens of meters away from the shuttle. He inserted a device called a "stinger" into the satellite's rocket motor. Thus locking himself to the satellite, he used his MMU to counter the spin of the satellite. Stabilizing the entire configuration in space, Gardner then used the thrusters on his MMU to tow the satellite to the shuttle's payload bay, where it was secured. Another satellite, Palapa B2, was also later recovered on the mission, and both were returned to Earth for repairs.

Knowledge Gained

The historic photographs of Edward White outside his Gemini IV spacecraft show the astronaut gripping his HHMU. In retrospect, when placed alongside images of Dale Gardner strapped to his MMU and maneuvering toward the disabled Westar, the HHMU seems primitive. Eugene Cernan's frustrated attempts to don his AMU stand in sharp contrast to the relative luxury of the MMU's flight support station and illustrate well the successful changes in design.

Edward White's mission demonstrated that the HHMU was extremely sensitive and its propellant supply far too limited. Yet the HHMU did provide positive control and maneuvering capability necessary to perform fundamental tasks in free space. After White's spacewalk, the HHMU was given larger propellant tanks and attached to the outside of the spacecraft's cabin to allow for more propellant volume, thus ensuring a greater maneuverability.

When the AMU was tested in space, designers realized that merely adding more parts would not solve the problems of the system. The Gemini program was branching into more fields than there were time and money to support. Since the United States' goal was a flight to the Moon — and since the AMU would not be useful there — the AMU project was put on hold. More thought would have to be put into donning methodology, self-propulsion, in-orbit storage, and the resupply of consumables and power.

These problems were addressed when the AMU was redesigned and flown on Skylab. The Skylab M509 experimental unit solved some basic problems: how to interface a large collection of equipment, power, and plumbing with the human form. Its chair-type design and handheld controls would become lasting contributions to the technology. Ideas on the placement of gas thrusters, propellant storage, and the translational/rotational controls were tested and found to be ideal design characteristics. During Skylab, an effective combination of design modes for storage, doffing, and donning was also verified. Although the M509 never actually saw free space, it represented an important link that would lead to the design of the Manned Maneuvering Unit.

By the time Bruce McCandless strapped himself into his MMU in orbit in February, 1984, he was to test a system that

the manned maneuvering unit

had evolved from several generations of space maneuvering technology. McCandless verified that the whole system was operational and performed as a unit. The ease of donning and doffing, the operability of control systems, the "invisible" heaters and thermal control elements that maintained the system's environment, and the ability to fly slightly beyond the expected distance from the spacecraft convinced technicians that the MMU was a success.

STS 41-B and STS 51-A would confirm that the United States had a superior, functioning manned maneuvering capability in space. Useful work was performed to capture satellites, while additional information was being amassed to enable the next generation of MMU to be used to construct large segments of the space station.

Context

From the beginning, no one underestimated the importance of man's ability to work in space outside the confines of his space vehicles and orbital shelters. From the first EVA on the Soviet mission of Voskhod 2 on March 18, 1965, and Edward White's spacewalk some three months later, the importance of this capability was clear.

With Gemini, the useful work that could be performed in free space was largely experimental. White realized the importance of some kind of maneuvering capability with his very limited HHMU. As the program progressed, leading to the AMU, it became evident that almost nothing useful in space could be performed without a maneuvering capability.

From the start, "work in space" was defined as the ability to move from one point to another and, upon arriving, to stabilize and use tools. This fundamental concept was first tested on the flight of Gemini IX-A. Although the EVA on that mission had its share of problems, it is a milestone in space history: the first time man, outside the confines of his space shelter, attempted useful work in space.

From Gemini through Apollo, work in free space had consisted of crawling around the external surface of a spacecraft and recovering items. On Skylab, a thermal shield was installed during an EVA, marking the first time that an EVA was accomplished to save a mission. Yet again, it was accomplished without the aid of a maneuvering unit. It was becoming clear, however, that the need to maneuver efficiently and safely in space was basic to the more ambitious space shuttle missions.

The stranded Westar 6 satellite provided the first opportunity to test the full capabilities of the MMU. The MMU performed flawlessly, so perfectly that it appeared nearly invisible in the drama that unfolded around it. The massive satellite dwarfed the astronaut who was approaching it in free space. The MMU provided the stable platform he needed to pinpoint his rendezvous, mate with the satellite, slow its spin, and return it to the shuttle — an astonishing first trial.

The Simplified Aid For EVA Rescue (SAFER) was designed and developed by the Johnson Space Center in a team project led by the Automation and Robotics Division. It is a small, self-contained, propulsive backpack device that can provide free-flying mobility for a spacewalker in an emergency. It is designed for self-rescue use by a spacewalker in the event the shuttle is unable or unavailable to retrieve a detached, drifting crew member. SAFER is a scaled-down, miniature version of the Manned Maneuvering Unit. It is designed for emergency use only, but without built-in backup systems. SAFER's propulsion is provided by twenty-four fixed-position thrusters that expel nitrogen gas. SAFER was not designed as a replacement for the MMU. A newer version of the MMU is being proposed for the construction of the International Space Station.

Bibliography

Allen, Joseph P., and Russell Martin. *Entering Space: An Astronaut's Odyssey.* New York: Stewart, Tabori and Chang, 1984. A heavily documented essay describing the U.S. space shuttle program. It describes the shuttle's development from processing at the Kennedy Space Center to launch, orbital activities, and landing. The book is aimed toward all readers and is probably the most beautifully photographed and illustrated of all shuttle books.

Baker, David. *The History of Manned Space Flight.* New York: Crown Publishers, 1982. This work offers a precise chronology of the history of manned spaceflight, oriented toward the United States' effort. It is quite a detailed work, tracing the U.S. manned space program from its beginnings to the dawn of the space shuttle program. Provides detailed analyses of all the Gemini flights. Suitable for all audiences.

De Waard, E. John, and Nancy De Waard. *History of NASA: America's Voyage to the Stars.* New York: Exeter Books, 1984. A pictorial essay on the history of NASA. It does not go into great detail on the Gemini and space shuttle programs, but it does present a large collection of color photographs from various missions.

Hartmann, William K., et al. *Out of the Cradle: Exploring the Frontiers Beyond Earth.* New York: Workman Publishing, 1984. This well-illustrated book concentrates on the future of manned spaceflight. It presents, with dramatic insight, space exploration in the twenty-first century and the benefits humankind will reap.

Joels, Kerry M., and David Larkin. *The Space Shuttle Operator's Manual.* New York: Ballantine Books, 1982. This manual is a detailed space shuttle system reference work. It gives specifics of the space shuttle system, from weights and sizes to operational characteristics. It describes countdown procedures, emergency instructions, and standard operational modes. It is thoroughly illustrated and offers a unique approach to the complex shuttle system. An entertaining as well as educational reference work.

Dennis Chamberland

THE MANNED ORBITING LABORATORY

Date: December 11, 1963, to June 10, 1969
Type of program: Manned space station

The Manned Orbiting Laboratory (MOL) was the first space station project that evolved beyond the study stage. It also represented the Pentagon's attempt to establish a manned military presence in space at a time when Cold War tensions between the Soviet Union and the Free World were at their greatest.

PRINCIPAL PERSONAGES:

LYNDON B. JOHNSON, President of the United States
ROBERT S. MCNAMARA, Secretary of Defense under the Johnson Administration
CYRUS R. VANCE, Deputy Secretary of Defense under the Johnson Administration
BERNARD A. SCHREIVER, Commander of the Air Force Systems Command
JAMES E. WEBB, Chief Administrator of NASA
JOSEPH S. BLEYMAIER, Titan 3 program director
RUSSELL BERG, initial deputy MOL program director
HARRY EVANS, initial MOL program vice director
PETER LEONARD, MOL program science adviser
MELVIN R. LAIRD, Secretary of Defense under the Nixon Administration

Summary of the Program

In the early 1960's, U.S. Air Force plans for establishing a military manned presence in space were based on the X-20 Dyna-Soar project. This was a ten-ton "mini-shuttle" spaceplane that was to be launched by an Air Force Titan 3C booster. As NASA's Gemini two-man spacecraft program evolved in the early 1960's, civilian defense analysts concluded that the Gemini might offer a means of performing military manned space functions better and more cheaply than the X-20.

Secretary of Defense Robert McNamara suggested that the Air Force be given an equal management role with NASA in the Gemini program, thus creating a parallel military "Blue Gemini" program. This drew a heated response from NASA administrator James Webb, who viewed the proposal as a militarization of NASA that would jeopardize its presidentially mandated goal of landing men on the Moon by the end of the decade.

In July, 1963, NASA began to look seriously at the concept of future space stations. It was suggested that an orbiting laboratory based on the Gemini spacecraft would offer the military a separate means to explore its role in manned space flight. This idea marked the birth of what was to become the Air Force's Manned Orbiting Laboratory (MOL) program. The resulting 11,340-kilogram cylindrical space station concept looked like a giant thermos bottle the size of a house trailer.

In October, 1963, McNamara requested that the X-20 be terminated and replaced with the Gemini B, a military version of the NASA two-man spacecraft. When McNamara met with the newly installed President Lyndon Johnson on December 10, 1963, the X-20 was cancelled. The next day, the Gemini B/MOL program was officially announced.

The MOL (pronounced "mole") program inherited some funds from the cancelled X-20 and benefited from the ongoing development of the X-20's Titan 3C booster. The original goals of the space station project were modest to provide a shirtsleeve environment to see how a crew could live and function in space for extended periods. Plans called for initial mission durations of thirty days.

One of the early surprises in manned space flight was the ability for astronauts to easily spot the track targets on the ground and to secure amazingly detailed photographs of the Earth with ordinary hand-held cameras. Thus, as the design stage of the program progressed, MOL evolved into space-based reconnaissance platform to develop surveillance techniques using both advanced sensors and the on-board astronauts. It was assumed that human observers in orbit could selectively choose the military and political targets to be photographed and studied. This would eliminate the need for blanket photographic coverage of vast areas of foreign territory and greatly simplify the interpretation of the resulting intelligence. In keeping with this surveillance role, the MOL project is designated in some references as the KH-10, part of a long series of "Keyhole" military space reconnaissance craft. Experiments with communication, meteorology, navigation, and electronic surveillance were also planned.

The project was first envisioned as a $1.5 billion program with the first manned flight scheduled for 1968. Five orbital laboratories were to be built and launched from Vandenberg Air Force Base, California. MOL was primarily an Air Force project, but plans called for the Navy to have dedicated use of two of the laboratories for the perfection of ocean surveillance techniques. The crew was to ride into space aboard the Gemini spacecraft, then transfer into the MOL laboratory once safely in orbit. At the conclusion of the mission, the crew would enter the Gemini, separate from the MOL, and return to Earth. The laboratory would then be deliberately deorbited into the ocean to prevent its secret reconnaissance equipment from falling into unauthorized hands.

The MOL design evolved into a polar orbiting cylindrical station 3.3 meters in diameter and 16.8 meters long with the Gemini B spacecraft attached. The laboratory was divided into different sections. The pressurized portion housed two spaces: the experimental laboratory where the surveillance equipment was to be operated, and a living compartment with consumables to sustain the two-man crew for up to thirty days. The attached unpressurized compartment opposite the Gemini contained housekeeping equipment such as power and oxygen supplies.

Official development approval for MOL was given by President Johnson on August 26, 1965. The timing of the approval is significant because it came while the eight-day mission of NASA's Gemini 5, carrying Gordon Cooper and Pete Conrad, was in progress. The success of the NASA Gemini missions were crucial to the MOL project, as the Gemini spacecraft was a key element in the system.

Douglas Aircraft won the competition to build the laboratory while the Gemini B spacecraft was to be produced by McDonnell Aircraft, the same company that built the Gemini for NASA's manned space program. Coincidentally, both prime contractors for the MOL project merged midway through the program's development and became the McDonnell Douglas Aircraft Company.

The MOL was to be launched by the Air Force Titan 3C booster. This was a modified two-stage Titan 2 ICBM clustered with twin 3.05-meter diameter solid-fuel rocket boosters. The solid boosters were unique in that they were assembled from separate "segments," which were stacked and bolted together. Their performance could be increased or decreased by adding or subtracting segments. For the Titan 3C, the solid boosters each contained five segments. The solid-fuel boosters were designated the "Zero stage" and each provided more than 453,600 kilograms of thrust. Both were ignited at liftoff along with the 213,190-kilogram thrust twin liquid fueled first stage engines, providing a combined liftoff thrust of 1,133,922 kilograms. This made the Titan 3C the most powerful booster in the Free World at the time. The second stage was powered by a single 45,360-kilogram thrust liquid fueled engine. The launcher was topped with a restartable "transtage," which provided an additional 7,257 kilograms of thrust.

One of the vexing problems facing engineers was how to move the crew from the Gemini B into the pressurized MOL laboratory once in orbit. A number of schemes were investigated, including transfer by Extravehicular Activity (EVA) or through an inflatable external tunnel connecting the two components of the station. The final solution was to modify

NASA's Gemini craft by adding a hatchway through the heat shield between the Gemini and MOL laboratory. This modification was the major difference between NASA's civilian Gemini and the Air Force Gemini B.

The heat shield is a critical item of any manned spacecraft. If it fails during the return to Earth, the crew is doomed to an incendiary death. Many engineers therefore considered the heat shield as an inviolable part of the spacecraft. Cutting a hatch through it was regarded by some as a dangerous joke. McDonnell Aircraft engineers in charge of designing the modified shield considered it to be theoretically usable. Only an unmanned space test would validate the design.

By the summer of 1966, the MOL concept grew more complex and it accordingly gained weight, swelling to a 13,608-kilogram station. This prompted the switch from the Titan 3C to the Titan 3M. The new rocket used more powerful seven-segment solid-fuel boosters and had a liftoff thrust of 1,363,636 kilograms.

As the program progressed, it became apparent that earlier cost estimates were not realistic. By early 1967, the total program cost had grown to $2.2 billion, with the initial manned launches slipping until 1970. Then the Vietnam War began to take its fiscal toll on the high-technology program. Funds were diverted from MOL to help finance the war effort and the first launch slipped further, to 1971. By 1969, the total program costs had expanded to a projected $3 billion, with the first launch still three years in the future.

Shrinking military budgets and increased Vietnam War costs eventually overtook the MOL project. On June 10, 1969, President Nixon's Secretary of Defense, Melvin Laird was forced to reluctantly cancel the MOL program. After a decade of effort with both the X-20 and the MOL program, America's military manned space program also ended.

Knowledge Gained

Had it been completed, the MOL would have been the world's first space station. While the goal of long-duration space operations was not achieved, the design and construction of the MOL components by Douglas Aircraft Corporation initiated an experience base for the building of large space structures. At the time of its design and initial fabrication, the MOL laboratory was the largest free-flying single-component manned space structure yet built.

Although the project was prematurely terminated, the MOL progressed through the development and validation of the Titan 3C booster and the initial development of the seven-segment solid booster for the Titan 3M. The seven-segment booster design was also shelved when MOL was cancelled, but its development was not wasted. In 1985, the design was resurrected for the even more powerful Titan 4 booster, which is the mainstay heavy-lift booster of the current American military space effort.

An unmanned MOL development launch designed by the Air Force as OV4 3 was flown on November 3, 1966. The payload consisted of a 3.3-meter diameter, 11.7-meter long simulated MOL fabricated from a Titan 2 ICBM oxidizer tank that was topped with a conical Gemini spacecraft adapter. The resulting vehicle was unusually long, 48.2 meters, and one of the design goals of the flight was to gather "long shape" aerodynamic data during the Titan 3C launch.

The Gemini B spacecraft launched atop the dummy MOL was in fact the same spacecraft used by NASA in the Gemini-Titan 2 suborbital flight on January 19, 1965. The NASA flight validated the craft's heat shield for manned use. In the case of the MOL launch, the same spacecraft was again used to validate the heat-shield design, the difference being the second flight used the MOL heat shield, which had a hatch cut into it for the transfer tunnel between the Gemini B and the laboratory.

The OV4 3 launch was, in all respects, a success. The Titan 3C performed nominally, flying an unusual "roller-coaster" ascent trajectory. Upon reaching an altitude of 210 kilometers, the first burn of the transtage pitched the vehicle downward as it accelerated to 28,255 kilometers per hour. Upon descending to 167 kilometers, the Gemini B spacecraft was released into a reentry trajectory that targeted it into the South Atlantic ocean near Ascension Island, where it was recovered. The transtage then reignited two more times to place the simulated MOL into a 295 by 296 kilometer orbit. The transtage remained attached to the payload, resulting in the largest military-oriented payload yet placed into orbit.

The OV4 3 mission provided not only operational experience with the flight of the simulated laboratory, but was also the first reflight of a spacecraft that had previously flown in space. Many respected engineers were doubtful the Gemini B heat shield would be structurally sound once the MOL crew hatch had been cut through it. The reentry test showed the modified heat shield did indeed work. Instead of destroying the shield, the heat of reentry "welded" the hatch shut and solidified the heat shield, verifying its design intent.

Although the program was cancelled, much of the reconnaissance sensor technology developed for the MOL project was later applied to unmanned spy satellites, such as KH-11.

Context

Had the project been concluded, MOL would have been the first manned space launch to use solid-fueled boosters. Instead, this legacy fell to the Space Shuttle, which itself owes the use of solid-fuel boosters to the experience gained with the MOL's Titan 3 booster. The reliability and simplicity of the solid-fueled boosters developed for the Titan 3 was instrumental in their acceptance into NASA's Space Shuttle program. Their initial success in the Titan program led to a false sense of security with the solid boosters, which contributed to the fatal Space Shuttle *Challenger* accident in 1986.

While the United States did not fly the hardware developed for MOL, the program's existence inspired the Soviet Union to initiate its own manned space-based reconnaissance system called Almaz (diamond in Russian). The Almaz was developed in the 1960's by the Soviet design bureau headed by Vladimir Chelomei. It was a twenty-ton cylindrical space station similar to MOL.

The Almaz program suffered from delays much as the MOL program did, and it was not ready to fly even when MOL was cancelled in 1969. However, in response to the planned 1972 launch of NASA's Skylab space station, the Soviet Almaz was "civilianized" and transferred to the Energia Design Bureau. Launched in 1971 as the Salyut 1 space station, the converted Soviet counterpart of MOL became the prototype of a twenty-five-year long series of space stations, which has culminated with the present day Russian Mir space station.

The MOL program evolved during a difficult era in American history. The bold space plans outlined by President Kennedy at the beginning of the 1960's clashed with the economic realities of massive government spending on both expanding domestic social programs and the burgeoning war in Southeast Asia. At the same time President Lyndon Johnson's "Great Society" programs absorbed more of the nation's fiscal resources, the Vietnam War was eating up more of a shrinking military budget. This placed a great strain on the development of new technologies such as the MOL. It became the target for continued budget cutting and saw its funding allocation continuously reduced until it could no longer maintain a realistic development schedule.

As MOL faced extended delays, the program had the public appearance of being outdated before it even flew. Based on the Gemini spacecraft that NASA retired in 1966, MOL would seem obsolete by the time manned launches began in 1972. Other detractors argued that the program was wasteful military duplication of the civilian space station program, which was being advanced as the Skylab program, also slated for its initial launch in 1972.

The reality behind MOL's eventual cancellation in 1969 goes beyond obsolescence or duplication of NASA efforts. The fact was that the operational requirement for a manned military reconnaissance platform in space had changed. The Air Force had experienced unexpected high success with its unmanned space reconnaissance programs. The KH-4 through KH-8 spy satellite series were far more productive than their planners dreamed. The complex and risky manned MOL missions were simply not needed any more for effective space-based reconnaissance.

There were thirteen military pilots selected as astronaut candidates for the MOL program. When the space station was cancelled, the seven pilots who fell within NASA's cutoff age of thirty-six were allowed to enter the NASA astronaut corps. Although it would be more than a decade before any of them would fly in space, all seven eventually flew on the Space Shuttle.

Significant among those who did not transfer to NASA

were Jim Abrahamson and Robert Herres, both with the Air Force. Too old to become NASA astronauts, they continued their military careers. Abrahamson earned three stars as a general and became first the head of the Space Shuttle program, then the first director of the Strategic Defense Initiative, or "Star Wars" program. Herres earned four stars and became vice chairman of the Joint Chiefs of Staff under General Colin Powell.

Bibliography

Bronner, Fritz. *Buzz Aldrin's Race Into Space Companion.* Berkeley, Calif.: Osborne McGraw-Hill, 1993. Though billed as a companion volume for the Buzz Aldrin's Race Into Space computer simulation, the first 183 pages of this book provide an excellent non-technical review of the politics and conduct of the Russian and American civil and military space programs.

Gatland, Kenneth. *The Illustrated Encyclopedia of Space Technology.* New York: Orion Books, 1989. This volume is noted for its concise yet detailed descriptions of the world's space programs. It is lavishly illustrated with color photographs and contains many drawings and charts that explain past and current space systems to the layman in a visually interesting manner.

Hacker, Barton C., and Grimwood, James M. *On the Shoulders of Titans.* NASA SP-4203. Washington, D.C.: U.S. Government Printing Office, 1977. This official NASA history of the Gemini manned space program is rich in technical detail and personal anecdotes about the development and conduct of the program. It is relevant to the Manned Orbiting Laboratory in that the Gemini spacecraft was to be used by that program.

Hobbs, David. *An Illustrated Guide to Space Warfare.* London: Salamander Books. This profusely and colorfully illustrated book details the development of military space systems from the birth of the Space Age to the modern "Star Wars" era. Non-technical text and excellent diagrams describe various military space programs for the lay reader.

Klass, Philip J. *Secret Sentries in Space.* New York: Random House, 1971. This book was one of the first to detail the Soviet and American military space reconnaissance programs. Although recent declassification has shown that many speculated details about covert programs were in error, the details of publicly discussed military space programs remain historically useful.

McDougall, Walter A. *...The Heavens and the Earth.* New York: Basic Books, 1985. This non-technical historical narrative deals exclusively with the politics of the Space Age and details how the civil and military space programs of both the United States and the Soviet Union evolved from the Cold War political struggle between the two superpowers.

Peebles, Curtis. *Guardians.* Novato, Calif.: Presidio Press, 1987. Written for the laymen, yet still detailed and fact filled, this book reviews the military space reconnaissance programs of both the United States and the Soviet Union. Technical details are presented in easy to understand context. Well illustrated with charts, diagrams, and photos.

Slayton, Donald K., and Michael Cassutt. *Deke!.* New York: A Tom Doherty Associates Book, 1994. An engaging, well written autobiography of one of America's first astronauts who, by a twist of fate, was thrust into a pivotal leadership role in the U.S. manned space flight program. This book gives a good insight into the politics and conduct of the American manned space program.

Von Braun, Wernher, and Frederick I. Ordway. *History of Rocketry and Space Travel.* New York: Thomas Y. Crowell Company, 1969. This fully illustrated volume details the development of rockets and the evolution of the Space Age. The historical narrative describes to the lay person how rocketry advanced from Chinese fireworks to the Apollo landings on the Moon.

Robert V. Reeves

MARINER 1 AND 2

Date: July 22, 1962, to January 3, 1963
Type of mission: Unmanned deep space probes

Mariners 1 and 2 were the first interplanetary spacecraft directed to study Venus. Although Mariner 1 suffered a launch failure, Mariner 2's flyby initiated an era of robotic spacecraft returning detailed data about the planetary system. Although it returned no photographs, Mariner 2 revolutionized knowledge of the conditions on Venus.

PRINCIPAL PERSONAGES

JACK N. JAMES, Mariner 2 Project Manager
DAN SCHNEIDERMAN, Spacecraft Systems Manager
EDWARD J. SMITH, magnetometer experiment
LEWIS D. KAPLAN AND
GERRY NEUGEBAUER, infrared experiment
J. COPELAND AND
D. E. JONES, microwave experiment

Summary of the Missions

The first interplanetary mission began developmental work even before the first Ranger spacecraft was ready for shipment to the Cape Canaveral launch site. (Ranger spacecraft were designed to be launched toward the Moon, taking photographs from a specified altitude until impacting the lunar surface.) The National Aeronautics and Space Administration (NASA) delegated primary responsibility for Mariners A and B to the Jet Propulsion Laboratory (JPL), in Pasadena, California. Mariner A was to be launched to Venus and Mariner B to Mars. Both spacecraft were to be in the 452-to-566-kilogram class. They were to be launched by a new booster consisting of an Atlas rocket with a Centaur upper stage. Centaur was designed to provide higher energy and a better performance than previous upper stages. During the 1962–1964 time period, desirable launch windows existed to both Venus and Mars. Program planners expected Centaur to be available to support these launch opportunities.

It became evident by August, 1961, however, that Centaur would not be ready to send Mariner A to Venus for the third quarter 1962 inferior conjunction, when Venus would be in a position between the Sun and Earth. Existing boosters were incapable of sending Mariner A to Venus, as the spacecraft was too heavy. Clearly a problem existed, but the tremendous pressure to encounter Venus during this inferior conjunction prevailed. Under the advice of JPL, NASA developed the compromise Mariner R program, wherein an Atlas-Agena booster would launch a lighter, Ranger-class spacecraft containing modified Mariner A components and experiments.

Mariner R would weigh less than 210 kilograms and carry between 11 and 18 kilograms of experiments. Two spacecraft were to be dispatched from the same launch facility during the 56-day launch window from July to September, 1962. Mission planners, scientists, and engineers had to redesign, build, and test a brand new spacecraft, devise a flight plan, and arrange twin launches in only eleven months. The decision to abandon Mariner A and commit to the new Mariner R spacecraft, redesignated Mariner 1 and 2, came in September, 1961. Test assemblies and design modification in the spacecraft and boosters were hurried along. After only nine months, project equipment was shipped to Cape Canaveral.

The Mariner R spacecraft was slightly more than 3 meters tall and 1.5 meters wide. As soon as the solar panels were deployed and the directional antennae unfurled, these dimensions became 3.5 and 5 meters, respectively. The final design weighed only 202 kilograms, including 18.6 kilograms of scientific equipment for Venus and interplanetary medium studies. The basic structure of the Mariner R spacecraft consisted of a hexagonal frame with superstructure to which a small liquid-fueled rocket motor, attitude control system gas jets, two folding solar panels, an infrared and microwave radiometer, numerous antennae, and other scientific experiments were attached.

Sixty minutes after launch, Mariner R's central computer and sequencer automatically activated the attitude control system to orient the spacecraft's directional antennae toward Earth and the solar panels toward the Sun, using nitrogen gas thrusters to make periodic adjustments to maintain this attitude to within 1 degree. In this attitude, the solar panels provided the spacecraft with 222 watts of power and permitted the telecommunications system to operate on a mere 3 watts of power (even at a distance of 54 million miles from Earth).

The small rocket engine in the hexagonal base contained sufficient fuel for only one midcourse correction firing. Mariner R's speed could be altered by 0.2 to 61 meters per second as a result of burning this engine from 0.2 to 57 seconds.

Three spacecraft (Mariners 1, 2, and a backup) were shipped to Cape Canaveral in late May, 1962. Final flight preparations for Mariner 1 were completed by mid-July. Thus, after only 324 days from program inception, a spacecraft was poised to provide the first closeup examination of another planet. The Atlas-Agena/Mariner 1 vehicle began its final countdown at 11:33 P.M. on July 20, but almost immediately, difficulty in the range safety command system forced a hold. (This system had already delayed the initiation of the 176-minute countdown.) The countdown resumed at 12:37 A.M. on July 21, but a blown fuse in the range safety circuit forced cancellation of the launch about two hours later.

The following day, the recycled countdown resumed just before midnight. Difficulties with both ground equipment and the vehicle's radio guidance system plagued the countdown with several unscheduled holds. At 4:16 A.M. on July 22, the countdown resumed after a recycle to T minus 5 minutes. The remainder of the countdown proceeded without incident, and Mariner 1 was lifted from the pad at 4:21:23 A.M. Although initial flight telemetry looked promising, the range safety officer noticed an unscheduled vehicle yaw two minutes after lift-off. The vehicle threatened populated areas and the North Atlantic shipping lanes. When corrective commands sent to the vehicle's guidance system were ineffective, the range safety officer was forced to destroy the vehicle, after only 293 seconds of powered flight. For more than a minute as the Mariner 1 spacecraft fell to the ocean, its transponder transmitted data.

The cause of the Atlas-Agena failure was not immediately clear. Before a commitment to proceed with Mariner 2 could be made, the cause had to be determined and corrective measures taken. After a detailed accident analysis, a simple fault was uncovered in the guidance system programming: A bar (or hyphen) above the element R in the guidance program had been overlooked. This programming error was shared by all previous Atlas launches, but booster performance was not affected by it during these launches because the remainder of the guidance systems functioned properly. On Mariner 1's Atlas, an antenna performed inadequately, causing the guidance signal received by the launch vehicle to be weak and noisy. Eventually, the launch vehicle lost its lock on the ground guidance signal, and when the signal was lost, the ground computer initiated a programmed search for the Atlas radio beacon to restore the proper signal lock. As this was in progress, the Atlas computer was suppressing information from this search, according to designed programming, and attempting to restore the signal lock on its own. Unfortunately, the command for that action was bar R. Thus, because the booster's computer was incapable of restoring the lock by itself, a result of the programming error, and was not responding to the ground search efforts, the Atlas was not properly supplied with steering commands and veered off course.

Thirty-five days after the destruction of Mariner 1, Mariner 2 began its final launch countdown. Like its predecessor, Mariner 2 was plagued by countdown difficulties. The first launch attempt had to be aborted because of a problem with the command-destruct system. The countdown resumed the following evening. During one unscheduled countdown hold, the primary Atlas booster battery had to be replaced; a planned hold then had to be extended to monitor the battery lifetime. When the countdown reached the scheduled T-minus-5-minute hold, technical difficulties in the radio guidance system forced a further delay. The countdown was able to proceed only to T minus 60 seconds before the radio guidance troubles forced another hold; it also caused a hold at T minus 50 seconds. After a recycle to T minus 5 minutes, the countdown was able to continue without mechanical trouble, but now the Atlas booster battery had only three minutes of pre-lift-off life remaining. To conserve power, flight planners decided not to switch on this battery until T minus 60 seconds. The test director pressed the fire button at T minus zero (1:53:14 A.M., August 27, 1962) and the Atlas-Agena/Mariner 2 rose from the pad and followed the desired trajectory. The Agena/Mariner 2 separated from the spent Atlas, and after coasting briefly, the Agena's restartable engine was fired to achieve the desired orbital velocity and a 186-kilometer parking orbit.

Just 16 minutes after orbital insertion, the Agena engine was restarted to increase spacecraft velocity to more than 40,000 kilometers per hour (escape velocity). Mariner 2 separated from the Agena upper stage two minutes after engine shutdown. Forty-four minutes after launch, Mariner 2's central computer and sequencer issued the commands to unfold the solar panels and radiometer dish, thus assuming the in-flight configuration. Sixteen minutes later, the attitude control system began the Sun orientation sequence to align Mariner 2's attitude properly.

Early projections indicated that Mariner 2 was on a flight path that would miss Venus by 373,000 kilometers. In order to obtain meaningful scientific data at closest approach, the spacecraft had to pass within 12,800–64,000 kilometers of the planet's surface. On September 4, a series of commands was transmitted to Mariner 2 to perform a required midcourse correction burn. The spacecraft engine burned for 27.8 seconds. As a result, Mariner 2's velocity decreased by 94.4 kilometers per hour with respect to Earth and increased by 72 kilometers per hour with respect to the Sun, causing the spacecraft to spiral in toward the Sun, passing Earth on mission day 65, and crossing the orbit of Venus on mission day 109.

During its interplanetary cruise, Mariner 2 suffered a series of nagging problems, but none that crippled the spacecraft. As Mariner 2 approached its target, spacecraft temperatures began to rise steadily. In fact, twelve hours before the Venus encounter, it was feared the central computer and sequencer would be unable to command Mariner 2 to initiate its encounter sequence properly because of overheating. As a check, a ground command was radioed to Mariner 2; the spacecraft, then 57.6 million kilometers away, responded by initiating the encounter sequence.

Mariner 2 Venus spacecraft P-1953. (NASA)

Mariner 2 approached Venus on its night side, making three scans of the planet: one on the night side, one along the terminator, and one on the daylight side. At its closest approach, Mariner 2 was only 34,800 kilometers above the surface of Venus.

Knowledge Gained

Mariner R carried six scientific instruments; two were primarily intended for Venus studies and the remaining four for interplanetary medium studies. Mariner's scientific objectives were twofold: measurement of cosmic dust, solar winds, high-energy cosmic rays, charged solar particles, and magnetic fields in the interplanetary medium; and measurement and investi-

gation of the magnetic fields, radiation belts, temperature, surface conditions, and cloud composition of Venus. Data from the scientific instruments were processed for transmission to Earth receiving stations by a data conditioning system.

For the study of the interplanetary medium, the spacecraft contained a cosmic dust detector, a solar plasma experiment, a magnetometer, and a high-energy radiation experiment. The cosmic dust detector measured the flux, direction of origin, and momentum of interplanetary dust and investigated the nature of any hazards this dust could pose to manned spacecraft. The energy and flux of low-energy positive ions streaming from the Sun were measured by the solar plasma experiment, which consisted of curved electrostatic deflection plates and collector cup, a sweep amplifier, programmer element, and an electrometer. The plasma experiment worked in a manner similar to a mass spectrometer, except that the instrument measured the energy of ions (in the range of 240 to 8,400 electron volts) rather than the mass. Mariner's magnetometer took readings of the magnetic fields in the vicinity of Venus and in interplanetary space with a lower sensitivity limit of 5 gammas. (One gamma is approximately one thirty-thousandth the intensity of Earth's magnetic field at the equator.) High-energy charged particles consisting of cosmic ray protons, alpha particles, electrons, and heavy nuclei were studied by the high-energy radiation experiment. To penetrate the walls of the instrument's ionization chamber, electrons required energies greater than 0.5 million electron volts, alpha particles more than 40 million electron volts, and cosmic ray protons of more than 10 million electron volts. Particle fluxes were measured in a detector consisting of three Geiger-Müller tubes.

For the study of Venus, the spacecraft contained a microwave radiometer and an infrared radiometer. The microwave radiometer scanned the planet at 19-nanometer and 13.5-nanometer wavelengths. Activated ten hours before closest encounter, the instrument had two scan-rate modes, 120 degrees at a rate of 1 degree per second for pre-encounter and 0.1 degree per second at closest approach. Designed to investigate the origin of the high surface temperature emanating from Venus and measure the surface temperature at encounter, the instrument had to be calibrated automatically twenty-three times during flight using a reference antenna system that scanned deep space for an absolute zero reference temperature. The infrared radiometer was attached directly to the microwave radiometer antenna so it would scan the exact area scanned by the microwave experiment at the same scan rate. This instrument scanned Venus at wavelengths in the 8- to 9- and 10- to 10.8-micron ranges of the infrared. A rotating disk alternately passed and blocked planetary emissions to compare the readings of Venus against the background empty space readings.

Context

Mariner 2 was the first successful interplanetary spacecraft, continuing beyond Venus and reaching solar perihelion (the point in its orbit that it was nearest to the Sun) on December 27, 1962, at a point 104 million kilometers from the Sun. Mariner 2 entered an orbit, sending it around the Sun and coming as close as 40 million kilometers to Earth and as far as 181 million kilometers from the Sun. Spacecraft signals continued to be received until January 3, 1963, when the Johannesburg, South Africa, tracking station shut down the spacecraft following a 30-minute data transmission. A few

days later, the Goldstone station searched for a weak spacecraft signal, but Mariner 2 had gone silent, becoming another relic of humankind's presence in outer space.

Scientists required several months to sift through Mariner 2's data to construct a new picture of the planet Venus. Rich in carbon dioxide and poor in oxygen and water vapor, the Venusian atmosphere consists of a continuous thick cloud layer beginning 72.5 kilometers above the surface and continuing for 24 -32 kilometers. This thick cloud layer allowed solar energy to reach the surface but prevented most of the heat from radiating back into space, creating a greenhouse effect. Before Mariner 2, the surface temperature of Venus was estimated to be about 320 degrees Celsius. Mariner data suggested a hotter planet, slightly more than 425 degrees Celsius at the surface. The thick atmosphere conducted heat well and explained why the spacecraft observed no temperature variation between the planet's daylight and dark sides. Mariner 2's magnetometer failed to detect a magnetic field and any belts of trapped charged particle radiation. These observations tended to confirm ground-based data indicating an extremely slow or nonexistent planetary rotation. Perhaps the most important result of the Mariner 2 mission was the inescapable evidence that interplanetary space was neither empty nor field-free, but full of particles having a wide range of energies. It was learned that these particles also contained time-dependent periodic fields with characteristic oscillation periods ranging from 40 seconds to several hours.

With the immense success of Mariner 2, subjective plans for a repeat mission were canceled. Plans began to take shape for a mission to Mars. The earliest "least-energy" opportunity for a Mariner Mars flyby was during 1964 – 1965. With these seemingly modest efforts, an era of planetary exploration had begun. The Mariner program concluded after ten attempted planetary missions. Two more Mariner spacecraft visited Venus. Mariner 5 flew 3,391 kilometers above Venus (October, 1967) and Mariner 10 flew within 6,000 kilometers of Venus (February 5, 1974) while on its journey toward Mercury. The United States sent Pioneer Venus probes in 1978 to orbit Venus and plunge through its atmosphere. The Soviet Union also investigated Venus, sending Venera probes through the atmosphere, attempting soft landings on the surface. Venera 7 (1970) was the first to transmit data from the surface. Several times the Russians repeated this feat with varying degrees of success; probes, however, cannot survive long in the harsh Venusian environment. Nevertheless, these spacecraft (Pioneer Venus, Mariner 10, and the Veneras) greatly increased scientific understanding of Venus, and the bold step of Mariner 2 paved the way for these more advanced scientific undertakings.

Bibliography

Fimmel, Richard O., et al. *Pioneer Venus*. NASA SP-461. Washington, D.C.: Government Printing Office, 1983. The official NASA history of the Pioneer Venus program. In addition to details about mission events and data, it provides a historical perspective of our scientific understanding of Venus as a result of earlier spacecraft missions (such as Mariner 2).

Fisher, David G. "Mariner: 25 Years of Interplanetary Space Flight." *Aerospace Educator* 4 (Fall, 1987): 33–37. This work provides details about the spacecraft systems and mission events. Suitable for general audiences.

Greeley, Ronald. *Planetary Landscapes*. Rev. ed. London: Allen and Unwin, 1987. This book presents pictorial geomorphology (study of the form and evolution of planetary surfaces) of all the planets visited by spacecraft prior to 1987 and the moons orbiting those planets. Suitable for advanced high school and introductory college-level audiences. Includes numerous references to more specific scientific papers.

Nicks, Oran W. *Far Travelers: The Exploring Machines*. NASA SP-480. Washington, D.C.: Government Printing Office, 1985. This work provides a perspective on planetary and lunar spacecraft programs unique to one who has participated in numerous programs. Illuminates the planning process and evolution of spacecraft programs. Suitable for general audiences.

Short, Nicholas M. *Planetary Geology*. Englewood Cliffs, N.J.: Prentice-Hall, 1975. This book details knowledge of the planets and their moons. Suitable as a text for planetary geology courses and includes photographs, charts, and diagrams. Questions are included at the end of each chapter.

David G. Fisher

MARINER 3 AND 4

Date: November 5, 1964, to July 24, 1965
Type of mission: Unmanned/Mars probes

Mariner 4 was the first U.S. unmanned probe to Mars. It provided data that greatly increased man's understanding of Earth's neighboring planet. The probe revealed a harsh, cratered wasteland rather than an environment likely to be capable of supporting life, as well as data on Mar's radiation density, cosmic-ray and cosmic-dust bombardment, characteristics of the "ion wind," and properties of the Martian atmosphere.

PRINCIPAL PERSONAGES

WILLIAM H. PICKERING, Director of the Jet
    Propulsion Laboratory
ORAN W. NICKS, Director of Lunar and Planetary
    Programs, NASA
DAN SCHNEIDERMAN, Mariner Project Officer at the
    Jet Propulsion Laboratory
JOHN R. CASANI, Mariner Spacecraft Systems
    Manager
ROBERT B. LEIGHTON,
BRUCE C. MURRAY, and
ROBERT P. SHARP, television subsystem experimenters
    at the California Institute of Technology
JAMES A. VAN ALLEN, the radiation and charged parti-
    cles physicist at the University of Iowa after whom
    the Van Allen radiation belts are named

Summary of the Missions

The Mariner missions were intended to advance man's knowledge of Mars and of the solar system in general by investigating certain physical aspects of the planet and its environs. After the success of Mariner 2, sent to investigate Venus in 1962, the National Aeronautics and Space Administration (NASA) contracted with the Jet Propulsion Laboratory (JPL) at the California Institute of Technology in Pasadena, California, to engage in a series of unmanned investigations of Mars over a period of many years. Mariner 3 was to be the first mission in that series.

The Mariner "family" of spacecraft had begun with a highly successful spacecraft, or "bus," on which numerous scientific instruments could be hung. This Mariner system was part of a broader system referred to as the overall "Project," which consisted of all the ground-support, testing, fabricating, launching, telemetry (radio signal data), and data-recep-

tion processes required to send the spacecraft to its target, correctly gather data, and return the data to Earth for scientists to interpret. In the case of Mariners 3 and 4, the Project involved a worldwide network of antennae stretching from California to Australia, South Africa, and Spain, as well as many subsystem contractors in the United States.

The Mariner/Mars Project established in 1962 succeeded within a short time in preparing two Mariner craft for launch. At midday on November 5, 1964, the Mariner 3 payload was lofted from Launch Complex 13 at Cape Kennedy by an Atlas rocket. Within a few minutes, it became apparent that the craft was not functioning properly. The "shroud," or protective covering that surrounds the payload during launch (when the possibility of damage is greatest), is normally ejected after the rocket passes through the atmosphere. At that point, the payload can ride safely on the nose of the rocket until it separates from the rocket and is sent on its journey. Mariner 3's shroud, however, would not disconnect and could not be ejected. Unable to shed itself of the weight of the shroud, the rocket was unable to reach sufficient velocity to push its payload far enough out into space. At the appropriate time, Mariner 3 separated from the rocket and tried to extend its solar panels to collect sunlight to convert into energy for powering the on-board equipment. Because the shroud still surrounded the Mariner payload, however, this step could not be achieved; after a short time the on-board backup batteries failed, leaving the spacecraft dead.

After studying the problems with the Mariner 3 craft, the engineering team experimented with a fiberglass payload shroud, redesigning the entire shroud in only three weeks. Toward the end of November, Mariner 4 stood on the launchpad; it was launched on November 28 at 9:22 A.M. eastern standard time by an Atlas-Agena rocket. The Atlas rocket separated, the Agena continued to push the Mariner 4 payload to an orbital altitude of 185 kilometers and an orbital velocity of 28,157 kilometers per hour. Mariner 4 was on its way. After a few minutes, the Agena rocket shut down and the Mariner craft coasted for 41 minutes in Earth orbit, preparing to swing outward toward Mars. At a predetermined time, the Agena reignited to boost the craft to escape velocity. Then Mariner 4 separated from the rocket, opening its four solar panels. Mariner tipped itself toward the Sun and received energy-giving sunlight, thus activating solar cells in the solar panels, generating 700 watts of electrical power.

The journey to Mars required crossing 523 million kilometers of open space over a period of almost eight months. During this time, the spacecraft used Canopus (one of the brightest stars in this part of the Milky Way galaxy) as its reference point for navigation. A spacecraft in Earth orbit or near Earth can use the planet for fixing its position. At great distances, however, Earth becomes a faint and unreliable reference; thus, Canopus was used.

During the long journey, Mariner continually checked its own performance and status as well as the condition of the space environment. On-board attitude control thrusters released puffs of nitrogen gas to keep Mariner upright and the craft was continually balanced and rebalanced against the pressure of the solar wind (the ionized particles that flow out from the Sun) by special vanes attached to the ends of the solar panels. To regulate the temperature of the spacecraft's instruments, special "window shades" fashioned like louvered slats opened or closed to let out or hold in the heat generated by the working instruments.

A solar plasma probe was designed to measure the charged particles making up the solar wind. A Geiger-Müller tube measured ionization caused by charged particles; unfortunately, a massive solar flare on February 5 damaged the instrument and rendered it useless. A cosmic-ray telescope detected protons and alpha particles, while a high-sensitivity helium magnetometer measured magnetic fields. The cosmic-dust detector consisted of a flat plate facing the direction in which the spacecraft was moving, and two surface penetration detectors were fixed with a microphone to record the impact force and the number of hits from micrometeoroids.

The imaging technology used by Mariner is known as remote sensing — acquiring data at a distance by electronic or mechanical means. Mariner's camera system is a vidicon television tube mounted behind a telescope; for this mission, it was also fitted with red and green filters that could be used interchangeably. The image falling on the tube is about 0.55 centimeter square. The camera scans the image in 200 lines of 200 dots for each line, producing a digital signal of 240,000 bits per image, which is recorded on a quarter-inch magnetic tape loop 91 meters long. The recorder stops automatically between images to save tape. It is incorrect, therefore, to say that the Mariner 4 spacecraft (or any interplanetary unmanned craft) "took pictures." The craft recorded images, which were converted to digital data so that the data could be radioed back to Earth.

Mariner 4 first encountered Mars on July 14, 1965, after traversing almost 523 million kilometers along the spacecraft's arcing trajectory (215 million straight-line kilometers). In its pass across Mars, Mariner recorded twenty-one images and a small portion of a twenty-second, starting at 18 minutes after midnight Greenwich mean time on July 25. The imaging sequence took only 26 minutes. The first image was recorded when Mariner was 16,895 kilometers from its target, and the last was recorded at 9,844 kilometers; at that time, Mars was moving in its orbit around the Sun at a speed 17,700 kilometers per hour faster than Mariner was moving as it crossed in front of the planet.

The digital data recorded on magnetic tape on board the spacecraft were released in a steady stream and rerecorded on magnetic tape by Mariner Mission Control at JPL in Pasadena. The series of Earth antennae, which spread from Goldstone, California, across Australia, South Africa, and Spain, recorded every bit of digital data, and the data then were converted into black-and-white photographs.

Knowledge Gained

It is an accepted truth in space science that the Mariner 4 mission completely changed humankind's concept of Mars and the solar system — and probably the universe. The startling images received from the craft, when compared to even the best photographs taken through Earth-based telescopes, immeasurably increased knowledge of planetary science. The other instruments, each conducting its own separate electronic investigations, also provided amazing information. Mariner 4 was an unqualified success.

Densely packed craters, both large and small, were revealed, as well as an impact basin. In all, more than seventy craters appear in the imagery. The surface of the planet was found to be rough and irregular, not unlike the topography of the Moon. Only in one instance did the Mariner 4 images hint at the presence of water, and that was in the ground haze rimming some of the deeper craters along the northern border of Phaethontis.

Carbon dioxide in considerable amounts was detected in the thin Martian atmosphere, although little nitrogen was discovered. Indeed, a thicker atmosphere might have protected the surface from the meteorite bombardment that produced the cratered surface. The absence of radiation belts and magnetic shock layers around Mars indicates that Mars has little or no magnetic field surrounding it. The Mars that greeted Mariner 4 has probably looked the same for millions and millions of years.

During Mariner's 227-day odyssey, about two hundred micrometeorites struck the craft. The hits apparently increased as the craft approached Mars but then decreased suddenly in the vicinity of the planet; this decrease suggests that Mars sweeps its path clear of cosmic dust as it orbits the Sun. The Mariner mission was undertaken during the period of the "quiet Sun," that time in the eleven-year sunspot cycle when the Sun's activity is at a minimum. At the same time, however, Mariner recorded violent solar eruptions in February, April, May, and June. Some of these flares were seen by telescope from Earth.

In a planned spacecraft occultation, the Mariner craft swept behind the planet and sent back to Earth careful readings of the thin atmosphere surrounding Mars. In fact, the ionosphere is virtually undetectable.

The sweep of Mariner 4's observations began at Phlegra, near the region of Cebrenia in the northern hemisphere; moved across the Mesogaea area near Amazonis at the Martian equator and Mare Cimmerium in the southern hemisphere, through Phaethontis and back up across Aonius Sinus; and ended just below and near Xanthe and the Chryse basin. By far the most spectacular image captured by Mariner 4 was image picture number 11 at 34° south latitude and longitude 199° east. The image showed a large crater in Cimmerium with newer and smaller impact craters overlaying the older large crater. The image also suggests ground haze or possibly frost rimming the large impact crater.

Context

The mission of Mariner 4 took place in that especially delightful era of scientific discovery known as the golden age of planetary exploration, made possible by breathtaking new technologies in space science research. For the first time, human eyes could see — considerably more precisely than Galileo had — the landscape of an alien world. Since 1960, when the world's first weather satellite (TIROS) looked at Earth for the first time, scientists have learned more about the solar system and the universe than had been learned during the preceding five hundred years. Mariner 4 was on the leading edge during that great epoch of lunar and planetary exploration — an exciting time of new discoveries, new theories, new possibilities, and new inventions.

Before Mariner 4, popular conceptions of Mars were determined by theorists who envisioned a planet that was more attractive and infinitely more appealing. Astronomer Percival Lowell had taken up the Martian canal debate and had tried to enlarge the Martian fantasy of earlier astronomers such as Giovanni Schiaparelli, who had claimed to see *canali*, or channels (which were mistakenly translated as "canals"). This controversy had raised the exciting possibility of intelligent beings who constructed the canals to transport or control the flow of enormous amounts of water.

The comic-strip adventures of Buck Rogers, piously waging war against dictators encamped on Mars, further added to the perception of Mars as a place inhabited by some kind of "people." On the heels of Buck Rogers, H. G. Wells wrote a novel about Martian monsters who came to Earth, destroying cities in their search for a nesting place. The fiction was enlarged upon by Orson Welles, then a radio broadcaster, who added sound effects to the story and used his creation as a huge Halloween prank on the American people. On October 30, 1938, Welles aired his fantasy as *The War of the Worlds*. Thinking that what they heard on radio was really happening, people panicked — some even committed suicide.

Given the prevailing myth of a life-supporting Mars, people the world over were surprised by the findings of Mariner 4. The night the first images came back to Earth, a televised newscast was held at JPL. One by one the images were shown; disappointment was evident as, not the lush landscape of a beautifully adorned planet, but a dry, barren, crater-pocked wasteland was revealed.

Although Mariner 4 brought about an end to a dream, it offered the beginning of another. Every planetary investigation provides new information with which to compare Earth. After the explorations of Venus and Mars by American and Soviet spacecraft, a new Venus-Earth-Mars scenario has slowly emerged, suggesting that the present appearances of these three planets are somehow linked in an evolutionary way. Perhaps at one time Earth looked like Venus; if the degradation of Earth continues, someday it may become like dusty, dry, and deadly Mars. Perhaps there is some kind of lesson to be learned from this trio of inner planets; perhaps continued explorations of Mars will reveal clues to the future of Earth.

The discoveries of Mariner 4 proved enticing; several more unmanned expeditions to Mars resulted, including Mariners 6, 7, 9, and 10, Vikings 1 and 2, and the Soviet Phobos probes. Manned landings remain a possibility. The success of Mariner 4 as a spacecraft led to the use of variations on the Mariner bus — the same type of frame was employed on subsequent missions. Mariner 5 went to Venus and Mercury in 1967, Mariners 6 and 7 to Mars in 1969, Mariner 9 to Mars in 1971, and Mariner 10 to Venus in 1973. The Mariner design blazed a path for other unmanned planetary explorers; the remote-sensing system used a flyby technique which was later used on all the Mariner-Mars missions, as well as on Pioneer missions to Jupiter and Voyager missions to Jupiter, Saturn, Uranus, and Neptune. The remote-sensing technique proved itself on the Mariner 2 and Mariner 4 missions.

Bibliography

Clarke, Arthur C. *The Promise of Space*. New York: Harper and Row, Publishers, 1968. An amazing survey of the human reach into space, this book contains dozens of charts, diagrams, tables, photographs, and illustrations. An engineer as well as a thoughtful writer, Clarke provides a careful overview of space technology, from Sputnik to the Apollo program.

Gatland, Kenneth. *The Illustrated Encyclopedia of Space Technology: A Comprehensive History of Space Exploration*. New York: Crown Publishers, 1981. The entire history of spaceflight, including the Mariner expeditions, is contained in this compendium of facts and figures. Key illustrations and color photographs make this book a reliable and easy-to-use reference for anyone interested in space.

Hartmann, William K., and Odell Raper. *The New Mars: The Discoveries of Mariner 9*. NASA SP-337. Washington, D.C.: Government Printing Office, 1974. One of the best research works extant for the reader who wants to understand the past, present, and future of Mars and Martian explorations. "Early Mariners and the Profile of the Mariner 9 Mission," as well as the first two chapters of historic data, presents a sweeping view of Mars as the planet of human hopes and fantasies.

Jet Propulsion Laboratory, California Institute of Technology. *The Mariner 6 and 7 Pictures of Mars*. NASA SP-263. Washington, D.C.: Government Printing Office, 1971. The official chronicle of the results of Mariners 6 and 7, this volume helps bring the contributions of Mariner 4 into perspective; the first chapter provides a thumbnail sketch of the Mariner 4 mission and other background information.

————. *Report from Mars: Mariner IV, 1964–1965*. NASA EP-39. Washington, D.C.: Government Printing Office, 1966. This illustrated booklet is a concise, accurate, well-written account of the entire mission, from first contact to imagery and photograph interpretation. An excellent reference source.

Koppes, Clayton R. *JPL and the American Space Program: A History of the Jet Propulsion Laboratory*. New Haven, Conn.: Yale University Press, 1982. A volume that must be considered the definitive work on the subject, this book covers the U.S. planetary exploration during its golden age.

Lewis, Richard S. *Appointment on the Moon: The Inside Story of America's Space Venture*. New York: Viking Press, 1968. Despite its misleading title, this storehouse of information about the exploration of space contains all that one needs to understand the technology, politics, and social issues inherent in the United States' first decade in space. Brilliantly and accurately written, with numerous historic photographs, the book is highly recommended.

Thomas W. Becker

MARINER 5

Date: June 14, 1967, to November 5, 1968
Type of mission: Unmanned/Venus probe

The primary scientific mission of Mariner 5 was to gather data on the atmosphere and the ionosphere of Venus and to study the interaction of the solar plasma with Venus' environment. A secondary engineering objective was to gain experience in converting a spacecraft designed to study Mars into one suitable to study Venus.

PRINCIPAL PERSONAGES
GLENN A. REIFF, Mariner-Venus 1967 Program
  Manager
DAN SCHNEIDERMAN, Mariner-Venus 1967 Project
  Manager
ALLEN E. WOLFE, Mariner-Venus Spacecraft System
  Manager
D. W. DOUGLAS, Mariner-Venus Missions Operation
  System Manager
NICHOLAS A. RENZETTI, Mariner-Venus Tracking
  and Data System Manager
CONWAY W. SNYDER,
ARVYDAS J. KLIORE,
CHARLES A. BARTH,
V. R. ESHLEMAN,
HERBERT S. BRIDGE,
EDWARD JOHN SMITH,
JOHN D. ANDERSON, and
JAMES A. VAN ALLEN, Principal Investigators

Summary of the Mission

The Mariner program was an early attempt by the National Aeronautics and Space Administration (NASA) to explore the environment of nearby planets using unmanned flyby space vehicles. The program began in the early 1960's, and the Jet Propulsion Laboratory of the California Institute of Technology was charged with the mission. The first two Mariner missions were to return data on Venus. The first Mariner launch, in 1962, was aborted soon after lift-off because of a malfunction. Mariner 2 was launched on August 27, 1962. It encountered Venus on December 14, 1962, and relayed valuable information to Earth. It established the presence and the properties of the solar wind (the plasma from the Sun) but did not provide information regarding its interaction with the environment around Venus, since its orbit did not pass through the zone of interaction.

The Mariner 3 and 4 missions were designed to gather data on Mars. Mariner 3 was disabled because its protective nose shield failed as the probe passed through Earth's atmosphere. Mariner 4, launched on November 28, 1964, performed flawlessly and operated for more than three years.

In December, 1965, NASA authorized two new Mariner programs for the Jet Propulsion Laboratory: one (Mariner 5) to investigate Venus further in 1967 and the other to return to Mars in 1969. A close look at the atmosphere and ionosphere of Venus was intended, and the interaction of solar plasma with the environment of Venus was to be a principal focus of Mariner 5. It was decided that the spare spacecraft designed for the Mariner 4 mission to Mars would be modified for the Venus mission. Seven scientific experiments were planned, and the spacecraft was fitted with appropriate equipment. A solar-plasma probe, a helium magnetometer, and an energetic-particle detector were designed to study the interaction between the planet and the interplanetary medium (the environment of Venus). An ultraviolet photometer was included to measure the properties of the planet's topmost atmospheric layers. S-band radio occultation and dual-frequency radio propagation experiments were to probe Venus' atmosphere at various heights from the surface, and precise measurements of the mass of Venus and the shape of its gravitational field were the aims of a celestial mechanics experiment.

The solar-plasma probe, helium magnetometer, ultraviolet photometer, and trapped-radiation detector experiments were slightly modified from earlier Mariner 4 experiments, and the dual-frequency radio experiment was fitted with a newly modified instrument package. The celestial mechanics and the S-band radio experiments involved no new instruments, since they utilized the tracking and communications systems. The instruments were fitted within an octagonal spacecraft bus. Since Mariner 5 would fly closer to the Sun than had Mariner 4 in its mission to Mars, the four solar panels that powered the probe's instrument package were reduced from 21.5 to 13.3 meters and were isolated and spaced 0.6 meter away from the bus to reduce heat transfer by reflection. The bus was thermally insulated and was fitted with a folding, octagonal, aluminized-Mylar sunshade facing the Sun. The bus was also fitted with thermostatically controlled louvers, which opened when the temperature in the bus exceeded a limiting value and closed when the temperature decreased. At the encounter point with Venus, the total heat leakage to the bus was expected to be about 50 watts.

Mariner 5 was equipped with sensors and an attitude control system that oriented the spacecraft to within half a degree with respect to the Sun and the star Canopus. A central computer and sequencer fulfilled timing, sequencing, and other computational functions for the spacecraft equipment. The spacecraft was aimed to encounter Venus on October 19, 1967, at an approximate distance of 8,165 kilometers from the planet's center. It had the capability of two midcourse maneuvers to avoid accidental impact on Venus.

The spacecraft, weighing 245 kilograms, arrived at Cape Kennedy on May 1, 1967. The Atlas-Agena, a multistage rocket with which the probe was to be fitted, was being readied at Launch Complex 12. The Atlas rocket was the main launch vehicle to lift the load through Earth's atmosphere. It was nearly 20.5 meters tall and 3 meters wide, and contained kerosenelike fuel and liquid oxygen feeding two 733,920-newton-thrust booster engines for lift-off, a 253,536-newton-thrust sustainer engine, and two 2,557.6-newton-thrust vernier control engines to stabilize the rocket. The Atlas rocket was to burn for about five minutes and then separate. The Agena D rocket, fitted atop the Atlas, was 6.1 meters long and 1.52 meters wide and was powered by a 71,168-newton-thrust engine. The Agena rocket was to be fired twice — first after the separation from the Atlas rocket, to provide sufficient velocity to place the spacecraft in orbit, and again at an appropriate time to power the spacecraft for its journey to Venus. The Agena rocket was fueled by dimethyl-hydrazine and nitric acid oxidizer.

The launch date was selected in advance to optimize the value of the scientific mission. Mariner 5 was launched at 0601 Greenwich mean time (GMT) on June 14, 1967. Soon after the probe's separation from the Agena rocket, the solar panels and the sunshade were opened and the scientific equipment was powered as Mariner 5 sailed toward Venus. The magnetometer returned measurements of the magnetic field of Earth and the interplanetary field of the Sun, and the ultraviolet photometer detected the Lyman-alpha glow caused by atomic hydrogen in the upper layers of Earth's atmosphere. The direction of approach to Venus was controlled by the spacecraft sensors, which aimed constantly toward the Sun and Canopus. A mid-course maneuver was made at 2308 GMT on June 19.

The instruments' data acquisition continued during the four-month flight to Venus. Approximately 850 hours of data (at 33.33 bits per second) were received by the Canberra station of the Deep Space Network during the first forty days. An additional 1,670 hours of data were received before Mariner's encounter with Venus, and 600 hours of data were received after the encounter, at a rate of 8.33 bits per second. Venus blocked Mariner 5 from Earth during the satellite's closest approach. This inhibited communications, and 648,000 bits of scientific data had to be stored in the spacecraft's tape

recorder and relayed back to Earth twice during the week following the closest encounter.

During its flight to Venus, an important experiment was conducted with Mariner 5. Ranging experiments had been performed earlier using Earth- and Moon-orbiting satellites, but the first ranging experiment at planetary distance was performed with Mariner 5. By accurately determining the frequency shift (Doppler shift) of a coded radio signal returned by the spacecraft, and measuring the total transit time (at approximately 300,000 kilometers per second), it was possible to determine the velocity and the distance of the spacecraft quite precisely. Near its approach to Venus, the accuracy of locating Mariner 5 was about 6 meters in 80 million kilometers.

As Mariner 5 neared Venus, various instruments started monitoring the planet. The encounter sequence with Venus started at 0249 GMT on October 19, 1967, in response to a command sent by ground controllers. Changes in the readings of the on-board magnetometer and the plasma probe indicated Mariner's passage through the zone of plasma shock. The closest approach to Venus occurred at 1734 GMT at a distance of 10,151 kilometers from the center (approximately 4,094 kilometers above the surface) of Venus. The planet's strong gravitational pull modified the flight path of Mariner 5 into a tight arc of more than 100 degrees and hurled it toward the Sun at a velocity of 30,650 kilometers per hour. During the three-hour period of the encounter with Venus, the thickness of the plasma shock and the boundary of the planet's influence on solar plasma were determined, ultraviolet photometric observations were completed, an ionosphere was detected, and observations of the atmospheric effect on S-band signals were made. Contact with Mariner 5 was re-established as the probe emerged from behind Venus.

As Mariner 5 approached the Sun, its telemetry signal grew weaker; the signal was eventually lost on November 21. At that point, Mariner was about 117 million kilometers from Earth and about 98 million kilometers from the Sun. From launch to this point, the spacecraft had performed flawlessly and exceeded all expectations. A large amount of scientific data was collected, and from every perspective the mission was deemed a complete success.

Mariner 5 attained a perihelion (closest distance to the Sun) of about 86 million kilometers on January 4, 1968. The spacecraft was traveling at the remarkable speed of 149,000 kilometers per hour. The strong gravitational attraction of the Sun placed it in an elliptical orbit around the Sun, with an orbital period of 195 days. Mariner 5 would be in the neighborhood of Earth (just outside the orbit of Venus) about every fourteen months.

Since the Mariner 5 mission was such a success, it was decided to extend the project through January 22, 1969. During this period, interplanetary space would be examined at short wavelengths, the nature of the space around the Sun would be studied during the period of high solar activity, and the understanding of the constants of celestial mechanics would be refined. Researchers also wished to gain knowledge of the effects on scientific equipment of the spacecraft's close passage around the Sun.

Attempts to reestablish radio contact with Mariner 5 began on April 26, 1968. After numerous attempts, contact was made on October 14, 1968, but the spacecraft behaved abnormally and did not respond to ground commands. Finally, on November 5, 1968, the project was terminated without achieving its extended objectives.

Knowledge Gained

Venus has been termed a twin planet of Earth. It is the closest planet to Earth, it has an atmosphere, and its size (radius about 6,053 kilometers) is comparable to that of Earth (radius 6,371 kilometers). In other ways, however, Venus is very different from Earth.

The atmosphere of Venus was found to be denser and more compressed than Earth's. The ultraviolet photometric experiment indicated the presence of atomic hydrogen, but the upper boundary of atomic hydrogen is only 19,000 kilometers above Venus' surface, or about one-fifth the height for Earth. Consequently, the temperature at this altitude is much lower than at the same altitude above Earth, suggesting that the thermal loss of hydrogen from the Venusian atmosphere is much lower.

Although Mariner 5 was not designed to make direct measurements of the composition, density, and temperature of Venus' atmosphere, estimates were made from various data that it did provide. S-band occultation measurements provided temperature and pressure profiles for Venus' atmosphere in the altitude range of 25 to 90 kilometers. The downward extrapolation indicated the surface temperature and pressure to be in the range of 430 degrees Celsius and 100 bars, respectively. The top of the cloud cover is located at an altitude of 65 to 70 kilometers from the Venusian surface. The dense lower atmosphere acts as a lens to refract light around the surface. Consequently, the nights on Venus are expected to be brighter than those on Earth, with sunlight forming smeared rainbows along Venus' horizon.

Mariner's ultraviolet photometer recorded a weak ultraviolet airglow on the nightside of Venus. The origin of this airglow is not known. The instrument failed to record the presence of atomic oxygen in the upper atmosphere, but verified the detection by the Soviet Venus probe, Venera 4, of very small quantities of molecular oxygen and water in the lower atmosphere. The dominant atmospheric constituent is carbon dioxide, which must constitute at least 85 percent, and perhaps much more, of the Venusian atmosphere. Nitrogen concentration is quite low. These observations confirmed the "greenhouse" model of Venus suggested in the early 1960's. Since carbon dioxide is transparent to sunlight but opaque to longer wavelengths (infrared) reradiated from the surface, heat energy builds up in the lower atmosphere. The thick atmos-

phere prevents cooling of Venus' nightside, and hot winds keep the surface temperature uniformly redistributed over the planet's surface.

Measurements of the magnetic field and ionosphere and their interaction with the solar plasma of Venus indicate considerable differences between Venus and Earth. Venus' magnetic field was undetected by Mariner's instruments and so must be at least one thousand times weaker than that of Earth. As a result of Earth's geomagnetic field, the solar plasma is deflected and flows around it, forming a cavity, known as the magnetosphere, within which the plasma cannot enter. The closest approach of the plasma on Earth's sunlit side is 50,000 kilometers from the surface. The deflected plasma causes a bow shock similar to one created by a supersonic plane. An analogous effect has been observed in Venus. In the case of Venus, however, the plasma is deflected not by the magnetic field but by a dense ionosphere on the sunlit side of the planet. The ionosphere prevents the penetration of plasma, piling it up in front and deflecting it on the sides to form a bow shock. The upper boundary of the ionosphere is, however, extremely low, not higher than 500 kilometers above Venus' surface.

The celestial mechanics experiment provided an accurate value for the mass of Venus. Analysis of the range and Doppler radio tracking data indicated, with an accuracy of one part per million, that the mass of Venus is 0.8149988 that of Earth. The radius of the planet could not be determined through the S-band occultation experiment, as the electromagnetic waves were deflected by the dense lower atmosphere of Venus.

Context

The early 1960's marked the beginning of planetary exploration. The first launch of an unmanned spacecraft, Venera 1, was in 1961 by the Soviet Union. Both Veneras 1 and 2 (the latter launched in 1965) flew past Venus without relaying any scientific information. Venera 3 landed on Venus in 1966 but failed to operate. Venera 4 was launched on June 12, 1967; it encountered Venus a day before Mariner 5 did. As Venera 4 descended through Venus' atmosphere, it transmitted valuable information on the temperature, density, and composition of the Venusian atmosphere; radio transmission from the spacecraft stopped, however, at an altitude of 25 kilometers.

Mariner 2 was the first United States probe to fly past Venus. Launched on August 26, 1962, it encountered Venus at a distance of 35,000 kilometers on December 14, 1962. Data from Mariner 2 confirmed the existence of solar plasma, which traveled with a velocity between 350 to 800 kilometers per second; the probe's instruments also measured the density of the plasma. It also returned data showing that the temperature of the Venusian atmosphere was high (around 130 degrees Celsius). The on-board magnetometer failed to detect any magnetic field around Venus.

The interaction of plasma with Earth's Moon has been

studied in detail. Since the Moon has no magnetic field, solar plasma strikes its surface and is absorbed, while the magnetic field associated with the plasma passes through the Moon without any interference. Measuring the interaction of the solar plasma with Venus was one of the primary objectives of Mariner 5, the second American probe to Venus. It discovered the bow shock of the plasma and the presence of an ionospheric layer around Venus; it also confirmed the existence of high temperatures, high pressure, and a high concentration of carbon dioxide on Venus.

Subsequent missions to Venus were Veneras 5 and 6, launched in 1969. Like Venera 4, both succumbed to the high atmospheric pressure of Venus and ceased functioning before they landed. Venera 7 operated for only 23 minutes. Venera 8 was more successful; it transmitted valuable information about temperature and wind velocities on Venus.

Mariner 10 represented the second generation of American probes to Venus. It took close-up photographs of the planet's cloud cover and identified weather patterns. Pioneer Venus (1978) and Veneras 15 and 16 (1983) mapped the surface of Venus using radar.

Since the days of Mariner 5, much has been learned about the atmosphere of Venus. Ninety-seven percent of Venus' atmosphere is composed of carbon dioxide. Sulfuric acid has been detected, as well as small amounts of hydrochloric and hydrofluoric acids. Most of the molecules of water are locked into the ions of these acids, making the surface of Venus much more arid than deserts on Earth. The concentration of water vapor is observed to be less than 0.1 percent in Venus' lower atmosphere.

Bibliography

Hunten, D. M., et al., eds. *Venus*. Tucson: University of Arizona Press, 1983. This book is a collection of scientific papers detailing the results of studies on the atmosphere, ionosphere, and surface of Venus. The treatment is at a university level.

Kaufmann, William J., III. *Exploration of the Solar System*. New York: Macmillan, 1978. This book summarizes the important observations made on planets and their moons. Suitable for a lower-level undergraduate audience.

Morrison, David, and Tobias Owen. *The Planetary System*. Reading, Mass.: Addison-Wesley Publishing Co., 1988. This volume provides short summaries of what is known of the planets, moons, and comets. Suitable for high school and college levels.

National Aeronautics and Space Administration. Science and Technical Information Office. *Mariner-Venus, 1967: Final Project Report*. NASA SP-190. Washington, D.C.: Government Printing Office, 1971. This report describes the complete project of the Mariner 5 mission. It contains details of the instruments, describes the observations, and analyzes the results. It is appropriate for university level audiences.

Snyder, Conway W., et al. "Mariner V Flight past Venus." *Science* 158 (December 20, 1967): 1665 – 1690. This collection of scientific articles by the principal investigators summarizes the preliminary findings of the Mariner 5 project. Suitable for university level readers.

D. K. Chowdhury

Mariner 6 and 7

Date: February 24, 1969
Type of mission: Unmanned /Mars probes

Mariners 6 and 7, the second and third unmanned missions to Mars, provided valuable photographic information on the surface characteristics of the planet, along with data on temperatures, atmospheric composition, and densities that were vital in planning and executing subsequent missions to Mars.

Principal personages

HARRIS M. "BUD" SCHURMEIER, Mariner 6/7 Project
    Manager at the Jet Propulsion Laboratory
JOHN A. STALLKAMP, Mariner 6/7 Project Scientist
HENRY W. NORRIS, Mariner 6/7 Spacecraft Systems
    Manager
W. RUSSELL "RUSS" DUNBAR, Launch Vehicle Systems
    Manager at Lewis Research Center
MARSHALL S. JOHNSON, Mariner 6/7 Mission
    Operations System Manager
ROBERT B. LEIGHTON, Principal Investigator, visual
    imaging experiment
GEORGE C. PIMENTEL, Principal Investigator, infrared
    spectrometry experiment
GERRY NEUGEBAUER, Principal Investigator, infrared
    radiometry experiment
CHARLES A. BARTH, Principal Investigator, ultraviolet
    spectrometry experiment
ARVYDAS J. KLIORE, Principal Investigator, S-band
    occultation experiment
JOHN D. ANDERSON, Principal Investigator, celestial
    mechanics experiment

Summary of the Missions

The Mariner 6 and 7 missions were undertaken to acquire information on the Martian surface and atmospheric characteristics in order to provide a firm basis for an effective long-range program of exploration. The selected instrument payload emphasized this purpose.

In late 1965, the National Aeronautics and Space Administration (NASA) had approved the Mariner Mars 1969 project, and the project management responsibility was assigned to the Jet Propulsion Laboratory (JPL), along with the responsibility for providing the spacecraft, mission operations, and Tracking and Data Acquisition (TDA) systems. The Lewis Research Center was assigned the responsibility for providing the launch vehicle system. The spacecraft subsystems were acquired from a number of industrial suppliers and were assembled and tested at JPL prior to being shipped to Cape Kennedy for launch.

The spacecraft system design was derived from the successful Mariner 4 spacecraft, with several significant improvements: The instruments were mounted on a two-degree-of-freedom platform, which made it possible to point the instruments in desired directions during the Mars encounter; dual tape recorders were used to store thirty-five times the amount of data that had been stored on Mariner 4; electronic equipment increased the rate at which data could be transmitted from the spacecraft to Earth to a maximum of 16,200 bits per second (Mariner 4 had transmitted data at a comparative snail's pace, 8.33 bits per second); and a central computer and sequencer, which could be reprogrammed from Earth, allowed information acquired from Mariner 6 to enhance Mariner 7 science data.

The spacecraft was octagonal in structure (as the previous Mariners had been), with electronic equipment mounted on its inner walls; the propulsion subsystem and its hydrazine tank were mounted on one of the faces of this octagon to provide trajectory corrections. The structure also supported the instrument platform; four solar panels that extended like wings to provide electrical power for the spacecraft; and a dish-shaped high-gain antenna, which was part of the radio subsystem used for receiving commands from Earth and transmitting data back to Earth. The radio subsystem was also used for navigating the spacecraft. The science payload consisted of two television cameras, two infrared instruments, and an ultraviolet spectrometer.

The two spacecraft could be launched during a "window" between February 24 and April 7, 1969. It was planned to launch the Mariner 6 as early as possible during this period to allow for the initial midcourse trajectory corrections and, if necessary, any unplanned modifications of the second spacecraft, which had to be launched during the same window. Two Atlas SLV-3C Centaur launch vehicles were prepared for the Mariner 6 and 7 missions.

One and a half weeks prior to the initial scheduled launch, an accidental depressurization of the Atlas booster occurred; the substitution of the launch vehicle designated for the second launch became necessary, and a new Atlas booster was obtained to launch Mariner 7. Rescheduling and extra shifts made it possible for the launch team to launch Mariner 6 at the desired early point in the window, on February 24, 1969. Roughly fifteen minutes into the scheduled one-hour launch window, Mariner 6 separated from its launch platform, at 8:29 P.M. eastern standard time. The launch vehicle was required to change its course following the burnout of the Atlas booster to ensure that debris did not fall on the Bahamas or the Leeward Islands. The launch vehicle's performance was excellent, and the spacecraft was safely on its way to Mars. Mariner 7 lifted off at 5:22 P.M. eastern standard time on March 27, 1969. A course change following the Atlas booster burnout was not necessary for Mariner 7, and the launch vehicle performance was equal to that for Mariner 6.

The launch trajectories selected for both Mariners were designed to ensure that the unsterilized Centaur stage could not strike Mars. Midcourse maneuvers were required to position the spacecraft roughly 3,200 kilometers above the planet's surface for their closest observations of the planet. These maneuvers were conducted on March 1 and April 8, 1969, respectively.

Several problems occurred during the missions that required the engineers and scientists to utilize great skill in evaluating problems, with limited data, and then design new operational sequences to overcome and reduce the risks of losing scientific data. The first of these problems occurred when the scan platform (the platform to which all the instruments were attached) was unlocked from the fixed position that prevented it from being damaged during launch. When the platform of Mariner 6 was unlocked, several small particles of dust were dislodged from the spacecraft; they drifted in front of the star tracker, which keeps the roll axis of the spacecraft fixed as it flies through space. As a result, Mariner 6 rolled away from Canopus (a bright star near the southern celestial pole), which was being used at that time as its reference. Standard commands were sent to the spacecraft to lock it back on Canopus.

The flight team recognized that a similar problem could occur during the Mars encounter, when valves would be activated to release the two gases used to cool the detector of the infrared spectrometer. As a result, a change in the encounter sequence was made; the spacecraft would be kept under gyroscopic control during the critical sequence.

The encounter sequences were designed to store the primary science data on on-board tape recorders; these data would be played back to Earth at a conservative rate of 270 bits per second so that the primary data would be recovered. An engineering experiment would allow a faster playback when the large Goldstone, California, antenna was tracking the spacecraft.

On July 29, 1967, the far-encounter sequence — which acquired 33 television pictures of Mars as Mariner 6 raced toward the planet — began. The next day, 17 images were scheduled as Mariner 6 came even closer. Fifteen minutes

before Mariner 6 took the last of these images, alarms rang in the operations center, indicating that radio contact with Mariner 7 had been lost. The emergency occurred when the staff was concentrating on recovering the primary science data on Mariner 6 and most of the tracking facilities were committed to that spacecraft. A small team of engineers was assigned to the Mariner 7 problem. After several hours, Mariner 7 was directed to switch to its omnidirectional antenna. After eleven minutes, signals were received on Earth and two-way communications with the spacecraft again were possible.

Obtaining science data during Mariner 6's near-encounter phase on July 31 now became the goal. High-rate data (16,200 bits per second) were received from the spacecraft. All went well, excepting some difficulties occasioned when the infrared spectrometer's low-temperature detector failed to cool, probably because a particle lodged in the fine tubing of its refrigerator.

Mariner 7 followed its predecessor five days later, and although considerable data were received on Earth, the navigation team determined that the craft's early failure had changed its flight path by the planet. The team heroically revised all the pointing angles to compensate for this change, sending them up to the spacecraft before its close encounter. The low-temperature refrigerator on the infrared spectrometer also worked on Mariner 7; thus, its information complemented that sent back by Mariner 6.

Following an extensive review of the lost radio signal, analysts finally concluded that the storage battery had failed. One or more of the eighteen sealed battery cases must have ruptured, causing severe power changes, and upsetting electrical circuits on the spacecraft. The liquid discharge from the battery cells also resulted in a small change to the spacecraft trajectory.

In spite of these problems, the flight team was able to reprogram the flight computers. As a result, the mission acquired more scientific data than needed to meet the established mission objectives. This improvement in data acquisition was most notable with the television experiment, in which 16 far-encounter and 48 near-encounter images were required; 2,561 far-encounter and 577 near-encounter images were actually recovered. Six hundred spectral pairs of data, compared to 288 pairs required for the infrared spectrometer, were acquired. Additionally, the infrared radiometer and the ultraviolet spectrometer also modestly increased the amount of information recovered.

Following the encounters with Mars, several engineering experiments and a few scientific measurements were made, but since the payload was entirely planet-oriented, few data were obtained. Toward the end of 1970, the altitude control gas on each spacecraft ran low; the missions were concluded when the transmitters were turned off before the end of December, 1970.

Knowledge Gained

Information from the ultraviolet and infrared spectrometers determined that carbon dioxide and its disassociation products were dominant in the Martian atmosphere. A high, thin cloud of atomic oxygen surrounds the outer atmosphere of Mars, just as it does on Earth and Venus. No nitrogen, however, was detected. Nor was any high-altitude ozone. Traces of water vapor and ice were found, and undetected amounts of inert gases such as argon and neon are believed to complete the constituents of the atmosphere.

The S-band occultation measurements made at four locations allowed the determination of pressure and temperatures at various altitudes. Thus, the ionospheric layer was detected at an altitude of about 140 kilometers. Surface pressures between 6 and 7 millibars were determined for three of four locations; the fourth had a pressure of 3.8 millibars, and it is believed that the surface altitude for this measurement has an elevation of 5 to 6 kilometers. The clouds of Mars are thin, occur mainly near the polar caps, and are made up principally of dry-ice crystals, water crystals, and dust. It does frost and snow on Mars.

Martian surface temperatures were measured by the infrared radiometer with two instruments compatible with the narrow-angle television cameras. A high temperature of 16 degrees Celsius was measured in equatorial regions and low temperatures of −120 degrees Celsius were found near the south pole. The temperatures measured by Mariner 7 as it crossed the southern polar cap are consistent with the frost temperature of carbon dioxide at a vapor pressure of 6.5 millibars; this information confirms that the polar caps are principally frozen carbon dioxide. The dark features on Mars absorb more solar energy during daylight and are warmer than the light areas.

The Mariner television cameras acquired roughly two hundred times the imaging data that were obtained from Mariner 4. The near-encounter pictures substantially increased resolution, allowing an improved understanding of the surface features and the geologic history of the planet. The cratered terrain is certainly Moon-like, but weathering has occurred, particularly along the edges of the polar cap. The carbon dioxide snow on the edges of the cap is generally thin; it disappears on the surfaces of craters while remaining on the crater walls, where the Sun does not shine as brightly.

Two new types of terrain were observed on these missions. The Mariner 7 television cameras crossed a southern circular desert area where there are few craters. This featureless terrain must have been significantly eroded, but liquid water cannot exist on Mars because of the planet's low atmospheric pressure. In another area, near the equator, the surface has a series of ridges and depressions that seemingly were created after the craters were produced. This area was identified by the television experiment team as chaotic terrain.

Context

Mariners 6 and 7 were the second and third spacecraft to visit Mars. The Russians had tried several times to visit Mars but were unsuccessful until the 1970's. Mariner 4, the first spacecraft to visit Mars, had discovered the cratered surface of the planet. That spacecraft also was able to determine that Mars has no significant magnetic field; during that mission, the planet's atmospheric density, temperature, and mass were measured. Mariners 6 and 7 were designed only to study the planet and make measurements that would be of value in subsequent long-range explorations of Mars.

Mariners 6 and 7 were able to photograph roughly twenty times the surface area covered by Mariner 4. Three types of terrain were observed: cratered, chaotic, and featureless. The featureless terrain indicates that significant weathering has occurred on Mars, eroding the impact craters that once existed in these areas. The measurement of surface temperatures confirmed those determined by Earth-based observations at the equator and the edges of the polar caps.

Two identical spacecraft were authorized by NASA to ensure that these important data were returned to Earth; this information would be used for future missions to Mars. Only one channel on the infrared radiometer on Mariner 6 worked, but the two working channels on Mariner 7 provided the missing data for the mission. The incorporation of an on-board, reprogrammable computer allowed information obtained from Mariner 6 to enhance that of Mariner 7.

The measurements collected by Mariners 6 and 7 do not encourage the belief that life exists on Mars, although the possibility of life is not excluded. Nevertheless, if life does exist, it probably is microbial. The absence of atmospheric nitrogen, coupled with ultraviolet rays penetrating the atmosphere, and the scarcity of water, make life-forms such as those found on Earth unlikely. The scientific and engineering data received from these missions, along with the successful demonstration of computer and communications technologies, were of significant value to the planning of the subsequent Mariner 9 orbiting mission of Mars, launched in 1971.

Bibliography

Chapman, Clark R. *Planets of Rock and Ice*. New York: Charles Scribner's Sons, 1982. This readable, well-illustrated book describes what is known of the solid planets and moons. Text includes material acquired by the Mariner 4, 6, 7, and 9 and Viking 1 and 2 missions to Mars.

Collins, Stewart A. *The Mariner 6 and 7 Pictures of Mars*. NASA SP-263. Washington D.C.: Government Printing Office, 1971. A brief background of previous investigations is included along with technical descriptions of the two cameras used for this investigation. Black-and-white prints of each of the images recorded on the spacecraft are included.

Scientific and Technical Information Division, Office of Technology Utilization. *Mariner-Mars 1969: A Preliminary Report*. NASA SP-225. Springfield, Va.: Clearinghouse, Department of Commerce, 1969. The initial report prepared by members of the project team and the scientific investigators, it describes the program objectives and the spacecraft that were built for this dual-spacecraft mission. Brief descriptions of each experiment and the preliminary results of each experiment are included with photographs and charts of the recovered data.

Wilson, James H. *Two over Mars: Mariner VI and Mariner VII, Feb –Aug, 1969*. NASA EP-90. Washington, D.C.: Government Printing Office, 1971. This report, with good pictures and charts, describes the missions of Mariners 6 and 7 and provides a brief summary of the scientific findings and their potential value to subsequent missions to Mars.

Henry W. Norris

MARINER 8 & 9

Date: May 8, 1971, to October 27, 1972
Type of mission: Unmanned/Mars probes

As the first artificial satellite of another planet, Mariner 9 did the work planned for both Mariners 8 and 9. It experienced a huge Martian dust storm, measured the two Martian moons, mapped a complete planet in less than a year, made preliminary determinations of the surface areas upon which landers could be set down, and revolutionized the understanding of a very old "New Mars."

PRINCIPAL PERSONAGES

DAN SCHNEIDERMAN, Mariner-Mars 1971 Project
  Manager at JPL
CARL W. GLAHN, Mariner-Mars Program Manager at
  NASA headquarters
ROBERT A. SCHMITZ, Manager of the Viking-
  Mariner-Mars 1971 Participation Group
HAROLD MASURSKY and
BRADFORD A. SMITH, Mariner 9 science experiment
  team leaders and principal investigators, television
  and orbiter imaging
CHARLES A. BARTH, Principal Investigator, ultraviolet
  spectrometer
R. A. HANEL, Principal Investigator, infrared interfer-
  ometer spectrometer
ARVYDAS J. KLIORE, Principal Investigator, S-band
  occultation
J. LORELL, Principal Investigator, celestial mechanics
GERRY NEUGEBAUER, Principal Investigator, infrared
  radiometer, thermal mapping

Summary of the Mission

Conceptually, the Mariner-Mars 1971 mission — the technical designation for the combined launching of Mariners 8 and 9 — began under the ambitious but premature first Voyager program, which was fiscally eliminated on October 26, 1967. The initial Mariner projects were approved on July 15, 1960, by T. Keith Glennan, the first Administrator for the National Aeronautics and Space Administration (NASA). At first, planners intended to send payloads of between 400 and 700 kilograms to Venus and Mars within the following two years, using Atlas-Centaur launch vehicles, but when the upper-stage Centaur booster continued to be unavailable, the Venus flyby flight (Mariner A) was canceled and that for Mars was successively redefined and redeferred.

In the interim, Jet Propulsion Laboratory (JPL) reconsidered several smaller varieties of the Atlas-Agena launch vehicle, proposing the pairs Mariner-Venus 1962 (only Mariner 2 completed its Venus flyby) and Mariner-Mars 1964 (only Mariner 4 completed its Mars flyby) and the single remodeled Mariner-Venus 1967 (which, as Mariner 5, completed a Venus flyby). As the Voyager program was phased out, reconsiderations permitted the final flights past Mars with more intensive optical equipment on the pair Mariner-Mars 1969 (which performed successfully as Mariners 6 and 7).

The upper-stage Centaur Satellite Launch Vehicle 3C became available in 1966. The Office of Space Science (OSS) proposed in November, 1967, that a pair of orbiters record with television relay systems the surface of Mars in anticipation of a new Viking project. The Mariner-Mars 1971 flights were approved August 23, 1968, to send orbiters with landers to Mars (Vikings 1 and 2), ostensibly to search for extraterrestrial life. Photographs from Mariners 4, 6, and 7, while suggesting a lunarlike, cratered surface, had shown a more dynamic planet than terrestrial observation had anticipated.

The Mariner-Mars 1971 mission, with identical spacecraft, was complementary during the ninety days of reconnaissance. Mission A (the prelaunch designation for Mariner 8) was to orbit Mars once every 12 hours, scanning most of the planet's surface, including the polar regions, from an orbit inclined 80 degrees to the Martian equator. The elliptical orbit was to have a periapsis (the closest approach to the gravitational center of the planet) of 1,250 kilometers, thus providing overlap in consecutive images taken directly downward by a wide-angle camera with a resolution of 1,000 meters.

Mission B (the prelaunch designation for Mariner 9) was to orbit Mars once every 20.5 hours, focusing every five days on details of the variable surface features within its equatorial belt from an orbit inclined 50 degrees. It was to come within 850 kilometers of the surface, from which its high-resolution camera would identify features as small as 100 meters.

Each was to carry three other scientific instruments. The infrared interferometer spectrometer, previously employed on Nimbus weather satellites, was redesigned to measure water vapor in the Martian atmosphere as well as the profile of atmospheric temperature from the surface outward into space. The ultraviolet spectrometer and the infrared radiometer had been employed by Mariners 6 and 7. The ultraviolet spectrometer was to identify the components of the Martian atmosphere and measure atmospheric pressure at the surface of the planet. The infrared radiometer would compare the background coldness of deep space (4 Kelvins) and the known temperature of an internal source in the instrument package.

An occultation experiment was to analyze the distortion of radio signals passing through the Martian atmosphere. With orbital repetition, the atmosphere could be measured according to latitude, season, or time of Martian day, and the shape of the planet could be determined. A celestial mechanics experiment, using the spacecraft's radio signals, was to determine the size, distance, and position of the planet and detect any large concentrations of mass.

As orbiters, Mariner-Mars 1971 required propulsion subsystems exerting 1,600-meter-per-second velocity changes to inject the craft into orbit. A 1,340-newton engine with restart capability, deriving its propulsion from monomethyl hydrogen and nitrogen tetroxide, was developed from components used on various other spacecraft.

The Mariner craft retained the basic shape and size of Mariners 6 and 7: an octagonal magnesium frame, measuring 138.4 centimeters diagonally and 45.7 centimeters deep. They required larger solar panels, still four in number; these measured 215 by 90 centimeters and spanned 68.9 centimeters, with a total surface area of 7.7 square meters, and included 14,742 solar cells, which were capable of generating 800 watts on Earth and 500 watts on Mars and of maintaining the nickel-cadmium battery at a charge of 20 ampere-hours. The restructuring did not increase the overall size of the spacecraft; the launch weight increased from 412.8 kilograms for Mariners 6 and 7 to 997.9 kilograms for Mariner-Mars 1971.

The launch window of May 6 to June 3, 1971, provided a basic time frame for both U.S. and Soviet efforts to investigate Mars at optimum distance. On August 10, 1971, because of the eccentricity (out-of-roundness) of the Martian orbit, the opposition (closest approach) of Mars and Earth was 56,166,105 kilometers.

Mariner 8 was launched with an Atlas-Centaur booster at Cape Kennedy's Eastern Test Range, Launch Complex 36, Pad A, at 9:11 P.M. eastern daylight time on May 8, 1971. Following a normal countdown and lift-off, a malfunction in the Centaur main engine occurred after its ignition, and the Centaur stage tumbled out of control, shutting down the engines. Centaur and spacecraft separated, reentering Earth's atmosphere 400 kilometers north of Puerto Rico.

On May 19, the Soviet Union launched the Mars 2 probe from its Baikonur cosmodrome near Tyuratam; the probe was placed into Earth parking orbit before being injected into flight trajectory for Mars. The probe had a weight of 4,650 kilograms. On May 28, the Mars 3 probe was launched.

The loss of Mariner 8 resulted in a revised orbital plan for Mariner 9; it was to accomplish the work of both with mini-

mum loss in data accumulation. A new orbit was assigned with an inclination of 65 degrees, a twelve-hour period, and a 1,200 kilometer periapsis. Mariner 9 was scheduled to arrive by November 14, ahead of both Soviet craft.

At 6:23 P.M. on May 30, Mariner 9 was launched from the Eastern Test Range into a direct-ascent trajectory along a 398-million-kilometer path toward Mars. The spacecraft separated from its Centaur booster and deployed its solar panels at 6:40. At 7:16, out of Earth's shadow, the craft's Sun sensor was set; at 10:26, the spacecraft locked its star sensor onto the star Achernar.

On June 4, a midcourse maneuver with a 5.11-second engine burn increased velocity by 6.7 meters per second, adjusting the trajectory and ensuring that the Centaur stage would not impact on Mars and thus potentially negate the results of the anticipated exploratory landings of the later Viking project. The correction aimed the craft so close to Mars (within 1,600 kilometers) that the anticipated second maneuver was unnecessary. The radio transmitter was switched from low-gain omnidirectional antenna to high-gain, narrow-beam (increased strength) antenna on September 22/23. On November 2, an unexpected anomaly shifted the navigational lock of the star sensor, requiring commands to the craft to search and resume orientation.

Earth-based astronomers were watching the Martian atmosphere. On September 22, a brilliant, whitish cloud appeared, moving rapidly over the Noachis region; the cloud was first photographed by G. Roberts of the Republic Observatory in South Africa. Trackers at the Lowell Observatory in Flagstaff, Arizona, watched the cloud spread from the initial 2,400-kilometer streak; dust-storm clouds obscured nearly the whole planet by the end of the month. Dust storms, known since 1892, show an intensity coincident with Martian perihelion (the planet's closest approach to the Sun) and are most visible when perihelion coincides with Earth-Mars opposition. Nevertheless, the intensity, speed, and spread of this storm were the greatest ever observed.

Calibration photographs of Mars were received on November 8, and the significance of the storm became apparent. Mariner 9 began regularly photographing the planet on November 10 through 11 as it approached from 860,000 to 570,000 kilometers. These images were transmitted through the Goldstone Tracking Station to JPL and broadcast live on national television.

On November 13, after a 15-minute, 23-second engine burn, Mariner 9 was inserted into an elliptical orbit: 64.4-degree inclination, 12-hour, 34-minute, 1-second period, 1,398-kilometer periapsis, 17,927-kilometer apoapsis (the farthest distance in its orbit from Mars). The initial orbit was intentionally long to allow optimal coincidence between the probe's periapsis pass over the planet and the Goldstone viewing period. By the fourth orbit, synchronization was achieved. An orbit trim maneuver was accomplished on November 15, yielding a 64.34-degree inclination, an 11-hour, 57-minute,

12-second period, a 1,394-kilometer periapsis, and a 17,048-kilometer apoapsis. The orbiting spacecraft was affected by an equatorial irregularity of the Martian gravitation, and a second trim maneuver was required on December 30 with a 17-second engine burn giving the craft a 64-degree inclination, an 11-hour, 59-minute, 28-second period, a 1,650-kilometer periapsis, and a 16,900-kilometer apoapsis.

The dust storm persisted through January, 1972, though by late November some clearing had begun; the rate again slowed in late December. On March 17, cameras were turned off to allow engineers to analyze a malfunction in the on-board computer; photography resumed until March 30, when all science instruments were turned off. The craft had taken 6,876 photographs.

Once each orbit between April 2 and June 4, Mariner 9 passed through the shadow of Mars; during these periods, the craft required battery power. The power shutdown during this interim not only saved the solar cells but also freed the Goldstone Tracking Station to give its attention to the flight of Apollo 16.

Mariner 9 resumed photography on June 8, giving special attention to the two Martian poles — which were not clearly visible from Earth and previously had been obscured by lingering clouds. From early August, 1972, until October 12, Mars was behind the Sun; Mariner 9 could not be commanded from Earth. The low-resolution camera had mapped most of the surface. The high-resolution camera had covered specially targeted areas over about 2 percent of the surface. The shrinking of the southern polar cap was examined in detail. When operations resumed, map coverage of the northern pole was completed.

Mariner 9 sent its last picture on October 17. By October 27, the spacecraft's mission ended with depletion of its attitude control gas. Final command 45,960 was given; the telemetry signal ceased during orbit 698. Scientists estimated that the spacecraft would remain in Martian orbit for fifty to one hundred years.

Knowledge Gained

The photographic mapping of Mars provided detail superior even to that of the Moon as viewed from Earth; moreover, the mapping covered the entire surface and displayed changes through one-half of a Martian year. The great dust storm provided an opportunity for photographic flexibility— repeated views of areas of interest, attention to Martian satellites, and measurements of the storm itself.

Mariner 9 indicated wind velocities as great as 180 kilometers per hour— an extreme required in low atmospheric pressure to raise dust 50 to 70 kilometers above the surface. Particles ranged from two to fifteen micrometers in size with a silicon content of about 60 percent.

Mariner 9 measured the shape of Mars. Rather than an oblate spheroid (flattened sphere) like Earth, Mars is a triaxial ellipsoid: 3,396 by 3,394 by 3,376 kilometers. The longer axis

passes through the highly elevated Tharsis Ridge, a Martian bulge 6,000 kilometers in surface diameter and 7 kilometers high.

While Mariner 9 did away with the notion of "canals" on Mars, it produced images of giant volcanoes even before the dust cloud settled. The large light spot known since the advent of telescopic observation as Nix Olympica (the snow of Olympus) was reidentified as Olympus Mons, the largest volcanic or mountainous mass known: 500 kilometers wide and rising 29 kilometers above its surroundings. Olympus lies on the western flank of a broad ridge, running diagonally northeast to southwest, which is marked by three great shield volcanoes — North Spot, Middle Spot, and South Spot — each about 400 kilometers across. These, and other smaller volcanoes, have been active recently enough areologically (areology is the Martian study equivalent of geology) that the plains on Mars appear as volcanic regions covering older cratered areas.

If the volcanoes were not impressive enough, as the dust subsided a spectacularly immense canyon within the equatorial belt was identified by Mariner 9. Valles Marineris, named for the spacecraft, stretches from the southeast flank of Tharsis Montes eastward more than 4,000 kilometers; its vertical edges descend up to 9 kilometers deep, and many minor side canyons are 250 kilometers wide, some extending back into surrounding uplands for 150 kilometers.

Numerous channel networks, some broad and sinuous, others narrow and dendritic (branched like trees), occur over widespread areas. Some of the largest, 30 to 60 kilometers wide and up to 1,200 kilometers long, originate in northern plateaus and flow (slope) northward into the Chryse region. Networks of narrower channels come off the sides of craters northwest of the Hellas Planitia. Other varieties demonstrate the exotic character of the planet.

The polar ice caps have a permanent core containing some frozen water. Their enormous expansion in the winters, extending to within 40 degrees of the equator, shows the deposit of frozen carbon dioxide from an atmosphere which, while thin, is about 90 percent that gas. Pitted plains and layered terrains awaited Viking Orbiter photography for further clarification. Carbon dioxide clouds formed and drifted about major topographical features.

While Mariner 9 did not obtain improved photographs of Phobos, the largest of the Martian moons, it did get preorbital pictures, including Phobos silhouetted against Mars. The smaller satellite, Deimos, was photographed while Mariner awaited the settling of the dust storm. Both were measured. Each is a triaxial ellipsoid with axes of 28, 23, and 20 kilometers for Phobos, and 16, 12, and 10 kilometers for Deimos. Both have nearly circular orbits about the Martian equator, keeping the same face (their longest axis) toward the planet. Deimos, at 23,400 kilometers, revolves in 30 hours, 18 minutes; Phobos, at 9,270 kilometers, in 7 hours, 39 minutes. Both are heavily cratered, the largest on Phobos some 8 kilometers wide. Their lack of crater erosion permitted some comparative consideration with craters on Mars. Both are among the darkest objects in the solar system.

Context

Space exploration and interplanetary travel were initiated in pre-World War II experimentation and post-World War II (Cold War) competition. The schemes proposed in the 1950's went beyond engineering capacities and budgeting possibilities. Mariner-Mars 1971, reduced to the actual success of Mariner 9, is no exception, as the effort to obtain its real cost illustrates. Figures from $76 to $115 million have been stated, but the reality is lost within a Voyager program that never occurred and a reorganized Viking mission that landed on Mars.

Mariner 9 was the fourth spacecraft to reach Mars of the first ten; six of these had only flyby-status missions. The Soviet Mars 1 spacecraft, launched on November 1, 1962, lost radio contact on March 21, 1963, some 106 million kilometers from Earth. Mariner 3 failed on November 5, 1964, shortly after its launch, when the shroud for the spacecraft did not open; it went into solar orbit. Mariner 4 was successfully launched on November 28, 1964, and took twenty-two photographs. Of them, nineteen were usable and two were beyond the Martian terminator (the darkening line of solar light); they revealed the cratered character of the Martian surface along a single sweep path. The Soviet Zond 2, launched November 30, 1964, failed when its transmission ceased at the beginning of May, 1965.

Mariners 6 and 7 were launched on February 24 and March 27, 1969. Mariner 6 sent 50 far-encounter and 25 near-encounter photographs over July 29-30; Mariner 7 sent 91 far-encounter and 33 near-encounter over August 2 -4. These hinted at the variety of Martian surface features.

Mariner 8 was a launch failure on May 8, 1971. The Soviet pair Mars 2 and 3, launched May 19 and 28, 1971, successfully reached Mars and went into orbit on November 27 and December 2. Each carried a sterilized scientific lander package. The Mars 2 lander, ejected before orbit, crashed on the Martian surface. The Mars 3 lander landed safely and relayed a television signal, which stopped after 20 seconds.

Their primary mission was to measure solar wind and cosmic radiation between Earth and Mars. They measured Mars's atmospheric humidity and surface temperature, ending their missions in August, 1972.

One significant result of these several efforts was the increased communication between U.S. and Soviet space scientists. The Working Group on Interplanetary Exploration was established under an agreement of January, 1971, by NASA and the Soviet Academy of Sciences. Meeting in Moscow in 1973, the group agreed to exchange on April 15 data from Mars 2 and 3 for pictures and maps made by Mariner 9. During 1973, the Soviets launched four more spacecraft of their Mars series. Mars 4 malfunctioned in one of its onboard systems and was not braked for orbit. It flew by on February 10, 1974, at 2,200 kilometers, taking pictures, and left to continue transmittal of outer space information. Mars 5 entered orbit on February 12, 1974. Mars 7 neared on March 9; its lander separated but malfunctioning of its onboard system caused it to miss Mars by 1,300 kilometers. Mars 6 approached on March 12, separated its lander at 48,000 kilometers, and continued into heliocentric orbit, passing within 16,000 kilometers. The lander transmitted for 148 seconds.

The U.S. Viking orbiters were launched August 20 and September 9, 1975, and inserted into respective orbits on June 19 and August 7, 1976. Care was taken to obtain higher-resolution photographs before permitting the landers to descend. Each was successfully landed: Viking 1 on July 20, Viking 2 on September 3. The prodigious output of photographic images (51,539 from the orbiters and more than 4,500 from the landers) was spectacular, while those from the surface had the intrigue of new but limited vision. The life experiments performed by the landers proved inconclusive, though essentially negative for the immediate regions. Viking Orbiter 2 ceased operating on July 25, 1978. Its lander was shut down on April 12, 1980. Viking Orbiter 1 was silenced on August 7, 1980. Viking Lander 1, with the capacity for direct communication to Earth, continued to supply weekly weather information or the occasional landscape image to indicate any change of surface conditions until contact was lost on January 11, 1983.

Bibliography

Cortright, Edgar M., ed. *Exploring Space with a Camera*. NASA SP-168. Washington, D.C.: Government Printing Office, 1968. As spacecraft of various types entered Earth, lunar, or solar orbits, experiments with camera observations, machine or handheld, produced one of the great achievements of the space age. Samples, arranged historically and topically and including those from the first lunar lander and the Mariner 4 flyby, receive interpretive commentary.

Ezell, Edward Clinton, and Linda Neuman Ezell. *On Mars: Exploration of the Red Planet, 1958 – 1978*. NASA SP-4212. Washington, D.C.: Government Printing Office, 1984. This excellent example of critical scholarship, demanding an attentive and knowledgeable reader, covers the technical details of the various Mariner and Viking missions and reports the complex web of political, financial, and managerial problems that eventually resulted in a series of firsts on Mars: the first photographing (Mariner 4), the first orbiting (Mariner 9), and the first landing with observational and biological results (Viking). Photographs, drawings, tables, appendices of data, bibliography, notes, and indexes make this the official NASA history.

Hartmann, William K., and Odell Raper. *The New Mars: The Discoveries of Mariner 9*. NASA SP-337. Washington, D.C.: Government Printing Office, 1974. A generous sample of Mariner 9 photographs, usually of adequate scale, supplemented by historical materials including drawings, telescopic observations, and comparative examples from Mariners 6 and 7, placed within a narrative context that interprets the diverse and novel features from the perspective of new discoveries. A concluding summary itemizes Mariner 9's achievements. For a broad audience.

Moore, Patrick, and Charles A. Cross. *Mars*. New York: Crown Publishers, 1973. A British effort in the wake of Mariner 9's apparent mapping triumph to provide any interested reader with the first real *atlas* of another planet. Aside from its maps, the volume is splendidly illustrated, carefully written, and indicative of the development of knowledge about Mars.

National Aeronautics and Space Administration. Scientific and Technical Information Office. *Mars as Viewed by Mariner 9: A Pictorial Presentation by the Mariner 9 Television Team and the Planetology Program Principal Investigators.* Rev. ed. NASA SP-329. Washington, D.C.: Government Printing Office, 1976. The largest and best collection of photographs, with captions expressing an interpretation by one of the members of the imaging team. The intelligent commentary reflects the excitement of the initial impact.

Spitzer, Cary R., ed. *Viking Orbiter Views of Mars*. NASA SP-441. Washington, D.C.: Government Printing Office, 1980. A carefully selected sample of photographs is presented in a topical arrangement with introductory comments, followed by examples with explanatory captions of sufficient detail to make each photograph more revealing for any audience. Several stereographically matched pairs allow three-dimensional effects when the enclosed viewing device is used. Greater clarity compared to the Mariner 9 equivalents illustrates how much more there was, how much more there remains, to be learned from Mars.

Clyde Curry Smith

MARINER 10

Date: November 3, 1973, to March 24, 1975
Type of mission: Unmanned deep space probe

Mariner 10 collected vital data on the inner solar system, including detailed photographs of Venus and Mercury. The first probe to use the gravitational pull of one planet to help it reach another, Mariner 10 proved the utility of the "gravity-assist" technique for interplanetary travel.

PRINCIPAL PERSONAGES

WALKER E. "GENE" GIBERSON, Project Manager
JAMES A. DUNNE, Project Scientist
JOHN R. CASANI, Spacecraft Systems Manager
BRUCE C. MURRAY, Principal Investigator, television experiment
NORMAN F. NESS, Principal Investigator, magnetic field experiment
VICTOR C. CLARKE, JR., Mission Analysis and Engineering Manager
JOHN A. SIMPSON,
HERBERT S. BRIDGE,
STILLMAN C. CHASE, JR.,
A. LYLE BROADFOOT, and
H. TAYLOR HOWARD, other Principal Investigators

Summary of the Mission

The Mariner 10 mission performed the first close flyby of the planet Mercury, using the gravity of Venus to divert the spacecraft's trajectory toward Mercury. While multiple-planet swing-bys had been anticipated since the 1920's, it was not until the 1960's, with advanced trajectory-computation techniques, that such methods of interplanetary navigation became feasible.

In 1969, the National Aeronautics and Space Administration (NASA) approved a Mariner Venus/Mercury mission for 1973 (MVM73). In February, 1970, it was discovered that the gravity of Mercury could be used to provide a "free return"— a second encounter with Mercury after the probe's initial flyby — using a minimum of spacecraft propulsion. The Jet Propulsion Laboratory (JPL) in Pasadena, California, was selected to implement the mission, and the project began in 1970 under the management of Walker E. "Gene" Giberson. John R. Casani was selected to manage spacecraft development, Victor C. Clarke, Jr., became Mission Analysis and Engineering Manager, and James A. Dunne was

Project Scientist. The Boeing Aerospace Corporation of Seattle, Washington, was selected to build JPL's Mariner spacecraft design.

The spacecraft itself was modeled on the Mariner series: a standard octagonal structure with electronic instrumentation installed in the inner walls, and within, hydrazine fuel tanks to feed the propulsion units, which handled trajectory corrections. Supported on this body were twin television cameras; two panels of solar cells extending like wings, which powered the spacecraft; and a dish-shaped high-gain antenna, which was part of the radio system used for relaying data, navigating the spacecraft, and making scientific measurements. Extensive modifications to the design of previous Mariner probes were necessary, such as changing the fuel tanks and attitude control system; others were necessary to protect the probe from the intense heat of the Sun, such as designing the photovoltaic solar panels to rotate and tilt away from the Sun and using new thermal protection techniques. Furthermore, the payload included instruments to take pictures in wavelengths from ultraviolet through infrared and others to measure electromagnetic and charged particle characteristics of the planets and interplanetary space.

The feasible launch period to Venus was open from October 16 to November 21, 1973. Normally, spacecraft are launched as close as possible to the beginning of the period in order to provide a margin for error. NASA, however, approved a launch at the optimal time of 12:45 A.M., eastern standard time, November 3, and the launch (achieved by the Atlas SLV-3D/Centaur D1-A launch vehicle) took place within a few thousandths of a second of that desired time.

Instruments for specific experiments were turned on soon after launch and began gathering data on the "solar wind" (the solar plasma that flows out from the Sun) and on the interplanetary magnetic fields. The television cameras were also engaged at that time to calibrate the cameras for Venus by taking pictures of the similarly cloudy Earth and for Mercury by taking pictures of the Moon, whose surface was suspected to be like Mercury's.

MVM73, now called Mariner 10, was — despite its ultimate success — plagued by failures of the spacecraft and of experiments throughout the mission, taxing the ingenuity of engineers and scientists to the utmost. Shortly after launch, a protective shield became stuck and caused much data to be lost in the plasma science experiment. Ten days after launch, during a series of roll maneuvers, the spacecraft's star tracker detected a bright particle (a minute flake of dust or paint from the spacecraft) and followed the particle instead of the intended star, Canopus, thus causing major losses of attitude control gas.

When employing an alternative method of flying the spacecraft, however, the JPL team found that the interaction between the gyroscope control system and the flexible body of the spacecraft produced unforeseen vibrations, again causing the loss of attitude control gas. A further threat to the mission occurred a few months later, when the spacecraft automatically and irreversibly switched from its main power system to a backup system, leaving no room for power failures without jeopardizing the mission. The high-gain antenna and television cameras also experienced recurrent failure, threatening to reduce drastically the amount of data transmitted to Earth.

Fortunately, the JPL team found ways to compensate for these difficulties. The solar panels were tilted so that the pressure of the solar wind could be used to balance the spacecraft and reduce waste of attitude control gas, a technique called solar sailing. After the Venus encounter, it was discovered that the trajectory could be corrected by "sunline" maneuvers, which do not require the spacecraft to roll or pitch, thereby avoiding use of the gas-wasting gyroscopes. The high-gain antenna and television camera heaters warmed up and repaired themselves before the Mercury encounter. In addition, the Deep Space Network (DSN), which received the spacecraft transmissions on Earth, made improvements in their large antennas.

On February 5, 1974, Mariner 10 began taking history's first close-up photographs of Venus, beginning with an image of a thin lighted crescent and gradually revealing the entire planet as the spacecraft swung around Venus from its night to its day side. Ultraviolet images showed swirling cloud patterns. At the same time, the high-gain antenna was used to reveal characteristics of both the Venusian atmosphere and the mass and shape of the planet.

On March 16, a sunline maneuver was performed to correct Mariner's trajectory, heading it toward Mercury. The first encounter with the planet occurred on March 29. For the eleven days surrounding this closest approach, Mariner's cameras transmitted more than two thousand high-resolution pictures back to Earth, displaying a cratered, Moon-like surface. Also, particles and fields measurements revealed an unexpected magnetic field.

On May 9 and 10, Mariner was readied to attempt a second, and possibly a third, encounter, using the "free return" gravity-assist trajectory. Two days after the first encounter, however, another power system failure occurred as a result of overheating. In August, 1974, the spacecraft's tape recorder failed, preventing pictures from being obtained when the spacecraft was out of contact with Earth. Part of the computer system also failed, making spacecraft control more difficult. The spacecraft team compensated for these problems, and the mission continued.

On September 21, Mariner 10 passed the day side of Mercury on its second encounter, at a distance of about fifty thousand kilometers. Three antennas of the DSN were arrayed to increase their ability to detect the spacecraft's radio transmissions, resulting in the receipt of about one thousand high-resolution pictures of Mercury from a new southerly viewing angle.

After the second encounter, the spacecraft was placed in a hibernation mode and allowed to roll slowly through space with its cameras deactivated. This move was done to save attitude control gas and to avoid exercising the unreliable power system. A "back-off" sunline maneuver was required at one point to prevent the spacecraft from actually hitting Mercury on its third encounter. One final malfunction occurred that resulted in the spacecraft rolling into an attitude that blocked radio signals from Earth. By using a combination of DSN antennas, educated guesswork, and analysis of the high-gain antenna transmission power pattern, scientists were able to stop the roll and the spacecraft was correctly oriented — only thirty-six hours before the third encounter.

On March 16, 1975 — nearly one year after its initial encounter with Mercury — Mariner 10 made its final and closest flyby and took its final installment of pictures, with resolutions to 140 meters. Eight days later, its supply of attitude control gas finally exhausted, the spacecraft's radio transmitter was turned off, leaving a silent spacecraft to orbit the Sun. A short time later, the United States Postal Service issued a stamp commemorating the mission.

Knowledge Gained

Ultraviolet pictures of Venus revealed a swirling pattern of clouds rotating rapidly (once every 4 Earth days) around the planet. The clouds move in a direction opposite to the planet's rotation (which had been discovered by Earth-based radar measurements of the surface to be very slow — once every 243 Earth days). Another experiment determined that the shape of Venus is one hundred times closer to a sphere than is Earth's.

The particles and fields experiments verified that Venus has a weak magnetic field but also confirmed that Venus' ionosphere interacts with the solar wind to form a "bow shock" (a shock wave similar to that produced by a supersonic airplane), which keeps the solar plasma from hitting the planet directly. The ultraviolet experiment found in Venus's upper atmosphere a great quantity of hydrogen, which is thought to control its chemistry, forming sulfuric acid clouds. The infrared experiment determined that the temperature of the cloudtops remained constant −23 degrees Celsius) between the day and night sides of the planet.

As Mariner 10 approached Mercury, the planet, as expected, showed a Moon-like appearance (heavily cratered with

Mariner 10 spacecraft. (NASA)

large, flat circular basins), suggesting the early bombardment of the inner solar system by meteors. On the second encounter, the back side of Mercury revealed a huge circular feature approximately 1,290 kilometers across, subsequently named the Caloris Basin for its heat. The infrared experiment determined that the surface temperatures — given the long Mercury days and nights and the planet's proximity to the Sun — range from −183 degrees Celsius on the night side to +187 on the day side.

The radio and ultraviolet experiments confirmed that Mercury, like Earth's Moon, lacks an atmosphere. Unlike the Moon, however, Mercury has large cliffs that are up to 3.2 kilometers high and 480 kilometers long. These are believed to have been caused by the shrinking of the planet during cooling of a hot central core. Mercury is also much denser than the Moon and, like Venus, much more spherical than Earth. Estimates of the mass of Mercury were improved one hundred times by the encounter of Mariner 10.

The particles and fields experiments discovered, quite unexpectedly, that Mercury has a significant magnetic field, sufficient to produce a bow shock between the planet and the solar wind. Whereas Venus's bow shock is produced primarily by charged particles in its ionosphere, Mercury, with no atmosphere, must have an intrinsic magnetic field. Mercury's field is Earth-like in its shape, though less intense.

Context

Mariner 10 was the fourth spacecraft to visit Venus and the first to take close-up pictures. The first interplanetary probe, Mariner 2, had made a flyby of Venus in 1962 and used infrared radiometry to determine that Venus had a very high temperature. Earth observations of Venus had previously established that the planet showed no surface features, implying an opaque atmosphere. In 1967, the combined measurements from the U.S. Mariner 5 flyby and the Soviet Venera 4, which sent an entry probe into the atmosphere, proved that the Venusian atmosphere was extremely hot and dense and composed primarily of carbon dioxide. None of these missions, however, carried cameras. Mariner 10 identified weather patterns on Venus that have produced data for the prediction of weather patterns on Earth. It confirmed the heat and density of Venus' atmosphere, which are attributed to a "greenhouse" effect resulting from the carbon dioxide atmosphere. This finding shed light on the greenhouse effects that could result on Earth from atmospheric changes caused by industrial pollution.

Later missions to Venus included the U.S. Pioneer Venus (in 1978) and the Soviet Veneras 15 and 16 (in 1983), which used radar to map surface features. Soviet Venera spacecraft were landed on Venus in 1975 and 1982, and the Soviet Vega spacecraft (which encountered Halley's comet in 1986) dropped balloons and landers on Venus as it flew by. Between 1989 and 1994, the U.S. Magellan mission produced a detailed map of 99 percent the Venusian surface.

Mercury observations are important in understanding the overall history of the solar system. Scientists speculate that Mercury and Venus formed originally in a zone around the Sun where there was little water, in contrast to Earth, which formed in a water-rich zone. Comparative studies of Mars, the Moon, and Mercury show that the three bodies were all subjected to a similar meteoric bombardment, including asteroid-sized bodies, between 4 billion and 3.3 billion years ago. The Mercury data suggest that the bombardment could have originated outside the solar system rather than in the asteroid belt, as previously thought.

Earth is in a period of tectonic action — creating volca-noes — and some scientists have speculated that Mars may be entering an active phase of tectonic activity leading to an Earth-like future. Mercury data, however, seem to strengthen a contrary theory, that Martian volcanism took place hundreds of millions of years ago and that the internal heat of Mars is too low to drive tectonic processes. Such findings, including Mercury's still-unexplained magnetic field, have raised new questions and changed scientists' perceptions of the solar system.

From the perspective of space technology, Mariner 10 established the viability of the gravity-assist technique, which was subsequently used for the 1977 Voyager 2 encounters with Saturn and Uranus via Jupiter. The gravity-assist technique will undoubtedly play a major role in future interplanetary navigation.

With Mariner 10, for the first time a group of scientists were selected to assist in mission planning early enough to influence the spacecraft design. JPL agreed to implement the mission for the very low budget of $98 million. The mission cost through the first encounter was $1 million less than the $98 million promised by JPL to NASA, including inflation. Costs for the extended mission were less than $3 million, so that the entire mission was achieved for $100 million, about sixty cents for each American.

Bibliography

Chapman, Clark R. *Planets of Rock and Ice: From Mercury to the Moons of Saturn.* Rev. ed. New York: Charles Scribner's Sons, 1982. Describes the characteristics of solid bodies of the solar system, as well as the people and missions that identified those characteristics. Draws comparisons between these planets and moons and draws conclusions about Earth in a planetary context. Includes some photographs of planets. Suitable for general audiences.

Dunne, James A., and Eric Burgess. *The Voyage of Mariner 10: Mission to Venus and Mercury.* NASA SP-424. Washington, D.C.: Government Printing Office, 1978. The official history of the Mariner 10 mission, written by the Project Scientist. Suitable for high school and college levels, it describes the mission in detail from early Earth observations of Venus and Mercury through the third Mercury encounter of Mariner 10. Contains numerous illustrations and photographs of the spacecraft and the pictures returned from the mission. Among the appendices is a list of the participants in the mission.

Greeley, Ronald. *Planetary Landscapes.* London: Allen and Unwin, 1985. A pictorial atlas describing the geomorphology (determination of the form and evolutionary history of planetary surfaces) of all the rocky planets and rock/ice moons from Mercury through Saturn. Suitable for advanced high school and college levels, this volume includes many references to books and scientific articles about the planets.

Hunten, D. M., et al., eds. *Venus.* Tucson: University of Arizona Press, 1983. A collection of scientific essays describing all aspects of Venus. The authors are primarily from the United States and the Soviet Union, and findings from all the Venus missions of both countries are described. Includes a chapter on the history of Venus studies, as well as maps, charts, equations, and calculations describing the geology, atmosphere, and magnetic environment of Venus. College-level material.

Murray, Bruce C., and Eric Burgess. *Flight to Mercury.* New York: Columbia University Press, 1976. A description of the Mariner 10 mission from the personal viewpoint of the Principal Investigator of the television experiment team. Written chronologically, it refers to world events of the period, including the Arab oil crisis and the Watergate hearings, to provide historical context for the mission. Suitable for general audiences; includes more than one hundred photographs of Venus and Mercury.

Murray, Bruce C., Michael C. Malin, and Ronald Greeley. *Earthlike Planets: Surfaces of Mercury, Venus, Earth, Moon, Mars.* San Francisco: W. H. Freeman and Co., 1981. This well-illustrated volume includes many black-and-white and some color photographs of the planets. It describes and compares the characteristics of the inner planets, Mercury, Venus, Earth, the Moon, and Mars and deduces facts about the origin and evolution of the solar system. Suitable for general audiences.

Strom, Robert G. *Mercury, the Elusive Planet.* Washington, D.C.: Smithsonian Institution Press, 1987. This readable, well-illustrated book provides a complete description of Mercury, including the history of observations of the planet. Describes the Mariner 10 mission in the context of the subsequent twelve years of interpretation of data from the mission, the geology and magnetic characteristics of the planet, and the theory behind scientific conclusions about Mercury. Includes many photographs, a table of definitions of technical and scientific terms, and a list of the names and locations of surface features.

Donna Pivirotto

MARS OBSERVER

Date: September 22, 1992, to August 21, 1993
Type of mission: Unmanned deep space probe

Mars Observer was to have photographed the Martian surface using the most sophisticated camera ever carried on a civilian spacecraft and to have carried out various scientific experiments concerning the Martian atmosphere.

PRINCIPAL PERSONAGES
WILLIAM PANTER, Mars Observer Program Manager
SAM DALLAS, Mars Observer Mission Manager
GLENN CUNNINGHAM, Mars Observer
    Project Manager
ARDEN ALBEE, Project Scientist
FRANK PALLUCONI, Deputy Project Scientist
MICHAEL C. MALIN, Principal Investigator,
    Mars Observer camera
MARIO H. ACUNA, Principal Investigator,
    magnetometer electron reflectometer
DANIEL J. MCCLEESE, Principal Investigator,
    pressure modulator infrared radiometer
DAVID E. SMITH, Principal Investigator,
    Mars Observer laser altimeter
PHILIP R. CHRISTENSEN, Principal Investigator,
    thermal emission spectrometer
WILLIAM V. BOYNTON, Team Leader,
    gamma-ray spectrometer

Summary of the Mission

Mars Observer lifted off from Cape Canaveral, Florida, attached to a Titan III rocket on September 25, 1992. After separation from the Titan, an on-board rocket burn of two-and-a-half minutes sent the tiny spacecraft toward a conjunction with Mars to be completed after a deep-space journey of just under one year.

Scientists modeled the Mars Observer on Earth-orbiting weather and communications satellites. The rectangular body of the spacecraft (the "bus") carried seven scientific instruments wrapped in thermal blankets to maintain their operating temperatures in the hostile environment of space. The instruments included a gamma-ray-spectrometer, a thermal emission spectrometer, a laser altimeter, a pressure modulator infrared radiometer, a magnetometer/electron reflectometer, the Mars Observer camera, radio-communications equipment, and computer equipment to control the remote actions of the satellite once it achieved Mars orbit. Mars Observer also carried radio

equipment provided by French and Russian scientists.

The gamma-ray-spectrometer was to detect and analyze gamma rays emitted by the Martian surface. Such analysis would provide an understanding of the mineral composition of Martian rocks and soil. The thermal emissions spectrometer, by measuring the heat produced by different areas of the Martian surface, was to provide information about weathering and mineral composition. The laser altimeter was to measure the topography of Martian surface features. The pressure modulator infrared radiometer was to measure the temperature, water content, and pressure of the Martian atmosphere throughout an entire Martian year. The magnetometer/electron reflectometer was to send back to Earth information concerning the Martian magnetic and gravity fields. The radio equipment aboard the spacecraft was to assist in the study of Martian magnetism, as well as relay information from all the instruments back to Earth. The Mars Observer camera represented the most sophisticated camera ever carried on a civilian spacecraft. It was to have photographed much of the Martian surface with very high resolution. The Franco-Russian equipment was to retrieve and store data supplied by a Martian lander scheduled by scientists in France and Russia to touch down on the Martian surface in 1994.

The mission objectives of Mars Observer, had they been realized, would have furnished American space scientists with much more detailed information about the planet Mars than had previously been available. The Observer accomplished none of its objectives. Almost from liftoff, foreshadowing of failure and controversy accompanied the Observer on its long path toward Mars.

In June, 1993, American space scientists had to delay the launch of the NOAA-I spacecraft when its Redundant Crystal Oscillator (RXO, a clock required for the operation of the spacecraft's central computers) failed. Mars Observer utilized the same type of RXO as did the NOAA-I spacecraft, making its reliability suspect. At the same time, the principal investigator of Mars Observer's photographic mission was coming under increasingly bitter criticism for his decisions concerning which areas of Mars' surface should have priority for photographic investigation.

In 1987, former National Aeronautics and Space Administration (NASA) consultant Richard C. Hoagland published a study of photographs of the Martian surface sent to Earth by NASA's own Viking space missions in the 1970's. Hoagland's book argued that several topographical features in an area of Mars called Cydonia revealed by Viking mission photographs could be artificial. The features included the "Face" (which NASA spokesmen insisted was nothing more than a hill with unusual geological formations which, under certain lighting conditions, made the hill look like a humanoid face surrounded by a vaguely Egyptian-looking headpiece), and the nearby "City" which Hoagland suggested might be a complex of huge buildings, some of them pyramidal in shape.

Between 1987 and 1993, a number of researchers (including several reputable scientists) examined Hoagland's evidence and arguments and concluded that there existed a good possibility that the unusual surface features photographed by the Viking mission might well be artificial. Hoagland and his supporters barraged NASA and many U.S. Congressmen with letters urging that any new NASA space missions to Mars include a closer examination of the Cydonian area as a top priority. When the principal investigator of the photographic mission of Mars Observer declined to accede to their urgings, the rhetoric of the researchers became increasingly acrimonious. In an exchange of words well publicized in the popular press, NASA officials and the Cydonia researchers leveled serious charges at each other.

Shortly after Mars Observer team members lost contact with the spacecraft, a philosopher published a book siding with those who thought that the Cydonia features might be artificial. The book charged that NASA was ignoring what the author termed an ethical and moral responsibility to ascertain the nature of the Cydonian topographical features revealed in the Viking photographs. The book also contained an account of the bitter exchanges between NASA officials and Cydonia researchers. NASA officials had dismissed the researchers as amateurs and UFO/ancient astronaut buffs whose ideas could not be taken seriously. The researchers strongly suggested that NASA officials might be deliberately suppressing what could be proof of extraterrestrial intelligence. NASA's supposed motives in the suggested coverup derived from a Brookings Institute report prepared for NASA in 1960 concerning the ways in which the discovery of extraterrestrial intelligence might affect human society. The report warned that such a discovery would inevitably have serious repercussions, including even possibly wide-spread political revolution. The resulting publicity brought little credit to either side in the debate.

The exchanges between NASA officials and the Cydonia researchers raged throughout Mars Observer's eleven-month journey to the red planet, with several U.S. Congressmen becoming involved. Ignoring the controversy as best they could, project scientists continued their mission. In January, 1993, NASA scientists successfully tested the focus of the Observer's camera. Less than a month later other scientists successfully calibrated the thermal emission spectrometer. In

April, those scientists responsible for the magnetometer part of the Observer's mission calibrated their instrument. In March and April, Mars Observer carried out the only successful experiment during its mission. In conjunction with two other spacecraft (Galileo and Ulysses), the instruments aboard the Observer gathered data concerning low-frequency gravitational waves. Scientists hoped the data might offer confirmation for Albert Einstein's general theory of relativity.

During the flight, project scientists decided to alter the original orbital insertion plan. Taking advantage of larger than expected fuel reserves on the spacecraft, the scientists moved ahead the planned date of insertion by three weeks. The earlier entry of the spacecraft into orbit would have allowed photography of Mars' surface to begin before the dust-storm season began. The maneuver required shutting down all communications between NASA and the spacecraft for seventy-two hours prior to orbital insertion. The shutdown came on August 21, three days before orbital insertion. After the systems shutdown, scientists were unable to reestablish communication with Mars Observer. This embarrassing failure coincided with the publication of the book mentioned above, which suggested wrongdoing at the highest levels in NASA.

During the following months, project scientists tried many expedients in attempts to salvage the mission. The last likely chance for reestablishing communications with the satellite failed in late September when scientists could not detect the radio beacon aboard the craft. When none of the expedients proved successful, NASA scientists decided to terminate the failed mission. A news release from the Jet Propulsion Laboratory in California dated November 24, 1993, quoted Mars Observer Project Manager Glenn Cunningham as saying that the majority of the mission team had been reassigned to other projects. Cunningham said that although a few scientists would continue ground-tracking the Mars Observer, little hope of salvaging the mission remained.

Cunningham's announcement brought to a sad conclusion a much-heralded and triumphant "return to the red planet." If accomplished, the mission's objectives would have provided significant advances in human knowledge of the Martian surface and atmosphere. NASA would have benefited from the favorable publicity generated by a successful reexamination of Mars, always a favorite with the public. Ending as it did in failure and controversy, the Mars Observer mission became a major setback for American efforts to explore the solar system in general and Mars in particular.

Knowledge Gained

Aside from the data concerning low-frequency gravitational waves in interplanetary space, the Mars Observer mission did little to increase our knowledge about Mars. Two official inquiries into the failure do provide some negative knowledge: Scientists may have gained a better understanding of what not to do on future space missions.

The two committees that investigated the Mars Observer

mission identified several possible causes for its ultimate failure. Although the speculation of mission scientists immediately after losing contact with the spacecraft centered around the possible failure of the RXO, the investigatory committees concluded that other malfunctions were just as likely to have caused the failure. The Independent Mission Failure Review Board appointed by NASA chief administrator Daniel Goldin agreed, and Jet Propulsion Laboratory's own Review Board identified several possible causes of the loss of contact with Mars Observer. These potential malfunctions included: (1) Electrical power loss due to a massive short in the power subsystems; (2) Loss of function that prevented both the spacecraft's main and backup computers from controlling the spacecraft; (3) Loss of both the main and backup transmitters due to the failure of an electronic part, and (4) A breach of the spacecraft's propulsion system.

The investigatory committees identified three possible causes of potential malfunction: (1) Liquid oxidizer (nitrogen tetroxide) of the rockets may have seeped past a check valve in the pressurization lines, causing the lines to burst; (2) The pressure regulator of the fuel tanks may have failed, causing the tanks to explode, and (3) A small pyrotechnic device called a squib, which was fired to open a valve in one of the pressure tanks, could have burst the tank. These latter three possibilities all resulted from the in-flight decision to attempt to insert the Observer into Mars orbit three weeks ahead of the original schedule. Members of both investigative committees concluded that the exact cause of the mission failure would probably never be exactly identified.

Context

The Mars Observer mission was not the first failed attempt to investigate the red planet. Some space scientists even speak of a "Mars Jinx," which has dogged both American and Russian missions to the planet since the early 1960's. Soviet space scientists lost contact with Mars 1, launched on November 1, 1962, at some 65.9 million miles from Earth. NASA's first attempt to retrieve data from Mars also ended in disappointment in 1964 when Mariner 3's shroud failed after the craft had been successfully launched on November 5, 1964. Mariner 3's sister ship Mariner 4, launched on November 28 the same year, did a successful flyby of Mars and sent back photographs on July 14, 1965. Several unannounced Soviet missions to Mars reportedly failed in the late 1960's.

In 1969, two NASA flyby missions to Mars experienced success. Mariner 6 (launched February 24) and Mariner 7 (launched March 27) both returned photographs of the red planet as they passed in its vicinity on July 31 and August 5, 1969, respectively. Mariner 8 fared less well, its launch vehicle failing shortly after liftoff on May 8, 1971. Soviet scientists sent two missions to Mars in 1971, both designed to orbit and photograph the planet, and to soft-land instruments packages on Mars' surface. Mars 2, launched on May 19,

arrived in Mars orbit on November 27, but sent back no useful data due to systems failures. Mars 3, launched on May 28, was more successful. It managed to send back a few photographs and some data.

The first major success in retrieving useful information from Mars through an unmanned space probe came with NASA's Mariner 9 mission. After liftoff from Earth on May 30, 1971, Mariner 9 arrived in Mars orbit on November 13. Mariner 9 orbited the planet and photographed the Martian surface for almost a year, until October 27, 1972. The quality of the scientific data and photographs relayed to Earth far surpassed anything from previous missions.

Soviet Mars missions continued to disappoint the scientists who designed them. In 1973 Russian scientists launched Mars 4, Mars 5, Mars 6, and Mars 7 on July 21, 25, August 5, and 9, respectively. Mars 4 and 5 were orbiters designed to relay signals from the landers carried by Mars 6 and 7. Mars 4 failed to achieve orbit. Mars 5 arrived in orbit on February 12, 1974, but transmitted data for only a few days. Mars 6 and 7 arrived in orbit on March 9 and 12, 1974, respectively, but their landers lost contact with the orbiters before actually touching down on the Martian surface.

American space scientists followed up their Mariner 9 success with the spectacular Viking missions of 1975 -1982. Viking 1 and Viking 2 left Earth bound for Mars on August 20 and September 7, 1975, respectively. Both carried landing craft and on-board relay equipment to send signals and data from the landers to NASA scientists and a world television audience. Viking 1 achieved Mars orbit on June 19, 1976, and successfully deployed its lander on July 20. Viking 2 achieved Mars orbit on August 7, 1976, and successfully deployed its lander on September 3. The photographs from the landers — the first from the surface of another planet — captivated people around the world as they were broadcast on television.

The Russian space scientists have not attempted any new missions to Mars since the disappointing Soviet attempts in the early 1970's. Three United States Mars missions since the Viking triumphs, up to and including the Mars Observer tragedy, have all ended in failure. Phobos 1 and Phobos 2, launched on July 7 and 12, 1988, respectively, both experienced signal malfunction before achieving Mars orbit. Both carried landers and equipment packages more sophisticated than those of the Viking missions. Despite these setbacks NASA scientists announced new Mars missions almost simultaneously with finally admitting that the Mars Observer mission had failed.

The Jet Propulsion Laboratory announced several new Mars missions shortly after Mars Observer's loss. The most intriguing of the announced missions, Mars Pathfinder, calls for a soft landing, a six-wheeled vehicle that will collect soil samples from different areas of the Martian surface, and a high-resolution camera that will photograph large areas of Mars from orbit. The mystery and fascination of Mars seems destined to attract the attention of Earth space scientists in spite of any "Mars Jinx."

Bibliography

French, Bevan M. *Return to the Red Planet: The Mars Observer Mission*. Washington, D.C.: United States Government Printing Office, 1993. A comprehensive description of the equipment and mission of the Mars Observer, prepared by one of the mission scientists. Relates the history of the mission from its inception until shortly before NASA scientists lost contact with the spacecraft. Includes photographs of the Mars Observer and many of its predecessors.

Hoagland, Richard C. *The Monuments of Mars: A City on the Edge of Forever*. Berkeley, Calif.: North Atlantic Books, 1987. An exhaustive analysis of several photographs taken by Viking 1 of the Cydonia region of Mars. The author, a former NASA consultant, argues passionately that several surface features of the Cydonian plain are artificial in origin. While not entirely convincing, the evidence he presents for his thesis is certainly strong enough to warrant closer investigation of the area in question.

Jet Propulsion Laboratory. *Mars Observer Loss of Signal: Special Review Board Final Report*. Pasadena, Calif.: Jet Propulsion Laboratory, 1993. Report of independent investigative committee retained by JPL to ascertain the causes of the Mars Observer mission failure. Concludes that many factors could have caused the mission failure, and that the true cause will probably never be known.

McDaniel, Stanley V. *The McDaniel Report: On the Failure of Executive, Congressional, and Scientific Responsibility in Investigating Possible Evidence of Artificial Structures on the Surface of Mars and in Setting Mission Priorities for NASA's Mars Exploration Program*. Berkeley, Calif.: North Atlantic Books, 1993. The author, a professor emeritus of Philosophy, castigates the United States Congress and NASA scientists and officials for failing to make the investigating of anomalous Martian surface features on the Cydonian plain a top priority in the Mars Observer mission. Makes a persuasive case that the possibility that these features are artificial should place them high on NASA's priority list for investigation by future Mars missions. Includes several intriguing photographs and a bibliography of scientific works analyzing the Cydonian surface features.

National Aeronautics and Space Administration. *International Exploration of Mars: A Special Bibliography*. Washington, D.C.: National Aeronautics and Space Administration, 1991. Comprehensive list of publications relating to Mars exploration space missions derived from STI database. Most complete reference available.

National Aeronautics and Space Administration. *Mapping the Red Planet: The Mars Observer Global Mapping Mission*. Pasadena, Calif.: National Aeronautics and Space Administration, Jet Propulsion Laboratory, 1992. A detailed explanation of the precise global mapping mission of the Mars Observer. Contains specifications and capabilities of the photographic equipment carried by the Observer.

National Aeronautics and Space Administration. *Mars Observer Mission Failure Report: A Report to the Administrator, National Aeronautics and Space Administration on the Investigation of the August 1993 Mission Failure of the Mars Observer Spacecraft*. Washington, D.C.: National Aeronautics and Space Administration, 1993. An internal NASA report detailing the possible causes of the loss of contact with the Mars Observer. Concludes that while there exist many possible causes for the failure of the mission, the exact cause cannot be ascertained.

Sheehan, William. *Worlds in the Sky: Planetary Discovery from Earliest Times through Voyager and Magellan*. Tucson: University of Arizona Press, 1992. Provides a brief account of the various space missions to Mars from 1962 to the Mars Observer mission. Puts Mars exploration into the broader perspective of the efforts of Earth nations in space exploration.

United States General Accounting Office. *Space Exploration: Cost, Schedule, and Performance of NASA's Mars Observer Mission*. Washington, D.C.: GAO, 1988. Exhaustive analysis of the cost estimates for the proposed Mars Observer mission, NASA's ability to administer the project, and the performance estimates of the craft and its equipment.

Paul Madden

Mars Surveyor Program

Date: Beginning February, 1994
Type of program: Unmanned Mars probes

Mars Surveyor is a planned ten-year effort to explore the planet Mars. A series of robotic orbiters and landers will be launched at twenty-five month intervals to study the surface chemistry and mineralogy, topography, atmosphere, polar cap composition, magnetic field, and seasonal variation of the climate of the planet.

Principal personages

MARIO ACUNA, Principal Investigator, magnetometer on Mars Global Surveyor

RAYMOND ARVIDSON, Principal Investigator on Mars Global Surveyor

JACQUES BALMONT, Principal Investigator, Mars relay experiment on Mars Global Surveyor

MICHAEL CARR, Principal Investigator on Mars Global Surveyor

PHILIP CHRISTENSEN, Principal Investigator, thermal emission spectrometer on Mars Global Surveyor

ANDREW INGERSOL, Principal Investigator on Mars Global Surveyor

MICHAEL MALIN, Principal Investigator, Mars orbital camera on Mars Global Surveyor

DANIEL MCCLESE, Principal Investigator, infrared radiometer on 1998 Mars Surveyor mission

DAVID SMITH, Principal Investigator, Mars orbiting laser altimeter on Mars Global Surveyor

LAWRENCE SODERBLOM, Principal Investigator on Mars Global Surveyor

LEONARD TYLER, Principal Investigator, radio science experiment on Mars Global Surveyor

BRUCE JAKOWSKY, Principal Investigator on Mars Global Surveyor

Summary of the Program

The Mars Surveyor program is planned as a series of spacecraft to explore the planet Mars. The program will include robot spacecraft that will orbit Mars, mapping the planet geologically, topographically, chemically, and mineralogically, as well as a series of landers that will conduct detailed scientific investigations at selected spots on the surface of Mars. The National Aeronautics and Space Administration (NASA) plans to launch one or more spacecraft at the planet Mars every twenty-five months for about a decade. The broad scientific objectives of the Mars Surveyor program are to collect the data necessary to understand how Mars has evolved since its formation, and to better understand the formation and evolution of planets in the Solar System. Mars Surveyor is also intended to conduct the initial experiments necessary for an eventual human mission to Mars, sometime in the twenty-first century.

The first launching in the Mars Surveyor program will be the Mars Global Surveyor, built by Martin Marietta Technologies Inc. located in Denver, Colorado. The spacecraft is scheduled for launching from Cape Canaveral, Florida, by a McDonnell Douglas Delta 2 rocket in November, 1996. The Mars Global Surveyor weighs about 1,075 kilograms, including about 75 kilograms of experiments and 380 kilograms of fuel. The spacecraft will reach Mars, after a 770 million kilometer journey, in September, 1997. When it arrives, the spacecraft will fire its main rocket engine to slow down enough to be captured into a highly elliptical orbit passing almost directly over both poles of Mars. This initial orbit is unsuitable for the science experiments that are planned. During the next four months the orbit will be adjusted by aerobraking, a procedure by which the spacecraft dips so low into the atmosphere of Mars that it is slowed by atmospheric drag, the same force you feel when you put your hand out the window of a moving car, reducing the high point of the orbit. Eventually the orbit will be adjusted by a combination of aerobraking and additional rocket firings so it is nearly circular at a height of about 378 kilometers above the surface of Mars. The final orbit will be Sun-synchronous, so the Sun will be at a standard angle above the horizon in each image and allow the midafternoon lighting to cast shadows in such a way that surface features will stand out.

The scientific instruments on Mars Global Surveyor will be turned on in April, 1998, and should continue to observe the planet until April, 2000. Since the Martian year lasts 687 Earth days, Mars Global Surveyor will observe the surface of the planet over a period a little longer than one full Martian year, allowing measurements over a full cycle of the seasons.

Mars Global Surveyor carries many instruments similar to those lost when the Mars Observer spacecraft failed, on August 21, 1993, just three days before it was scheduled to go into orbit around Mars. These scientific instruments were selected to determine the global topography, map the gravita-

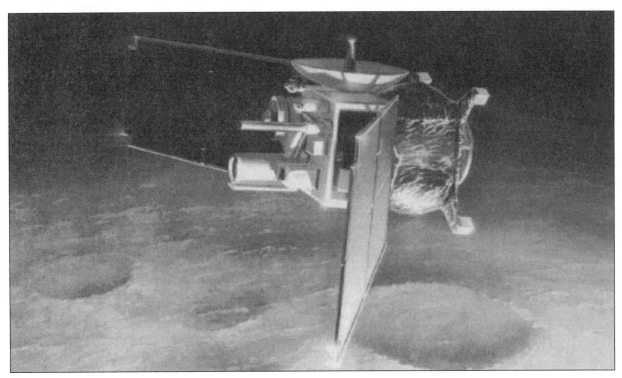

Mars Global Surveyor. (Lockheed Martin)

tional field, establish the nature of the magnetic field, monitor the global weather, determine the temperature profile of the atmosphere, and monitor the changes in the polar caps over a seasonal cycle.

The Mars Orbiter laser altimeter will fire a pulsed laser beam at the surface ten times a second. By carefully measuring how long it takes for each laser pulse to travel to the surface and back to the spacecraft, the height of each spot on the surface can be determined and detailed topographic maps can be prepared.

The magnetometer experiment carried on Mars Global Surveyor will determine if Mars currently has a magnetic field and will look for evidence of a stronger field that might have been present in the past. Meteorites that are believed to have come from Mars have strong remnant magnetic fields. This suggests that Mars may once have had a much stronger magnetic field, at a time when the interior of the planet was hotter than it is at present.

The Mars orbital camera will provide photographs of the surface of Mars. This camera will use a wide-angle lens to produce low resolution images of the entire planet, while a narrow-angle lens will take high-resolution images, seeing features as small as 1.5 meters in size, of selected areas of the surface.

Mars Global Surveyor also carries a thermal emission spectrometer, which will measure the infrared radiation emit-

ted by the surface. Using these measurements, scientists can infer the types of minerals that are present on the surface, how hot or cold the surface gets during the day/night cycle, and the size distribution of rocks and grains on the surface.

The radio transmissions from Mars Global Surveyor will be monitored by a radio science team who will be looking for changes that occur as Mars comes between the spacecraft and the Earth. Just before the spacecraft disappears from view the radio signals pass through the atmosphere of Mars, and careful monitoring provides information on the density, temperature, and pressure of the atmosphere. In addition, as the spacecraft orbits Mars it will speed up or slow down in response to deviations in the gravitational field of Mars from that of a perfect sphere. By measuring these very small changes in the speed of the spacecraft, the radio science team will map the gravity field of Mars.

After completion of its global mapping mission, Mars Global Surveyor will assume a new role as a communications relay satellite, receiving data transmitted by light-weight, low-power transmitters on future Mars landers and sending this data on to the Earth using its higher power transmitter and directional antenna. Mars Global Surveyor is expected to continue to serve as a radio relay through at least January 2003. Following its role as a relay satellite, the Mars Global Surveyor will be boosted into a quarantine orbit, about 405 kilometers above the surface of Mars. This higher orbit will place the spacecraft well above the bulk of the atmosphere, ensuring that it will not impact the planet for many years. This will allow future missions to search for possible life on Mars without the confusion of microbes transported from the Earth by the Mars Global Surveyor.

The second spacecraft in the Mars Surveyor program is scheduled for launching in December, 1998. This spacecraft, also an orbiter, is designed to observe the weather and climate on Mars. It will carry a small camera equipped with a wide-angle lens, which will transmit daily weather maps of the planet with a resolution of about 7 kilometers, giving about the same picture quality as the weather maps of Earth used in television weather reports. This spacecraft will also carry a pressure modulator infrared radiometer, which will monitor the time variation of the water vapor, carbon dioxide, and dust contents as well as the temperature and pressure of the atmosphere of Mars. This spacecraft is designed to operate for at least one Martian year to provide data over a full seasonal cycle, but the ultimate objective is to use these measurements to understand how the climate of Mars may have changed over a longer time period in the past.

The first lander in the Mars Surveyor program, called the Mars Volatile and Climate Surveyor, will be launched in January, 1999. The Mars Volatile and Climate Surveyor will be the first spacecraft to land in the polar regions of Mars. Using a low-mass television camera, this spacecraft will begin its photographic observations at an altitude of about eight kilometers above the surface, as it drifts down under its descent parachute. This series of photographs will allow scientists to place the landing site topography into the context of the regional topography. The lander contains a package of scientific equipment designed to investigate the unusual layered terrain in the polar regions, which appears to consist of alternating layers of clean ice and ice that is mixed with dust.

After landing on the surface, a robot arm will dig up samples of the surface and deliver them to an instrument that will determine the content of water ice and frozen carbon dioxide, two components of the polar cap of Mars. The experiment package will also include a weather station, with instruments to record the atmospheric temperature, pressure, and wind velocity and direction. The surface experiments are expected to operate for eighty-six days. The next spacecraft in the Mars Surveyor program will be an orbiter, planned for a launching in 2001, which will carry a gamma-ray spectrometer, designed to map the distribution of certain chemical elements on the surface of Mars. These element distribution maps will be important in understanding how Mars has evolved chemically and mineralogically. A Mars lander is also expected to be launched in 2001, but the scientific experiments and landing site will not be determined until after the results from the early missions are evaluated. Additional Mars Surveyor missions are planned for launching in 2003 and 2005, but their missions have not yet been determined. One proposal is that the mission in 2005 could return a sample of

the surface of Mars to the Earth for study in terrestrial laboratories.

Knowledge Gained

The Mars Global Surveyor is designed to perform many of the same scientific investigations that were planned for the Mars Observer spacecraft, which failed in 1993. This mission was designed to follow-up on a series of unexpected discoveries made by earlier spacecraft that explored the planet Mars. Among these discoveries are the layered terrain in the polar regions, surface features that appear to be flow channels, apparently indicating that rivers flowed on the surface of Mars at some time in the past, and the tallest volcanoes thus far discovered in the solar system.

Exploration of the layered terrain is a major objective of the Mars Volatile and Climate Surveyor, to be launched in January, 1999. This spacecraft will land near the North Pole of Mars and perform analyses of the ice and dust near the surface. The objective is to better understand how the layered terrain is deposited in the polar regions and to measure the contents of carbon dioxide ice and water ice in the North polar cap.

Detailed topographic maps of Mars will be produced by the laser altimeter carried on Mars Global Surveyor. At the initiation of the Mars Global Surveyor project, the heights of the major surface features on Mars were known to a precision of only +10 kilometers, which is insufficient for determining the geological context of some features. The laser altimeter on Mars Global Surveyor will significantly improve the precision of height determinations. This experiment will allow scientists to determine if the features that appear to be flow channels are consistent with water flow: for example, do they always flow downhill. The detailed topography of the volcanos will also be mapped.

High-resolution photography of the surface features will permit the determination of their ages, using observations of the density of impact craters on each feature. Features having a higher density of craters must have formed earlier, allowing more time for craters to accumulate. Since the rate of production of impact craters can be estimated, the time at which each major feature was formed can be inferred from its crater density. This should allow scientists to determine when the flow channels formed, dating the era when liquid water flowed on the surface of Mars.

Aside from Pluto, which has never been visited by a spacecraft, Mars is the only planet in the solar system whose magnetic field has not been characterized. The Mars Global Surveyor carries a magnetometer that will measure the strength of the magnetic field of Mars, if it has one. The results of this experiment will constrain models of how the interior of Mars has evolved over time.

The Viking landers measured the composition of the atmosphere of Mars. Subsequent to this, the gas content of glassy material extracted from a meteorite recovered from Antarctica was measured at the NASA Johnson Space Center.

The compositions of the two gases were so similar that scientists suggested this meteorite collected from the Antarctic, and several other meteorites similar to it, might be samples of Mars ejected into space by a major impact onto that planet's surface. More detailed measurements of the composition of the Martian atmosphere as well as detailed chemical analysis of surface rocks may provide a critical test of the Martian origin of these meteorites.

Context

The United States and the former Union of Soviet Socialist Republics (USSR) each launched a series of unmanned spacecraft to conduct extensive explorations of Mars. The first mission, Mars 1, launched by the USSR on November 1, 1962, was not successful. It was not until the Mariner 6 spacecraft, launched by the United States on February 24, 1969, flew by Mars that the first high-resolution photographs of the planet were obtained. Both the United States and the USSR launched spacecraft to orbit Mars in May, 1971, and each returned global mapping data. A lander launched by the USSR at the same time was not successful.

In 1975 the United States launched two spacecraft, Viking 1 and Viking 2, towards Mars. Each Viking spacecraft included both an orbiter, which undertook global mapping of the planet, and a lander, which undertook detailed analyses of its landing site. The major objective of the Viking mission was to determine if life ever existed on Mars. However, the highly reactive surface chemistry on Mars, resulting from the exposure of the surface to ultraviolet light from the Sun, gave rise to ambiguous results from some of the biological experiments on Viking. Nonetheless, the Viking orbiters provided detailed photographic maps of the surface, while the landers performed chemical analyses of the soils, measured the physical properties of the surface, and monitored the weather at the two landing sites.

Two Phobos spacecraft, to explore Mars and its largest moon Phobos, were launched by the USSR in 1988. Although one Phobos spacecraft failed before arrival at Mars and the second one ceased operation shortly after arrival in orbit around Mars, the initial measurements from the Phobos spacecraft confirmed many of the observations made earlier by the Viking spacecraft.

On July 20, 1989, on the twentieth anniversary of the first human landing on the Moon, U.S. President George Bush announced ambitious plans for American exploration of space, including the human exploration of Mars. Before humans can safely be sent to explore Mars, engineers need to know much about the surface and atmosphere to ensure that the spacecraft will not sink into the soil and that vital components will not be attacked by the atmospheric or soil chemistry. The Mars Surveyor program is designed to provide answers to those questions.

Bibliography

Barlow, Nadine G. *"NASA's Return to the Red Planet."* Ad Astra 8 (January/February, 1996): 39- 42. A discussion of the objectives of the Mars Pathfinder and Mars Global Surveyor missions, including detailed descriptions of each scientific instrument and the experiments it is expected to perform. Includes a brief discussion of the Mars Surveyor missions in 2001, 2003 and 2005. Suitable for general audiences.

Dasch, Pat, and John Kross. *"Mars Surveyor 98."* Ad Astra 8 (January/February, 1996): 12-13. A description, suitable for general audiences, of the mission objectives and scientific instruments to be flown on the orbiter and lander of the Mars Surveyor 1998 project.

Ezell, Edward, and Linda Ezell. On Mars: Exploration of the Red Planet. Washington, D.C.: NASA, U.S. Government Printing Office, 1984. A 535-page account, in the NASA history series, of the discoveries of the Viking missions to Mars, and the questions remaining for scientific investigation in the future.

French, Bevan M. Return to the Red Planet: The Mars Observer Mission. Washington, D.C.: U.S. Government Printing Office, 1993. A well-illustrated, fifty-nine page account of the scientific objectives of the Mars Observer mission including descriptions of the instruments carried by that spacecraft and the measurements they would have made. Many of these instruments will be flown on Mars Global Surveyor.

Jet Propulsion Laboratory. The Mars Global Surveyor: Return to the Red Planet. Pasadena, Calif.: NASA, Jet Propulsion Laboratory, 1995. A well-illustrated, thirty-six page description of the Mars Global Surveyor spacecraft, its scientific instruments, the project objectives, and the reasons for exploring Mars.

Miles, Frank, and Nicholas Booth. Race to Mars: The Harper & Row Mars Flight Atlas. New York: Harper & Row Publishers Inc., 1988. An extremely well-illustrated account of past missions to Mars, including detailed descriptions of the various types of surface features, a discussion of the types of scientific experiments to be performed on the surface, and a description of future plans for human missions to Mars.

Murray, Bruce. Journey into Space: The First Thirty Years of Space Exploration. New York: W. W. Norton, 1990. A first-hand account, by the former director of NASA's Jet Propulsion Laboratory, of NASA's efforts to explore the planets, including an in-depth account of the Viking mission and its accomplishments, and a concluding chapter on possible cooperative efforts between the U.S. and Russia to explore Mars in the future.

George J. Flynn

MARSHALL SPACE FLIGHT CENTER

Date: Beginning July 1, 1960
Type of facility: Research center

Marshall Space Flight Center (MSFC) is the National Aeronautics and Space Administration's principal center for development of large launch vehicles and propulsion systems. It is also one of four leading centers for space station research and development and one of the principal centers providing expertise in the physical space sciences and space technologies.

PRINCIPAL PERSONAGES

> WERNHER VON BRAUN, Director, 1960 –1970
> EBERHARD REES, Director, 1970 –1973
> ROCCO A. PETRONE, Director, 1973 –1974
> WILLIAM R. LUCAS, Director, 1974 –1986
> JAMES R. THOMPSON, JR., Director, beginning 1986
> T. J. LEE, Deputy Director
> WILLIAM R. MARSHALL, Space Shuttle
>> Projects Manager
> FRED WOTAJLIK, Hubble Space Telescope
>> Project Manager
> RAY TANNER, Space Station Project Manager
> JAMES A. DOWNEY III, Payloads Project Manager
> CHARLES R. DARWIN, Director,
>> Program Development
> JERROL WAYNE LITTLES, Director,
>> Science and Engineering

Summary of the Facility

Marshall Space Flight Center serves the National Aeronautics and Space Administration (NASA) as the leading center for development of large space launch systems, about a third of the space station, and a number of major space systems and payloads. The center's history dates to 1940, when a chemical munitions plant was located in Huntsville, Alabama. After World War II, the plant would have been shut down and sold as surplus, but the U.S. Army relocated a team of Army rocket engineers and scientists from Fort Bliss, Texas, to Huntsville, in 1950. The team was built around a core of 118 German engineers and scientists who had built V-2 rockets for Nazi Germany before surrendering to the U.S. Army in 1945. Popularly known as the von Braun rocket team, the group was led by the youthful Wernher von Braun, a charismatic engineer and scientist who would do much toward popularizing space travel in the United States. After early tests

and development work at the White Sands Missile Range in New Mexico, the team sought larger facilities for production of a nuclear-tipped tactical ballistic missile later known as Redstone.

For the first eight years in Huntsville, the team, which evolved into the Army Ballistic Missile Agency (ABMA), developed the Redstone and larger Jupiter missiles and started design work on the supersized Juno 5 missile for space exploration. Some effort was directed to development of a modified Redstone as a satellite launcher for the International Geophysical Year, an eighteen-month period dedicated to space exploration. After failure of the primary launcher, the U.S. Navy's Vanguard, the Army team was told to prepare its Jupiter C (later called Juno 1) for launch.

On January 31, 1958, the Army fired Explorer 1, the United States' first satellite, into orbit. The following October, NASA was formed around the old National Advisory Committee for Aeronautics, and nonmilitary space activities were transferred from the Department of Defense. The Army resisted, but in 1959 it was ordered to transfer the von Braun team, made up of 4,670 federal employees, to NASA. The transfer took place in October, 1959, and on July 1, 1960, ground was broken for its new headquarters building. The center was named in honor of George Catlett Marshall, the wartime chief of staff and postwar secretary of state, in the spirit of furthering the peaceful exploration of space.

The center's attention was initially focused on the development of large space launchers to overcome the apparent Soviet lead in this area. In the 1950's, the two superpowers had taken different paths in the nuclear arms race: The Soviets developed boosters that could launch the massive nuclear weapons common at that time, while the United States concentrated on making the bombs smaller to fit the mid-size missiles believed to be easier to build. The first of these large boosters was called the C-1, later known as Saturn 1.

In addition to the Saturn 1, the new center inherited the U.S. Air Force's huge F-1 engine, which generated 6.67 million newtons of thrust alone, and was working on designs for large boosters that could place almost 200,000 kilograms of mass into orbit. President John F. Kennedy's commitment, declared in 1961, to send men to the Moon brought this expertise and planning to the fore and soon turned the C-5 program into the Saturn 5 booster for the Apollo program. In

1963, Marshall was given the authority to proceed with the Saturn 5 rocket as the Moon launcher.

The magnitude of the Apollo program required construction of many complex facilities to support development and testing of new technologies that would be used in the Saturn rockets. The style of the German team was to carry this work through production of the first few flight units, then to turn the work over to a contractor that would build the production models. In this manner, the initial stages of the Saturn 1 and Saturn 5 vehicles were built in Huntsville by the Marshall team, then turned over to the Chrysler Corporation and Boeing Aerospace Company for production outside New Orleans, at a government-owned plant that built tanks and patrol boats during World War II. Subsequent stages of the Saturns, however, were built at contractor plants in California.

Important facilities built at Marshall include the F-1 test stand (completed in 1963), the vibration test stand for the complete Saturn 5 vehicle (completed in 1964), several smaller propulsion test facilities, a J-2 test stand (completed in 1965), and major machine shops.

As the Apollo program progressed, Marshall spawned two new NASA centers. The first, soon known as Kennedy Space Center, was built to launch the Saturn 5 Moon rocket. The other facility was the Stennis Space Center, built inland of the Gulf Coast near Bay St. Louis, Mississippi, to remove Saturn testing from the Redstone Arsenal, for it had been causing broken windows and crockery in downtown Huntsville.

As it was building the rockets that would send Americans to the Moon, MSFC figured at the center of several controversies regarding the management of the national space effort. With a manpower reduction in force effected in 1967, a federal judge ruled that mission support contracts were illegal. An eleven-year battle with the American Federation of Government Employees pushed the case to the Supreme Court, which ruled that the contracts were legal. The court ruled that NASA's charter and, by extension, the charter of any federal agency could be interpreted broadly.

Another MSFC project led to a revision of procurement practices. In 1969, the Senate Aeronautics and Space Science Committee learned that Marshall had built a Neutral Buoyancy Simulator (NBS) using more than $1 million of research funding after having been denied a construction of facilities request. The NBS is a water tank 22.9 meters wide and 10 meters deep in which spacewalks can be simulated; by using special spacesuits and flotation devices on heavy equipment, astronauts neither sink nor float and can move as if they were in space. The NBS has proved highly valuable in testing spacecraft designs and spacewalk procedures before flight.

The question whether MSFC would stay open was raised during the 1960's and well into the 1970's. "The Marshall problem" became a term that described the dilemma of any center that had facilities and manpower heavily vested in a single mission or line of work. One NASA official in the 1960's viewed Marshall as "that source for manpower needed elsewhere, and the place where surplus manpower would occur as the Apollo program phased down."

Indeed, the first of many manpower reductions occurred in 1966; these cutbacks continued into the 1970's. Marshall's employment fell from a peak of 7,272 in July, 1967, to 5,851 by mid-1970. Its strength would be whittled even further in the 1970's, leveling off at 4,000 in the 1980's. More serious than the raw numbers and their implications for the local economy was the loss of skilled management through layoffs, early retirement, or relegation to lesser positions. Many space veterans and one published history viewed a major portion of the layoffs as an effort "to get" the Germans. All but four or five of the original German rocket scientists were gone by the late 1970's because they did not have the rights held by most civil servants. A lawsuit was filed by two Germans alleging racial discrimination. In fact, there was resentment by many officials within NASA (and within MSFC) against the Germans and what was viewed as their autocratic, clannish management style.

Management styles changed radically at Marshall during the decade of the 1970's. At the urging of NASA officials, von Braun left MSFC for a post at NASA headquarters, where he was to lead advanced planning for the United States' future in space. His patron, Administrator Thomas Paine, left NASA soon after, and White House interest in the space program declined. Von Braun left NASA for private industry in 1975. At Marshall, he was succeeded by another member of the German team, Eberhard Rees. Rees, too, retired and was replaced by Rocco Petrone, an American who had been with the Army missile program at Kennedy Space Center. Petrone stayed only a year before being transferred to NASA headquarters.

Petrone was succeeded by William Lucas, a longtime civil servant who headed the program development branch from its inception. Lucas would have the longest tenure of any of Marshall's directors — twelve years — and his name would figure at the center of the postaccident investigation of the *Challenger* space shuttle disaster in 1986. Lucas resigned in 1986 and was replaced by James R. Thompson, the former manager of the main engine program, who was heading a fusion project at Princeton University.

Diversification and the space shuttle program kept the gates open at MSFC during the 1970's and returned it to a central role as one of NASA's most vital centers in the 1980's. Earlier, in the mid-1960's, NASA officials had started the Apollo Applications Program, which would use the Apollo and Saturn spacecraft in missions outside the scope of a simple manned lunar landing. It was an ambitious effort that was soon limited to the single Skylab launched in 1973 and manned through 1974. Yet this project gave Marshall experience in developing a space station and a major science payload, the Apollo Telescope Mount (ATM), which carried a cluster of eight ultraviolet and X-ray telescopes to study the Sun.

Although a manned space station seemed markedly different from large rockets, both involved integration and simultaneous operation of many complex systems. In this area, Marshall had excelled with the Saturn rockets. The center had also started developing a science capability in support of the engineering work on the rocket vehicles. This scientific expertise, in turn, expanded into experiments for these and other spacecraft.

Marshall's capability and innovation were clearly displayed in two projects in the early 1970's. The first was the Lunar Rover Vehicle, an electric "Moon buggy" that allowed astronauts to drive several kilometers from their lunar module. Marshall and the Boeing Aerospace Company developed the rover in a short period of time and saw it operate successfully on the last three Apollo missions.

When Skylab was launched in 1973, it was headed for a failure; Marshall was able to save the mission — even make it a success — in less than a month. During ascent, a combined sunshade and micrometeoroid shield tore, eliminating one solar power wing and jamming the other shut. Marshall engineers were able to modify a full-scale mock-up in the neutral buoyancy simulator to duplicate the damage and allow astronauts to practice repair procedures a few days before launch. Marshall's final involvement in Skylab came in 1978 – 1979 as the station started its spiral to reentry. MSFC engineers initially tried to develop a small, shuttle-launched robot stage to reboost the station. Yet Skylab would be on the ground again before the shuttle was launched, and by late 1978 the effort was abandoned. By the time Skylab reentered, much new experience had been gained in attitude control by momentum wheels (on Skylab's ATM), and the robot stage effort led to development of an orbital maneuvering vehicle that would act as a short-range space tug returning satellites to the shuttle in orbit.

Marshall's work on Skylab led to the High-Energy Astronomical Observatories (HEAO) program, which culminated in launches of three highly successful X-ray and cosmic-ray astronomy satellites from 1977 to 1979. The HEAO program substantially expanded the number of known X-ray sources and showed the universe to be more violent (that is, energetic) than previously thought. Skylab work also resulted in Marshall having lead responsibility for the docking module used on the joint Apollo-Soyuz Test Project (1975) with the Soviet Union.

Concerns about how to restart rockets and how to weld structures in space led to experiments in the fundamental behavior of fluids in a weightless environment. Experiments to test such behavior were carried aboard the Apollo spacecraft in 1971, and a larger suite of experiments was launched during Skylab and the Apollo-Soyuz Test Project. To span the gap between Skylab and Spacelab, a series of ten space processing applications rockets was flown from 1975 to 1980 to conduct basic research in the field. That, in turn, led to experiments aboard Spacelab 3 and the formation of a space commercialization office at MSFC.

In the late 1960's, Marshall and the Manned Spacecraft Center (later Johnson Space Center) were involved in research and development for a space shuttle. After several study cycles, which progressively reduced the size of the vehicle, NASA was authorized to proceed in 1972. MSFC was assigned to develop the high-performance main engines, their external fuel tank, and the twin solid-fueled rocket boosters. Johnson Space Center would manage the winged orbiter and the overall program. Several facilities at Marshall were modified to support shuttle testing. The Saturn vibration test stand was widened so that the complete shuttle "stack" could be assembled to simulate launch vibrations. The test stand built by the Army to fire the Saturn 1 first stage was modified to accommodate structural and pressure testing of a three-quarter-length solid-fueled rocket booster. All engine testing was conducted at the Stennis Space Center under Marshall supervision. The need for advanced engine development, however, had led NASA officials to renovate one of the old Saturn test stands.

Marshall soon acquired oversight responsibility for the European Space Agency's Spacelab program, which would be a major shuttle payload. MSFC was designated lead agency for the first three Spacelab missions. This work broadened the center's systems capabilities and increased its reputation as a science center rather than solely a builder of large rockets. Marshall was also charged with developing the Inertial Upper Stage, which boosts shuttle payloads to geostationary orbit, and advanced studies on reusable space tugs that would provide greater capability in the future. This work on Spacelab, and the center's record with Skylab, finally led to the assignment of more than 30 percent of NASA's space station program to MSFC in 1984. In the interim, between Apollo and the shuttle, Marshall's future again was in doubt, and rumors were often circulated about its closing.

Context

Marshall Space Flight Center is a strong, diverse facility that will play a major role in the U.S. space program into the next century. The center's staff is organized into four major groups. The programs and projects group covers the space shuttle, space station, Hubble Space Telescope, and payload projects. Program development includes preliminary design and analysis of advanced programs. Science and engineering provides research and technology, analytical, scientific, and other support to all project offices. Institutional and program support is the source of managerial and facilities support to all offices.

This organizational structure gives the MSFC flexibility in supporting a wide range of projects. For example, work done by the program development office on an unmanned cargo version of the shuttle was supported by the space shuttle projects office and various laboratories in the science and engineering office. One element of the Structures and Dynamics Laboratory has supported research on atmospheric science and on shuttle engine hot gas flow, since both involve modeling complex fluid flows. Marshall's Space Science Laboratory has achieved international recognition for work in solar-terrestrial physics, X-ray astronomy, and materials sciences and has developed several payloads for the shuttle and other spacecraft.

Chief among the new payloads are the core module, which will be used as laboratory and living quarters for the crew, and the resource nodes, which will join the modules. Marshall will also supply the life-support systems, outfit the laboratory module, and manage science operations aboard the manned U.S. space station. By overseeing the space laboratory, Marshall will play a large role in materials and life science experiments to be conducted aboard the space station.

Its experience with large projects also led to MSFC's being designated the lead center for the Hubble Space Telescope and the Advanced X-Ray Astrophysics Facility (AXAF). The Hubble project contains a 2.4-meter-aperture reflector telescope, which, though smaller than many ground-based telescopes, performs at least ten times more effectively because it is designed to function above Earth's atmosphere. Although the Hubble Space Telescope is operated by the Space Telescope Science Institute in Baltimore, Marshall retains responsibility for any necessary maintenance work in orbit. AXAF will be an X-ray complement to the Hubble observatory and a high-resolution continuation of the HEAO program. Marshall is also developing a so-called Tethered Satellite System to explore the upper atmosphere and remains the lead center for several Spacelab missions.

Marshall's history of excellence in designing space propulsion systems has not been overlooked. In addition to improving the shuttle main engine and fixing the solid-fueled rocket boosters, Marshall sponsors and conducts research in advanced propulsion areas such as the Space Transportation System booster and main engines, the joint NASA/Air Force advanced launch system, and an advanced solid-fueled rocket motor.

Bibliography

Baker, David. *The Rocket: The History and Development of Rocket and Missile Technology*. New York: Crown Publishers, 1978. A highly detailed and comprehensive technical survey of the development of major and minor launch vehicles. Written for the technically oriented reader.

Belew, Leland F. *Skylab: Our First Space Station*. NASA SP-400. Washington, D.C.: Government Printing Office, 1977. First of a series of books reporting Skylab results and experience. An introductory text written for the lay reader, with many details about MSFC. Well illustrated.

Bilstein, Roger E. *Stages to Saturn*. NASA SP-4206. Washington, D.C.: Government Printing Office, 1980. A highly detailed history of the development of the Saturn family of rockets and the legacy it left for the space shuttle program. Written for the educated lay reader.

Braun, Wernher von, et al. *Space Travel: A History*. New York: Harper & Row, Publishers, 1985. A revision of *The History of Rocketry and Space Travel*, which appeared during the 1960's. This book covers the entire space program for people unfamiliar with space travel. Includes surveys of the German rocket team and the Apollo-Saturn program.

Compton, David W., and Charles D. Benson. *Living and Working in Space: A History of Skylab*. NASA SP-4208. Washington, D.C.: Government Printing Office, 1983. An official NASA history of the Skylab program. Covers development work and management at Marshall as well as at other centers and contractor facilities. Includes discussion of the repair required after the sunshade was lost.

Levine, Arnold S. *Managing NASA in the Apollo Era*. NASA SP-4102. Washington, D.C.: Government Printing Office, 1982. A highly detailed history of the difficulty of and lessons learned from managing a large, growing agency as it developed a complex program to place Americans on the Moon. Activities and problems at MSFC are discussed in the context of the agency as a whole.

McConnell, Malcolm. *Challenger: A Major Malfunction*. Garden City, N.Y.: Doubleday and Co., 1987. An incisive account of how the *Challenger* accident came about and the roles played by people at Marshall and other centers. Written for a general audience.

National Aeronautics and Space Administration. *Marshall Space Flight Center, 1960–1985: Twenty-fifth Anniversary Report*. Washington, D.C.: Government Printing Office, 1985. Celebratory booklet on Marshall's history. Includes many photographs of MSFC employees at work, contrasting early work with more recent activities. Included is a three-page time line showing activities on major projects, the management, and the supportive Huntsville community.

Ordway, Frederick I., III, and Mitchell R. Sharpe. *The Rocket Team*. Cambridge, Mass.: MIT Press, 1982. A well-researched history of the origins and fates of the German rocket team that led the development of U.S. Saturn rockets in the 1960's. The bulk of the book deals with the prewar history and the V-2 program, but the latter chapters describe the history of MSFC and how the rocket team left NASA.

Report of the Presidential Commission on the Space Shuttle Challenger Accident. Washington, D.C.: Government Printing Office, 1986. The summary of the activities of the commission organized to determine the cause of the Challenger accident. This thorough report provides technical and managerial details about Marshall and NASA and speculates about failures to monitor the booster and other problems on the shuttle. Recommends changes in procedure.

Dave Dooling

MATERIALS PROCESSING IN SPACE

Date: Beginning January 31, 1971
Type of technology: Materials processing

Materials processing on Earth has resulted in improved tools and devices and sophisticated composites that have contributed to higher standards of living worldwide. Materials processing in space, where gravitational effects are eliminated, will lead to even more dramatic technological gains.

Summary of the Technology

Materials processing in space was first attempted in January of 1971, by the astronauts on Apollo 14. The term "materials processing" means changing the characteristics of materials. For example, by freezing water it is changed from liquid to a solid. The process of changing from a liquid to a solid is called solidification, and the characteristics of the material are altered.

Materials are usually processed to obtain a desired end product. Metal alloying, for example, is the process of mixing different metals together while they are in a liquid state, usually at a very high temperature, and then solidifying the mixture by cooling to obtain a metal alloy. Iron is a strong and useful metal; yet, it will oxidize (rust) easily. When iron is alloyed with chromium, manganese, and carbon, a new metal alloy called stainless steel, which does not rust easily, is produced. Therefore, the process of alloying these materials produces a new, improved end product.

There are many processes in which materials are combined to obtain different end products. On Earth, these processes can be adversely influenced by the force of gravity. Gravity can play a counterproductive role in processing. Sedimentation, buoyancy, and convection are the main gravity-driven effects that produce undesirable results in certain kinds of processing on Earth.

Sedimentation and buoyancy are really the same observable phenomenon described from different reference points. On Earth, if a material placed in a fluid, which is designated as the reference fluid, is more dense than the fluid, it will settle to the bottom. This process is called sedimentation. If the material is lighter than the fluid, it will float to the top and is said to be buoyant. Both sedimentation and buoyancy result in the separation of materials. On Earth, water is accepted as a common reference material, and the terms sedimentation and buoyancy are used to describe what happens when other materials are placed in water. Generalization of these terms, however, allows other reference materials to be designated.

Convection is a stirring action within a fluid. Usually the term "convection" is used to describe stirring which is induced by temperature differences within a fluid. Placing a heating source at the bottom of a container expands the molecules closest to the heat source and makes them buoyant with respect to the rest of the molecules in the gravitational field. The buoyant molecules move to the top of the container, inducing a stirring action. Placing a cooling source at the top of a container in a gravitational field will result in a similar stirring action. The stirring action, called convection, is usually caused by a temperature-induced movement of molecules in a fluid.

In space, these gravity-driven effects can be minimized. Theoretically, these effects can be eliminated, because they depend on the density (weight per unit volume) of one material versus that of another. Weight is a measure of the gravitational force on a material or object. In the weightlessness of space, there is no gravity-induced sedimentation/buoyancy or convection.

Materials processing in space covers a wide range of activities across the processing spectrum. At one extreme the objective is to produce completely homogeneous products. At the other extreme the objective is to separate and isolate specific materials from others in a mixture. Materials processing in space is particularly applicable at these two extremes because in space the effects of sedimentation/buoyancy and convection will not adversely influence the processing.

Sedimentation/buoyancy can be counterproductive when obtaining homogeneous mixtures is the primary goal. In order to obtain homogeneous mixtures, the materials must be mixed in such a way that one unit volume of the mixture is the same as any other unit volume within the mixture. Any sedimentation/buoyancy effect would counteract this desired goal. This is especially true when the mixture needs to be solidified as a homogeneous solid, because most solidification processes require relatively long quiescent periods.

The absence of sedimentation/buoyancy in the microgravity environment of space allows attainment of higher levels of homogeneity in metal alloys and other solids. Lightweight, strong metals could be formed by mixing air with a metal that is in a liquid state, allowing bubbles to form in the metal, and then solidifying the end product. On Earth, this cannot be done easily or uniformly because of the differ-

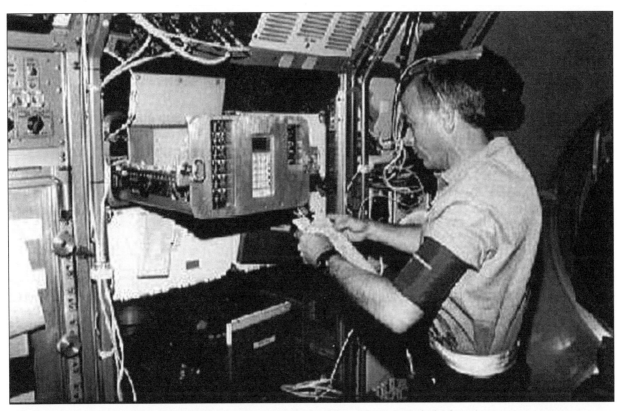

Crew member working on the spacelab Generic Bioprocessing Apparatus (top); United States Microgravity Payload 1 (bottom). (NASA)

ence in densities between air and metal. In space, metal and air can be mixed and solidified because sedimentation/buoyancy is absent and therefore does not contribute to separation. New products formed in this way are called foamed metals, and are strong yet lightweight.

In contrast to the process of mixing to obtain homogeneity, the process of separation on Earth is aided by sedimentation/buoyancy. When biological substances have weight, however, small differences in density and electrical charge create small-scale convection, which counteracts their separation. When measures are taken to minimize convective effects on Earth, the devices and mechanisms used create other problems. In the microgravity environment of space, where convection can be minimized, separation of biological substances that would be difficult on Earth can be achieved. Methods used to separate biological substances include electrophoresis, isoelectric focusing, chromatography, and filtering.

Electrophoresis is the process of separation using the differences in electrical charge between molecules. In this process, an electric potential is created across the material. With a positive charge on one side of the material and a negative charge on the other, molecules with a positive charge will migrate toward the negative side and molecules with a negative charge will migrate toward the positive side. This process requires that the materials being separated have the necessary charge differences to separate.

Isoelectric focusing also uses an electric potential across the material but incorporates the use of a solution with a pH (hydrogen-ion concentration) gradient. First, a pH gradient is established by "stacking" a series of carrier molecules (ampholytes) in the solution. Each ampholyte has the same molecular weight but a different stationary position (isoelectric point) within the fluid column in a specific electric field. Under the electric potential field, these ampholytes will form layers according to their isoelectric points. The result is a column with a natural pH gradient. Molecules introduced to this column will migrate to a corresponding stationary position. Sharp, concentrated, stable zones of like molecules are formed. This process of separation is useful for separating materials with isoelectric points as little as 0.01 pH apart.

Chromatographic methods such as ion exchange, molecular sieve, and affinity all use the principle of attraction. The molecules to be separated are attracted to a material that they will cling to called a "getter" material. The associative forces common to biological systems such as coulomb forces, van der Waals forces, and hydrophobic attractions are utilized to hold the molecules of interest while other, undesirable molecules are allowed to pass. Then the molecules of interest are removed from the "getter" and collected. Filtration methods involve forcing a mixture of molecules through a membrane of filtering material with a rigorously defined pore size; the molecules are separated according to molecular size.

Separating and sorting biological substances allows scientists to obtain ultrapure materials for study. For example, using ultrapure materials, they can grow very pure crystals and extract information about the molecular structures of substances. Crystal growth is the most common approach for obtaining structural information about biological substances.

Understanding the molecular structure of a substance allows scientists to manipulate it or vary its application. The more knowledge there is about the molecular structure of a substance, the less expensive and time-consuming it is to obtain the desired end product. Drug development, treatment of disease, and a host of other expensive, time-consuming problems can be minimized by a good understanding of the molecular structures involved.

Crystal growth is one of the best ways to obtain an understanding of molecular structures. The process involves several steps. First, an ultrapure sample of the material is needed. Such a sample is made through the separation and sorting techniques discussed earlier. Second, the substance is subjected to one of several methods of crystallization. After a crystalline form of the substance is obtained, the third step is to analyze the structure. Currently, the most common method of analyzing the structure is X-ray diffraction.

Crystallization is one process in which the systematic arrangement of identical molecules is accomplished. This orderly arrangement results in sufficient material for pattern recognition, which makes structural definition possible. To achieve the orderly arrangement of identical molecules on a microscopic level, it is necessary to allow enough time for the natural and appropriate orientation of each molecule as it takes its place in the structure. Consider this analogy. If 100,000 football fans had only five minutes to find their seats, there would be mass confusion. In many cases fans would occupy seats not legitimately belonging to them, and after they were in place it would be difficult, if not impossible, to extract them. Given more time, the process of achieving the proper seating arrangement would be accomplished more smoothly and acceptable positioning would result. Similarly, most macromolecules, especially mammal proteins, are made of 50,000 to 100,000 basic units of inheritance, or genes. Allowing enough time, without convection, for the natural and proper orientation of each molecule to take its proper place in the crystalline structure is vital.

There are several methods used to grow crystals, but the time consideration is common to all of them. There are other parameters that are important to the formation of crystals — such as the thermal environment, solution concentration, and externally imposed forces. Each of the crystallization methods provides advantages in controlling one or more of these parameters and therefore has its benefits. The most popular growing method for biological crystals is vapor diffusion. This method is used by more crystal growers in the biological field than any other.

There are many methods associated with crystal growth. Bulk crystallization occurs when a large quantity of solution is prepared and then given sufficient time for crystals to form.

Batch crystallization occurs when a batch, or small quantity, of solution is prepared and given time to form crystals. Dialysis involves a semipermeable membrane used to separate the solution to be crystallized from the solution of the precipitating agent; dialysis of solute through the membrane is used to bring about nucleation conditions (conditions under which crystals will begin to form). Liquid/liquid diffusion occurs when two previously separated liquids are brought in contact with each other in a closed container and allowed through the process of liquid-through-liquid diffusion to reach nucleation conditions. Vapor diffusion, yet another process, occurs when a droplet of the solution to be crystallized is allowed to reach the nucleation condition by vapor diffusing through an air space between the droplet and a reservoir of diluted precipitating agent.

After a crystal has been grown, the next step is to analyze its pattern and derive the molecular structure. The method most commonly used is X-ray diffraction. In this approach, the crystal is placed in a beam of X rays that are scattered in all directions by the electron complement of each atom in the crystal structure. The degree of scatter is proportional to the size and spacing of the atoms. When the lattice of the crystal is symmetrical, the distribution of the scattered X rays will form a pattern that can be analyzed to deduce the arrangement of the atoms.

Processing in space clearly offers unique advantages. On Earth, materials that are processed at high temperatures may become contaminated by the very containers in which they are processed. The weightless environment of space affords the possibility of high-temperature processing without a con-

tainer. In addition, ultrahigh processing, a method which would be useful for purifying materials, is possible in space. Because a constant stream of ions (an ion is an atom or group of atoms that carries an electric charge) is created by the orbital velocity in space, this process would probably require the use of a shield, or wake, to protect the materials being processed from the contamination of ions.

Knowledge Gained

Materials processing in space is a very young space technology. It has been shown, however, that processes with homogeneity or separation as their goal are greatly facilitated when compared to similar processing on Earth. In addition to metals and biological substances, insulators (such as glass and ceramics) and semiconductors have also been processed successfully in space.

The knowledge about materials processing in space comes from a smattering of sources. While there are no quantified analyses on the subject, the results obtained clearly show improvements in achieving materials homogeneity, materials separation, and crystal growth.

Context

It is doubtful that manufacturing facilities will be built in space to process large amounts of materials. It would not be economical to build manufacturing facilities in space simply to gain a minimal improvement in materials unless the materials were unique and valuable. It is more likely that the experiments involving materials processing in space will provide knowledge that will enable the processes on Earth to be improved.

Bibliography

Bugg, Charles E., et al. "Preliminary Investigations of Protein Crystal Growth Using the Space Shuttle." *Journal of Crystal Growth* 76 (1986): 681- 693. A summary of the results of four different shuttle flights of a vapor diffusion experiment to obtain protein crystals. It describes the advantages, techniques, strategy, and hardware design used for these four flights. Also describes the results of the experiments.

Chassay, Roger P., and Bill Carswell. *Processing Materials in Space: The History and the Future.* AIAA-87-0392. New York: American Institute of Aeronautics and Astronautics, 1987. An AIAA paper presented at the twenty-fifth Aerospace Sciences Meeting held in Reno, Nevada, in January, 1987. It gives a historical overview of materials processing in space and a description of the experiments performed. Describes the work being done at NASA centers and speculates on the future facilities available for materials processing in space and the technologies needed.

Doremus, Robert H., and Paul C. Nordine, eds. *Materials Processing in the Reduced Gravity Environment of Space.* Pittsburgh: Materials Research Society, 1987. Materials Research Society symposia proceedings were held in December, 1986, in Boston, Massachusetts. This is a compilation of the papers presented, which describe the highlights of ground-based experiments and plans for future experiments.

Feuerbacher, Berndt, Hans Hamacher, and Robert J. Naumann, eds. *Materials Sciences in Space: A Contribution to the Scientific Basis of Space Processing.* Berlin: Springer Verlag, 1986. This book has three objectives: to stimulate new scientific experiments in space in order to expand the knowledge gained from microgravity research, to provide industry with the information obtained from space experiments, and to contribute to the scientific background for commercial space utilization.

Hazelrigg, George M., ed. *Opportunities for Academic Research in a Low-Gravity Environment.* Vol. 108, *Progress in Astronautics and Aeronautics.* New York: American Institute of Aeronautics and Astronautics, 1986. A series of formal presentations on eight topics followed by a panel discussion on the policy implications of the identified research opportunities. The eight topics covered are infrastructures for low-gravity research, critical phenomena, gravitation, crystal growth, metals and alloys, containerless processing, combustion, and fluid dynamics.

McPherson, Alexander. *Preparation and Analysis of Protein Crystals*. New York: John Wiley and Sons, 1982. Basically a textbook describing the techniques and procedures used in the preparation and analysis of biological substances such as proteins, this book provides a working knowledge for the nonspecialist who is familiar with protein crystalization.

Naumann, Robert J., and Harvey W. Herring. *Materials Processing in Space: Early Experiments*. NASA SP-443. Washington, D.C.: Government Printing Office, 1980. A photographic presentation of the theories, applications, evolution, processes, results, and probable future of materials processing in space. Emphasis is on the theories and how they are supported, suggested by, and developed from the experiments conducted on the ground and in space up to and including the Apollo-Soyuz mission. It summarizes the state of the art at the time of publication and speculates on future facilities available for supporting materials processing in space.

Weinberg, Robert A. "The Molecules of Life." *Scientific American* 253 (October, 1985): 48-57. This special article is dedicated to introducing the new biology, which seeks to explain the molecular mechanisms underlying a new way of thinking about life. Provides an in-depth discussion of DNA, RNA, and proteins and molecules of the cell membrane, cell matrix, and immune system. There is a discussion of the molecular basis of communication between cells and within the cell, development of cells, and evolution.

Yates, Iva C., Jr. *Apollo 14 Composite Casting Demonstration Final Report*. NASA technical memorandum, TM X-64641. October, 1971. A description of the casting furnace experiment on Apollo 14. This experiment was one of three materials processing experiments performed on the mission. The other two were experiments investigating separation using electrophoretic techniques and fluid flow in selected configurations.

David W. Jex

PROJECT MERCURY

Date: October 7, 1958, to May 16, 1963
Type of program: Manned Earth–orbiting space program

Project Mercury was the United States' first manned orbital space program. In all, six manned missions were flown, two suborbital and four orbital, yielding important data on human adaptability to space travel.

PRINCIPAL PERSONAGES

ROBERT R. GILRUTH, Manager, Project Mercury
MAXIME A. FAGET, Chief, Flight Systems Division,
 Space Task Group
ALAN B. SHEPARD, the first American in space, on the
 Mercury-Redstone 3 mission
VIRGIL I. "GUS" GRISSOM, commander of the
 Mercury-Redstone 4 mission
JOHN H. GLENN, the first American to fly in Earth
 orbit, on the Mercury-Atlas 6 mission
M. SCOTT CARPENTER, the second American in orbit,
 on the Mercury-Atlas 7 mission
WALTER M. "WALLY" SCHIRRA, commander of the
 Mercury-Atlas 8 mission
L. GORDON COOPER, commander of the Mercury-
 Atlas 9 mission
DONALD K. "DEKE" SLAYTON, one of the original
 seven U.S. astronauts, replaced as commander of
 MA-7 as a result of a heart murmur

Summary of the Program

Project Mercury was officially born on October 7, 1958. Its stated goals were threefold: to place a manned spacecraft in Earth orbit, to investigate man's performance and his ability to function in space, and to recover the pilot and spacecraft safely. Beginning fifteen months after the first orbital satellite (Sputnik 1) was launched, Mercury would last nearly five years, send six men into space, cost an estimated $400 million, and employ approximately thirty-three hundred engineers and scientists.

The prime spacecraft contractor, McDonnell Aircraft Corporation, was chosen in January of 1959, and the first full-scale mockups were ready for inspection a mere two months later. The final vehicle was the classic cone-shaped "space capsule" which would become the traditional symbol of the United States' progress in space exploration. Mercury was a stubby black cone only 2.9 meters long and 1.89 meters in diameter at its base. The spacecraft tipped the scales at 1,355 kilograms and was small enough to impose a height limitation on the astronauts of 1.82 meters (5 feet, 11 inches).

Two booster rockets were selected; the smaller one was the Redstone, a direct descendant of the German V-2 designed by Wernher von Braun. Also designed by von Braun, the Redstone stood 21.5 meters high, was 2 meters in diameter, and produced a sustained thrust of slightly more than 34,000 kilograms. It would be used for high-altitude structural tests and the first suborbital manned flights. The second and larger booster was the venerable Atlas. With development beginning in 1951, the Atlas would be the longest-lived of all American missiles, still launching satellites up through the late 1980's. This "workhorse" of the American space program measured 23 meters long and carried a thrust of more than 250,000 kilograms.

With the official go-ahead, it was now necessary to concentrate on the third important element in this project: the astronauts. It was decided early in the project that military test pilots would be the perfect candidates. The original list consisted of 110 men: 5 Marine, 47 Navy, and 58 Air Force. By early March, 1959, the list had been whittled down to 32. The 32 became 18, and the 18 became "The Seven." In mid-April, The Seven were introduced to the public, becoming instant idols to millions of young boys, subjects for harassment by the press, unofficial ambassadors of the United States, and "the best pilots in the world."

The original, and rather optimistic, flight schedule proposed in January, 1959, would have had the first manned mission take place by January, 1960. Afterward, men would fly every three to four weeks, completing the project by September of that year. It was planned that six suborbital and six orbital missions were to fly with men aboard. Primarily because of the high failure rate of the Atlas rocket, however, Project Mercury got off to a slow start. Tests of both the Redstone and Atlas rockets were at first discouraging. Mercury-Atlas 1 and Mercury-Redstone 1 both failed in 1960. The first Mercury-Redstone test was finally successful just before the end of the year. In January, 1961, Ham the chimpanzee flew a suborbital lob on MR-2, experiencing six minutes of weightlessness. Despite problems with electronics aboard the Mercury spacecraft and much joking from their test-pilot friends about astronauts being no better than apes, The Seven were heartened by Ham's success. Mercury-Atlas tests 2, 3, and 4, all unmanned, were completed by September, 1961. Meanwhile, the first manned U.S. spaceflight was made on May 5, 1961, at 9:34 A.M. eastern standard time. At that moment, the name of Alan B. Shepard joined those of Charles Lindbergh and the Wright brothers, as he became the United States' first man in space. Riding his spacecraft, *Freedom 7*, for a modest 15 minutes and 22 seconds, Shepard flew twenty-two days after Yuri Gagarin's stunning single-orbit mission. The secrecy surrounding the Soviets' Vostok flight, however, had made Shepard a hero to most of the world. The small Redstone rocket made orbital flight impossible; still, *Freedom 7* reached an altitude of 187.5 kilometers and a maximum velocity of more than 8,000 kilometers per hour.

Gus Grissom became the next hero, taking *Liberty Bell 7* on a repeat of Shepard's flight, lifting off at 7:20 A.M. on July 21, 1961. Flying 9 seconds longer and 3 kilometers higher, Grissom's flight went as planned, until the landing. One major modification in Grissom's spacecraft was the use of explosive bolts on the hatch to make emergency escape much easier. Shortly after landing, while Grissom was waiting for the recovery crew, the hatch blew open. *Liberty Bell 7* quickly filled with water, weighing 400 kilograms more than what the recovery helicopter could handle. The capsule was dropped and swiftly sank from sight into 4,600 meters of water. Meanwhile, Grissom was struggling to stay afloat while his spacesuit filled with water as a result of an open valve. After an "eternity"— about five minutes — he found himself safely on board the helicopter returning to the carrier.

On February 20, 1962, an estimated 100 million people watched the televised broadcast of the first American to go into orbit. After a monthlong series of frustrating launch delays, John Glenn soared toward the heavens in *Friendship 7*. For the next four and one-half hours, Glenn would take hundreds of photographs, observe African dust storms, describe mysterious "fireflies," and test dozens of intricate spacecraft systems. A potentially serious problem, however, came at the start of the second orbit. An indicator on the ground said that the heatshield had come loose and was being held in place only by the retropackage straps. Glenn was not told of the problem until the very end, however, and reported that everything was fine. He was finally instructed to leave the retropackage on during reentry to hold the heatshield in place. Nearing the end of the third orbit, Glenn fired the retro-rockets to begin his slow glide toward Earth. Four hours and fifty-five minutes after launch, *Friendship 7* settled into the waters of the Atlantic Ocean.

With the success of *Friendship 7*, the next mission was to concentrate on science. Scott Carpenter was the pilot in the spacecraft dubbed *Aurora 7*. Like Glenn's, Carpenter's mission would take three orbits, lifting off on May 24, 1962. His many tasks included zero-gravity observations of liquid, weather photography, and the release of a 76-centimeter-diameter balloon to study air drag. As a result of his heavily loaded schedule of experiments, fuel consumption was

unusually high, such that nearly 50 percent was used in the first orbit. Because of this and a malfunction in the spacecraft control systems, Carpenter had to effect a manual retrofire. Being distracted from his busy final preparations he ignited the rockets three seconds too late, while the spacecraft was at the wrong attitude. As a result, *Aurora 7* landed several hundred kilometers away from the intended splashdown site and required more than four hours to be rescued.

October 3, 1962, was the day on which Walter Schirra lifted off at 7:15 A.M. in *Sigma 7* for his six-orbit mission. The one remarkable thing about Schirra's flight was just how unremarkable it was. Some called it perfect; Schirra agreed, calling it a textbook flight. Unlike Carpenter, Schirra managed to conserve fuel well, using in six orbits what Carpenter had used in the first half orbit. Schirra performed more photographic and systems experiments than were done on previous missions. He also tried extended periods of freedrift, allowing the spacecraft to drift unguided. Reentry and landing were likewise uneventful, with Schirra hitting the ocean a scant seven and one-half kilometers from the planned site. Unlike Glenn and Carpenter, he elected to ride *Sigma 7*, dangling from a helicopter's tether, all the way to the rescue carrier's deck. About forty-five minutes after splashdown, and 9 hours, 59 minutes after launch, Wally Schirra blew open the hatch to the jubilant shouts of the ship's crew.

On May 15, 1963, at 8:04 A.M., the last Mercury spacecraft lumbered upward toward the sky. L. Gordon Cooper, in *Faith 7*, was on his way toward a twenty-two-orbit, thirty-four-hour mission. The primary goal was to test man's endurance in zero gravity. With a full day in space, Cooper was given eleven experiments to occupy his time, and like the earlier astronauts, he saw the "fireflies." One experiment had him launch a small beacon to test visual perception; other tasks were medical in nature. Cooper performed tests in transferring liquids from one container to another. He took radiation measurements and searched for Earth details, claiming that he could see roads, villages, and individual houses. Near the end of the flight, Cooper made observations of "airglow" and the "zodiacal light," mysterious dim glows seen in the night sky. Photographs were taken to help in navigation studies for the Apollo program.

Another addition to this mission was that of a slow-scan television camera, which transmitted still cockpit pictures back to the ground every few seconds. On board, television had very little utility for the weight it took up and would not be used again until the first Apollo mission. (It is interesting to note that the Soviets used television from their very first flight.) The only problems in the mission involved a cranky pressure suit coolant system that alternated between extremes of heat and cold. More serious was an electrical short circuit that disabled Cooper's automatic control system. The latter problem required a manual re-entry similar to Carpenter's. Unlike Carpenter, however, Cooper landed only 6.6 kilometers from the aircraft carrier *Kearsarge*. Like Schirra, Cooper elected to stay with his spacecraft as it was hoisted up to the deck. Forty minutes after landing, he blew the explosive bolts on his hatch. Project Mercury had come to a close.

Knowledge Gained

The Mercury program proved that man could be sent into space, perform useful duties, and be recovered safely. These events alone fulfilled the primary objectives. It provided a test bed for spaceflight management, communications, planning, training, and basic spacecraft engineering, proving that all worked well. In-flight experiments were divided into several areas: visual perception, photography, radiation studies, tethered balloon, aeromedical, and miscellaneous zero-gravity studies.

During several missions, flares were launched from the ground or intense beacons activated for the astronaut to sight. Poor weather hampered all attempts except on the last flight, when Cooper reported seeing one of the lights. The light was first visible from 530 kilometers away and appeared as a "star" of average brightness. Also on this same flight, a small beacon was launched from the retropack. This light flashed brightly about once a second and was used to test the use of beacons for identifying space targets in later programs. Cooper sighted it on two orbits — once when it was about twenty-one kilometers away, and again when it was more than twenty-eight kilometers in the distance. While the beacon was rather dim, Cooper stated that the flashing made it easily distinguishable from the stars.

In other simple visual studies, the astronauts were able to see ships' wakes, airplane contrails, railroad tracks (one actually saw a train), and small villages. All agreed that ground details were amazingly clear. Photography experiments revealed that Earth is best viewed in near-infrared light. This information would be of value for weather satellite design. Hundreds of ground photographs were taken, supplying useful geological and topographical data.

The radiation experiments showed that there were very low levels of radiation inside the spacecraft. In another test, readings were taken on both Schirra's and Cooper's flights of radiation remnants from a July, 1962, atomic explosion. The results showed that the radiation decayed by several orders of magnitude over the seven-month period between the missions. The balloon experiment failed on both missions. As a result of the medical experiments on Cooper, it was shown that he withstood the stresses of the flight situation with no evidence of degradation of his performance. He did, however, suffer some dehydration. His pulse and blood pressure returned to normal between nine and nineteen hours after splashdown. Sleep in flight appeared to be relatively normal, and there was no evidence of psychological disturbances during the thirty-four-hour flight. All evidence pointed to the ability of humans to handle flights of a much longer duration, both physically and psychologically. This was welcome information, since the Apollo Moon missions would last up to ten days.

On Schirra's mission, a number of panels made of differing ablative materials were attached to the outside of the spacecraft. These were to test the behavior of various materials to the reentry heat. None of the panels was clearly superior to the others. In-flight television transmissions were attempted from Cooper's mission, but the poor quality of the system rendered it of little value.

Context

As the United States' first manned space program, Project Mercury served to pioneer the fundamentals of spaceflight. It answered the questions, "Could a man be sent into orbit and recovered successfully, and while in space, could he perform useful work?" These missions laid the groundwork for the management, control, and training techniques required in later and more complex programs. Mercury provided training for both astronauts and ground-control personnel. Three of the original seven astronauts were to go on to fly the two-man Gemini spacecraft: Gus Grissom in the first Gemini mission, Gemini 3, Schirra in Gemini 6, and Cooper in Gemini 5. Three would join Apollo crews: Schirra on Apollo 7, the first manned Apollo flight, Shepard, who became the fifth man to walk on the Moon on Apollo 14, and Deke Slayton, who never flew in Mercury as a result of a minor heart problem, but was finally given his chance during the Apollo-Soyuz Test Project in 1975.

The more specific flight elements required for lunar missions were not explored until Gemini. These included spacewalks, rendezvous and docking, navigational techniques, and missions of extended duration. The Mercury flights also helped to galvanize the American public and government into accepting President John F. Kennedy's goal of a manned lunar landing by the end of the decade. This declaration was made only fifteen days after Shepard's flight, when the National Aeronautics and Space Administration (NASA) had a mere total of 15 minutes and 28 seconds of manned spaceflight time.

When compared to the Soviet's first manned program, Vostok, Mercury seems rather primitive. Vostok landed on land, the Mercury space capsule in water. Vostok flew six orbital missions, the longest being five days, compared to Mercury's four orbital missions, with thirty-four hours being the longest flight duration time. Vostok cosmonauts accumulated a total of 16.46 days in space, compared to the minuscule 2.2 for The Seven. The Soviet spacecraft weighed a frightful five thousand kilograms compared to the Mercury's featherweight fourteen hundred kilograms. What the Mercury lacked in size and weight, however, it made up for in technology. While the Soviets were the undisputed leaders in the space race during this time, much of this success was a result of the brute-force nature of their technology. Their rockets were larger than those of the United States, but this was out of necessity; they lagged behind in the important technology of miniaturization, and therefore manufactured heavier spacecraft. Furthermore, the early Soviet program was propelled with the need to "show off," giving science a back seat. For example, no repeat missions were allowed; each had to accomplish a clear-cut first for the history books. Mercury, by contrast, was a clearer and more methodical flight test program. This model presented by Mercury served the Gemini and Apollo programs well, leading them to take a more cautious step-by-step approach.

Bibliography

Baker, David. *The History of Manned Space Flight*. New York: Crown Publishers, 1982. A detailed look at "how it all began." This book is a large coffee table model, more than 530 pages long and containing hundreds of photographs. Unlike many coffee table books, however, this volume has a generous amount of text. Recommended for anyone with even a passing interest in space history. Contains many tables.

Gatland, Kenneth. *The Illustrated Encyclopedia of Space Technology: A Comprehensive History of Space Exploration*. New York: Crown Publishers, 1981. Aimed at a general audience, this book stands out for its numerous photographs, highly detailed drawings, and charts. The twenty-one sections are divided among different contributors of note and cover such areas as space pioneers, space centers, and man on Mars.

Gunston, Bill. *The Illustrated Encyclopedia of the World's Rockets and Missiles*. London: Salamander Books, 1979. A remarkable reference book, its sections are divided among the various kinds of missiles, including surface-to-air and air-to-air, and is then subdivided into countries. An attempt was made to provide photographs, drawings, and text for every known missile. Nearly all illustrations are in color. Highly recommended to those interested in model building.

National Aeronautics and Space Administration. *Mercury Project Summary Including Results of Fourth Manned Orbital Flight, May 15-16, 1963*. NASA SP-45. Washington, D.C.: Government Printing Office, 1963. This work is the official NASA summary of the project. It contains twenty sections detailing Project Mercury's medical experiments and booster and spacecraft performance. It is fairly technical in nature but should be suitable for general audiences. This volume also contains the official flight report for the last mission, *Faith 7*, along with transcripts of air-to-ground communications.

Swenson, Loyd S., Jr., James M. Grimwood, and Charles C. Alexander. *This New Ocean: A History of Project Mercury*. NASA SP-4201. Washington, D.C.: Government Printing Office, 1966. This publication is part of the NASA historical series and one of the "Special Publications" made available to the public. The history of Project Mercury is traced up through Cooper's flight. Includes material on both the technical and management sides of the program. Many rare photographs and diagrams are included. Suitable for ages twelve and above.

Michael Smithwick

THE DEVELOPMENT OF
PROJECT MERCURY

Date: January, 1958, to September 9, 1959
Type of program: Manned Earth-orbiting space program

The technical development of the Mercury spacecraft culminated in the Big Joe *launch of September, 1959, when the feasibility of the ablative heatshield and the survivability of a spacecraft during reentry and splashdown were demonstrated and the use of an Atlas booster to launch Mercury spacecraft was proved effective.*

PRINCIPAL PERSONAGES

MAXIME A. FAGET, Chief, Flight Systems Division,
 Space Task Group
ROBERT R. GILRUTH, Manager, Project Mercury
JAMES A. CHAMBERLAIN, Program Manager, Project
 Mercury
CHARLES W. MATHEWS, Operations Manager, Project
 Mercury
JOHN YARDLEY, the manager of the Mercury program
 at McDonnell Douglas
ANDRE J. MEYER, JR., a developer of heat shielding
 mechanisms for the Mercury project
WALTER C. WILLIAMS and
CHARLES J. DONLAN, Associate Directors, Project
 Mercury
ALEC BOND, *Big Joe* Project Engineer

Summary of the Program

The launching of the "boilerplate" Mercury capsule, named *Big Joe*, on September 9, 1959, capped almost two years of technical developments that ultimately resulted in the manned spacecraft of the Mercury program. *Big Joe's* successful twenty-minute flight proved that a frustum-shaped ("Coke bottle," in the designers' vernacular) capsule with an ablative heatshield would align itself properly, with its blunt end forward, during reentry and would land safely. The *Big Joe* flight also showed that efforts to make the Mercury compatible with existing rockets, such as the Atlas, had been successful.

The "father" of the Mercury capsule— and, to some extent, the later Gemini and Apollo craft— was Maxime Faget. In the early spring of 1958, he and his coworkers at Langley Memorial Aeronautical Laboratory in Virginia were already designing features for the manned spacecraft that were to become the mainstays of the U.S. space program for the next decade and more.

Faget's group first worked for the National Advisory Committee for Aeronautics (NACA). In July of 1958, NACA was merged into the new National Aeronautics and Space Administration (NASA). In September, 1958, plans for the program that was to be known as Project Mercury began to take shape and the scope of the necessary technology became clear. The mission objectives and vehicle configuration were drawn up by a panel of experts from NACA and the Advanced Research Projects Agency (ARPA), and a document was developed that outlined the retrograde firing rocket systems and the life-support, attitude control, recovery, and emergency escape systems. This panel, called the Space Task Group, also developed plans for a girdle of tracking stations around Earth, craft-to-ground communication methods, and ground support and test program technology.

The objectives of the project were to "achieve at the earliest practicable date orbital flight and successful recovery of a manned satellite, and to investigate the capabilities of man in this environment." The principles of the program were as follows: The simplest and most reliable approach was always to be used; there was to be a minimum of new technical developments; and there was to be a steady progression of tests.

Debates had raged for years within NACA and ARPA and among aerodynamic theorists as to how it might be possible to retrieve a capsule safely, whether manned or unmanned, from Earth orbit. Since a spacecraft orbits Earth at speeds of about 30,000 kilometers per hour and must slow to a relatively safe landing speed of less than 100 kilometers per hour, there is a tremendous loss in kinetic energy. Most of this energy is released as heat. Theorists had predicted that reentry vehicles would experience temperatures ranging from 1,000 degrees Celsius to as high as 3,000 degrees Celsius and that these temperatures, since they are well above the thermal limits for most metal structures, would kill the capsule's occupants and destroy the capsule itself. Three solutions had been proposed by various aerodynamicists in the United States, and each had advocates. They were an ablative heatshield, a beryllium heatsink, and a slow glider reentry.

The slow glider mechanism was rejected rather early in the NACA studies, although parts of the idea survived in the Rogallo wing, which was proposed for the final landing of the craft. It was believed that the first shock of the craft's hitting the air stream at the beginning of reentry would make deployment of glider wings a very unsound procedure.

Andre J. Meyer was detailed to keep tabs on the development of the beryllium heatsink and the ablative heatshield. In the late 1950's, fiberglass, ceramics, and other thermally resistant materials were in the earliest stages of development, and it was not clear whether a heatsink made of beryllium would be able to carry heat away from the critical areas fast enough. Beryllium, like titanium and vanadium, is a refractory metal; it maintains its structural integrity at relatively high temperatures. It was theorized that beryllium would get hot during reentry but would be able to spread the heat over its entire volume quickly; thus, no part of the structure would get too hot to cause mechanical damage.

The ablative heatshield, in contrast, would consist of a very poor heat conductor, or insulator. The theory, as developed by investigators at the Langley and Ames research centers, was that most of the heat would be dissipated by shock waves and that the shield would not conduct the remaining heat to the titanium shell of the spacecraft or to the occupants within.

There were no accurate ways to evaluate shielding devices of the size necessary to protect a manned spacecraft. Wind tunnels could not produce air speeds high enough, and earlier rockets had relied on rather small nose cones for reentry. Beryllium, moreover, is a difficult metal to work, and in the late 1950's no beryllium structures comparable to the required device had been built. There were beryllium coatings on engines and mirrors but nothing similar to a rather thick, well-formed, pure metal heatsink. Meyer soon formed the opinion that successful development of the mechanism would require a breakthrough in beryllium technology.

With the flight of *Big Joe* and the success of the ablative heatshield, the beryllium heatsink idea was dropped from all subsequent manned space programs. An Atlas 10D modified Intercontinental Ballistic Missile (ICBM) launched *Big Joe* successfully. Even though the outboard engines had failed to detach correctly and the additional weight had caused *Big Joe* to fail to reach the desired speed, it did reach a speed of almost 25,000 kilometers per hour, which was acceptable for the heat studies.

The ablative heatshield worked effectively; only one-third of the heatshield melted during reentry. The fringe areas, where the heatshield was connected to the capsule, were hotter than expected, and thicker beryllium shingles were placed there in subsequent Mercury flights.

Big Joe demonstrated that Mercury had the proper aerodynamic configuration and that upon reentry it would turn so that its blunt end was leading. Had the retro-rockets not fired to initiate reentry, it was calculated, the heatshield would have kept the capsule intact for a natural decay orbit. The launch also proved that the fabrication techniques used on the titanium outer shell worked, that the Atlas could be used as a launcher, that the capsule could be landed safely on the ocean with the help of a parachute, and that recovery procedures could be effective 500 kilometers from the anticipated landing site. *Big Joe's* success made NASA personnel confident that the Mercury program could be manned, and a second test flight was deemed unnecessary.

Knowledge Gained

The Mercury project's development was an organizational and technological triumph. Titanium welding came of age at the McDonnell Douglas plant in St. Louis, where the titanium shell of the Mercury spacecraft was welded in an inert gas atmosphere. The layers were only 25 millimeters thick.

The success of *Big Joe* demonstrated that the heatshield approach would work even better than had been expected. The heat experienced by *Big Joe* was more intense than predicted, but it did not last as long as predicted. The paint on the titanium shell identifying it as a U.S. spacecraft sustained no damage during the flight, and a piece of tape left on the outside surface returned undamaged. It was necessary to increase the thickness of the beryllium heat shingles at the interface of the craft with the heatshield. The only damage — and it was very slight — occurred at this interface. Some studies had suggested that a gradually decaying orbit might seriously damage the heatshield, but such reentries would be highly unlikely in future manned flights. It was the flight of Mercury-Atlas 7 several years later that proved how efficacious the heatshield design really was; this capsule had a sustained-heating reentry trajectory.

The Atlas 10D, since it did not "stage" (detach from the capsule) properly, was classified as a failure by the Air Force. To NASA officials, however, the failure was almost irrelevant, since the capsule landed in such good shape.

After *Big Joe*, a series of launches using "Little Joe" rockets were conducted to test launch staging, escape and abort mechanisms, and other specialized features of what was to be the manned Mercury craft. A rhesus monkey named Sam rode Little Joe 2 for four minutes and survived reentry and the shock of impact. Another monkey, dubbed Miss Sam, survived a gravitational force of 10 and was able to perform several functions during her ride.

Big Joe and its thermal monitors showed that a manned spacecraft could survive reentry and that its occupants would experience no deleterious thermal effects. It also showed that the spacecraft's design was sound and that the capsule could reorient itself for reentry and survive the shock of a splash landing.

Context

During the early technical development stages of Mercury, the U.S. space program underwent several reorganizations. NACA and ARPA were competing agencies of sorts in early 1958, and there was considerable tension as to the role the Air Force might play in directing future space efforts. The early Mercury-Redstone series tests were partially under control of the U.S. Army. The various organizations were looking out for their own interests.

The successful Soviet launchings of Sputniks 1 and 2 in October and November of 1957 caused consternation within the space and aeronautical communities. The Ames and Langley research centers, however, were succeeding in their technological investigations. Even though the Atlas was designed to launch thermonuclear warheads, it could be modified for use as a launch vehicle for manned spacecraft. Titanium airframes were being built by the aircraft industry, miniaturization of electronics was also beginning, and most of all, the country was developing a desire to get into space in response to what it saw as the Soviet challenge.

Basically, the Mercury capsule was a hull that could surround a pilot. The feasibility of using the Atlas booster with the Mercury capsule was demonstrated by *Big Joe*, as was the recoverability of a returning spacecraft after a splash landing. The Mercury techniques and design were carried over to the Gemini and Apollo programs; the spacecraft that put the first humans on the Moon a decade later had the same basic features as *Big Joe*.

Within the space community, particularly at McDonnell Douglas and NASA, the idea of staged program management and meticulous testing of each feature systematically gained acceptance. The management teams that developed the early Mercury craft were talented, and their organizational abilities, high motivation, and clear decision making saw the U.S. space program off to an excellent start.

Bibliography

Caidin, Martin. *The Astronauts: The Story of Project Mercury, America's Man-in-Space Program*. New York: E. P. Dutton Co., 1960. This book is a bit dated, but it contains a very good discussion of the problems antedating the manned Mercury flights. It focuses primarily on the astronauts but provides several very good insights into the technical problems associated with the Mercury capsule and the Atlas launch vehicles.

Emme, Eugene M. *Aeronautics and Astronautics: An American Chronology of Science and Technology in the Exploration of Space, 1915 – 1960*. Washington, D.C.: Government Printing Office, 1961. Emme's books on the early history of Mercury and later Gemini are very informative. His section on the late 1950's is very relevant to the technical development of Mercury and *Big Joe*. It is detailed and is for the spaceflight devotee rather than the interested amateur.

Faget, Maxime A. *Manned Space Flight*. New York: Holt, Rinehart and Winston, 1965. Faget was known as "the father of the U.S. space capsules." This brief volume details the Mercury program's problems and solutions.

Gatland, Kenneth. *Illustrated Encyclopedia of Space Technology: A Comprehensive History of Space Exploration*. New York: Crown Publishers, 1981. This is a comprehensive book on all aspects of the space program, including the efforts of many different countries. Includes detailed sketches of both the Soviet Vostok capsule and the U.S. Mercury capsule. These sketches are necessary for readers who wish to compare the U.S. and Soviet space programs at the time when the Mercury program was being developed.

Rosholt, Robert L. *An Administrative History of NASA, 1958 – 1963*. NASA SP-4101. Washington, D.C.: Government Printing Office, 1966. This book focuses mainly on the administrative tangles in the aeronautics and space community at the time Mercury's technical developments were under way. The conflicts between the Air Force and the civilian space community are discussed, and the rationale for the national organization that was to emerge is well presented.

Sobel, Lester A., ed. *Space: From Sputnik to Gemini*. New York: Facts on File, 1965. This small book contains a remarkably detailed account of all launches of the U.S. and Soviet space programs from 1957 to 1966. It includes brief statements about each Mercury flight, including technical development efforts, and gives a good account of the early days of the program. It is ordered chronologically.

Swenson, Loyd S., Jr., James M. Grimwood, and Charles C. Alexander. *This New Ocean: A History of Project Mercury*. NASA SP-4201. Washington, D.C.: Government Printing Office, 1966. This book is part of the NASA Historical Series. It is a complete and well-written account of all aspects of the space program until 1962 as they influenced the development of Mercury. It is a bit weak in outlining rivalries among the military and the civilian authorities, but it gives a good account of *Big Joe* and the early technical problems, such as those with the heatshield.

Thomas, Shirley. *Men of Space: Profiles of the Leaders in Space Research, Development, and Exploration*. 5 vols. Philadelphia: Chilton, 1960. This book contains the biographies of several of the key personnel associated with the early days of Mercury, such as Robert Gilruth and other members of the Space Task Group.

John Kenny

MERCURY-REDSTONE 3

Date: May 5, 1961
Type of satellite: Manned suborbital spaceflight

Alan Shepard became the first American to reach space on Mercury-Redstone 3 in a fifteen-minute suborbital mission. In his Freedom 7 *spacecraft, Shepard experienced five minutes of weightlessness. Although the United States was not the first country to put a man in space, Mercury-Redstone 3 gave President John F. Kennedy confidence to commit NASA to a lunar landing.*

PRINCIPAL PERSONAGES
ALAN B. SHEPARD, *Freedom 7's* Pilot
JOHN H. GLENN, JR., *Freedom 7's* backup Pilot
CHRISTOPHER C. KRAFT, JR., Assistant Chief, Flight
    Operations Division, NASA
JAMES E. WEBB, NASA Administrator
HUGH L. DRYDEN, NASA Deputy Administrator
ROBERT R. GILRUTH, Director of the Space Task
    Group and Mercury Program Manager
MAXIME A. FAGET, the spacecraft's designer
WILLIAM K. DOUGLAS, the Flight Surgeon
VIRGIL I. "GUS" GRISSOM and
DONALD K. "DEKE" SLAYTON, astronauts

Summary of the Mission

On February 22, 1961, the Space Task Group revealed that three Mercury astronauts had been selected to begin concentrated training for manned Mercury-Redstone (MR) suborbital flights. These three astronauts were Virgil I. "Gus" Grissom, Alan Shepard, and John H. Glenn, Jr. Although the identity of the prime pilot for MR-3 was kept secret, the astronauts themselves already knew who had won the competition to be first.

While the National Aeronautics and Space Administration (NASA) pressed forward to send one of these three men into space, the Soviet Union orbited Yuri Gagarin in Vostok 1 on April 12, 1961. Losing the race into space to the Soviets was a bitter defeat for Project Mercury, but efforts were redoubled to proceed. NASA would not fly MR-3 until it was reasonably certain that both the booster and the spacecraft were ready to support a manned mission safely.

Redstone number 7 was selected to support MR-3. The Mercury spacecraft to be used for this flight would not, however, incorporate all the changes the astronauts had requested. Spacecraft number 7 would have only a small porthole and

periscope through which the astronaut would be able to view. The hatch was sealed by seventy bolts once the astronaut entered. Redstone number 7 arrived at Cape Canaveral in March, 1961; the spacecraft had been there since December 9, 1960.

The United States stood poised to enter space on May 2, 1961. Unlike the Soviets, the Americans planned to carry out their launch attempt before the eyes of the world. Newsprint, radio, and television reporters converged on the Cape to cover MR-3. The question asked most often concerned which astronaut would be making the flight. The announcement had been withheld in the event of a last-minute change. The weather was poor, and a hold was called at T minus 290 minutes when a storm broke out. Shepard and Glenn were awakened at 2:00 A.M. to eat breakfast, have a medical examination, and prepare for the flight. The press learned that Shepard had been selected as pilot and that Glenn was to be his backup. Shepard awaited a decision on weather for more than three hours while fully suited.

Shepard was born November 18, 1923, in East Derry, New Hampshire. He received his bachelor of science degree from the United States Naval Academy in 1944. As a naval aviator, Shepard ultimately achieved the rank of rear admiral in the Navy, and NASA selected him as one of the original seven Mercury astronauts in 1959. Following his MR-3 suborbital flight, Shepard served as backup pilot for Mercury Atlas 9. He was then grounded from flight status because of an inner ear ailment. While grounded, until May 7, 1969, Shepard served as chief of the Astronaut Office. A successful ear operation restored Shepard to flight status, and he was assigned as commander of Apollo 14, becoming the only Mercury astronaut to walk on the Moon. From June, 1971, to August 1, 1974, Shepard resumed his post as Chief Astronaut. Then he retired from both NASA and the Navy.

The May 2 launch attempt was scrubbed after a report of rain near the recovery area was made after 7:00 A.M. A forty-eight-hour recycle was announced, but poor weather foiled a May 4 launch attempt. Countdown for the next attempt began at 8:30 P.M. on May 4. Shepard was awakened at 1:10 A.M. on May 5. He showered and shaved before sharing a breakfast of orange juice, filet mignon, bacon, and scrambled eggs with Glenn and physician William K. Douglas. After a medical examination, biomedical sensors were attached to Shepard,

Launch of the MR-3 spacecraft. (NASA)

and he donned his pressure suit. A transfer van transported Shepard, carrying a portable air conditioner, to the launchpad. At 5:15 A.M., Shepard exited the van, briefly looked up at the Redstone booster, and entered the elevator to rise to the level of the spacecraft.

Shepard entered the spacecraft, which he had named *Freedom 7* (making allusions to the number of the booster, spacecraft, and Mercury astronauts), the hatch was bolted shut, and he began breathing pure oxygen. Weather conditions 15 minutes before scheduled launch were inadequate for proper photographic coverage, and a hold was called to await clearer skies. Meanwhile, a 115-volt, 400-hertz inverter was replaced on the Redstone. This hold ended after 52 minutes, and the countdown picked up at the T-minus-35-minute mark. At T minus 15 minutes, a hold was again called, this time because of computer deficiencies at the Goddard Space Flight Center. This hold lasted 154 minutes. When the countdown resumed, it continued down to T-minus zero. Shepard had been inside *Freedom 7* for 4 hours and 14 minutes awaiting lift-off.

Astronauts Gordon Cooper and Donald Slayton communicated with Shepard from the pad blockhouse and Mercury Control Center, respectively. Walter Schirra flew overhead in an F-106 chase plane, ready to follow after MR-3. Glenn had helped prepare the spacecraft controls for Shepard, and Grissom served as the reserve pilot. Thus, although Shepard would fly the United States' first manned spaceflight, MR-3 was an astronaut team effort.

As the firing command was given, Shepard's heart rate rose from 80 to 126 beats per minute. *Freedom 7* rose from the launchpad at 9:34 A.M., and for the first 45 seconds of powered flight the ride was smooth. As *Freedom 7* passed into the transonic zone, aerodynamic buffeting vibrated the vehicle. Maximum dynamic pressure (3.8 pounds per square inch) was encountered 88 seconds after lift-off at an altitude of 10.6 kilometers. Shepard was shaken so violently that he was unable to read instruments. After becoming supersonic, the vibration levels quickly diminished.

Shepard experienced 6 gravitational loads two minutes after lift-off. T plus 142 seconds and at an altitude of 59 kilometers, the Redstone engine shut down. Ten seconds later, the spacecraft/booster clamp ring was released. A trio of solid-fueled posigrade rockets (rockets that increase the speed of a satellite, thus placing it in a higher orbit) at the back of *Freedom 7* (each packing 160 kilograms of thrust) pushed it free of the booster. Simultaneously, the launch escape tower's escape and jettison motors fired, pulling the tower safely away from *Freedom 7*. Shepard attempted in vain to watch the separation through the spacecraft porthole.

The Redstone booster accelerated *Freedom 7* to a speed of 8,214 kilometers per hour. Outside the spacecraft, the temperature was 104 degrees Celsius; inside it was 32.8 degrees, and Shepard's suit temperature was a comfortable 24 degrees. Cabin pressure stabilized at 5.5 pounds per square inch.

Shepard extended the periscope while the Automatic Stabilization Control System (ASCS) turned *Freedom 7* around so that it was flying heatshield first; oscillations were damped by ASCS thruster firings. Shepard performed pitch, yaw, and roll movements of the hand control stick, while *Freedom 7* was still under automatic control, to familiarize himself with the handling characteristics. He then switched to manual control and performed these movements one axis at a time, at no time moving more than 20 degrees on a single axis.

Shepard had become weightless at booster burnout, but he experienced no debilitating symptoms, and could read instruments and make limited movements allowed by his restrictive spacesuit. Shepard's vision was unimpaired; he was able to view Earth through his periscope. Unfortunately, Shepard had forgotten to remove a gray filter (used prior to lift-off to minimize glare) from the periscope, although he was able to distinguish clearly between cloud cover and land masses, seeing first the west coast of Florida and the Gulf of Mexico and later Lake Okeechobee and the Bahamas.

Shortly after *Freedom 7* had achieved a maximum altitude of 186.4 kilometers, the retrofire sequence was initiated. Shepard had pitched the spacecraft up to the 34-degree retro-attitude, and a trio of retro-rockets fired in turn for 10 seconds apiece, each burn overlapping the previous one by 5 seconds, to reduce spacecraft speed by 560 kilometers per hour. The retropack was held in place by a trio of steel straps equipped with explosive bolts so that the pack could be jettisoned. When retropack jettison was executed, Shepard heard a thud but saw no pack jettison light indication on the instrument panel. Astronaut Slayton informed Shepard that telemetry indicated retropack jettison. Shepard could see pieces of the pack and straps passing by his porthole, and switching to manual override, Shepard got a green retropack jettison light.

The periscope was retracted when *Freedom 7* was restored to automatic control for the reentry phase. Shepard performed a dexterity test and vainly attempted to see stars through the porthole. At T plus 490 seconds, Shepard set up a two-revolution-per-minute roll to minimize landing dispersions. Deceleration was rapid and loads quickly reached a gravitational force of 11.6. Shepard strained to make his voice communications readable but withstood the crushing force of reentry.

The drogue parachute mortar fired at an altitude of 6.4 kilometers at T plus 570 seconds. Four seconds later, the ambient air snorkel opened, and through his periscope Shepard could see the drogue parachute. Passing 3 kilometers altitude at T plus 600 seconds, the main parachute deployed. When fully blossomed, the parachute lowered *Freedom 7* to the water at a gentle 10 meters per second. *Freedom 7* splashed down 128 kilometers east-northeast of Grand Bahama Island, only 11.2 kilometers off target, at 9:49 A.M. Within five minutes a Marine helicopter recovered *Freedom 7* from the water and transported Shepard and the spacecraft to the deck of the USS *Lake Champlain*.

Knowledge Gained

Gathering biomedical data on human response to space-flight stresses and engineering data on the Mercury-Redstone spacecraft-booster combination was the primary goal of MR-3. The ability of astronaut Shepard to perform specific tasks during the flight and his usefulness as a pilot were carefully monitored and assessed after the flight. On June 6, 1961, a conference was held in the U.S. Department of State Auditorium by NASA in cooperation with the National Institutes of Health and the National Academy of Sciences to report the results of MR-3 investigations. This scientific meeting was open to both U.S. and foreign scientists.

Bioinstrumentation for real-time monitoring of pilot physiological responses during test flights had to be specially developed for Project Mercury. Biosensors to measure and record body temperature, respiratory activity, and heart rate were perfected for use on MR-3. (Blood pressure measurements were taken on later Mercury flights.) These sensors, attached directly to Shepard's body, did not interfere with his responsibilities as a pilot.

Shepard's vital signs were determined eight hours before flight. His body weight was 76.9 kilograms, rectal temperature was 37.2 degrees Celsius, respiration rate was 16 breaths per minute, heart rate was 68 beats per minute, and blood pressure was 120 over 78 while he was seated. Shepard was given another physical examination on the USS *Lake Champlain* immediately following recovery, at which time his body weight was 76 kilograms, rectal temperature was 37.9 degrees Celsius, pulse was 100 beats per minute, and blood pressure was 130 over 84 (seated). Three hours after the mission had ended, Shepard's status was measured on Grand Bahama Island. At this time Shepard's body weight was 75.6 kilograms, oral temperature was 36.7 degrees Celsius, respiration rate was 20 breaths per minute, heart rate was 76 beats per minute, and his blood pressure was 102 over 74 while he was standing and 100 over 76 while he was supine. Shepard performed programmed activities during the flight without major difficulty. He suffered no apparent physiological or psychological abnormalities as a result of his exposure to launch and reentry stresses and a brief spell of weightlessness (4 minutes and 45 seconds). There were no medical indications from MR-3 that would preclude American manned orbital missions. Shepard's only medical complaint was that his biomedical sensors caused minor skin irritation. Different adhesives were used on subsequent flights.

Spacecraft environmental control systems maintained comfortable suit and cabin conditions. Shepard's suit temperature varied between 23.6 and 27.8 degrees Celsius; the cabin temperature, between 33.3 and 37.8 degrees. Only a portion of the available oxygen supply had been used, and there was an adequate cooling supply.

Shepard exercised all three methods of spacecraft control: automatic, manual, and fly-by-wire (a combination of manual and automatic). Manual control was very responsive, behaving similarly to procedures during training simulations except for a tendency toward a slight roll clockwise, which was caused by a tiny thrust expelled from leaky hydrogen peroxide thrusters. The ASCS consumed more fuel than desired when Shepard apparently forgot to turn off a manually controlled fuel-flow valve after he had switched to the automatic control mode. Attitude control and stabilization was maintained as desired with interactions between spacecraft, Mission Control, and the astronaut pilot.

Context

On May 5, 1986, five of the original seven Mercury astronauts and the wife of the late Virgil Grissom were reunited in Los Angeles, California, for a celebration of the twenty-fifth anniversary of *Freedom 7*. The celebration overshadowed the difficulties faced by NASA in the wake of the explosion of the space shuttle *Challenger*. Shepard, then fifty-two and a successful Houston businessman, was especially honored as the first American astronaut into space. Looking at the flight of MR-3 retrospectively after more than a quarter century of manned spaceflight, which has included such impressive space achievements as Apollo Moon landings, Skylab and Salyut long-duration flights, and space shuttle missions, Shepard's 15-minute suborbital arc down the Atlantic Missile Range at first glance seems more like a spectacular stunt than an important milestone in space travel. Such an appraisal, however, clearly fails to recognize the proper place that *Freedom 7* deserves in the history of both the United States and manned spaceflight.

The *Freedom 7* mission occurred on the heels of several politically embarrassing incidents for the fledgling Kennedy Administration, including the Bay of Pigs fiasco. While the United States was still suffering fear and humiliation over the Soviet launching of Sputnik forty-two months earlier, the Soviets put the first man into space, Yuri Gagarin. The *Freedom 7* mission had several profound effects within the United States. First, it provided a renewed sense of national pride. Second, it served to crystallize support for NASA's plans for goals beyond Project Mercury. Third, it was used to illustrate the difference between the open society of the United States and the secrecy surrounding the Soviet space program. Fourth, it demonstrated to the nation and the rest of the world that the American program was not discouraged by its late entry into the space age and was capable of competing with the Soviets in the area of spaceflight.

Shepard was honored in Washington, D.C., on May 8, 1961, with a parade along Pennsylvania Avenue and a ceremony in the White House Rose Garden, where he received NASA's Distinguished Service Medal from President Kennedy. *Freedom 7* and Shepard drew headlines all over the world, much to the surprise and chagrin of Soviet premier Nikita Khrushchev, considering that the American suborbital mission occurred weeks after Gagarin had actually orbited

Earth. *Freedom 7* gave the Kennedy Administration confidence to propose a bold new initiative in an attempt to elevate the United States into a position of leadership in spaceflight, as well as a strong indication that both the American people and Congress would support and finance the effort.

President Kennedy spoke before Congress on May 25, 1961, and publicly committed NASA to a manned lunar landing before the end of the decade — effectively challenging the Soviets to a race to the Moon. This bold step was made before an American had even orbited the Earth.

Bibliography

Carpenter, M. Scott, L. Gordon Cooper, et al. *We Seven.* New York: Simon & Schuster, 1962. Written by the astronauts in conjunction with the staff of Life magazine, this book portrays the original Mercury astronauts the way they were, and still are, revered. Fascinating reading. Includes photographs.

Grimwood, James M. *Project Mercury: A Chronology.* NASA SP-4001. Washington, D.C.: Government Printing Office, 1963. This book chronologically traces the general developments of manned spaceflight capability, research and development of Project Mercury, and the operational phase of the Mercury program. Includes numerous diagrams, photographs, charts, and eight appendices that are key to a complete understanding of Mercury's achievements.

Link, Mae Mills. *Space Medicine in Project Mercury.* NASA SP-4003. Washington, D.C.: Government Printing Office, 1965. Prepared by the Office of Manned Space Flight, this text traces aviation medicine through Mercury bioscience and biotechnology. Provides actual flight biomedical data; photographs and references abound.

McDonnell Astronautics Company. *Project Mercury Familiarization Manual.* SEDR 104. St. Louis: Author, 1961. Once classified, this contractor document provides the most detailed description available of the Mercury spacecraft and its systems. Contains structural diagrams, block diagrams, wiring diagrams, and photographs. For the serious researcher.

National Aeronautics and Space Administration. *The Mercury Redstone Project.* TMX-53107. Huntsville, Ala.: Marshall Space Flight Center, 1964. Reports in a single volume a complete history of Redstone's development for unmanned, chimpanzee, and manned Mercury suborbital spaceflights. Highly technical, providing flight data in chart, graph, and tabular forms.

National Aeronautics and Space Administration, with the National Institutes of Health and the National Academy of Sciences. *Proceedings of the Conference on the Results of the First U.S. Manned Suborbital Spaceflight.* Washington, D.C.: Government Printing Office, 1961. Proceedings of a conference held on June 6, 1961. Provides a summary of the Mercury program, *Freedom 7*'s biomedical data, pilot training and evaluation of pilot in-flight performance, and Shepard's flight report. Includes charts, diagrams, tables, and photographs.

Olney, Ross R. *Americans in Space: A History of Manned Space Travel.* Rev. ed. Nashville: Thomas Nelson, 1973. This volume includes general summaries of American spaceflights from *Freedom 7* through Apollo 13. Perfect for younger readers and novices. Contains photographs, appendices, and a glossary, as well as numerous quotations from mission air-to-ground conversations.

Silverberg, Robert. *First American Into Space.* Derby, Conn.: Monarch Books, 1961. Published shortly after Mercury-Redstone 3, this paperback book demonstrates the American people's reaction to Shepard's historic flight, one of relief and pride. Details the flight of *Freedom 7* and provides personal biographies of astronauts Shepard, Grissom, and Glenn.

Swenson, Loyd S., Jr., James M. Grimwood, and Charles C. Alexander. *This New Ocean: A History of Project Mercury.* NASA SP-4201. Washington, D.C.: Government Printing Office, 1966. An excellent NASA History Series text, providing a fascinating report of research and development, management, and flight operations of the Mercury program. Extensive photographs, appendices, and references are included. An important reference for serious spaceflight researchers.

David G. Fisher

MERCURY-REDSTONE 4

Date: July 21, 1961
Type of mission: Manned suborbital spaceflight

The flight of Liberty Bell 7, *Mercury-Redstone 4, ended the Mercury-Redstone program. Astronaut Gus Grissom duplicated the successful suborbital profile flown by Alan Shepard in* Freedom 7. *Although* Liberty Bell 7 *sank after the spacecraft hatch blew off prematurely, Grissom was safety recovered.*

PRINCIPAL PERSONAGES

VIRGIL I. "GUS" GRISSOM, *Liberty Bell* 7 Pilot
JOHN H. GLENN, JR., *Liberty* 7 backup Pilot
CHRISTOPHER C. KRAFT, JR., Assistant Chief,
    Flight Operations Division
JAMES E. WEBB, NASA Administrator
HUGH L. DRYDEN, NASA Deputy Administrator
ROBERT R. GILRUTH, Director of the Space Task
    Group and Mercury Program Manager
MAXIME A. FAGET, spacecraft designer
WILLIAM K. DOUGLAS, astronaut Flight Surgeon
ALAN B. SHEPARD and
DONALD K. "DEKE" SLAYTON, astronauts

Summary of the Mission

The National Aeronautics and Space Administration (NASA) announced in early July, 1961, that Air Force Captain Virgil I. "Gus" Grissom had been selected as the primary Pilot for Mercury-Redstone 4 (MR-4), the United States' second suborbital flight. Marine Lieutenant Colonel John H. Glenn, Jr., would serve as backup Pilot for Grissom, as he had for Shepard's Mercury-Redstone 3 flight in May, 1961. Launch of MR-4 was originally scheduled for July 18, 1961.

Grissom was born April 3, 1926, in Mitchell, Indiana. He earned a bachelor of science degree in mechanical engineering from Purdue University in 1950. As a military pilot, Grissom ultimately achieved the rank of lieutenant colonel in the Air Force. NASA selected Grissom as one of the original seven Mercury astronauts in 1959. Grissom, in addition to his *Liberty Bell* 7 suborbital flight, later flew as Command Pilot of Gemini 3 and served as backup command pilot for Gemini VI. Assigned as Commander of the first manned Apollo flight, Grissom perished in the Apollo 204 flash fire on Pad 34 at Cape Kennedy on January 27, 1967.

Meteorological forecasts for July 18 predicted cloud coverage that would have severely hindered optical coverage of the launch vehicle during ascent. A one-day delay in launch was ordered. On July 19, however, clouds and precipitation postponed the launch once more. MR-4 was rescheduled for July 21. The two-day delay was ordered because the Redstone booster's liquid oxygen supply had to be removed and its cryogenic tanks purged and dried before the next launch attempt.

Grissom was awakened at 1:10 A.M. eastern standard time on the morning of the launch. Within an hour, he had eaten breakfast and had been given a physical examination. After suiting up, Grissom was driven to the launchpad. He climbed into Mercury spacecraft number 11, which he had named *Liberty Bell 7*, at 3:58 A.M. Grissom would wait nearly three and a half hours inside the spacecraft before lifting off.

Mercury spacecraft number 11 incorporated a number of new innovations, encouraged by the astronauts themselves, that Shepard's spacecraft did not have. The two most important changes were the inclusion of a new trapezoidal window, which greatly increased the pilot's viewing ability, and an explosive hatch, which greatly reduced the problem of exiting the spacecraft after splashdown. Originally, egress from the Mercury spacecraft was through the antenna compartment, a procedure requiring removal of a pressure bulkhead. This maneuver proved difficult for the astronauts and would be disadvantageous in the event the astronaut was injured or disabled, for the only alternative access to the pilot required external removal of seventy bolts around the hatch.

Weather was favorable during the early morning hours of July 21. The countdown proceeded without delay from T minus 180 minutes (3:00 A.M.) to T minus 45 minutes. Proper installation of a misaligned hatch bolt forced a 30-minute hold, and at T minus 30 minutes, the countdown held for 9 minutes while pad searchlights, which interfered with Redstone telemetry during launch, were turned off. The final hold, 41 minutes in duration, was called at T minus 15 minutes as Mission Control waited for better cloud cover conditions to prevail. The countdown resumed, and MR-4 lifted off the pad at 7:20 A.M.

Proper flightpath angle was maintained by the Redstone control system during ascent. As the vehicle entered the transonic region at an altitude of 11 kilometers, Grissom experienced less noise and vibration than Shepard reported at the similar point in his flight. (Damping material and alternate aerodynamic bearings had been added to *Liberty Bell 7* to reduce these vibrations.) Grissom was able to read his instruments clearly and communicate with Mission Control. Pausing briefly to peer out the window, he saw nothing but blue sky above him. The vehicle had just passed through a thin layer of clouds. Grissom later reported that the sky had turned from a beautiful blue to pitch black in a matter of seconds; shortly after, he was able to see a star.

The Redstone engine cut off at T plus 2 minutes, 23 seconds. At the same time, the escape tower clamp ring was released and both the escape and tower jettison rockets fired. The gravitational force (g) of six times Earth's gravity was suddenly replaced by weightlessness. Grissom saw two streams of smoke pass by his window and the launch escape tower off to the right of the spacecraft. Ten seconds after booster burnout, the spacecraft's booster adapter clamp ring separated, and posigrade rockets fired to separate *Liberty Bell 7* from the spent Redstone. The spacecraft automatic stabilization and control system provided five seconds of rate damping and turned the spacecraft around to its flight attitude of −34 degrees with respect to the horizon.

Grissom assumed manual control of *Liberty Bell 7*'s attitude at T plus 3 minutes, 5 seconds. He performed pitch and yaw movements to check out the manual proportional control system, dividing his attention between looking out the window and controlling the spacecraft. Grissom yawed to the left to view the coast of Florida. Alan Shepard, serving as Capsule Communicator, reminded Grissom to return to retro-attitude. Retrofire sequence was initiated by a timer, and thirty seconds later, *Liberty Bell 7* reached its maximum altitude, 188.8 kilometers.

When the retro-rocket firing sequence began, Grissom keenly perceived their effect. Until retrofire he had had the distinct feeling that he was flying backward. (The heatshield pointed near the direction of the spacecraft velocity vector.) After retrofire he experienced a sensation as though he had changed direction and was now flying forward. Actually, what Grissom felt was a reduction in his retrograde motion. He controlled spacecraft attitude within proper limits during retrofire, and after T plus 5 minutes, 43 seconds, he manipulated the spacecraft attitude by using the manual rate command system. *Liberty Bell 7*'s retropack was jettisoned at T plus 6 minutes, 7 seconds.

Grissom switched from ultrahigh-frequency to high-frequency radio systems. He attempted to inform Mission Control about the tremendous view he saw, but was unable to establish radio contact. Failing on the high-frequency wavelength, Grissom returned to the ultrahigh-frequency wavelength; glancing down into the periscope, he was able to see the retropack floating away from the spacecraft. The spacecraft was pitched to the 14-degree reentry attitude. Loads on *Liberty Bell 7* increased as it plunged through the upper atmosphere. Eventually Grissom experienced 10.2 g, slightly less than Shepard had been subjected to aboard *Freedom 7*.

At an altitude of 20 kilometers, Grissom reported being able to see white, wispy contrails. The drogue parachute deployed to stabilize the spacecraft. The main parachute was deployed 33 seconds later at an altitude of 3,300 meters. Grissom watched the main parachute blossom into its full shape, noting a small triangular rip and a quarter-sized hole as its only imperfections. Through his periscope, Grissom could see the Atlantic Ocean's surface rapidly approaching and prepared for impact. Splashdown came at T plus 15 minutes, 37 seconds. Grissom had traveled 488 miles down the Atlantic missile range from lift-off to splashdown. He had been weightless for 5 minutes and 18 seconds, 37 seconds longer than Shepard.

Splashdown imposed no excessive loads on the spacecraft, and *Liberty Bell 7* tipped over on its left side. The parachute compartment, heavily laden with Grissom's reserve parachute, dropped below the water level. Grissom waited until the reserve chute package popped above the water and then ejected it. Thirty seconds later, the capsule turned itself to an upright position. The waves were gentle, and Grissom was in good shape and in contact with recovery helicopters. Meanwhile, Grissom disconnected his suit attachments, except for an inlet hose that provided cooling, disconnected his helmet, and rolled up the suit's neck dam. He then began reading the items on his postsplashdown switch checklist. As a recovery helicopter closed in on *Liberty Bell 7*, Grissom removed the cover on the detonator for the explosive bolts on the hatch and pulled out the safety pin. (To activate the explosive hatch, a plunger off the astronaut's right shoulder had to be pulled.) Grissom sat back and awaited instructions from the recovery helicopter.

Unexpectedly, the spacecraft hatch blew off. Grissom, seeing water pouring in through the open hatch sill, instinctively removed his helmet, grabbed the instrument panel, and thrust himself into the water. With his suit neck dam up, Grissom would have been able to float easily. Unfortunately, he failed to close one suit port, through which water entered. The helicopter crew grappled the loop on top of the spacecraft, but *Liberty Bell 7* filled with water. The recovery helicopter, unable to keep *Liberty Bell 7* afloat, cut its attachment line loose, allowing the spacecraft to sink. Now the recovery forces turned their attention to Grissom, who was struggling to stay afloat because of excess water in his suit. A horse collar was then dropped to Grissom, and he was extracted from the ocean and flown to the recovery vessel USS *Randolph*.

Knowledge Gained

Because of the brevity of the mission, there were no purely scientific investigations included in the flight plan; all knowledge gained on MR-4 was biomedical or engineering in nature. Grissom was assigned fewer spacecraft tasks to perform than Shepard so that he could use more time for observations through his trapezoidal window.

Apart from feeling somewhat tired and having a mild sore

throat, Grissom was in excellent shape after his flight. His first postflight examination was conducted on the USS *Randolph*, 15 minutes after splashdown. Grissom appeared tired and breathed heavily, which could have been attributed to his ordeal in the water following splashdown. After a brief nap and a breakfast meal, Grissom was flown to Grand Bahama Island for more tests. Three hours after splashdown, his appearance and vital signs were much improved, and he was subjected to more medical, psychological, and neurological examinations over the next 48 hours. As a result of the medical evaluations, a number of conclusions were reached. Grissom was in good health both before and after MR-4, and his immediate postflight examination results were consistent with exertion and exposure to salt water. No specific functional abnormality was found to be attributable to spaceflight stresses, and there were no biochemical alterations.

The flight profiles of MR-3 and MR-4 were quite similar, but Redstone performance differed slightly. Grissom's Redstone engine burned 0.3 second in excess of prediction, which pushed the craft 4.8 kilometers higher and 14.4 kilometers farther downrange than scheduled. Shepard's Redstone engine burned a half second short of schedule.

The loss of the spacecraft forced a review of Mercury recovery plans. Navy and Marine recovery forces expressed a need for checklists with specific detailed recovery functions and responsibilities. Use of larger, more powerful helicopters, availability of flotation gear for helicopters, and use of flotation collars to be attached by frogmen to the spacecraft were given consideration for future Mercury flights as a result of the MR-4 experience. *Liberty Bell 7* need not have been lost. The recovery helicopter crew cast the spacecraft loose when a warning light indicated overheating of the helicopter's engines. Later examination revealed no problem with the engines; the warning light was faulty. There was no official NASA or military recovery effort for the lost $2 million spacecraft. Although amateurs have on occasion attempted to arrange recovery efforts, *Liberty Bell 7* resides on the ocean floor in 4.5 kilometers of water.

Context

NASA's Project Mercury was a measured response to the perceived and very real potential of the Soviet Union to place the first man in space. This suspicion was largely based on the heavier lift capability of existing Soviet boosters, as proved by some of the early Sputnik launches. The Atlas Intercontinental Ballistic Missile (ICBM), already in development, was selected to place Mercury astronauts into orbit. Recognizing that the Mercury-Atlas (MA) rocket would not be safe and reliable until mid-to-late 1961, NASA pursued a parallel program to gain manned spaceflight operational status

as early as possible, in order to beat Soviet cosmonauts into space. This parallel program, Mercury Redstone (MR), used a modified Redstone Intermediate Range Ballistic Missile (IRBM) as a booster to propel the Mercury spacecraft at suborbital speeds on ballistic arcs originating at Cape Canaveral and terminating with a splashdown in the Atlantic Ocean. MR-4 was the last mission of the MR program. In mid-August, 1961, NASA announced that additional planned manned suborbital flights would be canceled. Spacecrafts number 15 and 16, originally manufactured for two extra MR flights, were diverted to the MA program. NASA said all MR program test objectives had been achieved with the conclusion of MR-4, *Liberty Bell 7*.

MR-1 was a miserably disappointing failure. When ignition was commanded on November 21, 1960, a cloud of smoke surrounded the launchpad and a piece of hardware accelerated rapidly through that cloud. Only the Mercury capsule's launch escape tower had lifted off. The parachute canister popped off and the parachutes were ejected, yet both the booster and spacecraft remained fixed to the pad. The capsule was repaired and given a new booster, parachute compartment, and launch escape tower. MR-1A was successfully launched unmanned on December 19, 1960. Although experiencing slight over-acceleration, the spacecraft demonstrated the entire suborbital flight profile.

MR-2 was launched on January 21, 1961, with a chimpanzee named Ham as its passenger. Ham was a biomedical test specimen to collect data on the effects of weightlessness and launch and reentry stresses. He was subjected to as much as 18 g when the launch vehicle followed a steeper flight path than desired, resulting in abort conditions. Ham reached an apogee of 251 kilometers and landed downrange 77 kilometers farther than planned; he was not injured during this severe Mercury spacecraft test.

These two MR flights experienced Redstone booster over-accelerations. At the suggestion of Wernher von Braun, another unmanned MR flight test using a boilerplate Mercury spacecraft was inserted into the program schedule before attempting a manned MR flight. Designated MR-BD, this test took place on March 24, 1961. The design changes to correct the over-acceleration problem also corrected booster performance, clearing the way for manned Mercury suborbital flights.

Before Project Mercury could launch astronauts, the Soviet Union had sent a man, Yuri Gagarin, into Earth orbit. Nevertheless, Project Mercury proceeded with a pair of manned suborbital flights, MR-3 and MR-4. Despite these successes, the disparity between Mercury and Vostok was underscored when cosmonaut Titov spent a full day in space later in 1961.

Bibliography

Carpenter, M. Scott, L. Gordon Cooper, Jr., et al. *We Seven*. New York: Simon & Schuster, 1962. Written by the astronauts in conjunction with the staff of *Life* magazine, this book portrays the original Mercury astronauts the way they were and continue to be revered. Includes photographs.

Grimwood, James M. *Project Mercury: A Chronology*. NASA SP-4001. Washington, D.C.: Government Printing Office, 1963. Chronologically traces general developments of manned spaceflight capability, the research and development of Project Mercury, and the operational phase of the Mercury program. Numerous diagrams, photographs, and charts. Includes eight appendices essential to a total understanding of Mercury's achievement.

Link, Mae Mills. *Space Medicine in Project Mercury*. NASA SP-4003. Washington, D.C.: Government Printing Office, 1965. Prepared by the Office of Manned Space Flight, this text traces aviation medicine through Mercury bioscience and biotechnology. Provides actual flight biomedical data. Photographs and references abound.

NASA Manned Spacecraft Center. *Results of the Second U.S. Manned Suborbital Space Flight, July 21, 1961*. Washington, D.C.: Government Printing Office, 1961. Provides a summary of the Mercury program, *Liberty Bell 7* biomedical data, pilot training and evaluation of pilot in-flight performance, and Grissom's flight report. Charts, diagrams, tables, and photographs.

Olney, Ross R. *Americans in Space*. Nashville: Thomas Nelson, 1973. General summaries of American spaceflights from *Freedom 7* through Apollo 13. Perfect for younger readers and less informed spaceflight aficionados. Contains photographs, appendices, and a glossary. Gives numerous quotations from mission air-to-ground conversations.

Silverberg, Robert. *First American Into Space*. Derby, Conn.: Monarch Books, 1961. Published shortly after Mercury-Redstone 3, this paperback book demonstrates the American people's response to the United States' first steps in space: one of relief and pride. Although it details the flight of *Freedom 7*, the book provides personal biographies of astronauts Shepard, Grissom, and Glenn.

Swenson, Loyd S., Jr., James M. Grimwood, and Charles C. Alexander. *This New Ocean: A History of Project Mercury*. NASA SP-4201. Washington, D.C.: Government Printing Office, 1966. Excellent NASA History Series text. Fascinating report of research and development, management, and flight operations of the Mercury program. Extensive photographs, appendices, and references. An essential reference for serious spaceflight researchers.

David G. Fisher

MERCURY-ATLAS 6

Date: February 20, 1962
Type of mission: Manned Earth-orbiting spaceflight

The Mercury-Atlas 6 (MA-6) mission, better known as the flight of Friendship 7, was the United States' first manned Earth-orbiting flight. John H. Glenn, Jr., the astronaut on board, proved that man could work in a zero-gravity environment and that the Mercury spacecraft design was sound.

PRINCIPAL PERSONAGES

JOHN H. GLENN, Pilot, the first American to fly in
 Earth orbit
M. SCOTT CARPENTER, backup Pilot
ROBERT R. GILRUTH, Manager, Project Mercury
CHRISTOPHER C. KRAFT, JR., Flight Director
WALTER C. WILLIAMS, Mission Director
WILLIAM M. BLAND, JR., the chief designer
 of the capsule

Summary of the Mission

Until 1962, the United States had been second to the Soviet Union in the international space race. The Soviets, after orbiting the world's first satellite and sending the first probe to the Moon, captured the imagination of millions with the first manned orbital mission. Yuri Gagarin became instantly famous after his one-orbit mission, which was soon upstaged by Gherman Titov's seventeen-orbit flight. By the end of 1961, the Soviets had accumulated more than twenty-seven hours of flight time, the Americans a scant thirty-one minutes. (Actually, the chimpanzee Enos flew a two-orbit Mercury mission in November, 1961. This event prompted one cartoonist to depict Enos walking away from the spacecraft, helmet in hand, saying, "We're a little behind the Russians and a little ahead of the Americans.") Clearly, 1962 was the year for Americans to come from behind, the year in which the United States could at last demonstrate its prowess in space.

Forty-one-year-old John H. Glenn would join the likes of Charles A. Lindbergh and the Wright brothers as one of the world's most popular adventurers. Serving as backup astronaut to both Alan B. Shepard and Virgil I. "Gus" Grissom in their brief suborbital flights, Glenn was the natural choice to make the first orbital mission. Scott Carpenter would be his backup pilot.

Mercury spacecraft number thirteen was not much different from that flown by Grissom in July, 1961. A classic cone-shaped capsule, the black spacecraft measured 2.9 meters high and 1.89 meters across the base. On its nose perched the launch escape system, a fiery red solid-fueled rocket balanced atop a ladderlike tower. This device was designed to propel the spacecraft away from the launch vehicle in case of an emergency. The spacecraft weighed 1,935 kilograms at lift-off. Once in orbit, its weight would drop down to 1,355 kilograms. It would weigh 1,099 kilograms upon landing. (In comparison, the Soviet's Vostok spacecraft weighed 5,000 kilograms.)

Glenn's launch vehicle would be the venerable Atlas. First developed in 1951, the Atlas would prove to be the longest lived of all American missiles, still launching satellites up through the late 1980's. This "workhorse" of the American space program was 23 meters long and had a thrust of more than 2.45 million newtons at lift-off. The Atlas is unique in that it is virtually an aluminum "balloon." To keep the booster lightweight, its metal skin was made so thin that it could not even support its own weight. Therefore, the rocket was pressurized at all times, inflated like a balloon, to keep it rigid.

The original — and optimistic — schedule for Project Mercury had the first orbital flight slated for April, 1960. This flight was rescheduled to December, 1961, a month after the successful mission of Enos the chimpanzee. Soon other problems caused the flight's postponement to January 16, 1962, and later to January 23. Weather problems continued to delay the flight. On January 27, all looked well, and Glenn suited up and climbed into the capsule. Five hours later he was instructed to climb back out. Questionable weather had caused another delay. More weather problems resulted in further postponements. Finally, almost one month later, on February 19, the countdown began.

Glenn's three-orbit flight would help verify and evaluate the integrity of the Mercury spacecraft's design. Glenn would also make astronomical observations while on the dark side of Earth. These data would give researchers an idea of how the stars might be used for navigational purposes. While in daylight, he was to observe weather patterns and take photographs. Also, many questions regarding man's capabilities in the weightless environment remained unanswered: Would it affect his judgment, his eyesight, his sense of balance?

Glenn was awakened at 2:05 A.M. on February 20, 1962. After donning his spacesuit, he slid into the couch of the spacecraft at 6:03 A.M. At 7:00 A.M., weather forecasters at Cape Canaveral gave a favorable report for the flight. Fifty-five minutes before launch time the tower was rolled back, exposing the gleaming rocket.

At 9:37 A.M. the Atlas engines came to life, propelling John Glenn into space. "The clock is operating, we're under way," Glenn called. At 2 minutes and 10 seconds before lift-off, the two outer engines shut down and dropped away while the center, so-called sustainer, engine continued firing for another few minutes. Glenn was pressed back into his seat with a force of nearly eight times Earth's gravity. In only five minutes he was in a 158-by-256-kilometer orbit, circling Earth.

Once in orbit, Glenn immediately began status checks and measurements. The astronaut tested the maneuverability of the spacecraft. As he quickly moved from one tracking station to another, he commented on the beauty of the sunset, the blackness of the sky, and the brilliance of the stars. Over the Indian Ocean he reported again on the stars, the constellations he could recognize, and a high-altitude band of light in the atmosphere. He watched the Moon rise above the darkened horizon, its soft white light illuminating the ground below him. During all this time, Glenn's heartbeat stayed a calm eighty to eighty-five beats per minute.

Starting a tradition which would continue for years, the residents of Perth, Australia, turned on all of their lights for the orbiting pilot. At fifty-five minutes into the mission, Glenn reported to capsule communicator Gordon Cooper, "Just to my right I can see a big pattern of lights."

It was during his first sunrise that Glenn reported one of the most intriguing observations of his flight, the mysterious "fireflies" that surrounded his spacecraft. These brilliant lights were later found to be particles of ice that had broken away from the spacecraft.

Beginning with the second orbit, Glenn reported a slow sideways drift at about 1 degree per second. This drift was the result of a failed thruster, a thruster that was to be used for fine motions. When Glenn reached a certain attitude, more powerful thrusters would automatically fire. Thus, he switched over to manual control, a method that consumed fuel much more quickly but that allowed for more precise navigation.

At about this time, a portion of telemetry code indicated a problem with the landing system. Because of the forces of splashdown, the spacecraft engineers designed an air bag to be used to cushion the impact. Tucked under the heatshield, this air bag would normally be deployed well after reentry. Yet the telemetry code suggested that the landing bag had already deployed, meaning that the heatshield was loose. A loose heatshield could have subjected Glenn to temperatures of up to 1,600 degrees Celsius.

The decision was made not to inform Glenn of this malfunction until later so as to avoid distracting him. Still, the test pilot's opinion was needed to diagnose the problem. He was asked whether he heard strange sounds (indicating the loose heatshield), whether the landing bag indicator light was on (it was not), and whether certain switches were in the right positions.

Christopher C. Kraft, Flight Director, immediately devised a plan to leave the rear-mounted retro-rockets on during reentry. Normally, the package of three rockets, strapped to the heatshield, is jettisoned directly after firing. In this case, Mission Control hoped that these retro-rockets would hold the heatshield in place. Of primary concern were the aerodynamics of the spacecraft as it hit the atmosphere. This configuration had never been tested.

Meanwhile, Glenn continued watching "fireflies," taking photographs of sunrises and sunsets, and reporting data to ground controllers. It was not until his last pass over Hawaii that Glenn was finally informed of the potentially life-threatening problem. He was given new landing instructions: He would need to use manual control for the reentry sequence.

The retro-rockets fired at 4 hours and 32 minutes after lift-off. Soon Glenn was hitting the upper atmosphere. He heard the roar of the air as it began to tear at the titanium shingles of the spacecraft, the noises that sounded like "small things brushing against the capsule." One of the straps from the retropackage broke loose, swinging around and slapping the spacecraft's window. Large chunks of the package flew by the window, and an orange glow crept up the side. Glenn noticed that the interior of the spacecraft was becoming hotter, and for a moment he thought that he might not make it. The ionized air prevented any communications with the outside. Glenn was completely alone in his white-hot spacecraft.

With two minutes left before the drogue parachute was to be deployed, the fuel was exhausted. At an altitude of 8.5 kilometers, the drogue was finally released, stabilizing *Friendship 7* for the main parachute, which was released soon afterward.

John Glenn landed eleven kilometers from the recovery ship, the destroyer *Noa*. Within seventeen minutes, *Noa* had pulled alongside the small bobbing capsule and the recovery crew quickly winched the spacecraft aboard. Glenn opened the hatch and stepped into the cool Atlantic air.

Knowledge Gained

John Glenn's mission again demonstrated that man could go into space, perform useful duties, and return with no ill effects. The spacecraft design was tested in real-life flight situations.

One of the most important aspects of any mission is the astronaut's ability to control his spacecraft. Glenn's flight showed that, by using external references, the pilot could yaw (move left and right) the vehicle more easily than he could pitch or roll it. Furthermore, it was easier to yaw in the daylight than in the dark.

The utility of having a pilot on board was clearly demonstrated when the thruster problem arose. Had the flight been unmanned, it would have been terminated early because of excessive fuel use. Furthermore, Glenn's on-site observations of the heatshield problem provided useful data that could not be gathered from the ground.

Glenn took six spectrograms of the Orion region of the sky. They were used to determine the kind of light that could pass through the window. This information would help compensate for spectral distortion in the future. Overall, Glenn took only two rolls of film and produced rather poor photographs. His photographs did reveal cloud patterns, ground details, sunrises and sunsets, and the so-called fireflies (later called the Glenn effect).

Glenn reported that clouds of all types were easily seen with the naked eye. Cities were clearly visible on both the dayside and nightside. Colors appeared much as they would from a high-altitude aircraft. On the dark side of Earth, Glenn could observe clouds, weather fronts, and land masses in the moonlight. Without the Moon, looking at Earth was like "looking into the black hole at Calcutta." Lightning was seen in storms over the Indian Ocean.

From the in-flight problems of the thrusters and heatshield much was learned about coordinating ground control with the pilot's observations. The decision to withhold information about the possible landing bag problem was later deemed imprudent. It was determined that for future missions, the pilot would be informed of problems as quickly as possible.

The medical team concluded that spaceflight had no ill effect on man, or that whatever physical effects the flight might have had on Glenn were so short-lived that they disappeared before the postflight medical examination.

Context

The flight of *Friendship 7* was the first of four manned Earth-orbiting Mercury flights. It proved that man had a place in space, and it disproved the theory that space travel is only safe for robots. Glenn's flight also demonstrated that the basic design of the spacecraft was sound and could be used to make future designs.

Because of the unusual environment, Glenn felt rushed to perform all of his duties. Subsequent flight plans would be reworked to allow the astronaut more time to perform his tasks. It was decided that the first orbit should be used both for systems checkout and to allow the pilot to get accustomed to the space environment. Air-to-ground communications were simplified based on Glenn's having to report all switch positions and gauge readings twice during each orbit. The mission flight plan would be finalized earlier so the pilot could become more familiar with the schedule.

The next flight, MA-7, would be a virtual repeat of Glenn's, used to gain additional experience for manned spaceflight operations and to test spacecraft modifications. MA-8 would orbit Earth six times instead of three and perform more of the same experiments. MA-9, the last mission, was thirty-four hours long; designed to test man's endurance in space, it lasted for twenty-two orbits.

In 1961, the McDonnell Douglas Corporation, the Mercury's manufacturer, was asked to begin development of the next generation of manned spacecraft. This two-man Mercury, later known as Gemini, would be built using the experience gained with the original Mercury.

John Glenn retired from the National Aeronautics and Space Administration (NASA) in 1964. Being the oldest of The Seven (the name by which the seven original Mercury astronauts were best known collectively), he had little prospect of any new flight assignments. The Gemini astronauts would be training for Apollo flights, and by that time he would be nearing fifty years old. Ten years after his retirement from NASA, John Glenn was elected senator from his home state of Ohio. In 1984, he ran an unsuccessful campaign for president on the Democratic ticket.

Bibliography

Baker, David. *The History of Manned Space Flight*. New York: Crown Publishers, 1982. This comprehensive book, more than 530 pages long, details all manned spaceflight programs up to 1982. Contains hundreds of photographs. Recommended for anyone with even a passing interest in space history.

Gatland, Kenneth. *The Illustrated Encyclopedia of Space Technology: A Comprehensive History of Space Exploration*. New York: Crown Publishers, 1981. A general reference book accessible to the nontechnical reader. Its many sections cover such areas as space pioneers, space centers, and manned missions to Mars.

Swenson, Loyd S., Jr., James M. Grimwood, and Charles C. Alexander. *This New Ocean: A History of Project Mercury*. NASA SP-4201. Washington, D.C.: Government Printing Office, 1966. This NASA publication traces the history of Project Mercury through Gordon Cooper's flight in *Faith 7*. Both the technical and managerial sides of the program are discussed. Many rare photographs and some diagrams are included.

Thomas, Shirley. *Men of Space: Profiles of the Leaders in Space Research, Development, and Exploration*. 5 vols. Philadelphia: Chilton, 1960. Covers the people who managed the early development of Project Mercury. Includes biographical material on Robert Gilruth and other members of the Space Task Group.

Wolfe, Tom. *The Right Stuff*. New York: Farrar, Straus and Giroux, 1979. This narrative about Project Mercury and the Mercury astronauts looks at how instant hero status affected both the pilots and their families. Investigates what kind of men would risk their lives to fly in space.

Michael Smithwick

MERCURY-ATLAS 7

Date: May 24, 1962
Type of mission: Manned Earth-orbiting spaceflight

Mercury-Atlas 7 was the first U.S. space mission devoted to manned scientific research in space. This second orbital mission for the United States, flown by Scott Carpenter, investigated the capability of man in space to perform various experiments, ranging from photographic trials to investigations of the characteristics of Earth's atmosphere at the edge of space for navigational purposes.

PRINCIPAL PERSONAGES
M. SCOTT CARPENTER, Pilot
JAMES E. WEBB, NASA Administrator
D. BRAINERD HOLMES, NASA's Director of Manned
  Space Flight
KURT H. DEBUS, Director of NASA's Cape Canaveral
  launch facilities

Summary of the Mission

American manned spaceflights that preceded the Mercury-Atlas 7 (MA-7) flight were conducted primarily to determine fundamental questions, such as vehicle performance and human ability to survive in space. Following the examples of Alan B. Shepard, Virgil I. "Gus" Grissom, and John H. Glenn, M. Scott Carpenter in MA-7 was to inquire into a whole new set of questions about manned spaceflight. For the first time, the agenda had been cleared for basic scientific research in an American manned spacecraft.

Carpenter was to be allowed three orbits, requiring an in-orbit flight time of approximately four and one-half hours. During this period, he was assigned an assortment of scientific experiments in addition to piloting and attending to the spacecraft.

During his mission, he would deploy an inflated, tethered balloon from the capsule to study the tenuous atmosphere at his orbital altitude as well as the reflectivity of sunlight on the balloon's surface. Carpenter would also be making weather observations from orbit; his camera was equipped with special filters and film types. His scientific itinerary also included the study of liquids in a gravitational force of zero (the apparent "weightless" environment of orbital free-fall), an investigation of the luminous band of light around Earth's horizon as seen from space (which would be used for future navigational references), looking for flares set off on the ground, testing the human ability to use star sights for space navigation, and

investigating a phenomenon discovered on Glenn's flight, "space fireflies." Carpenter's schedule was literally packed with tasks that needed attention every minute, a fact that would cause eventual problems during the flight.

The first launch date of MA-7 had been set for May 15, 1962. Problems with the Atlas booster rocket delayed the flight, however, and then the Mercury capsule (named *Aurora 7* by Carpenter both because it represented the "dawn of a new age" and in honor of Project Mercury, which he described as "a light in the sky") developed problems with its attitude control system. The difficulties were corrected, and at the same time a modification was developed and installed in the parachute deployment system. Finally, May 24, 1962, was established as the launch date.

Carpenter lifted off Launchpad 14 at Cape Canaveral, Florida, at 7:45 A.M. on May 24, 1962, in the United States' second orbital flight. Five minutes after leaving the Florida launch tower, the Atlas missile completed its task of placing MA-7 into orbit and separated, falling away from the capsule. After only five minutes of flight, the Mercury capsule had attained orbital velocity, flying at a speed of 28,000 kilometers per hour.

Carpenter immediately began to follow his flight plan, turning the spacecraft around. Then he felt compelled to maneuver his capsule in space, discovering that there was absolutely no sensation of speed or even disorientation as he inverted *Aurora 7* to look down at the world passing beneath him. Yet Carpenter had much work to accomplish, and with only three orbits in which to do so, he set about his assigned tasks.

He maneuvered his capsule busily, attempting to see the flares over darkened Australia, but the country was obscured by clouds. He was forced by the requirements of his many photographic assignments to place the capsule in a variety of positions — so many, in fact, that on several occasions he used excessive fuel in trying to maneuver the capsule into position as rapidly as possible. He also made a mistake on several occasions, activating the manual and automatic thruster modes simultaneously and thereby wasting even more fuel. In these terms, the first orbit was to be costly.

Carpenter got his first look at the firefly phenomenon on his first encounter with night. He described them as resembling snowflakes and hypothesized that they were "frost from a thruster"— a speculation that would prove surprisingly accurate.

Launch of the Mercury-Atlas 7 mission. (NASA)

A regulator inside his suit had not been able to keep up with its rising temperature. Although the heat in his suit never rose above 36 degrees Celsius (96.8 degrees Fahrenheit), the system designed to remove the humidity did not function properly. In addition, Carpenter's biological temperature monitor was malfunctioning, indicating to worried ground controllers a body temperature of 38.8 degrees Celsius (101.84 degrees Fahrenheit). A persistent problem throughout his flight, the heat would account for his losing seven pounds of body weight to perspiration.

Carpenter released the balloon at 1 hour and 38 minutes into the flight; he would later describe it as "a mess." It did not inflate properly, did not deploy behind the spacecraft as predicted, and in the end refused to release when commanded. Carpenter trailed the balloon until retrofire, "like a tin can attached to the bumper of a car." Although he confidently attempted some analysis of its deployment, it was impossible to obtain any meaningful drag data from it, and because of its partially inflated state, reflection data were for all practical purposes useless. Nevertheless, the unexpected results themselves would yield valuable scientific data.

By the final orbit, Carpenter was genuinely low on fuel. He set the spacecraft on an automatic control system to place it in position for firing of the braking rockets (retro-rockets). In doing so, he found that the automatic system was malfunctioning, wasting even more of his dangerously depleted fuel. He switched back to the manual mode. With the two systems operating at once, the fuel requirements were doubled. Carpenter would need what little fuel he had remaining to orient the capsule for reentry into Earth's atmosphere. If he ran out before this procedure was completed, he would enter at the wrong angle and could burn up. Worse, he might not be able to orient the capsule to fire the retro-rockets and become stranded in space as a result.

To complicate matters further, just as Carpenter was orienting the spacecraft for return, he bumped the side of Aurora 7 with his hand, and a series of the mysterious fireflies flew past the window. Then he bumped the walls again, releasing even more of them, which he identified conclusively as frost particles.

Nevertheless, the moment of reentry was near at hand, and as Carpenter, dwelling on the fireflies, completed the orientation of his capsule and fought the automatic control problem, the short countdown on the retrofire began. The firing was supposed to be automatic, but it never came. Waiting a few seconds, Carpenter finally hit the manual fire switch; the rockets fired three seconds late.

Retrofire was not only three seconds later than planned, but because of the problems with his automatic control system, the spacecraft was oriented 25 degrees off center and the rocket itself delivered less thrust than planned. These seemingly small factors magnified themselves at 28,000 kilometers per hour and acted to bring the spacecraft in 463 kilometers off target.

Though Carpenter had fired his retro-rockets and his craft had begun the long fall to Earth, he was not home yet. He still needed fuel to orient the capsule for the interface with the atmosphere so that the capsule would descend in a controlled manner. He used fuel from his automatic tanks to assist efforts at maintaining control just before he reentered the atmosphere.

The reentry itself was affected by the sparing use of fuel. Aurora 7 began to oscillate excessively, and as the spacecraft buffeted from side to side, Carpenter heard the craft "bang and whump." He finally used every last drop in fuel in an attempt to restrain these motions. Aurora 7 then began wide oscillations in an arc of about 270 degrees, and Carpenter began to fear that the capsule would invert and go completely out of control. Thus, at 7,900 meters, he deployed his drogue chute to control the motions of the spacecraft. Carpenter splashed down northeast of the Virgin Islands, five hours after lift-off.

Because he had overshot his targeted landing zone, Carpenter would have a long wait in the water. He exited his craft to a life raft, where he waited, in the company of two Navy recovery divers. He was transported safely to the deck of the USS *Intrepid* some four hours later.

Knowledge Gained

In no previous space mission had a manned space vehicle been maneuvered in space so extensively as had Carpenter's. Not only was *Aurora 7* loaded with experiments to perform, but Carpenter's automatic control system malfunctioned as well, requiring even more maneuvering. He literally pushed the tiny space capsule to its maximum abilities, providing practical knowledge on the limits and performance of the Mercury spacecraft system.

Carpenter and MA-7 represented the first manned space mission dedicated primarily to science. In past space voyages, tolerance of the space environment had been a fundamental area of inquiry with regard to both human and machine. MA-7 was the first mission, however, to place a man in space with the express aim of discovering, in a scientific sense, the effects of an unknown environment. MA-7 proved that humans could achieve Earth orbit, survive there, and maneuver a space vehicle to obtain useful scientific data.

Carpenter discovered that cloud cover obscured any attempt to communicate with him by flares from the ground. While seemingly a commonsensical discovery, it did prove that high altitudes of orbit affected visual observation in much the same way as lower ones. Until Carpenter's mission, that idea was only conjecture.

The mission of MA-7 also proved that the density of the atmosphere was so insignificant that when traveling at 28,000 kilometers per hour at an altitude of 268 kilometers there was not enough drag to stretch out the balloon. This was a surprise to the experiment's designers, who had wanted Carpenter to report on the effect of the drag. MA-7's finding would influence nearly all space missions to follow and would ultimately affect the design of future space stations.

Carpenter provided some of the first meteorological photographs of Earth taken from space. These images were taken with several filters that would allow later comparison of weather formations and would be instrumental in determining the design of satellite requirements ranging from weather satellites to the Earth resources satellites. The human ability to change photographic filters and aim at specific features was at that time not possible using unmanned vehicles. MA-7 provided badly needed yet otherwise unobtainable data.

Carpenter solved the substantial space mystery of the day. On an earlier flight, Glenn had described what appeared to be luminous fireflies dancing outside the window of his spacecraft. One of Carpenter's assignments was to find out more about these mystifying objects, which he discovered to be flakes of frost that had accumulated on the sides of the space vehicle. Again, the ability of a man in space to make these on-the-scene evaluations was well beyond the capability of any unmanned probe.

Finally, Carpenter was assigned two vital navigation tasks. He was to see if he could spot specific stars from the window of his spacecraft and to investigate the band of light at Earth's visible horizon. He was able to distinguish Earth's horizon clearly, and he was able to spot the target stars distinctly, proving that man would someday be able to use these guideposts to navigate in space.

Context

MA-7 was the sixth manned space mission flown in history, the fourth space mission launched by the United States, the United States' second orbital mission, and the first mission devoted substantially to scientific research in space. The second manned mission to fly into space atop an Atlas booster rocket, MA-7 still holds the record for missing its intended landing zone, by some 463 kilometers. Nevertheless, Carpenter's flight proved conclusively not only that man could achieve orbit and survive but also that he could work there and collect data.

By early 1962, there had been few spaceflights at all. The question of simple survival had been answered conclusively, first by the Soviet Union some thirteen months earlier with the flight of Vostok 1. Since the Soviets were not sharing their information, however, American scientists could rely only on their own data. The Cold War between the United States and the Soviet Union had been heating up, and the high ground of space was being eyed for military use. The U.S. Air Force had already initiated Dyna-Soar, a planned space laboratory in which it would conduct military exercises on a nearly continuous basis. Dyna-Soar was replaced by the Manned Orbiting Laboratory, which would be canceled some seven years later.

In 1962, when Carpenter flew on MA-7, the United States and the Soviet Union had just entered the so-called space race, initiated one year earlier by President John F. Kennedy's challenge to land a man on the Moon by the end of the decade. This proclamation gave the nation something for which to strive, a goal: to reach the Moon and return safely. Although this was the focus of the space program in the early 1960's, other vital space applications were being developed as well. The first meteorological, Earth resources, military, and navigation satellites were also being developed and launched.

Thus, the mission of MA-7 came at a time of energetic space science development, of which there were two primary focal points: manned spaceflight leading to a landing on the Moon, and unmanned investigations of many varieties, ranging from Earth resources to military surveillance. Carpenter's mission benefited both goals substantially by providing key pieces of information.

Bibliography

Baker, David. *The History of Manned Space Flight.* New York: Crown Publishers, 1982. This work offers a precise chronology of the history of manned spaceflight, although it is heavily oriented toward the American effort, to the beginning of the space shuttle program. It gives detailed analyses of all the Mercury flights.

Carpenter, M. Scott, L. Gordon Cooper, Jr., et al. *We Seven.* New York: Simon & Schuster, 1962. This book is a delightful, personal, exciting accounting of the first four U.S. manned spaceflights, written by the Mercury astronauts themselves. Among other things, it gives Carpenter's personal reflections on the events leading up to his flight on MA-7 and the flight itself.

Clarke, Arthur C. *The Promise of Space.* New York: Harper & Row, Publishers, 1968. This work is a thoughtful, developmental history of manned spaceflight, seeded with Clarke's ideas of humanity's future movements in space. It details some of the early days of manned flights with the purpose of accounting for time in space as a step toward eventual permanent colonization. It can be enjoyed by most readers with at least a high school background and can help in the understanding of long-term goals in space.

De Waard, E. John, and Nancy De Waard. *History of NASA: America's Voyage to the Stars.* New York: Exeter Books, 1984. Although this essay on the history of NASA does not go into great detail on the Mercury program, it does offer an abundant collection of color photographs taken during the Mercury and other flights.

Hartmann, William K., et al. *Out of the Cradle: Exploring the Frontiers Beyond Earth.* New York: Workman Publishing, 1984. This extensively documented and artistically illustrated book defines humanity's quest for the stars, concentrating on futuristic endeavors. It presents, with dramatic insight, the tomorrow of space exploration and all the benefits humankind will reap. Directed toward a general readership.

Dennis Chamberland

MERCURY-ATLAS 8

Date: October 3, 1962
Type of mission: Manned Earth-orbiting spaceflight

The flight of Sigma 7, the Mercury-Atlas 8 mission, was the United States' third manned Earth-orbiting spaceflight. Astronaut Walter M. Schirra demonstrated that man could work in a zero-gravity environment for an extended time.

PRINCIPAL PERSONAGES

WALTER M. "WALLY" SCHIRRA, Pilot
L. GORDON COOPER, backup Pilot
ROBERT R. GILRUTH, Manager, Project Mercury
CHRISTOPHER C. KRAFT, JR., Flight Director
WALTER C. WILLIAMS, Mission Director

Summary of the Mission

With the success of two three-orbit flights by mid-1962, the United States decided that its space program was ready for longer Mercury missions. Since the Mercury spacecraft was designed for only a three-orbit, five-hour-long flight, several modifications on its design would be required. There would no longer be space for extra supplies. The atmospheric leak rate of the vehicle had to be cut from 1,000 to 600 cubic centimeters per minute. A new fuel-saving plan was implemented to minimize the use of attitude fuel. Excessive fuel use had been a problem that had plagued the flights of both John Glenn and Scott Carpenter. Finally, a new plan was developed to conserve electrical power; the pilot would turn off all systems that were not actually in use.

Thirty-nine-year-old Walter M. "Wally" Schirra was the pilot selected for the third Mercury orbital mission. Gordon Cooper was his backup pilot. This mission was essentially an engineering flight to test the new conservation measures, paving the way for still longer missions. Schirra was given few scientific experiments; instead, he was to concentrate on observing the behavior of the spacecraft and its systems. Schirra named his spacecraft "Sigma 7." "Sigma" is an engineering term for "summation," and this flight was to be a summation of all previous ones.

While preparing for the flight of *Sigma 7*, National Aeronautics and Space Administration (NASA) officials were apprehensive about the next steps the Soviet Union might take. The Soviets' last mission had lasted a full day in orbit. Americans believed that, with some luck, it might be possible to launch a daylong Mercury mission before any more Soviet

spectaculars were reported, effectively overtaking them. This pressure gave added importance to the long-duration *Sigma 7* flight. On August 11, 1962, however, the Soviets launched Vostok 3, with cosmonaut Andrian Nikolayev at the helm. A day later, Radio Moscow announced that Vostok 4 had lifted off, piloted by cosmonaut Pavel Popovich. For three days, the two Soviets sent live television images back to Earth, chatted with each other, and stole the world's headlines. The United States was clearly trailing in the space race. (The two cosmonauts landed within minutes of each other, after four and three days in space, respectively.)

Because of the length of the mission, *Sigma 7* would fly over portions of Earth without the benefit of tracking stations. As Schirra's spacecraft traveled around Earth, the planet itself would revolve under him 27 degrees each orbit — causing his flight path to shift west. This shift would gradually take him away from the normal set of ground stations, leaving him with communications lapses of up to thirty minutes. Until Schirra's flight, ten minutes without ground communications had been the accepted maximum. Therefore, for this mission, three tracking ships were added to the network. The *Huntsville, Watertown,* and *American Mariner* were stationed close to Midway Island, near Hawaii. Another land-based station was added at Quito, Ecuador, to aid communications on the final orbit.

The *Sigma 7* would land in the Pacific Ocean, while all the previous missions had landed in the Atlantic Ocean. The Department of Defense assigned more than seventeen thousand personnel, eighty-three airplanes and helicopters, and twenty-six ships to recovery operations. The aircraft carrier USS *Kearsarge* was designated the primary recovery ship.

On the morning of October 3, 1962, Schirra climbed aboard *Sigma 7*. The automobile ignition key that had been hung inside the cabin by a good-natured technician added a playful spirit to the day's activities. Schirra made a quick inventory of the gear in the cabin: star charts, flightplan, camera equipment, and the like. Checking in the "glove compartment," where personal items were stowed, Schirra discovered a second gift from the ground crew: a steak sandwich.

The launch was scheduled for 7:00 A.M., but a radar malfunction at the Canary Island tracking station delayed the countdown by fifteen minutes. At precisely 7:15 A.M., the Atlas rocket was ignited. After only ten seconds had elapsed, however, the mission was nearly aborted. The Atlas had made an unexpected clockwise roll because of a misalignment of the booster engines. The vernier rockets (small attitude control motors) stopped the roll before it could trigger an abort condition. The remainder of the launch was uneventful, much less dramatic than Schirra had expected. The sustainer engine did, however, fire ten seconds longer than planned, putting *Sigma 7* in a slightly higher orbit than intended: some 167 by 283 kilometers, the highest orbit of all four Mercury orbital flights.

Schirra first cartwheeled the spacecraft around to watch the Atlas rocket drift into space. He then performed some manual attitude tests and began to monitor the spacecraft systems. While flying over Zanzibar, Schirra could feel his spacesuit become warmer. The suit's delicate environmental control system had sent its temperature soaring from 23.3 to 32.2 degrees Celsius. On the ground, Christopher Kraft, the flight director, suggested that the mission be terminated after one orbit. Schirra carefully adjusted the suit's control knob and managed to stabilize the temperature. Kraft gave permission for a second orbit. By the time Schirra was passing over Australia for the second time, the temperature had dropped to a comfortable 22.2 degrees Celsius.

Over Australia, Schirra was supposed to watch for flares that would be fired from the ground. As with past missions, the launching area was covered with clouds. Thus, nothing could be seen of the ground.

Another test was scheduled to verify the utility of a periscope mounted under the main instrument panel. Carpenter had claimed that it was useless, taking up much needed space and weight, but Schirra believed that it could be used for more accurate attitude alignment. At the post-flight press conference, however, it was revealed that Schirra had not been able to use the periscope effectively.

Over Muchea, Australia, on the second orbit, Schirra performed the night yaw experiment, designed to determine whether the spacecraft could be properly aligned by using the stars as a guide. The small window and incorrect star charts made it difficult to identify constellations. Nevertheless, ground telemetry showed Schirra's error to be no more than 4 degrees.

Approaching his flight over California for the second time, Schirra radioed to Capsule Communicator John Glenn that he saw the "fireflies," bright particles that Glenn had observed on his *Friendship 7* mission. Schirra also noticed that he could generate more "fireflies" by knocking the side of the spacecraft. He reported seeing the Salton Sea, Mount Whitney, and several roads in the Mojave Desert. Unlike most astronauts, Schirra was rather unimpressed with the view. Although he was ten times higher than he had ever flown before, the details and landscape looked to him like those viewed from high-altitude aircraft. "Same old deal, nothing new," he stated at the debriefing.

On the third orbit, Schirra performed an experiment sug-

gested a year before by his training officer. He closed his eyes and attempted to touch certain target points on the control panel. In nine trials, he made only three errors. Schirra concluded that weightlessness created no significant disorientation.

By the end of the third orbit, Schirra began to take photographs of various regions — from California to Cuba. He then measured radiation levels in the cabin, but the levels were so low that they were virtually unreadable. Kraft urged Schirra to look for the large silver Echo balloon satellite when he passed over Zanzibar, but he was unable to find it.

On Schirra's fourth orbit, he took more photographs, performed communications checks with the ships *Watertown* and *Huntsville*, and observed thunderstorms over Australia. Nearing California once again, at 6 hours and 8 minutes after lift-off, Schirra was given two minutes of live network radio time to report on his status to the entire world.

Flying over South America for the last time, Schirra bid the Quito, Ecuador, communications crew farewell as he prepared for reentry. The fuel conservation procedures had succeeded. *Sigma 7* still had 78 percent of its fuel left in its tanks. Less fuel had been used in six orbits than in the three revolutions on each of the previous two missions. Orienting for reentry consumed more fuel, and the supply dropped to 52 percent. With the blunt heatshield forward, the rockets fired one at a time at five-second intervals. The retropackage dropped away, and Schirra began his fiery fall toward the Pacific Ocean.

The friction of reentry heated the small capsule to more than 1,600 degrees Celsius. Schirra reported seeing a brilliant green glow outside the window. A retropackage strap still attached to the heatshield slapped against his window much as it had for John Glenn a few months before. The small drogue parachute opened at an altitude of 13 kilometers, stabilizing the spacecraft. The single main chute opened at 5 kilometers altitude. Some nine hours after lift-off, *Sigma 7* settled into the Pacific Ocean a mere 7.5 kilometers from the aircraft carrier *Kearsarge*. As had Glenn, Schirra elected to stay in the capsule until it was placed on the deck. The reporters shouted to him, "How do you feel, Wally?" He replied with a curt "Fine" and headed down to the ship's sick bay.

Knowledge Gained

The *Sigma 7* mission was intended to test different flight procedures so as to correct problems encountered during the previous two missions. The crowded flight plans during Mercury-Atlas 6 and 7 had rushed the astronauts too much, causing them to waste both fuel and energy. Schirra did not have to tend many scientific experiments, so he had more time for each flight test. Many of the new procedures on Mercury-Atlas 8 dealt with factors critical to long-duration space missions, specifically, the conservation of fuel and electrical power. Attitude changes were kept at a minimum, and those that were effected took more time than on previous flights. As a result, Schirra's first attitude change at 4 degrees per second (turning around to observe the detached booster), consumed only 0.14 kilogram of propellant, compared to 2.45 kilograms for the same maneuver on John Glenn's flight.

The bulky periscope was tested to see if it could aid in the adjustment of the spacecraft's attitude. When the periscope was used at high power, Schirra reported that it was only a bit faster, but not any more accurate than other methods. It was recommended that the periscope be dropped from future missions.

Various ablative panels were attached to the outside of the spacecraft. These were used to test the reactions of different materials to the reentry heat. None of the panels was clearly superior to the others.

Previous mission flight rules regarding a ten-minute maximum communications gap were changed, since the longer mission would carry Schirra farther from the prime tracking stations. The decrease in pilot communications time did not appear to hinder the mission in any way.

Psychomotor tests proved that an astronaut's spatial perception did not change significantly in weightlessness. In fact, only one significant observation concerning Schirra's physical condition was made. After the mission, Schirra's change in blood pressure and heart rate when standing and lying down was somewhat greater than it had been before the mission. This remained constant until one day after splash-down. In-flight exercise was recommended to counteract any such changes in heart rate caused by future long-duration missions.

Context

The *Sigma* 7 flight was called an "engineering evaluation" by Schirra. Drawing from the experience gained from the first two missions, NASA officials used *Sigma* 7 to verify new approaches required for long-duration flights. Many of these new approaches were tested further by Gordon Cooper in the final Mercury flight, the thirty-four-hour-long *Faith* 7 mission. The *Sigma* 7 experience would also affect the way flight plans were organized and pilots were trained. Furthermore, as a result of Schirra's mission, the pilot was given a more autonomous role, and the need for continuous communications became less urgent.

Compared with the Soviets, the Americans were still a distant second in the space race, with only a total of nineteen hours in space compared to the Soviets' eight days.

Wally Schirra would go on to command the fourth manned Gemini mission — Gemini VI with Thomas P. Stafford in December, 1965. The pair would perform the world's first rendezvous between two manned spacecraft. Three years later, Schirra would command the first manned Apollo flight, Apollo 7. Donn Eisele and Walter Cunningham accompanied him for the 163-orbit flight. With his participation in the Apollo program, Schirra became the only astronaut to fly in all three manned programs during the 1960's.

Bibliography

Baker, David. *The History of Manned Space Flight*. New York: Crown Publishers, 1981. A detailed reference work, more than 530 pages long and filled with photographs. Unlike many well-illustrated books, this source has a generous amount of text.

Caidin, Martin. *The Astronauts: The Story of Project Mercury, America's Man-in-Space Program*. New York: E. P. Dutton Co., 1960. Although dated, this book contains an excellent discussion of the early problems with Project Mercury. It focuses on the astronauts but also offers insight into the technical problems associated with the Mercury capsule and the Atlas launch vehicle.

Gatland, Kenneth. *The Illustrated Encyclopedia of Space Technology: A Comprehensive History of Space Exploration*. New York: Crown Publishers, 1981. A comprehensive book on all aspects of the space program, including the United States' efforts at international cooperation in space. Includes detailed sketches of the Mercury and Vostok spacecraft.

National Aeronautics and Space Administration. *Mercury Project Summary Including Results of Fourth Manned Orbital Flight, May 15-16, 1963*. NASA SP-45. Washington, D.C.: Government Printing Office, 1963. The official summary of Project Mercury, this book contains material on medical experiments and the performance of the booster and spacecraft. Technical in nature but suitable for a general readership.

Swenson, Loyd S., Jr., James M. Grimwood, and Charles C. Alexander. *This New Ocean: A History of Project Mercury*. NASA SP-4201. Washington, D.C.: Government Printing Office, 1966. Part of the NASA Historical Series. A complete and well-written account of all aspects of Project Mercury. Gives a good description of early technical problems, such as those with the capsule's heatshield.

Michael Smithwick

Mercury-Atlas 9

Date: May 15 to May 16, 1963
Type of mission: Manned Earth-orbiting spaceflight

The Mercury-Atlas 9 mission, which saw the flight of the Faith 7 *spacecraft, was the United States' fourth manned orbital flight. Astronaut L. Gordon Cooper demonstrated that man could work in a zero-gravity environment for more than one day.*

Principal personages

    L. Gordon Cooper, Pilot
    Robert R. Gilruth, Manager, Project Mercury
    Christopher C. Kraft, Jr., Flight Director
    Walter C. Williams, Mission Director

Summary of the Mission

On May 15, 1963, Leroy Gordon Cooper was positioned atop an Atlas booster in his Mercury spacecraft *Faith 7*. Cooper would shortly be launched on the most complicated Mercury flight ever. It would also be the last mission of the Mercury program.

Mercury-Atlas 9 (MA-9) got its start following the successful flight of MA-8 (*Sigma 7*), piloted by astronaut "Wally" Schirra. The Mercury spacecraft was originally built for a simple three-orbit, five-hour-long mission. With the success of the Soviet Union in long-duration flights lasting several days, not to mention delays with the United States' two-man Gemini spacecraft, it became apparent that a lengthy Mercury flight was in order. First, both the spacecraft and flight procedures had to be modified to support longer stays in space. These modifications were successfully tested by Wally Schirra in his six-orbit mission of October, 1962. Still, a new set of problems had to be solved for Cooper's flight, as it was to last for more than one day.

Faced with this task, engineers dissected the capsule, looking for every gram of excess weight that could be replaced with extra fuel, oxygen, and experiments. Based on Schirra's recommendations, the bulky periscope was the first item to be removed, saving 34 kilograms. Two backup radio transmitters and an extra automatic control system were also taken out of the spacecraft. These changes made room for more batteries, more cooling water, nearly double the amount of drinking water, and more oxygen. A total of 183 changes were made to the Mercury spacecraft. In the end, it weighed 1,963 kilograms — only 1.8 kilograms more than Schirra's *Sigma 7*.

Cooper's flight was a controversial one. Many wanted Mercury to end on the high note of Schirra's textbook-perfect mission. In fact, this feeling was so strong and the changes in Cooper's spacecraft so extensive that for a while MA-9 was considered a separate program. In a quarterly report on the Mercury program dated November, 1962, it was stated, "This report will be the final in the series of Project Mercury, as such, since the MA-8 flight was the last mission of Project Mercury." MA-9 was now called the Manned One-Day Mission, or MODM, project.

Another controversy arose when Cooper announced the name of his spacecraft. "Faith 7" was selected, symbolizing for Cooper his trust in God, his country, and his teammates.

Faith 7 would be a scientific mission. The previous Mercury flights had been much too short for serious scientific research. Cooper called *Faith* 7 a "flying camera," because it contained a 70-millimeter Hasselblad (the same camera used on the lunar missions), a 35-millimeter still camera, a 16-millimeter general-purpose motion-picture camera, and a television system, the first on an American manned flight. Experiments included the releasing of a flashing beacon from the rear of the spacecraft, which would test the pilot's visual ability to track other objects in space. There would be a tethered balloon to study air drag. Studies of the behavior of liquids in weightlessness would be made, and photographs were to be taken to aid in weather satellite design, celestial navigation, and studies of the upper atmosphere.

Cooper would experience long gaps in communications with ground controllers, much as Schirra had on his flight, because *Faith* 7 would be moving away from the established tracking network. Two new ships were added to fill in some of the gaps in communications: *Twins Falls Victory* and *Range Tracker*.

Thus, at 8:04 A.M. on May 15, the nearly $18 million Mercury-Atlas 9 spacecraft was launched. Five minutes later, *Faith* 7 was in an orbit of 267 by 161 kilometers. Cooper immediately began a slow turn to observe the Atlas booster tumbling behind his spacecraft. Using the fuel savings procedures that Schirra had pioneered, Cooper maneuvered his spacecraft. The *Faith* 7 consumed only 0.09 kilogram of fuel for that maneuver, compared to 0.14 kilogram on Schirra's flight and 2.45 kilograms on Glenn's flight.

Cooper was the most relaxed Mercury astronaut. He had been caught dozing in his capsule during countdown and was awakened several times from catnaps throughout the entire flight. On his first pass over Australia, he saw the lights of Perth and observed the so-called Glenn effect, the mysterious "fire-flies" first reported by John Glenn. He released the beacon on the third orbit but was unable to see it until the next orbit.

On the sixth orbit Cooper tried releasing the balloon, a sphere 75 centimeters in diameter attached to a 30-meter line. A strain gauge was to measure the differences of air drag between the highest and lowest points of the orbit, but the device failed for unknown reasons, just as it had when Scott Carpenter had tried to use it one year earlier.

Cooper performed the liquid experiments on the seventh orbit, transferring urine samples and water from tank to tank. The syringes he used were unwieldy and leaked. He observed that in microgravity the liquid "tends to stand in pipes, and you have to actually force it through." Next, he took mea-surements of the residual radiation from a July, 1962, atomic blast. These measurements would be compared with similar measurements taken by Wally Schirra seven months earlier.

Cooper spent much time performing vision tests. Early astronauts had reported the amazing clarity with which they could see objects on Earth. Schirra claimed to have seen indi-vidual roads and railroad tracks. Cooper caught sight of a 44,000-watt xenon light located at Bloemfontein, South Africa. Astronauts on previous missions had attempted similar tests, but clouds had always obscured their views. Other visual observations made by Cooper would stir much controversy. On the ninth orbit, while passing over the Himalayas and later over the southwestern United States, Cooper claimed that he could see individual houses with smoke coming from their chimneys. He reported seeing fields and roads, and he believed that he actually saw a vehicle on one of the roads. Over northern India, Cooper watched a steam locomotive, and he detected the wake of a boat in a Burmese river. These observations were made without any visual aids whatsoever. Even after long post-flight debriefings, many would maintain that the astronaut must have had some sort of visual halluci-nation. At this time Cooper also took some of the best pho-tographs ever taken during the Mercury program.

At 13 hours and 35 minutes into the mission, Cooper said good night to John Glenn, who was Capsule Communicator (CapCom) on board the ship *Coastal Sentry Quebec*. For the next eight hours Cooper dozed, waking periodically to record status reports into the on-board tape recorder. Later, Cooper recorded the first prayer from orbit. In his casual southern drawl, he gave thanks for the success of the mission and for the privilege of being on the flight, and he prayed for help in future space endeavors. On the fifteenth orbit, Cooper received greetings from the Salvadoran president. Later, as he soared above Zanzibar, Cooper sent greetings to African lead-ers currently meeting in Addis Ababa, Ethiopia.

The spacecraft had been performing nearly flawlessly until the seventeenth orbit, when the level of carbon dioxide began to climb. A higher level could slow Cooper's reactions and dull his senses (although he could activate his emergency oxygen supply if necessary). On the nineteenth orbit, Cooper had a more serious problem: The light signifying the begin-ning of reentry switched on, indicating that the spacecraft was experiencing a deceleration. This malfunction had the effect of triggering the automatic recovery sequence. Cooper quickly deactivated the system, but the power supply to the automatic stabilization control system short-circuited. He switched to the backup system, but it, too, had failed. Since all attitude readings were lost, Cooper would have to perform a manual reentry, a procedure attempted on Scott Carpenter's 1962 Mercury mission. As if this problem were not worri-some enough, Cooper noticed that the carbon dioxide levels in his suit and cabin were increasing again.

During this emergency, Cooper calmly photographed cloud formations and the Moon setting behind Earth. He also took time-exposure photographs of "dim light" phenomena, which included high-altitude "airglow" (a glow caused by solar radiation ionizing the thin gases in the upper atmos-phere) and the "Zodiacal light" (a dim glow from the Sun's atmosphere stretching millions of miles from the Sun).

On the twenty-second orbit, Cooper reported that he was in retrograde attitude, with all tasks on the checklist complet-ed. CapCom Glenn gave the ten-second countdown, and Cooper, keeping his pitch at 34 degrees, fired his retro-rock-ets. With the spacecraft traveling nearly 12 kilometers per sec-ond, the slightest error could throw him off course. His tim-ing was perfect, and the manual control during the reentry was flawless. Gordon Cooper landed nearest the recovery point of any Mercury astronaut, only 1.8 kilometers away from the target. He had flown for 34 hours and 21 minutes.

As with Glenn and Schirra, Cooper elected to ride his spacecraft all the way to the carrier. Forty minutes later, he opened the hatch. Project Mercury had ended, and the cele-brations began. President John F. Kennedy awarded all the astronauts medals, and a tickertape parade in New York City followed. Radio Moscow recognized Cooper's flight and the entire Mercury program, expressing hope for future Soviet-U.S. cooperation in exploring the universe.

Knowledge Gained

Perhaps the most controversial aspect of the *Faith* 7 flight had to do with Cooper's visual reports. He claimed to have seen individual houses, a locomotive on its tracks, roads, and even a vehicle on a road. Well after the mission, there was still much debate over whether he had truly seen these things or whether he had been experiencing visual hallucinations.

On the third orbit, Cooper released a flashing beacon. For two or three orbits' time, he was able to see the beacon, one time when it was estimated to be 28 kilometers away. Even though the beacon was rather dim, Cooper said that its flash-ing made it easily distinguishable from the stars. This informa-tion would aid spacecraft planners when they began designing rendezvous devices for the two-spacecraft Apollo missions.

Astronaut Gordon Cooper is assisted backing out of his spacecraft "Faith 7". (NASA)

Photography experiments performed aboard MA-9 revealed that Earth is best viewed in near-infrared light, an important piece of data for weather satellite designers. The value of space-based photographs was debated until Cooper returned with the hundreds that he took. These images showed that much could be learned from high-quality photographs. Photography from space would become an integral part of future missions.

The radiation experiments showed that there were very low levels of radiation inside the spacecraft. The readings taken onboard *Faith 7* were compared to those taken onboard *Sigma 7*, seven months earlier. Residual radiation from an atomic test of July, 1962, was seen to decrease by several orders of magnitude.

The medical experiments showed that Cooper "withstood the stresses of the flight situation with no evidence of degradation" of his performance. He did, however, suffer from mild dehydration. His pulse and blood pressure returned to normal between nine and nineteen hours after splashdown. There was no evidence of psychological disturbances during the thirty-four hours in orbit. All evidence pointed to the ability of man to handle, both physically and psychologically, missions of a much longer duration. This news was welcome, since the Apollo Moon flights would last up to ten days.

The liquid flow experiment gave only limited results; it essentially confirmed previous data gathered in ground-based laboratories. The experiment would, however, provide information needed to design weightless fuel systems required for transferring fuel during future missions.

The Zodiacal light photographs were underexposed and yielded little data as to the light's possible origins. The airglow photographs were overexposed but still useful. Scientists determined that the airglow appeared at an altitude of about 88 kilometers and that it was about 24 kilometers thick.

Context

Gordon Cooper's spaceflight was a milestone: It was the first scientific manned space mission for the United States. It was a fitting end to Project Mercury, which had started some five years earlier, had cost more than $400 million, and had involved 3,345 scientists and engineers. The flight of *Faith 7* combined everything that had been learned up to that time into a single, highly successful mission. Space was now actually being used, not simply explored. The mission proved that the long-duration procedures rehearsed by Wally Schirra during the previous flight were valid and that long-term weightlessness had no ill effect on man. This final determination was especially important, considering that the Apollo missions would last up to ten days.

Cooper's success with the manual reentry justified the expense of sending man along with machine into space. Had *Faith 7* been unmanned, it would likely have been stranded in orbit.

The use of an on-board television camera was the first in the American manned program, although the Soviets had used television beginning with their first manned Vostok flight. The poor quality of Cooper's camera rendered it nearly

useless, and television was not to be used again until the first Apollo flight, some six years later.

The interesting sightings made by Cooper of roads, houses, and the like prompted NASA to send him into space again, only two years later, to follow up on his reports. In 1965, he flew the longest flight up to that time, with Pete Conrad on Gemini V. It lasted more than eight days. Cooper later became Commander of the Apollo 10 backup crew, but he never flew again.

Up until the MA-9 mission, there was serious talk of having yet another Mercury flight, MA-10. That idea was rejected because of the tremendous success of *Faith 7*, although there were still several unused Mercury spacecraft (three of which were modified for long-duration flight). This proposed mission would have lasted three days and would have been flown by Alan B. Shepard, the United States' first man in space. In anticipation of this, a Mercury capsule even had "Freedom 7-II" painted on its side, after Shepard's first spacecraft, *Freedom 7*.

Bibliography

Baker, David. *The History of Manned Space Flight*. New York: Crown Publishers, 1981. This book details the evolution of space programs throughout the world. Well illustrated with both color and black-and-white photographs. Recommended for the general reader with an interest in space history.

Gatland, Kenneth. *The Illustrated Encyclopedia of Space Technology: A Comprehensive History of Space Exploration*. New York: Crown Publishers, 1986. A well-illustrated volume covering the early development of various space programs and detailing the goals of each. Includes drawings and photographs.

National Aeronautics and Space Administration. *Mercury Project Summary Including Results of Fourth Manned Orbital Flight, May 15-16, 1963*. NASA SP-45. Washington, D.C.: Government Printing Office, 1963. The official summary of the Mercury program, this book focuses on the *Faith 7* flight. Contains a discussion of scientific experiments and details the design of the capsule. The official flight report of this last mission is also included, along with transcripts of the ground-to-air communications.

Sobel, Lester A., ed. *Space: From Sputnik to Gemini*. New York: Facts on File, 1965. This small book provides a complete account of the pertinent data concerning launches from 1957 to 1966 in both the Soviet Union and the United States. Ordered chronologically, it contains details about each Mercury flight.

Swenson, Loyd S., Jr., James M. Grimwood, and Charles C. Alexander. *This New Ocean: A History of Project Mercury*. NASA SP-4201. Washington, D.C.: Government Printing Office, 1966. One of the special publications made available to the general public as part of the NASA Historical Series, this text traces Project Mercury through Cooper's flight. Includes material on both the technical and managerial aspects of the program. With photographs and diagrams.

Michael Smithwick

METEOROLOGICAL SATELLITES

Date: Beginning April 1, 1960
Type of satellite: Meteorological

Since 1960, meteorological satellites have helped predict the weather, saving lives in storms, protecting crops during droughts, and making life easier on Earth. These satellites have revolutionized meteorological science.

PRINCIPAL PERSONAGES

JOHN HERBERT HOLLOMAN, Assistant Secretary of
    Commerce for Science and Technology
DAVID S. JOHNSON, Director, National Weather
    Satellite Center
FRANCIS W. REICHELDERFER, Chief, United States
    Weather Bureau
S. FRED SINGER, Director, National Weather Satellite
    Center
WILLIAM G. STROUD, Chief, Aeronomy and
    Meteorology Division, Goddard Space Flight
    Center, and manager of the TIROS project
MORRIS TEPPER, Director of Meteorological
    Programs, Office of Space Science and
    Applications, NASA

Summary of the Satellites

The first successful meteorological experiment was flown in 1959 on the United States' Explorer 7 satellite, which measured reflected sunlight, infrared radiation emitted from Earth, and energy radiated from the Sun. These components of Earth's radiation are fundamental to deriving the atmospheric and oceanic conditions which give rise to daily changes in weather patterns.

In 1960, the Television Infrared Observations Satellite (TIROS) program began. With a special television camera called a vidicon, TIROS 1 was able to provide the first cloud images from space and thus become the first successful, dependable, long-term weather satellite. The TIROS series demonstrated that satellites could be used to survey global weather conditions and study other surface features from space. TIROS 1 transmitted to Earth more than twenty thousand high-quality cloud cover photographs from a 122.5-kilogram spacecraft.

TIROS 2 proved the accuracy and reliability of the experimental television and infrared equipment for global meteorological information data gathering. The TIROS series

proved the cornerstone of the effort to develop twenty-four-hour weather monitoring of Earth. The Improved TIROS Operational System (ITOS) and the Environmental Science Services Administration (ESSA) and National Oceanic and Atmospheric Administration (NOAA) systems that have followed simply improved on the basic design.

RCA was the prime contractor for the TIROS series, and the project was managed by the National Aeronautics and Space Administration (NASA) with data handled by the United States Weather Bureau (later renamed ESSA, still later titled NOAA). The ten TIROS satellites were all successfully launched, and only TIROS 5 experienced an infrared imaging system failure.

The satellites of the Environmental Science Services Administration were designed to follow up on the successful TIROS series. In this series, dependability increased to the point that, for the first time, weather forecasters could use the satellite data routinely in preparing daily weather forecasts, storm and marine advisories, gale and hurricane warnings, cloud analyses, and navigational assistance. Again, RCA was the prime contractor. These nine ESSA satellites were extensions of and improvements on the original TIROS design.

The ESSA spacecraft carried either the automatic picture transmission systems or the advanced vidicon camera system. These systems included two 800-line cameras with nearly twice the resolution of a normal television camera. They could photograph in real time a 3,000-kilometer-wide area with 3-kilometer resolution. The two television cameras were mounted 180 degrees apart on the sides, with a 40-centimeter receiving antenna on top, solar cells covering the sides and top, and four whip antennae extending from the baseplates. Generally, it was some 50 centimeters high and nearly 100 centimeters wide, shaped like an 18-sided cylindrical polygon. Two tape recorders could store up to 48 pictures for later transmission.

ESSA satellites provided global weather data to the United States Department of Commerce's command and data acquisition stations in Wallops Island, Virginia, and Fairbanks, Alaska, and then relayed it to the National Environmental Satellite Service at Suitland, Maryland, for processing and forwarding to the major forecasting centers in the United States and around the world. ESSA provided the world's first true meteorological satellites as well as the world's first operational applications satellite.

In 1969, an image from ESSA 7 made history by revealing that the snow cover in parts of the Midwest was three times thicker than normal. Measurements showed that there was the equivalent to 15 to 25 centimeters of water covering thousands of square kilometers. A disaster area was declared and before the floods came the area was prepared. By the time ESSA 9 was in orbit there were some 400 receiving stations in operation around the world, as well as 26 universities, up to 30 U.S. television stations, and an unknown number of private citizens receiving the photographs.

In 1970, a second generation, called the Improved TIROS Operational System (ITOS), was introduced. The National Oceanic and Atmospheric Administration had taken over the duties of the ESSA. Thus the second ITOS was also known as NOAA 1 when launched in December, 1970, from Vandenberg Air Force Base. Gradually the ITOS and the NOAA satellites began to replace the functions of the remaining ESSA satellites. These spacecraft were manufactured by RCA. ITOS-B failed to achieve orbit as a result of launch vehicle failure, and ITOS-C was never used as a designation. The NOAA series formally began with NOAA 6. Like the ITOS spacecraft, the NOAA satellites were developed by RCA.

The TIROS-type satellites had proved that an operational weather satellite system had considerable value, but still the series had limited capabilities. Modern weather forecasting is done numerically, with the aid of computers. Some of the most important data are the vertical temperature structure of the atmosphere, the motion of the winds (both speed and direction), and the atmosphere's moisture content. The ESSA system could provide no direct data about these factors. The United States was well aware of these limitations and eventually turned to the Nimbus system, which had remote sensors that could measure atmospheric temperatures and moisture levels.

The Nimbus series ("nimbus" is Latin for rain cloud) was originally conceived of as only a weather forecasting program, without experimental capabilities, but the later satellites, Nimbus 5 through Nimbus 7, were upgraded to provide data for a larger range of scientific programs — including agriculture, cartography, geography, geology, and oceanography. The Nimbus program was especially important for gathering weather data over the two poles. The Nimbus series was developed by General Electric Company and launched from Vandenberg Air Force Base. NASA's Goddard Space Flight Center managed the satellite series, using seven Nimbus satellites to test and improve measuring instruments. Noteworthy experiments were performed. For example, the cloud imaging vidicon camera on Nimbus 2 was equipped to transmit images of weather to local receiving stations by a very high-frequency (VHF) signal. Thus many low-cost receiving stations, which were often located in developing countries, could receive and use data from the weather satellites for the first time. In 1969, instruments that were able to probe the

atmosphere's temperature and moisture layers by means of a selected spectral apparatus were placed on Nimbus 3.

Of special significance in the late 1960's were the launchings of NASA's Applications Technology Satellites (ATS's) 1 and 3. These geosynchronous satellites were actually early communications satellites that appeared to "hang" in orbit over one spot on Earth. They carried sensors which, from a stationary position high above the equator, provided a constant view of changing weather patterns. For the first time meteorologists could track in detail and with precision the movement and growth of weather systems from space. The ATS satellites were thus the first series of meteorological functioning satellites in a geosynchronous orbit and were positioned to view 45 percent of Earth's surface. This program was managed by NASA, with Hughes Aircraft Company the primary spacecraft contractor. The second, fourth, and fifth ATS's failed to gain the proper orbit and were thus excluded from meaningful experiments. ATS 3 not only performed meteorological experiments but also performed experiments involving communications, navigation, and Earth reconnaissance. Later it was moved to a more favorable orbit to better monitor any storms developing over the United States.

The Synchronous Meteorological Satellite (SMS) series was the first series actually designed to be geosynchronous. The satellite was designed by Ford Aerospace and Communications Corporation and managed by NASA for NOAA. Parked over a part of the globe, they enabled meteorologists to make twenty-four-hour forecasts. So successful was this program that it was succeeded by NOAA's Geostationary Operational Environmental Satellite (GOES) program. SMS 1 (sometimes known as SMS-A) and SMS 2 (or SMS-B) were the developmental satellites leading up to the GOES program, and GOES-A also carried the designation SMS-C.

The first GOES spacecraft were produced by Ford, and after they were launched by NASA they were turned over to NOAA, which used them to provide meteorological data worldwide — including observations of ocean currents and river water levels. The GOES spacecraft provided pictures of approximately one-fourth of the globe at 30-minute intervals day and night with an infrared image maker. The GOES-D was the first United States spacecraft capable of near-continuous measurements of atmospheric water vapor and temperature. The spacecraft enabled scientists to provide more accurate weather predictions.

Paralleling these developments in the civilian space progam, the United States Department of Defense had developed and flown the unclassified Defense Meteorological Satellite Program (DMSP) satellites to forecast the weather for Air Force and Navy planes and ships. These eight satellites, the first of which was launched in October, 1971, included a precision sensor on top. Designed and built by RCA, the DMSP spacecraft served as the primary military meteorological reconnaissance satellites. Similar to the TIROS spacecraft, the DMSP satellites transmit weather data to the United States Air Force Global Weather Control at Strategic Air Command Headquarters at Offutt Air Force Base in Nebraska and to other military installations around the world. The sensors include an electron measurement instrument to evaluate atmospheric conditions that might affect military communication, an instrument to measure temperature, and a cloud cover imager.

Knowledge Gained

Basically, meteorological satellites are platforms for "topside observation," or electromagnetic scanning of the atmosphere of Earth from above. Their scanning is passive in the sense that satellites merely make use of existing radiation emitted or reflected from the atmosphere, without adding to it as radar does. Many experiments, especially those conducted with the Nimbus satellite series, improve and perfect weather forecasting.

Some satellites also do special experiments. For example, there have been many important experiments measuring both direct and reflected energy from the Sun. Orbiting satellites provide an ideal platform for such measurement. The first successful meteorological satellite experiment, on Explorer 7, directly measured the Sun's energy for the first time. Additional radiation experiments have been conducted on TIROS, Nimbus, and NOAA satellites.

In general, meteorological satellites perform five major functions. They are able to take the images of clouds and cloud patterns and relay them back to Earth, they can observe the movement of clouds and weather systems (this work was previously accomplished through the use of tracer balloons), they can measure the atmosphere's moisture content and temperature, they can help forecast storms, especially severe thunderstorms, through cloud patterns and upper-air movements, and they can help track tropical storms by observing cloud patterns.

The meteorological satellite utilizes three classes of instruments with which to gather data, forecast weather, and perform scientific experiments. Radiometers measure Earth's infrared and reflected solar radiance and use small, selected wavelengths (ultraviolet to microwave) to measure temperature, ozone, and water vapor in the atmosphere. The radiometers are sensitive to one or more wavelength bands in the visible and invisible ranges. If visible wavelengths are used, the satellite can detect cloud vistas from reflected sunlight, resulting in a slightly blurred version of those images photographed directly by astronauts. Using the invisible range, satellites can capture terrestrial radiation, producing images from Earth's radiant energy.

Such sensitive radiometers, combined with sensitive optical devices, can measure clouds as small as 1 kilometer. As the satellite spins, the radiometer scans parts of Earth. These radiometric data are then transmitted to each ground processing station or recorded for later replay. The ground processing station then produces a computer-generated image with landmark features, coastlines, and sometimes even artificial political boundaries.

Sounding sensors are more complex than radiometers. These sonarlike devices probe the atmosphere and aid in understanding temperature and moisture, among other things. For example, Nimbus 3 measured the global distribution of ozone in the atmosphere. Tracking and data-relay systems are special transponders that receive signals from Earth stations and relay back other data. These systems help identify wind patterns and ocean currents.

Context

The TIROS satellites first proved the feasibility of gathering meteorological data from space, and in 1962 and 1963 they helped save lives during a period of severe hurricanes. Meteorological satellites continually examine cloud formations, estimate atmospheric water vapor, and measure air and water temperatures. These data are then used routinely to forecast storms and provide basic information for modeling weather conditions. Aircraft and ship travel would be far more difficult without the use of meteorological satellites. Nearly all Americans are familiar with the work of these satellites from the weather forecasts on local television news programs.

The huge panoramic views of the atmosphere afforded by meteorological satellites confirmed the structures of large cloudy weather systems that had previously emerged only after painstaking assembly and analysis of synoptic data. Few can even appreciate the work that was required in the 1950's to produce a cloud map half as accurate as one poor satellite photograph. Satellite images were remarkably clear and precise, and with the continual improvement of equipment they became better and better. The relatively high resolution of the images (a few kilometers compared with tens of thousands of kilometers of the earlier, coarser maps) revealed a bewildering range of subsynoptic scale structures in cloud systems, many of which are still not well understood. For the first time there was regular uniform weather information. Meteorological satellites provide a consistent flow of images, covering oceans and landmasses heretofore impossible to observe.

In mid-1983, the nations of the World Meteorological Organization began a cooperative climatic experiment called the International Satellite Cloud Climatology Project. Using six satellites to cover the entire globe, the project sought to determine the quantity and height of clouds every three hours during a five-year period. Scientists have begun to use these data to test and study the natural variability of cloud patterns. Such experiments will become more sophisticated as computers enable meteorological satellites to gather and process even more information.

The development of meteorological satellites has been particularly advanced by international cooperation. For example, several different nations contributed the instruments that the United States used in its Nimbus and TIROS programs. Brazil, the Bahamas, and other countries in Latin America use the United States' GOES and NOAA satellites in daily forecasting and research.

In the longer run, the availability of years and years of daily coverage of Earth for rain, temperature, cloud behavior, and other meteorological data have enabled researchers to develop climatologies — that is, baselines of what is normal; thus, deviations can be compared. These baselines are especially important for detecting drought, temperature change,

and snow caps. Previously, scientists had small descriptions but lacked data for the entire planet. With the same complement of instruments over the entire planet, scientists will be able to attain a better understanding of what is normal weather for Earth. In addition, NOAA's advanced very high-resolution radiometer was developed in such a way that the infrared signals could be analyzed to determine the state of greenness of the vegetation on Earth. Used over a number of years, this instrument can monitor slow but significant changes in large-scale vegetation cover brought about by droughts and floods.

Bibliography

American Meteorological Association. *The Conception, Growth, Accomplishments, and Future of Meteorological Satellites.* NASA SP-2257. Springfield, Va.: National Technical Information Service. An important collection of articles about the history and practice of the United States meteorological program. Clearly written.

Barrett, E. C. *Climatology from Satellites.* London: Methuen and Co., 1974. A systematic look at the applications and uses of satellites in all phases of meteorology. Written for the advanced student, although the second chapter lays out in clear detail the extent of the program in the early 1970's.

Eagleman, Joe R. *Meteorology: The Atmosphere in Action.* New York: Van Nostrand Reinhold Co., 1980. This basic textbook provides a context for the importance of the meteorological satellites as tools of the modern-day scientist. Written in clear and understandable language.

Henderson-Sellers, Ann. *Satellite Sensing of a Cloudy Atmosphere.* London: Taylor and Francis, 1982. A collection of articles about the various technologies and uses of meteorological satellites. Written in conjunction with the Global Atmospheric Research Program.

McIlveen, Robin. *Basic Meteorology.* New York: Van Nostrand Reinhold Co., 1986. This introductory textbook provides a survey of the history and context of the meteorological satellite. Written clearly and concisely.

National Aeronautics and Space Administration. *Meteorological Satellites: Past, Present, and Future.* NASA SP-2227. Springfield, Va.: National Technical Information Service, 1982. An important collection of articles about how the United States' meteorological satellite program began. It covers the programs until the beginning of the 1980's. Accessible to the beginner.

Scorer, Richard S. *Cloud Investigation by Satellite.* New York: Halsted Press, 1986. A wonderful overview, filled with photographs illustrating the uses of meteorological satellites. The images provide a clear sense of the importance of the various advancements in the satellite meteorological program.

Smith, W. L., et al. "The Meteorological Satellite: Overview of Twenty-five Years of Operation." *Science* 231 (January 31, 1986): 455- 462. This is an important review of the contributions of the meteorological satellite programs. Written for the layman.

U.S. Congress. House. Committee on Science and Technology. *Space Activities of the United States, Soviet Union, and Other Launching Countries/Organizations: 1957 – 1984.* Report prepared by Congressional Research Service, the Library of Congress. 99th Cong., 1st sess. Washington, D.C.: Government Printing Office. This is a clear analysis of the various activities in space. Includes an interesting section on meteorological satellites.

Douglas Gomery

MILITARY METEOROLOGICAL SATELLITES

Date: Beginning April 1, 1960
Type of satellite: Military meteorological

Military meteorological satellites provide cloud-cover photographs and other weather data from orbit. These data are used for many purposes, from the scheduling of reconnaissance satellite launches to the planning of military operations. Much of this important information could not be obtained from ground instruments.

Summary of the Satellites

The weather is a determining factor in all human activities — including military activities. History is filled with examples of weather's impact on battles, and military strategists know that commanders must be able to predict the next day's weather. For example, General Dwight D. Eisenhower's decision to launch the invasion of Europe on June 6, 1944, was based on a forecast of marginally acceptable weather.

When the first studies of satellites were made in the late 1940's, it was quickly realized that these spacecraft could provide weather data on a worldwide basis. In the late 1950's, the U.S. Army considered the possibility of using a satellite equipped with a television camera for battlefield reconnaissance. By the fall of 1958, this project became a meteorological satellite and was transferred to the National Aeronautics and Space Administration (NASA) as a joint civil-military project. It was called the Television Infrared Observations Satellite (TIROS). TIROS 1 was launched on April 1, 1960, into a 740-by-692-kilometer orbit. Within days it was clear that the science of meteorology had been fundamentally altered.

It was during this time that the Discoverer reconnaissance satellites began returning the first satellite photographs of the Soviet Union. To support these satellites, reliable weather forecasts were needed of specific sites inside the Soviet Union, China, and Eastern Europe. The film supplies of the satellites were limited; none would be wasted on photographs of clouds. Even partly cloudy conditions result in poor photographs, because of large shadows cast by the clouds. (U-2 overflights of the Soviet Union and, later, Cuba often had to be delayed because of weather.) During the early years of the United States' reconnaissance satellite operations, weather support was provided by the civilian TIROS and Nimbus satellites. By early 1963, however, it was becoming clear that military and civilian requirements were too different; whereas civilian weather forecasters needed widespread coverage, the military needed information on specific areas. Accordingly, a separate military meteorological satellite program was approved.

The first of these satellites was launched on January 19, 1965, from Vandenberg Air Force Base in California. The Thor booster put it into an 822-by-471-kilometer orbit, inclined 98.8 degrees. The satellite passed over a given spot on Earth at the same local time each day. A total of six Thor launches were made until March, 1966. (One launch failed.)

On September 16, 1966, the satellite program became known as the Defense Meteorological Satellite Program (DMSP). The satellites were built by the Radio Corporation of America. Called the DMSP Block 4A satellites, they weighed 82 kilograms and were spin-stabilized. In 1968, the Block 4B was introduced, and the Block 5A followed in 1970.

While the first DMSP satellites were being launched, the United States was becoming increasingly involved in the Vietnam War. Knowledge of weather conditions was critical, particularly for air strikes. Air Force and Navy fighter pilots had to spot their targets visually from 3,660 meters. Tanker aircraft, which refueled the strike aircraft, needed to operate in areas that were free of clouds and turbulence. In South Vietnam, Laos, and Cambodia, close air support missions required good weather at low altitudes. It was difficult to provide the necessary forecasts for these activities: There were few weather stations in Southeast Asia, the North Vietnamese regularly falsified or coded their data, and the seasonal monsoon weather caused additional problems.

Many different types of satellites were used to forecast the weather in Southeast Asia during this time. Civil weather satellites, equipped with a feature called Automatic Picture Transmission (APT), transmitted images to ground stations constantly so they could be detected by any receiver within range. (The Chinese and the North Vietnamese may have also made use of this technology.) The civil satellites made their passes from 7:00 A.M. to 9:00 A.M. and from 11:00 A.M. to 1:00 P.M. The DMSP satellites provided photographs in the visible and infrared spectra at 7:00 A.M., 12:00 P.M., 7:00 P.M., and 12:00 A.M. (Unlike the civilian photographs, those from the DMSP were encoded.) The Air Force ground stations were located at an air base in South Vietnam, outside Saigon, and at a Thai air base. The South Vietnamese facility had a mobile van with an APT antenna, a large DMSP dish antenna, and two sandbagged DMSP trailers. The photographs were transmitted via microwaves to a U.S. Air Force headquarters facility. (Unfortunately, the Navy lagged behind the Air Force in readout equipment during the early years of the war. That was corrected by 1970. The USS *Constellation* was the first aircraft carrier to use shipboard readout equipment.)

The Soviets were slower to fly an operational weather satellite system. Early experimental weather equipment was carried on a number of Soviet photoreconnaissance satellites, starting with Kosmos 4 in April, 1962. The first true Soviet weather satellite is believed to be Kosmos 44. It was launched on August 28, 1964, by an A-1 booster from Tyuratam and reached an 813-by-604-kilometer orbit, inclined 65.1 degrees. Three more followed: Kosmos 58 (February 26, 1965), Kosmos 100 (December 17, 1965), and Kosmos 118 (May 11, 1966). All were thought to be test payloads. Kosmos 122 (June 25, 1966) was the first mission the Soviets acknowledged as a weather satellite. It went into an orbit similar to those attained by the four earlier satellites.

The shift to operational status was signaled with the launch of Kosmos 144 on February 28, 1967. This satellite established the basic pattern of operational flights: a 650-kilometer circular orbit, an 81-degree inclination, and an A-1 launch vehicle. Four more followed between April, 1967, and June, 1968. The operational system made its official debut on March 26, 1969, with the launch of Meteor 1. The early launches of the Meteor satellites followed the same pattern as the test launches. The fifth Meteor I launch (June 23, 1970) went into a 906-by-863-kilometer orbit. Starting in 1972 this 900-kilometer orbit would become standard. The Meteor satellite was cylindrical, with two solar arrays; it measured 4.88 meters in height and 1.5 meters in diameter. The two television cameras, for taking photographs during the daytime, and the infrared scanner, for both day and night measurements, were located in the base of the cylinder. The resolution of the Meteors was described as being three times that of the TIROS satellites. That would be sufficient to provide military support, but it was not until 1968-1969 that the Soviets took advantage of their capability. Three or four Meteor satellites were launched per year through the 1970's.

In 1971, the DMSP Block 5B satellite was introduced in the United States. It was a twelve-sided cone, 1.64 meters high and 1.31 meters wide. A 1.22-by-2.74-meter solar shade protected the sensors from glare. The satellite was powered by solar cells covering eleven of the sides. It weighed 195 kilograms and had a three-axis stabilization system. This increase in weight required an upgrading of the launch vehicle's upper stage. In early 1973, the Air Force announced the existence of the DMSP program, and the following year the satellite photographs were released for civilian use. Also in 1974, the improved Block 5C was introduced. The final DMSP Block 5C launch was made on February 19, 1976. The Thor booster malfunctioned, and the satellite was left in a 355-by-90-kilometer orbit that soon decayed.

Meanwhile the Soviets were also improving their weather satellites. The first Meteor 2 version was launched on July 11, 1975. It carried improved systems, with a 2-kilometer resolution for the television camera and an 8-kilometer resolution for the infrared scanner. On June 29, 1977, the Soviets launched their twenty-eighth Meteor 1 satellite into a 685-by-602 kilometer, Sun-synchronous orbit. It was the first time such an orbit was used by the Soviets. The Sun-synchronous Meteors were launched due south from Tyuratam. The spent stages made impact just short of the Soviet-Iranian border. Because of the weight of the payload, a lower altitude orbit had to be used. (Plesetsk could not be used for this launch; a launch toward the north could be mistaken for an attack on the United States, and a launch toward the south would pass over populated areas.) Because of these difficulties, only one Sun-synchronous Meteor launch is made every one or two years. In the 1980's, the role of these spacecraft was changed from that of Earth resources satellites to that of military meteorological satellites.

The DMSP Block 5D-1 version made its debut in 1976 in the United States. The 473-kilogram satellite was a goldfoil-wrapped box with a solar panel on a long boom. It was 5.2 meters long and 1.8 meters wide. The television and infrared devices had two resolutions: 0.56 kilometer and 2.78 kilometers. Atmospheric temperature and water vapor could also be measured. The first DMSP Block 5D-1 was launched on September 11, 1976, into an 834-by-806-kilometer, Sunsynchronous orbit. After launch, a high-pressure nitrogen leak caused the satellite to tumble. Because the solar panel was no longer directed toward the Sun, the satellite lost power and ceased to function. On October 5, the solar panel began to be illuminated by the Sun and generate power. Ground controllers were able to raise the altitude of the satellite and slow its spin. This effort was made more difficult when a gyroscope problem occurred in February, 1977. Finally, on March 24, the satellite was fully stabilized. Four more satellites from this series were launched.

The first of the improved DMSP Block 5D-2 satellites was launched on December 21, 1982. The booster was an Atlas E, an intercontinental ballistic missile converted for satellite launching. The Block 5D-2 carried a larger sensor payload and had a longer life span (thirty-six months versus eighteen months for the earlier Block 5D-1 series). A second Block 5D-2 was launched on November 18, 1983, and a third on June 20, 1987. The operating network consists of two satellites: the "early morning bird," which makes a south to north crossing of the equator at 6:30 A.M. local time, and one that makes its crossing at noon.

The Soviets also upgraded their satellites in the 1980's. The Meteor 1 satellites were gradually replaced, the F-2 became the preferred booster, and the launch rate dropped to one or two per year (indicating better reliability). The first Meteor 3 was launched on October 24, 1985. This type uses a higher orbit of 1,210 by 1,185 kilometers and is inclined 82.5 degrees.

Knowledge Gained

The DMSP satellites have provided weather data for the planning of reconnaissance satellite coverage, thus preventing the waste of valuable film. In Vietnam, these satellites proved critical. Air Force officials have said that the DMSP was among the most significant innovations during the war. For three years the Navy had made repeated attacks on the Thanh Hoa Bridge, which was one of the two main rail lines to the south. One of the problems was weather over the target. Only eighteen kilometers inland, it was covered with low clouds, fog, smoke from burning fields, and haze during much of the year. Sometimes there would be only two to four days of clear weather per month. The DMSP photographs indicated when the weather was about to change. In the late 1960's, improved sensors increased effectiveness. The infrared scanner revealed the burning rice paddies, and warnings could be given to the pilots.

During the November, 1970, raid on the Son Tay prisoner-of-war camp in North Vietnam, DMSP satellites gave a highly accurate three- to five-day forecast. In May, 1975, the Cambodians captured the freighter *Mayaguez* on the high seas. During the military action that followed, the DMSP provided most of the weather information. Air refueling areas were moved on May 14 based on DMSP data, and decisions were made to land several damaged helicopters in Thailand rather than abandon them.

The Soviet Union's Meteor satellites also contributed to that country's national safety. In July, 1970, an article described how Meteor photographs had helped save a large dry dock that was threatened by Typhoon Juliette. Sailing times of ships have been cut 5 to 7 percent as a result of weather forecasting provided by Meteors. Civil air transport has been aided by timely weather data, particularly cloud cover data. It is also probable that Meteor satellites supported Soviet military operations in Afghanistan, as strikes played a key role in the Soviet strategy.

Context

Military meteorological satellites provide a support function. They can make military operations more effective but they cannot undertake such activites as reconnaissance surveys or air strikes. Because of their limited role, they face budgetary problems that do not affect other military satellite programs. In the DMSP, these problems gave rise to a number of innovative solutions. During the early 1970's, the program was run by forty military and civilian personnel. Many of the Air Force officers were of junior rank but held doctorates or had specific technical qualifications. That meant that personnel with ranks as low as captain held senior positions. Their tours were also longer than normal, giving a cohesiveness to the program, while saving funds.

In addition, the Block 4 and early Block 5 satellites had no built-in redundancy, making the program even more cost-efficient. Spacecraft with built-in redundancy are equipped with duplicate systems to prevent failure of the mission in the case of one component's malfunction. Thus, the Block 5A and Block 5B spacecraft had a variety of life spans; they would function from three months to twenty-three months. Even with the increased priority in the early 1970's, the DMSP Block 5D had only "selective redundancy," that is, backup systems for certain critical areas. This modification was intended to ensure a longer lifetime. By using the satellite's stellar inertial platform as the guidance system for the rocket booster, designers were also able to reduce the weight of the DMSP satellites — making them less expensive to launch.

The design and contracting process was also innovative. The concept for the infrared scanner, used in all the Block 5 satellites, was proved by three Air Force and two Aerospace Corporation engineers. Hardware development was then undertaken by Westinghouse.

Despite its military origins, the DMSP has its civilian importance. The photographic data have been available to civilian weather forecasters since 1974. Meteorologists have been able to take advantage of the satellite's improved resolution. Moreover, the DMSP Block 5D has served as the basis of the civilian TIROS-N weather satellite.

Bibliography

Klass, Philip J. *Secret Sentries in Space*. New York: Random House, 1971. The first book on military satellites. It covers the historical background and development of reconnaissance satellites and briefly discusses U.S. military weather satellites.

Peebles, Curtis. *Guardians: Strategic Reconnaissance Satellites*. Novato, Calif.: Presidio Press, 1987. Covers the history and technology of reconnaissance satellites and the profound impact they have had on international relations. The book has a brief mention of U.S. military weather satellites. Suitable for high school and college readers.

Turnill, Reginald. *The Observer's Book of Unmanned Spaceflight*. New York: Frederick Warne, 1974. A brief overview of unmanned satellites. It includes a section on the Meteor satellites. Suitable for general audiences.

U.S. Congress. Senate. Committee on Aeronautical and Space Sciences. *Soviet Space Programs, 1971 – 1975*. Vol. 1. Report prepared by Congressional Research Service, the Library of Congress. 94th Cong., 2d sess., 1976. Committee Print. This volume is an earlier version of the source listed below. It includes material on the Meteor weather satellite. Recommended for high school and college readers.

U.S. Congress. Senate. Committee on Commerce, Science, and Transportation. *Soviet Space Programs: 1976 – 1980*. Part

3, *Unmanned Space Activities*. Report prepared by Congressional Research Service, the Library of Congress. 99th Cong., 1st sess., 1985. Committee Print. This book includes a section on the Meteor weather satellites, a table of launches, and illustrations of the satellites. Recommended for high school and college readers.

Wilding-White, T. M. *Jane's Pocket Book of Space Exploration*. New York: Collier Books, 1977. This source focuses on the achievements of space exploration. It provides a capsule summary of the history and statistics of satellite programs and includes discussion of both the Meteor and DMSP Block 5D satellites.

Curtis Peebles

Military Telecommunications Satellites

Date: Beginning December 18, 1958
Type of satellite: Military communications

Since the mid-1960's, communications satellites for the United States' military have provided reliable, worldwide communications between troops and decision makers; they have made possible a new form of conflict: advanced nuclear war.

Summary of the Satellites

Although the first artificial satellites were developed for nonmilitary, scientific purposes, it did not take long for the United States' military to understand and appreciate their value. Communications satellites in particular, it was realized, could aid the Army, Navy, Marines, and Air Force strategically and tactically. Communications satellites would enable a country to wage global war by maintaining links across the planet.

The Score (Signal Communication by Orbiting Relay Equipment) program of the United States Army was the first project of the world's first military communications satellite. The Score spacecraft was launched from Cape Canaveral around Christmastime, 1958; from its elliptical Earth orbit, it transmitted messages for two weeks, including a holiday greeting from President Dwight D. Eisenhower. The satellite weighed 70 kilograms and was launched atop an Atlas B rocket into a 185-by-1,470-kilometer orbit. Next came the Courier, a military communications satellite that made few advances. In fact, the experiment was a failure, and because of it, other military communications satellites were delayed for four years.

The first operational system was named the Interim (or Initial) Defense Communications Satellite Project (IDCSP), later referred to as phase 1 of the Defense Satellite Communications System (DSCS 1). This system made use of a total of twenty-six satellites, launched between June, 1966, and June, 1968. These were simple spin-stabilized spacecraft with no complicated moving parts; they were designed for the highest possible reliability, weighing 45 kilograms and laced with solar cells. Dispensed at six-hour intervals into slightly different orbital velocities to give them global coverage, the IDCSP satellites were placed in subsynchronous orbits at 33,915 kilometers, so for about four and one-half days they remained within view of an Earth ground station located at the equator. They were deployed in slightly subsynchronous orbits with an eastward drift in groups of seven at a time, using a massive Titan military rocket.

Since each satellite remained in view of an equatorial station for four or five days, several were accessible at any one time. Some thirty-five Earth terminals (including seven located on ships) linked with stations up to 16,000 kilometers apart using an FM band. Links between South Vietnam, Hawaii, and Washington, D.C., were used for high-speed digital data during the last seven years of the United States' involvement in the Vietnam War. Indeed, many of the sophisticated bombing raids would not have been possible without satellite communications to and from headquarters to manage the strategy. Moreover, this experience proved how valuable the satellite system could be during wartime. Communications involved single-channel voice relay, imagery (including photographs), teletype (written communications), and computerized digital data transmission. Although they were designed for a life of only eighteen months, many of the satellites lasted years.

The same technology was used in two other military communications satellite systems, including the United Kingdom's Skynet and a system for the North Atlantic Treaty Organization (NATO) countries. The Skynet system included large fixed ground stations as well as stations aboard ships and on mobile land units. Two satellites were built in the United States and launched by the United Kingdom under a cooperative program during 1969 and 1970. A second was launched in 1974, after an initial failure.

The Skynet satellites were designed to provide exclusive and highly secretive voice, telegraph, and facsimile links for both strategic and tactical communications between the United Kingdom and bases throughout the world. The drum-shaped cylinders, 137 centimeters in diameter by 81 centimeters in height, were spin-stabilized so that their antennae were always within sight of proper ground stations.

For NATO, two satellites were built in the United States and then successfully launched during 1970 and 1971. The original NATO operational system included twelve ground stations near the capitals of the twelve NATO countries. The NATO spacecraft were of the IDCSP design, only slightly larger, with an improved antenna system. They were placed in geostationary orbits so that they were above the same spot on Earth at all times. Later, the NATO system expanded the number of ground stations so the system could cover the Northern Hemisphere from Ankara, Turkey, to the suburbs of northern Virginia, home of the Pentagon.

The DSCS 2 followed the IDCSP. It consisted of spacecraft designed to provide the Department of Defense with a reliable network of strategic communications satellites for global coverage. Managed by the United States Air Force, the DSCS 2 satellites were developed by TRW as the primary contractor. The DSCS 2 program had its first launch in 1971 and featured much larger, cylindrical spacecraft, more than 3 meters in height and 3 meters in diameter. These were designed to function for five years and were placed in geosynchronous orbits. The system provided up to 1,300 duplex voice communication channels. The DSCS 2 was designed with a multiple-channel, wideband antenna system so it could link more easily to smaller ground stations. In this way, the design resembled that of Intelsat 4.

The DSCS 3 program operated on the same geostationary equatorial orbits as DSCS 2 but offered 50 percent greater communications capability and a promised ten-year life span. Like the DSCS 2, the third phase would be controlled by the United States Air Force Satellite Control Facility near San Francisco, with communications control built into selected Defense Communications Agency terminals. Through its larger antennae, DSCS 3 would provide Earth coverage as well as spot-beam transmissions to smaller, portable ground stations.

The first DSCS 3 was launched in October, 1982, and offered greater flexibility and increased channel capacity. It weighed 1,042 kilograms. The DSCS 3 satellites were to be used by the National Command Authority, the White House Communications Agency, and the Diplomatic Telecommunications Service. By 1984, their price had risen to $150 million each.

The United States Navy Fleet Satellite Communications, or FLTSATCOM, system and the Satellite Data System (SDS) are part of the total Air Force Satellite Communications System, or AFSATCOM. Both were designed to establish reliable networks for communication between decision makers and military forces in the field, on sea, or in the air. FLTSATCOM was designed to provide worldwide high priority ultrahigh frequency (UHF) communications between aircraft, ships, submarines, and ground stations — as well as between the United States military and presidential command networks. The communications provided more than thirty voice and twelve teletype channels designed to serve mobile as well as stationary centers.

The 1,860-kilogram FLTSATCOM spacecraft have twenty-three communications channels. Ten are allocated to the Navy for command of its air, ground, and ocean forces. Twelve are allocated to the Air Force as part of the AFSATCOM system of worldwide command and control, especially of nuclear weapons. One is singularly reserved for command authorities. The AFSATCOM system enables aircraft in flight to communicate with ships, Earth stations thousands of miles away, and even properly equipped submarines. Information from these satellites is used to control and monitor intercontinental ballistic missile and nuclear weapons sites. The SDS provides additional coverage in these areas.

The FLTSATCOM spacecraft is a hexagonal structure, nearly 2 meters wide and more than 1 meter high, weighing 1,005 kilograms. It is solar-powered and contains three antenna systems. Its launch vehicle was an Atlas-Centaur. Fully operational by January, 1981, the FLTSATCOM was built by TRW. It was set up to establish lines of communication among the president, troops, and nuclear weapons anywhere around the world. There are relay links among nine hundred Navy ships, submarines, and aircraft of the Navy and selected ground stations; among more than one thousand United States Air Force aircraft and ground stations; and for the Strategic Air Command. Only the polar regions are not covered.

The SDS system denotes satellites in highly elliptical orbits similar to those of the Soviet Molniya satellites. The SDS spacecraft were designed to cover the polar cap areas not covered by FLTSATCOM. They also are set to link with spy satellites directly. Weighing approximately 700 kilograms, they were launched by a Titan 3B/Agena D rocket from Vandenberg Air Force Base, with a planned life span of ten to one hundred years.

The Leasat system is also used for classified military work. It is intended to replace the FLTSATCOM system when the latter ceases to be operational. Designed to be launched from a space shuttle, each satellite is 426 centimeters in diameter and 617 centimeters in length. The total payload is 1,288 kilograms and includes two large UHF antennae.

Knowledge Gained

Generally, military communications satellites are not designed to gain new knowledge; they are merely expected to provide reliable service. Yet some appear to have had experimental components. For example, the Tactical Communications Satellite, or Tacsat, was first launched in 1967 as part of an experimental system. Tacsat went into orbit as the largest communications satellite in the West. Its size, 0.5 meter in diameter and 1 meter in height, enabled it to work with a network of very small ground stations. A cylindrical satellite, it was designed and built by Hughes Aircraft Company for a February, 1969, launch at the direction of the United States Air Force and Missile Systems Organization. It was placed in geosynchronous orbit and was the first used for the experiment between tactical communications among the Air Force, Navy, and Army ground stations and mobile field units, aircraft, and ships. Submarines, it was found, could be connected to command stations almost anywhere on the planet.

The knowledge gained by U.S. military communications satellites can best be summarized by examining the Military Satellite Communications (MILSATCOM) system, designed to serve both strategic and tactical forces. Labeled Milstar, it includes many improvements. It is designed to cover the entire globe and have the ability to detect both the deepest

Milstar Satellite Communications System. (USAF)

submarine and activity on the highest mountain. In order to achieve worldwide communications capability, Milstar must use the extremely high-frequency radio band. Milstar is also designed to be jam-resistant; with the ability to confuse an enemy's alternative signals and prevent interference. No military tool can survive all battle conditions, especially in a nuclear age, but planners would like satellites to be as survivable as possible. Milstar has built-in components to enable it to fend off missiles or enemy satellites. Researchers have designed the Milstar to stay in orbit for a decade or more to allow time to develop and perfect a replacement system.

Context

In the nineteenth century, ships carried messages from Europe to the United States, Asia, and Africa. A singular revolutionary change came in the early part of the twentieth century, with radio communications from land to land, ship to ship, ship to shore, and later aircraft to shore and ship and other aircraft. Radio communications permitted the rapid and reliable exchange of information and facilitated the logistics of managing a global war on the scale of World War II.

Another generation of revolutionary change came with radio communications by satellite. Such an advanced network permitted the increase of reliable communications worldwide. Within massive grids, planes, missiles, submarines, ships, ground troops, and bases could be linked. Fast, reliable communication has always been a military necessity, and the military communications satellites provide this. Because of their high costs, all spacecraft are designed to be as efficient as possible. Only military communications satellites are specifically required to perform uniformly well twenty-four hours a day throughout the year for a period of years.

The effect of satellites on the United States' military was evident by the late 1960's. The IDCSP, for example, provided a crucial link for the United States in its war in Vietnam. With IDCSP satellites, information and reports could flow from South Vietnam through Hawaii to headquarters in Washington, D.C. Thus it was possible to run the war from the United States, indeed from the White House itself. Still, the United States seemed not to have pushed as hard in this direction as the Soviet Union had. The Soviet Union's military launched almost twice the number of satellites as the United States' military. Virtually all unmanned Soviet spacecraft were launched into Earth orbit by the military, many for communications purposes. The military employs satellites for many noncommunications functions, including navigational assistance, weather forecasting, surveillance, nuclear testing and radiation detection, and research and technological development. Yet certainly one of the most valued benefits of these satellites has been improved communications.

Bibliography

Brown, Martin P., Jr. *Compendium of Communication and Broadcast Satellites: 1958 – 1980*. New York: IEEE Press, 1981. This reference guide provides a listing and accessible description and analysis of the various military satellite communications systems at the end of the 1970's. In diagrammatic fashion, it examines the inner workings of all satellites that were not classified. A wonderful guide to a secretive industry.

Chetty, P. R. *Satellite Technology and Its Applications*. Blue Ridge Summit, Pa.: TAB Books, 1987. This book provides a useful overview of the functions of communications satellites in space and includes sections on the workings of the military satellites.

Jarett, David, ed. *Satellite Communications: Future Systems*. Vol. 54 in *Progress in Astronautics and Aeronautics*. New York: American Institute of Aeronautics and Astronautics, 1977. Chapter 12, by William H. Curry, Jr., details the architecture of a military communications system. Curry offers the reader an interesting discussion of how military communications satellites differ from civilian satellites.

Porter, Richard W. *The Versatile Satellite*. New York: Oxford University Press, 1977. This short book provides a fine introduction to the various uses of satellites. Chapter 4 deals with communications satellites and covers their military uses.

Raggett, R. J. *Jane's Military Communications*. 7th ed. London: Jane's Publishing Co., 1986. This catalog examines all sorts of military communications hardware, including satellites. Updated annually, this fascinating guide is somewhat technical.

Turnill, Reginald, ed. *Jane's Spaceflight Directory*. London: Jane's Publishing Co., 1987. A comprehensive guide to spacecraft. Somewhat technical but still appropriate for the layman.

Yenne, Bill. *The Encyclopedia of U.S. Spacecraft*. New York: Exeter Books, 1985, reprint 1988. A fine, well-illustrated introduction to all spacecraft, with a discussion of major military communications space satellites. This volume is useful for the beginning student.

Douglas Gomery

MISSION TO PLANET EARTH

Date: Beginning 1990
Type of satellite: Unmanned Earth-orbiting space program

The Mission to Planet Earth (MTPE) is a long-term program within the National Aeronautics and Space Administration (NASA) to study how the global environment is changing. It uses the unique perspective available from space to observe, monitor, and assess large-scale environmental process, with an emphasis on climate change. Satellite data is complemented with data collected by aircraft and ground stations.

PRINCIPAL PERSONAGES

CHARLES F. KENNEL, NASA Associate Administrator
for the Office of Mission to Planet Earth
GHASSEM ASRAR, Earth Observing System (EOS)
Program Scientist
MICHAEL KING, Earth Observing System (EOS)
Senior Project Scientist

Summary of the Program

The Earth has always experienced changes in its environment and climate. Until very recently, however, this was viewed as being a gradual process and interested only scientists. But today, issues such as ozone depletion, global warming, deforestation, desertification, acid rain, rising sea levels, extreme weather conditions, and reduction in biodiversity have served as increasing evidence of rapid, large-scale changes in the global environment. There is a growing amount of evidence that the normal rate of global environmental change is being accelerated by human activities. Exhaust fumes from automobiles, factories, and power plants increase the level of greenhouse gases in the atmosphere; certain chemicals used in manufacturing deplete the ozone layer that protects us from ultraviolet light; stripping and burning of forests changes the overall balance of life on the planet. Within only a few decades, we have made significant changes in the world about us without understanding the long-term effects on the ability of the environment to sustain life.

In order to understand the global implications of our activities, we must observe and monitor the Earth as a whole. Scientific communities within the United States are working with international scientific communities to examine these issues on a sound scientific basis. These research efforts are being coordinated under the U.S. Global Change Research Program (USGCRP), the International Geosphere-Biosphere Program (IGBP), and the World Climate Research Program (WCRP). NASA's contribution to this effort is the Mission to Planet Earth (MTPE), which uses the unique perspective of space to understand how the Earth's environment is changing and man's role in that change.

MTPE is a long-term program to study the Earth using data from spacecraft, supplemented with data from aircraft and ground stations, and making that data available to scientists around the world. A research program supports scientific analysis of the data, and the results of these scientific investigations will be used by decision makers to help them make economic and environmental policy. Interdisciplinary research, education, and international coordination are all major elements of MTPE. In order to accomplish its mission, MTPE will use a constellation of satellites organized in two groups: Earth Probes and the Earth Observing System (EOS).

Most of the Earth Probes will be deployed during the Phase I Program from 1990 to 1998 and will consist of satellites and instruments addressing specific Earth science investigations. There are currently more than thirty such probes planned or operational. The motivation behind these probes is to provide focused missions in a faster, better, and cheaper manner. They are all small to moderate sized satellites with highly focused objectives. Some examples of Earth Probes are the Upper Atmosphere Research Satellite (UARS), the Ocean Topography Experiment (TOPEX/Poseidon), and the Total Ozone Mapping Spectrometer (TOMS/Earth Probe). In addition to Earth Probes already approved, small Earth Probes will be launched as particular observations are requested by the national and international science community or as data gaps develop.

While the Earth Probes are very important and will provide scientists with valuable data, the heart of MTPE is the fifteen-year long EOS program. EOS satellites will be launched during the Phase II Program from 1998 to 2014. These will be a series of polar-orbiting and low-inclination satellites for long-term global observations of the land surface, biosphere, solid Earth, atmosphere, and oceans. EOS satellites will carry instruments supplied by NASA as well as instruments from researchers selected through a competitive process.

Planning for the EOS mission began in the early 1980's and was included in the 1990 Presidential initiative, Mission to Planet Earth. Originally, there were three groups of instruments: EOS-A spacecraft, EOS-B spacecraft, and instruments to be attached as payloads on then planned Space Station Freedom. But in 1991, Congress directed the plans for EOS to be restructured to focus the science objectives of EOS on global climate change; increase the flexibility and survivability of the program by flying the instruments on multiple smaller platforms instead of a few large satellites; and reduce the cost of EOS from $17 billion to $11 billion through fiscal year 2000. In 1992, Congress further reduced the budget for EOS through fiscal year 2000 to $8 billion, and to $7.25 billion in 1994. These budget cuts required elimination of some satellite missions, consolidation of others, and a greater reliance on foreign contributors. This left EOS with three main satellite programs, in addition to several smaller ones, for a planned total of seventeen satellites. The main programs are EOS-AM, EOS-PM, and EOS-CHEM, with three satellites in each of these programs.

EOS-AM will study the land and ocean surfaces, clouds, aerosols in the air, and the radiative balance (the amount of heat coming into and leaving Earth). EOS-PM will study clouds, precipitation, the radiative balance, terrestrial snow, sea ice, sea-surface temperature, and atmospheric temperature. EOS-CHEM will study atmospheric chemicals and their transformations and oceanic surface stress.

All satellites in these programs are to be in Sun-synchronous polar orbits, meaning they orbit over both of the Earth's poles and they always cross the equator with the same angle relative to the Sun. The first launch in the series will be EOS-AM-1 in June, 1998, followed by EOS-PM-1 in 2000 and EOS-CHEM-1 in 2002. All satellites are designed to last five years, but because of the budgetary restraints, they are scheduled to be replaced every six years. This means that each satellite will have to survive one year longer than designed in order to have continuous observations. With the exception of EOS-AM-1, all satellites will be functionally identical. Using the same satellite design helps to reduce costs.

EOS-AM-1 will support an instrument mass of 1,155 kilograms, provide an average power of 2.5 kilowatts, and provide an average data rate of 18 Mbps (million bits per second). All other spacecraft will support an instrument mass of 1,100 kilograms, provide an average power of 1.2 kilowatts, and an provide an average data rate of 7.7 Mbps.

All satellites in the three series will be placed into orbits with altitudes of 705 kilometers and an inclination of 98.2 degrees. Since EOS-AM studies mainly surface features it will observe the Earth in the morning when cloud cover is at a minimum, hence its name. It will cross the equator going from north to south at 10:30 A.M. local Sun time. Likewise, EOS-PM will provide complementary observations of the Earth in the afternoon and will cross the equator from south to north at 1:30 P.M. local Sun time. EOS-CHEM will cross the equator from south to north at 1:45 P.M. local Sun time. These orbits will allow the satellite to view the same region on the surface of the Earth every 233 orbits, or about every 16 days.

Space-based observations are not sufficient to accomplish all mission objectives, and extensive ground and airborne based observations will be necessary to provide a complete picture. These campaigns are important components of MTPE. More than any other factor, the key to success for MTPE is the commitment to make Earth science data easily available to researchers everywhere. This will be accomplished with the EOS Data and Information System (EOSDIS). EOSDIS is a comprehensive data and information system designed to perform a wide variety of functions to support scientists everywhere. Some of the services provided by EOSDIS are for the casual user, while some are restricted and designed for select scientists chosen by NASA. Many other services will fall between.

NASA designed EOSDIS as a distributed, open system. This permits the distribution of EOSDIS components to various locations to take best advantage of different institutional capabilities and science expertise. There is an EOSDIS Core System (ECS) that provides centralized control and site-unique extensions. Although physically distributed, the components are integrated together and will appear as a single entity to users.

Most interaction with EOSDIS will occur through nine EOSDIS Distributed Active Archive Centers (DAACs). Most of this interaction will occur through human-computer interface over the internet, but each DAAC also has a User Support Services to assist users in data acquisition, search, access, and usage. EOSDIS will charge users not more than the costs of fulfilling an order. This cost may include charges for machine use to execute searches, electronic- or hardcopy delivery, and performing any unique processing of the data. EOSDIS may also charge for packaging and media used for data delivery but will not charge for the institutional costs of a DAAC. To promote the use of EOSDIS, first-time users will receive a credit balance. The user will be charged only after this credit has been used.

Ultimately, the product of MTPE is to educate decision makers and the public concerning the effects of man's activities on global climate change. One of the four goals of the program, as cited in the Mission to Planet Earth Strategic Plan, is to "foster the development of an informed and environmentally aware public." In support of the NASA education initiative, the Office of MTPE is working with NASA's Education Division to establish a focused education strategy. This strategy has the objectives of training the next generation of scientists; educating and training educators; raising awareness of decision-makers and the public regarding global climate change; improving science and math literacy; improving the interface between educators and scientists; increasing resources availability; increasing the knowledge base; and encouraging the development of external resources capable of translating scientific research into usable forms for a variety of customers.

Knowledge Gained

With most of MTPE's missions flying in the twenty-first century, we can expect to see results for several decades to come. But this does not mean that MTPE has not already had an impact on our understanding of global climate change. We can examine the results of just one satellite mission under MTPE to provide us with a preview of what we can expect. In 1995, TOPEX/Poseidon, a joint mission between NASA and the French space agency Centre Nationale d'Etudes Spatiales (CNES), successfully completed its three-year primary mission to help scientists study how the Earth's oceans affect our climate. By improving our understanding of how oceans circulate, information gathered by TOPEX/Poseidon is enabling oceanographers to study the way the oceans transport heat and nutrients and how oceans interact with weather patterns. This data allows scientists to produce global maps of ocean circulation. It also provided oceanographers with unprecedented global sea level measurements with an accuracy to less than 5 centimeters. This data has shown a global rise in sea level during the mission time frame of .3 centimeters per year. Further measurements will help determine if this rise is a short-term or a long-term event. TOPEX/Poseidon also detected data concerning the El Niño condition in the equatorial Pacific Ocean that has been linked to global weather patterns. TOPEX/Poseidon is expected to continue making measurements until at least 1999.

Context

Greater scientific understanding and research findings over the last thirty years reveal that the Earth's climate is not static but is actually continuously changing. Increasing evidence that mankind's activities contribute to this change has provided the incentive to launch a major, large-scale study of the global environment. Mission to Planet Earth is NASA's contribution to this worldwide effort. MTPE will study the global climate from space in an effort to increase understanding of many environmental issues facing mankind today.

One of these issues is global warming. Scientists know that the Earth's average global temperature has been increasing since the beginning of the Industrial Revolution. What is currently unknown is exactly what is causing this rise. It could be a natural, cyclical event, or it could be due to gases emitted by man's machines that trap heat in the atmosphere. Some of these gases are carbon dioxide and methane and are called greenhouse gases. Measurements of global levels of greenhouse gases show that these levels have steadily increased and in 1990 were more than 14 percent higher than in 1960. Increases in temperature could alter precipitation patterns, change growing seasons, melt polar ice resulting in coastal flooding, and create new deserts where none currently exists. Many of MTPE's investigations will be to study factors that contribute to global warming, to predict further trends, to determine man's role in this change, and to discover what effect changes in man's activities will have.

Another case of man's activities affecting the environment is the ozone depletion. Ozone is a molecule consisting of three oxygen atoms and exists high in the atmosphere in small amounts. But these amounts are great enough to absorb much of the Sun's deadly ultraviolet light and protect all of us on the Earth's surface. This gas is destroyed, though, when it reacts with chlorine, bromides, and other compounds in the presence of light. The main source of chlorine in the atmosphere is from chloroflourocarbons (CFCs), manmade chemicals used in air conditioning, refrigeration, industrial solvents, and fire fighting agents. Once released into the atmosphere, they gradually drift up through the atmosphere until they reach the ozone layer. Under the proper conditions these CFCs destroy ozone. Space studies have long documented the depletion of ozone, and this data has been supplemented with airborne measurements. As a result, most countries, including the United States, signed the Montreal Protocol in 1989 calling for the phaseout of production and use of CFCs by the year 2000. This phaseout has been accelerated and the United States no longer manufactures any CFCs. There is still much to learn, and MTPE will continue to study ozone depletion.

Another issue that will be studied by MTPE is deforestation. As our cities grow, we have cleared land to make room for structures and have harvested forests for timber or for agriculture. Deforestation has three major impacts on the environment: it reduces the planet's ability to absorb carbon dioxide, a major greenhouse gas; it influences local weather, especially precipitation and water storage; and it reduces biodiversity. Plants store carbon dioxide in their structure. By burning vast areas of timber, carbon dioxide is released into the air. At the same time, the amount of plants removing carbon dioxide from the air has been reduced. MTPE will study global vegetation and other land processes to better understand their role in the global environment.

These are just a sampling of the issues confronting society today and areas of research that MTPE will be dedicated to address. Great effort has already been made to understand our role in the global environment without being able to determine definitive answers. Mission to Planet Earth should extend our understanding greatly and let us know, once and for all, just how much our activities affect the environment we live in.

Bibliography

1995 MTPE EOS Reference Handbook. NP-215. This publication provides a wealth of data on MTPE in general and EOS in particular. Detailed descriptions of the satellite programs, their instruments, the research programs and interdisciplinary programs is provided. Some drawings of satellites are provided. Additional references are provided at the end of each chapter. Contact points for all programs and for EOSDIS is provided. This publication can be accessed over the World Wide Web via the MTPE Home Page, or it can be obtained by writing to EOS Project Science Office, Attn: Charlotte Griner, Code 900, NASA/Goddard Space Flight Center, Greenbelt, MD 20771, phone 301 286-3411, internet cgriner@ltpmail.gsfc.nasa.gov. Suitable for audiences from high school and up.

Committee on the Environment and Natural Resources Research(CENR). *Our Changing Planet — The FY Year U.S. Global Change Research Program*. This publication outlines all of the U.S. Federal Government activities in the area of global climate change. Ground, airborne, and space-based programs are all covered, as well as economic research. Budgets are presented by agency and by scientific element. This report is expected to be published annually. In the title, replace the word 'Year' with the year desired. Suitable for high school and college level audiences.

The Earth Observer. The bimonthly newsletter of the EOS Project Office provides current reports on the status of various elements of the EOS Program. Issues dating back to November/December, 1992, are available on the World Wide Web at http://spso.gsfc,nasa,gov/spso_homepage.html. Suitable for high school and college level audiences.

Christopher Keating

THE NATIONAL AERONAUTICS AND SPACE ADMINISTRATION

Date: Beginning October 1, 1958
Type of organization: Space Agency

The National Aeronautics and Space Administration was formed in 1958. Its purpose is to unite under one administration all U.S. space exploration activities. Although it is a civilian organization, it sometimes undertakes special military projects.

PRINCIPAL PERSONAGES

T. KEITH GLENNAN, NASA Administrator from
October 1, 1958, to January 20, 1961
JAMES E. WEBB, NASA Administrator from
February 14, 1961, to October 8, 1968
THOMAS O. PAINE, NASA Administrator from
March 21, 1969, to September 15, 1970
JAMES C. FLETCHER, NASA Administrator from
April 27, 1971, to May 1, 1977
ROBERT A. FROSCH, NASA Administrator from
June 21, 1977, to January 20, 1981
JAMES BEGGS, NASA Administrator from
July 10, 1981, to February 25, 1986
JAMES C. FLETCHER, NASA Administrator
from May 12, 1986 to 1989
RICHARD H. TRULY, NASA Administrator
from 1989 to 1992
DANIEL GOLDING, NASA Administrator
beginning 1992

Summary of the Agency

The National Aeronautics and Space Administration (NASA) came into being October 1, 1958, when President Dwight D. Eisenhower signed into law the National Aeronautics and Space Act of 1958.

On October 4, 1957, the Soviets launched the world's first artificial satellite, Sputnik 1. At that time, the United States had several space efforts spread across a large number of government agencies. Sputnik's launch convinced U.S. officials that in order to compete with the Soviet Union in space, the United States needed to establish one comprehensive space agency. To that end, NASA was founded. NASA absorbed the existing National Advisory Committee for Aeronautics (NACA) and acquired personnel and projects from other government programs, including Project Vanguard from the Naval Research Laboratory, lunar probe programs from the Army, and lunar probe and rocket engine programs from the Air Force.

Initially, NASA's resources included eight thousand employees, three laboratories, two flight stations, and an annual budget of $100 million. Later, the Jet Propulsion Laboratory, in Pasadena, California, and the Army Ballistic Missile Agency, in Huntsville, Alabama, were added to the facilities. Today, NASA has centers throughout the nation, including NASA Headquarters, Washington, D.C.; Ames Research Center, Mountain View, California; Dryden Flight Research Facility, Edwards, California; Goddard Space Flight Center, Greenbelt, Maryland; the Jet Propulsion Laboratory, Pasadena, California; Johnson Space Center, Houston, Texas; Kennedy Space Center, Cape Canaveral, Florida; Langley Research Center, Hampton, Virginia; Lewis Research Center, Cleveland, Ohio; Marshall Space Flight Center, Huntsville, Alabama; National Space Technology Laboratories, Bay St. Louis, Mississippi; and Wallops Flight Facility, Wallops Island, Virginia. In addition, Marshall Space Flight Center operates two more centers: the Michoud Assembly Facility, in New Orleans, and Slidell Computer Complex, in Slidell, Louisiana. The White Sands Test Facility, in New Mexico, is operated by Johnson Space Center. Kennedy Space Center is responsible for the operation of the Space Transportation System Resident Office at Vandenberg Air Force Base, California. Plum Brook Station, which provides large-scale, specialized research installations, is operated by Lewis Research Center.

NASA Headquarters manages the spaceflight centers, research centers, and other NASA installations. Headquarters officials direct the planning and management of NASA research and development programs, determine what projects and programs will be undertaken, and review and analyze all phases of the activities.

Ames Research Center performs research and development in aeronautics, space science, life science, and spacecraft technology. Ames has responsibility for the Pioneer series of spacecraft, and Ames scientists study the origins of life and provide medical support for manned missions. Ames contributes to the space shuttle program by researching heat protection systems and conducting wind tunnel experiments.

Dryden Flight Research Facility is administered by Ames Research Center. Its primary research goal is the study of high-speed aircraft, but its personnel also work on vertical takeoff and landing, low-speed flight, supersonic flight, hypersonic flight, and flight vehicle reentry.

Goddard Space Flight Center conducts automated spacecraft and sounding rocket experiments to provide data about Earth's environment, Sun-Earth relationships, and the universe. Goddard is the home of the National Space Science Data Center, the repository of data collected from spaceflight experiments. The center is responsible for the Delta launch vehicle and has the lead role in the management of the international Search and Rescue Satellite-Aided Tracking (SARSAT) program.

NASA's Jet Propulsion Laboratory (JPL) is a government-owned facility that is staffed and managed by the California Institute of Technology. The laboratory is active in deep-space missions and data acquisition and analysis for those missions. JPL scientists also study solid- and liquid-propellant spacecraft engines and space-craft guidance and control systems. The laboratory operates the Deep Space Communications Complex, which is one station of the Deep Space Network.

Johnson Space Center (JSC) was established in September, 1961. It is NASA's primary center for the design, development, and manufacture of manned spacecraft; selection and training of spaceflight crews; and ground control of manned flights. JSC is the lead NASA center for management of the space shuttle. The Mission Control Center is one of NASA's best-known facilities, since it is from there that manned flights, from Gemini IV to the space shuttle missions, have been monitored. Television broadcasts of spaceflights have featured the activity at Mission Control. JSC also operates the White Sands Test Facility, which is responsible for the space shuttle propulsion system and power system and for materials testing.

Kennedy Space Center (KSC) serves as the main facility for the testing and launch of space vehicles, manned and unmanned. It also oversees launches at the Air Force's Western Space and Missile Center at Vandenberg Air Force Base in California. KSC concentrates on the assembly, testing, and launch of space shuttle vehicles and their payloads. KSC operates the Space Transportation System Resident Office at Vandenberg Air Force Base in California. The office supports the Air Force in the design, construction, and operation of the space shuttle launch and landing site.

Langley Research Center's primary mission is to research and develop new aircraft and spacecraft. The center's scientists study ways to increase the performance and efficiency of air and space vehicles. Langley has a variety of wind tunnels to aid this research. The National Transonic Facility, a cryogenic, high-pressure wind tunnel, is also located at Langley.

Lewis Research Center is NASA's main center for research and development in aircraft propulsion, spacecraft propulsion, space power, and satellite communications. Power system research includes studies on the conversion of chemical and solar energy into electricity. Lewis manages the Centaur launch vehicle, a second-stage rocket used with the Atlas first stage. Lewis also operates Plum Brook Station, which provides specialized research installations.

Marshall Space Flight Center (MSFC) was established in 1960 by a team of former Army rocket experts headed by Wernher von Braun. MSFC is responsible for the development of the space shuttle's main engines, solid-fueled rocket boosters, and external propellant tank. MSFC is also responsible for the Hubble Space Telescope, the Spacelab research facility, and many other programs.

The Michoud Assembly Facility and Slidell Computer Complex are operated by MSFC. Michoud's primary function is the design, engineering, manufacture, assembly, and testing of the space shuttle's external tank. The computer complex is responsible for NASA's computational requirements. It also provides computational services for the National Space Technology Laboratories (NSTL).

NSTL became a separate field installation in 1974; it was previously called the Mississippi Test Facility. Since 1975, NSTL's function has been to support the development of the space shuttle's main engine. Between 1965 and 1970, NSTL oversaw the static test firing of the first and second stages of the Saturn 5, the launch vehicle used in the Apollo manned lunar landing missions and the Skylab program.

Wallops Flight Facility is part of Goddard Space Flight Center. Wallops' main responsibility is to manage NASA's suborbital sounding rocket projects, and its responsibility is complete; it covers this program from mission planning to landing and recovery. Wallops personnel design and develop payloads, control launches, track rockets, and acquire data. They also monitor, schedule, and control all NASA balloon activities, and to this end, they operate the National Scientific Balloon Facility at Palestine, Texas.

NASA has made a number of international agreements, results of the mandate given it by the 1958 National Aeronautics and Space Act, which required NASA to cooperate with other nations in peaceful aeronautical and space activities. Cooperation with other nations contributes to broad national goals by stimulating scientific and technical contributions from abroad, providing access to foreign areas for tracking stations and possible emergency landing sites, enhancing satellite experiments with foreign scientific data, increasing the possibility for the development of space technology, extending ties to foreign scientific and engineering communities, and supporting U.S. foreign policy.

One of the best-known agreements involved the Apollo-Soyuz Test Project. In July of 1975, three American astronauts in an Apollo capsule and two Soviet cosmonauts in a Soyuz capsule met in space by means of a docking module. The two capsules remained docked for forty-four hours while the astronauts and cosmonauts exchanged visits. It was hoped at the time that increased cooperation between the two countries would result from this joint mission.

Other international projects have included cooperative spaceflights involving American and foreign astronauts, cooperative satellite experiments in which all or part of a satellite has been built by foreign governments, foreign space missions that carried U.S. experiments, and solar energy projects. More such projects are under way, and at least seventy-two countries have participated in scientific and technical information exchanges with the United States.

NASA's budget is substantial. In 1965, it was $5.092 billion. It had dropped to $3.269 billion by 1975, but it rose to $7.251 billion over the next ten years.

Context

A nation receives a good return on the money it spends on a space program. One of the benefits of a space program is the information it can provide about the solar system and the universe. For example, the Voyager flybys of Jupiter and Saturn allowed U.S. scientists to study those planets' atmospheres in detail. Information obtained from such studies may appear to have no application, but, in fact, it sheds light on the structure of Earth's atmosphere.

Another space program benefit is improved military technology. In most space studies, rockets or missiles must be used to lift the payload or experiment, and the rockets' engineering must be reliable. Although NASA is a civilian agency, it contracts with the military for various projects. Also, any rocket that can put a satellite in orbit for peaceful purposes can put one in orbit for military purposes.

Perhaps most important are the space program "spin-offs."

A spin-off is the application of space-age technology to other uses. For example, when the question of humans entering space was first raised, NASA scientists quickly determined that medical data would have to be transmitted from the spacecraft back to Earth. Medical technology had to be developed which would transmit data on astronauts' heart activity, respiration rate, and temperature. When the devices to perform these functions were developed, they were quickly adapted for use in hospitals. Many patients may owe their lives to the fact that an electrocardiogram could be read at some distance from a hospital bed and thereby alert medical personnel to an impending heart attack. Other spin-offs include image processing, advanced aircraft, drag reduction techniques for racing yachts, eyeglass filters, solar water heaters, water filters, relaxation systems, speech aids, and a "cool suit." The same image processing techniques that remove interference from data returned by deep-space probes, such as Voyager, can be used to make ordinary X-ray radiographs clearer. The cool suit is based on technology developed at Ames Research Center; a liquid-cooled helmet liner for military pilots was invented after heat exhaustion appeared to be the cause of some accidents.

A national space program also has intangible benefits. Humans seem to need the kinds of challenges offered by space exploration.

Bibliography

Couper, Heather, and Nigel Henbest. *Space Frontiers*. New York: Viking Press, 1978. This book contains more than one hundred color illustrations. It begins with the early astronomy of the Egyptians and Greeks and continues through modern astronomy, including radio astronomy. NASA's role in space exploration is discussed in the chapter "Man Conquers Space." The possibility of space colonization is evaluated. Suitable for general audiences.

De Waard, E. John, and Nancy De Waard. *History of NASA: America's Voyage to the Stars*. New York: Exeter Books, 1984. Illustrated with many color and black-and-white photographs, this book begins with a discussion of mankind's early attempts to fly and proceeds through descriptions of NASA's programs. It also speculates about future space exploration. For general audiences.

Haggerty, James J. *Spinoff*. Washington, D.C.: Government Printing Office, 1985. *Spinoff* is one of NASA's annual reports. This issue contains approximately 130 pages and is illustrated throughout with color photographs. The publication covers the many technological developments for which NASA is responsible and which have applications in fields other than space science. For general audiences.

Koppes, Clayton R. *JPL and the American Space Program: A History of the Jet Propulsion Laboratory*. New Haven, Conn.: Yale University Press, 1982. This volume discusses research activities at JPL during World War II and the years that followed. Describes JPL's internal organization and the relationship between JPL and NASA. Most of the text is devoted to JPL space projects.

Levine, Arnold S. *Managing NASA in the Apollo Era*. NASA SP-4102. Washington, D.C.: Government Printing Office, 1982. A highly detailed history of the difficulties encountered and lessons learned in managing NASA, a growing agency, as it developed a complex program to place humans on the Moon.

National Aeronautics and Space Administration. *Marshall Space Flight Center 1960 – 1985: Twenty-fifth Anniversary Report*. Washington, D.C.: Government Printing Office, 1985. A celebratory booklet on Marshall's history and its role in the U.S. space program. Includes many photographs of Marshall personnel at work, contrasting early and later activities. Contains a timeline showing the dates for major projects.

————. *NASA: The First Twenty-five Years, 1958 – 1983*. NASA EP-182. Washington, D.C.: Government Printing Office, 1983. Designed to be a teacher's resource, this illustrated volume covers the history of NASA and its many programs. Includes sections on the pre-NASA U.S. space program, manned space-flight, and the uses of space technology. An appendix lists all major NASA launches through 1983. Suitable for high school and college students.

Time Life Books Editors. *Life in Space*. Boston: Little, Brown and Co., 1984. A volume in a Time-Life series, this work contains many photographs and illustrations. It covers the Mercury, Gemini, Apollo, Skylab, and space shuttle programs. Suitable for the nonspecialist.

T. Parker Bishop

NATIONAL AEROSPACE PLANE

Date: Beginning February 4, 1986
Type of technology: Reusable launch vehicle

The National Aerospace Plane (NASP) is a program to design and build a Single-Stage-To-Orbit (SSTO) vehicle that will land and take-off on conventional runways. The plane, traveling at hypersonic speeds, will require the development of new materials and technologies to become a reality.

PRINCIPAL PERSONAGES

RONALD REAGAN, former President of the United
States who established the Aerospace Plane as a
national goal

ROBERT HEAPS, U.S. Air Force Colonel and Program
Director for NASP

ROBERT BARTHELEMY, former Program Director for
NASP

PHIL AITKEN-CADE, U.S. Air Force Colonel and
Deputy Program Director for NASP

GEORGE BAIRD, Chief Program Engineer for NASP

FRANK BERKOPEC, chief of Space Vehicle Propulsion
at National Aeronautics and Space
Administration's (NASA) Lewis Research Center

STEPHAN WOLANCZYK, subsystem Technology
Development Manger for NASP

MARGARET WHALEN, Coleader of Slush Technology
Program at NASA's Lewis Research Center

Summary of the Technology

President Ronald Reagan in his February, 1986, State of the Union address asked for congressional support of a National Aerospace Plane that he referred to as the "Orient Express." Reagan's vision has become the X-30 National Aerospace Plane (NASP), a technological device capable of reaching Mach 25 (twenty-five times the speed of sound or 28,000 kilometers per hour) and delivering payloads into low Earth orbit. The NASP program, if successful, would lead to a new concept of aerospace vehicles including a commercial transport, a spaceplane replacing the shuttle, and a series of military Transatmospheric Vehicles or TAVs.

The NASP program was started secretly in the early 1980's by the Defense Advanced Research Projects Agency (DARPA). The engineering and scientific research tasks were allocated to various government and industrial labs across the country, with the main program office located at Wright Patterson Air Force Base in Dayton, Ohio. The military importance of TAVs is reflected in the NASP funding. The U.S. Air Force, U.S. Navy, DARPA, and the Strategic Defense Initiative Organization (SDIO) all contribute 80 percent to the costs, while NASA's share is only 20 percent. The total cost to build and fly the X-30 has been estimated to exceed $5 billion dollars.

The decision to build the NASP is based upon the view that less costly access to space travel will benefit both commerce and the military. Current methods of space travel are very expensive and require cumbersome vehicles. The launch of a space shuttle payload costs about $7,500 per kilogram with the time between launches often measured in months. As a contrast, there are significant savings to launch, land, and service a SSTO type craft. The reduction estimates vary from only one-tenth to only one-one-hundredth of the shuttle's cost per kilogram of payload placed in a low Earth orbit.

A radically new design concept like the NASP emerged from earlier research aircraft that were given an X classification. The North American X-15 was able to achieve Mach 6.7 (more than 8,000 kilometers per hour) and reach near-space altitudes in excess of 100 kilometers. The X-15 flights yielded valuable information on frictional heating from the atmosphere. A metal alloy known as Inconel-X was used in the metallic skin of the aircraft, which could withstand temperatures to 700 degrees Celsius. Other areas of the plane heated to 1,100 degrees Celsius but were coated with an ablative heat-shield.

The X-24s were a series of lifting-body aircraft with the purpose of demonstrating that high-speed craft with low lift and high drag coefficients could land on conventional runways. The X-24 lifting bodies attained Mach 1.8 (2,200 kilometers per hour) yet made unpowered, accurate, and precise landings. This wingless craft generated lift in the atmosphere through a streamlined but stubby shape. The shape and construction of both of these unusual aircraft influenced the design of the X-30.

The X-30, like the X-15, X-24, and the shuttle, will use rocket propulsion, but not for the entire flight. Unlike the shuttle, the X-30 will not discard its booster rockets but will retain them for reentry. This concept is contrary to conventional practice, which uses a series of rocket stages to decrease the vehicle's weight in order to increase thrust. A SSTO craft

An artist's conception of the National Aerospace Plane that would have been used for low cost transport into near Earth orbit. (NASA)

will require that the engine efficiency be as high as 95 percent to extract as much thermal energy as possible from each kilogram of fuel. Researchers also believe that the vehicle's empty weight must be less than 25 percent of its takeoff weight to achieve orbit.

The thrust required to deliver the X-30 into orbit will be produced from a propulsion system based on the ramjet and scramjet, as a single air-consuming engine type will not operate over the required range of altitudes and speeds. Ramjet engines have been in operation since the 1940's on missiles. A ramjet employs a cone-shaped structure in the front of the engine to reduce the speed of the incoming air from supersonic to subsonic range. The air becomes compressed and heated, passing over fuel injectors that ignite the mixture. Resulting hot gases expand out of the tube-shaped engine at a higher velocity than the incoming air, providing thrust.

Ramjet engines operate best from Mach 1 (the speed of sound) to Mach 6; at greater speeds they become overheated and ineffective due to the drag induced by the compressed

air. To exceed Mach 6, the ramjet must be able to effectively mix fuel with air moving through the engine at supersonic speeds. Engineers compare this task to that of igniting a match in hurricane force winds. The principle of producing combustion in a supersonic airflow is termed a supersonic-combustion ramjet or scramjet.

With air moving with such high velocities inside the scramjet, combustion must be successfully achieved within only one or two milliseconds. Hydrogen becomes the desired fuel due to the rapid burn-rate, low weight, and high thermal energy, but there are problems. The low density means that sufficient fuel to propel the SSTO craft will occupy a large volume. Hydrogen is an extremely inflammable gas, as aviation history reveals, and requires special facilities for storage and refueling.

An engine under serious consideration for the NASP is a long and rectangular hybrid of the ramjet and scramjet. This engine will be able to shift from the ramjet to scramjet mode at higher speeds without substantially changing the internal airflow path. Engineers indicate that a single engine combining both modes of operation is preferable to building and attaching the two separate engines because of the savings in overall weight. To allow the engine to run under such a range of conditions, fuel injection must be varied in response to the combustion region, which moves upstream with an increase in flight speed. Groups of fuel injectors will be installed at critical points inside the combustion chamber walls.

Ramjet and scramjet engines both require air to sustain combustion. The scramjet will be functional at very high altitudes, but just how high is unknown. At these high altitudes and speeds the engine's thrust will decrease at the expense of drag. When drag forces overcome the thrust, the aircraft must turn off the scramjet engines and rely on rocket propulsion to reach orbit. The X-30 will use the same rocket engines to gain reentry from orbit and enter a gliding trajectory back to Earth. The X-30, unlike the shuttle, will be able to restart its engines once it reenters the atmosphere, allowing a greater margin of safety for landings.

Five contractors working with NASP's Joint Program Office in Dayton, Ohio, have designed a composite configuration. The configuration that has emerged is a two-seat aircraft about the size of a Douglas DC-10 airliner. The craft, a massive structure of engines and cryogenic fuel tanks, would measure between 45 and 60 meters long and have a mass of between 100,000 and 140,000 kilograms. The fuselage of the craft is very broad, to provide lift and accommodate multiple scramjets underneath. The wings are very stubby, providing flight control but little lift.

The wider fuselage permits a more efficient air flow path, allowing air collection and compression prior to the engine inlets. The wide body design also accommodates a larger fuel tank without significantly increasing drag forces. The X-30 was designed so that at high speeds the thrust, drag, and lift forces all pass through the center of gravity. To achieve this, the wings must be movable and placed in the neutral position upon reaching hypersonic speeds. At lower speeds, the wings provide lift that acts as a control surface to stabilize the craft. At hypersonic speeds, control surfaces are not necessary since the forces are stabilized through the center of gravity

Actual testing of the X-30 design has been limited to wind tunnels that cannot attain the required airspeeds that the full-scale prototype will reach. Engineers have overcome some of these drawbacks by modeling airframes on supercomputers. Air flow down to as small as six one-hundredths of a square millimeter across the airframe can be modeled. Supercomputer time alone over the first five years of the project amounted to close to 50 percent of the national total. Flight materials and systems will also be tested when attached to conventional rocket launch vehicles.

A major consideration surrounding the X-30 design requires the development and testing of entirely new materials. Sections of this aircraft must be able to withstand temperatures as high as 2,800 degrees Celsius while resisting tremen-

dous pressures generated by hypersonic flight in the atmosphere and from orbital reentry. These materials must maintain high strength yet be very light in weight. Titanium aluminum boron alloys sustain strength to 750 degrees Celsius and carbon-carbon composites can resist temperatures to 3,000 degrees Celsius.

These materials alone will not be sufficient to absorb all of the heat generated by this vehicle. Active cooling systems utilizing the fuel as a coolant have been used on the SR-Blackbird reconnaissance plane. The hottest sections of the airframe, the nose, engines, and leading edges can be actively cooled with cold circulating fuel. The X-30's fuel tanks will hold liquid hydrogen as a slush cooled to minus 242 degrees Celsius. The semi-solid hydrogen slush that can be contained in a smaller volume than liquid hydrogen alone will be pumped through the hottest portions of the aircraft, absorbing heat.

The NASP program has yet to build a series to test devices that will demonstrate the capabilities of the various technologies. Large funding requests have delayed obtaining the required budget from Congress. As a result, the U.S. Air Force and NASA have delayed the development of SSTO demonstrators in favor of testing a series of experimental devices atop Minuteman rockets in the late 1990's. It is hoped that these experiments can help resolve a number of program issues resulting from the lack of data on the performance of scramjets as well as the absence of data on the control and flight characteristics of hypersonic vehicles.

Knowledge Gained

Research and development stemming from the NASP program have resulted in new devices, materials, and technologies. NASP contractors have produced a titanium alloy that will be used in the medical field and in chemical processing. The light weight and strength make the alloy suitable for hips and other human joints. The alloy's resistance to corrosion will find use in pipes and valves that transport corrosive gases.

Composites developed for aircraft engine components could be used in other types of power-generating turbine machinery that are exposed to high temperatures. Carbon-carbon composites have been developed that are able to resist temperatures as high as 1,700 degrees Celsius, which will be encountered on the leading edges of the flight surfaces of the NASP. If given an anti-oxidant coating of these carbon materials, power plant combustion chambers should become more efficient, with less pollution.

Texas Instruments' Metallurgical Materials Division, under contract to the NASP program, has succeeded in developing a titanium-aluminum alloy that can be rolled to foil thickness. Direct outcomes of the process were a higher yield and a cost savings of from $6,600 dollars per kilogram to only $1,500 to $2,000 per kilogram. The foils will be applied to metal matrix composites for gas turbine powerplants as well as to combustor and exhaust components. The foil will find other commercial applications in heart valve assemblies and pacemakers due to its high strength, low weight, and compatibility with human tissue.

Titanium foils are being manufactured into honeycomb materials and heat exchangers. Honeycomb materials of titanium are light in weight, strong, and resist high temperatures, and they find use in the aircraft industry. In the construction of heat exchangers for NASP, Texas Instruments bonds separate layers of titanium, copper, and nickel foil to a stainless steel sheet. Heating of the stainless steel sheets allows the foils to alloy into a brazing material. The titanium-copper-nickel brazing material forms joints of higher yield and better quality at a lower cost.

In order to model high-speed airflow, NASP researchers have modified fluid dynamics software to study structural and thermal analysis of airframes and engines. The huge volume of computational data needed to model these physical systems has produced new computer algorithms that can run at high speeds. This field known as computational fluid dynamics is expected to be applied to the study of commercial engine combustion and brake systems, as well as to wind and ocean current circulation patterns.

NASP's supercryogenic fuel requirements have led to the development of special fuel tanks constructed of graphite epoxy that provides high strength and low weight. Advanced turbomachinery pumps developed for the cooling systems will allow engineers to reduce the overall weight by as much as twelve times. The fuel under development for the NASP will be a slush, a mixture of both solid and liquid hydrogen, as more hydrogen can be carried in a fuel tank this way. Tests conducted by NASA have already demonstrated the feasibility of producing large amounts of slush hydrogen.

Context

Progress in the areas of propulsion, high temperature materials, and supercomputer modeling distinguishes the NASP program from earlier aerospace planes and vehicles. Kerosene, used for most conventional jet aircraft, is not a suitable fuel for hypersonic transports designed to travel two or more times faster than the speed of sound. Air-consuming propulsion systems, like the scramjet, by using atmospheric instead of on-board oxygen, will be able to carry much larger payloads. A scramjet engine alone, however, will not be capable of reaching orbital altitudes that are well above the atmosphere.

The NASP will also have rocket engines to operate in space and for reentry but, unlike the shuttle, will not discard them. A vehicle able to take off and land from a runway and attain an orbit without resorting to rocket staging offers many advantages, including more frequent return flights and cost effectiveness. A Single-Stage-To-Orbit (SSTO) craft like the NASP must burn a low-density fuel such as hydrogen to save weight.

To carry enough hydrogen for the flight requires very large fuel tanks, which contributes to a large, bulging fuselage. The fuselage must therefore be designed to produce both lift and directed airflow. One solution to the problem of large fuel tanks was resolved with the realization that a larger quantity of hydrogen could be carried in these tanks if at least past of it was in a solid state. The solution was to develop a slush hydrogen technology that could produce large quantities of this liquid-solid mixture.

A major challenge to production of the NASP has been the development of materials that can maintain strength under the extreme conditions of high temperatures and pressures encountered during hypersonic flight. A major breakthrough was preventing carbon-carbon composites, which are very light and strong, from burning in the presence of oxygen. Silicon-carbide ceramic coatings, for example, would seal out oxygen but were subject to cracking due to differences in the thermal expansion rates of these two materials. The answer was to layer the material with a silicon sealer and to add an oxygen inhibitor to the carbon matrix.

The ability to investigate fluid dynamics on supercomputers has enabled researchers to model characteristics of high-speed flight that would not otherwise be available experimentally. The basic design of the NASP airframe, as well as the ramjet and scramjet engine performances modeled by computer, have provided data that have been validated in supersonic wind tunnel testing. Computational modeling, when appropriate, can lead to results that are reasonably predictive and cost-effective.

When viewed only as applied research, the NASP, like other major government-industry joint ventures, involves considerable risks with no guarantees for successes or applications. The potential technology spin-offs from basic research, on the other hand, have prompted other countries to develop their own hypersonic vehicles. If the U.S. should drop the NASP program, then countries like Japan, Germany, and England would catapult to leadership roles in aerospace technology.

Bibliography

Boyne, Walter B. *The Leading Edge.* New York: Artabras Publishers, 1986. Written by the former Director of the National Air and Space Museum of the Smithsonian Institution, this is a fascinating account of breakthroughs in aviation history, with many full, color plates. Advanced aircraft incorporating new materials and designs are discussed in Chapter 9.

Brown, Stuart F. "X-30: Out of the World in a Scramjet." *Popular Science* 239 (November, 1991): 70 (11). A featured article covering the historical development of National Aerospace Plane. A good discussion of aerodynamics and propulsion systems including ramjets, scramjets, and rockets. Gives the general reader a feeling for the economics and global interest in the program.

Kandebo, Stanley. "TI Finding Commercial Uses for NASP Materials." *Aviation Week and Space Technology* 137 (September, 1992): 56-58. The author has kept the aviation reader well-informed on the progress of the NASP program in recent years. Discusses advancements in commercialized materials and processes developed in connection with the NASP program by Texas Instruments as a prime contractor. Applications for the thin metal alloy foils are listed.

Logsdon, John M., and Ray A. Williamson. "U.S. Access To Space." *Scientific American* 260 (March, 1989): 34–40. Primary focus is given to the shuttle and existing expendable launchers, but the NASP program is discussed as a space program goal. Excellent diagrams of space launch vehicles and space flights by category from 1989 to 1995. A chart of possible goals for a U.S. Space transportation system is also contained in this issue.

Murray, Charles J. "NASP, A Leap into the Unknown." *Design News* 47 (August, 1991): 70-74. An excellent review covering the complete history of the aerospace plane. Nontechnical and easy to understand by the lay reader. Problems and possible solutions encountered in the design and testing of this revolutionary craft. Materials, fuel systems, and industrial spinoffs are discussed. The author excerpts interviews from prominent individuals who are familiar with the program.

Reithmaier, Larry. *Mach 1 and Beyond: The Illustrated Guide To High Speed Flight.* New York: TAB Books, 1995. A short but informative chapter is provided on the NASP program. Both objectives and values of the overall project are listed. The reader is given information on the technical problems that must be overcome for the NASP to become a reality. Profile diagrams of the proposed design along with an artistic rendition of the completed craft highlight the chapter.

Taylor, Michael J. H. ed. *Jane's Aviation Review.* London: Jane's Publishing, 1987. A complete chapter written by Graham Warwick is devoted to trans-atmospheric vehicles (TAVs) along with a brief history of the NASP program. Includes five conceptual illustrations of proposed aerospace planes. Discussion of problems of obtaining high Mach numbers with ramjets and scramjets and efforts to overcome airframe temperatures caused by high-speed flight. The British version of the aerospace plane is also presented.

Michael L. Broyles

THE NATIONAL COMMISSION ON SPACE

Date: July 16, 1984, to August, 1986
Type of organization: Space agency

The National Commission on Space prepared Pioneering the Space Frontier, *a thorough look at long-term space goals which includes a "Declaration for Space," what many consider to be as important to space-age civilization as the Declaration of Independence is to the United States.*

PRINCIPAL PERSONAGES

THOMAS O. PAINE, NCOS Chairman and former
NASA Administrator

LAUREL L. WILKENING, NCOS Vice Chairman and a
planetary scientist at the University of Arizona

LUIS ALVAREZ, a Nobel-prizewinning physicist based
at the Lawrence Berkeley Laboratory

NEIL A. ARMSTRONG, the first man to walk on the
Moon

PAUL J. COLEMAN, President of the Universities Space
Research Association

GEORGE B. FIELD, Senior Physicist at the Smithsonian
Astrophysical Observatory

JACK L. KERREBROCK, Associate Dean of Engineering
at the Massachusetts Institute of Technology and
NASA Associate Administrator

GERARD K. O'NEILL, President of the Space Studies
Institute and Chief Executive Officer at Geostar
Corporation

KATHRYN D. SULLIVAN, the first American woman
to walk in space

DAVID C. WEBB, Chairman of the National
Coordinating Committee for Space

MARCIA SMITH, NCOS Executive Director
and a specialist on Aerospace Policy
at the Library of Congress

LEONARD DAVID, NCOS Director of Research and
editor of *Space World*

THEODORE SIMPSON, NCOS Director of Planning
and Project Leader of the Air Force Space Plan

Summary of the Agency

In 1984, the space shuttle and space station programs were billed by NASA as the next logical steps in space exploration. Some members of the U.S. Congress, however, believed that the U.S. space program had had no real objective or focus

since the Apollo program ended in the mid-1970's. The Senate Committee on Commerce, Science, and Transportation and the House Committee on Science and Technology created a private citizens' commission to determine the goals and the pacing of the American space effort, both manned and unmanned.

These committees held hearings. Testifiers included Thomas Paine, former National Aeronautics and Space Administration (NASA) Administrator, and James Beggs, the NASA Administrator at that time. Paine testified that, if a study were performed, it should be done by NASA, which would have to implement the recommendations. Beggs testified that NASA had file cabinets filled with earlier studies. Congress, however, decided that the earlier reports were unsatisfactory and stressed the need for a study by people outside government but familiar with the space effort, a study that would balance all scientific and political interests.

The result, Title II of Public Law 98-361, the National Aeronautics and Space Act of 1984, was approved July 16, 1984. It created the National Commission on Space (NCOS) and included requirements that the NCOS should plan for the next twenty years, "define the long-range needs of the Nation that may be fulfilled through the peaceful uses of outer space," and "articulate goals and develop options for the future direction of the Nation's civilian space program." It stated that the NCOS should consider a permanently manned space station a necessity. The act included a so-called sunset provision, requiring the commission to disband within sixty days of submitting its report to Congress and the President of the United States.

The act also required President Ronald Reagan to appoint fifteen people (of whom no more than three could be federal employees) to the commission within ninety days. The members would review classified documents; therefore, Federal Bureau of Investigation (FBI) clearance would be required. The time needed to process these clearances delayed the naming of the members. On March 29, 1985, Reagan appeared at the National Space Club in Washington, D.C., where he announced the names of fourteen of the fifteen members. He later named the fifteenth member, after an FBI clearance was received.

Nonvoting NCOS members included Senator Slade Gorton and Senator John Glenn, two members of the House of Representatives, and nine *ex officio* members (from the National Science Foundation, the Office of Science and Technology Policy, and the U.S. Departments of Agriculture, Commerce, Transportation, and State). Marcia Smith, Executive Director, Leonard David, Director of Research, and Theodore Simpson, Director of Planning, headed the staff of nine.

The commission began work quickly. It personally invited more than three hundred space experts, including many from NASA, to speak at eight hearings to convene between May and December, 1985, in various cities. (The five people on the commission staff loaned from NASA performed only administrative work. In line with Paine's 1984 congressional testimony, the commission invited NASA policy experts to the hearings because the NCOS believed that NASA would more readily implement directives that it had helped establish.)

Even before the first hearings, the commissioners recognized that planning for the next twenty years only, as Congress had requested, would be insufficient, because many programs have lead times of more than ten years. They decided to look at what the United States wanted to be doing in fifty years, and what would be needed in the next twenty years to lay the groundwork for the fifty-year goals.

Even with the quick start, the NCOS did not have a full year before the deadline because of the administrative and clearance delays in naming the commission members. Chairman Paine and Vice Chairman Laurel Wilkening returned to Congress, requesting a six-month extension of the deadline. They also asked for funding for a thirty-minute video. Congress agreed to both requests. The total appropriation came to $1.4 million.

Continuing its diligent approach, the National Commission on Space held fifteen public hearings to receive comments from as wide a cross section of the public as possible. These hearings — intentionally held around the country, from Honolulu, Hawaii, to Iowa City, Iowa, and Boston, Massachusetts — occurred between September, 1985, and January, 1986. More than six hundred people, of all ages and from all walks of life, presented their views. Besides these public hearings, the NCOS arranged to have a "Life in the Twenty-first Century Workshop" on October 18, 1985. It also asked the American Institute of Aeronautics and Astronautics to sponsor a "Workshop on the Commercialization of Space," which took place on October 31, 1985. In addition to organizing these many hearings and workshops, NCOS commissioners spoke individually to numerous groups, invariably requesting the members of the audience to contribute their ideas by sending either letters or electronic mail.

More ideas came from an unexpected direction. The astronomer Carl Sagan wrote an article for *Parade* magazine (a national Sunday newspaper supplement) in early February, 1986, asking American readers to write the NCOS outlining what they believed their nation's space goals should be. His request specifically pointed to a joint U.S.-Soviet manned Mars mission. By the end of February, when the NCOS was completing its report, it had received more than one thousand letters sparked by Sagan's article. Frank White, another concerned planner, worked closely with Commissioner George Field on the "Declaration for Space." NCOS member Gerard O'Neill commented that White's work "contributed freely to the National Commission on Space at a critical point in its deliberations, [and] became the organizing theme about which its report was written."

This reaching out for ideas on space goals went beyond the borders of the United States. The commission realized that the United States could not be a leader in space if it did not know what other nations were doing. Chairman Thomas Paine and Executive Director Marcia Smith visited Moscow to obtain Soviet input. Paine also went to Paris to determine what the European Space Agency (ESA) had planned.

The many suggestions received through the letters, electronic mail, workshops, hearings, and consultations helped the commissioners formulate their report. Very early in their planning, the commissioners recalled an illustrated report on the United States' future in space that had appeared in a national magazine in 1951. They realized that the illustrations that had accompanied that piece were still distinctly in their memory. Yet the text, which had been written by Wernher von Braun, was not as memorable. The commissioners decided that their report should be well illustrated. They clearly wanted the publication to be as accessible to the public as possible.

On May 23, 1986, a commercial publisher released the commission's 218-page report, *Pioneering the Space Frontier: The Report of the National Commission on Space*. The report recommended that NASA review the report's findings, and, by December 31, 1986, suggest a long-range implementation plan, including a specific agenda for the next five years.

The prologue to the report, entitled "Declaration for Space," states the pioneering mission for twenty-first century America:

> To lead the exploration and development of the space frontier, advancing science, technology, and enterprise, and building institutions and systems that make accessible vast new resources and support human settlements beyond Earth orbit, from the highlands of the Moon to the plains of Mars.

The report provides a rationale for exploring and settling the solar system. It states that the solar system is humanity's extended home; the human species is "destined to expand to other worlds." It adds that American freedoms must be carried into space, and individual initiative and free enterprise in space must be stimulated. Comparing the American frontier to the space frontier, the report advocates combining space's vast resources with solar energy to create new wealth. This must be done, it asserts, logically and wisely, sustaining investment at a small but steady fraction of the gross national product (GNP).

The report advises that the United States work with other nations in exploring the universe. It states that, with the United States' economic strength, technological ability, and frontier background, it will lead in the space effort and challenge and inspire individuals and nations to contribute their best efforts. It calls for common sense in settling space, in a sustained effort combining technology, scientific research, exploration, and development of space resources. Government's role, it states, is to support exploration, science, and critical technologies and provide the transportation systems and access to this new territory. The NCOS report affirms that Americans must not lose their values, such as peaceful intent, equal opportunity, planetary ecological considerations, and respect for alien life forms.

Part 1, "Civilian Space Goals for Twenty-first Century America," discusses the proposed long-range three-pronged interlocking program: stimulating space enterprise; increasing understanding of Earth, the solar system, and the universe; and exploring and settling the solar system. To accomplish these three goals, NCOS recommends in part 2, "Low-Cost Access to the Solar System," that the United States commit to providing low-cost access to space and advancing technology across a broad spectrum.

Part 3 proposes twenty-year civilian space programs, with budget planning five years in advance. It assumes that NASA's budget will grow at the rate the GNP grows, remaining below 0.5 percent of GNP (one-half the peak percentage during Apollo). The projected fifty-year total is $700 billion. Among the twenty-year goals stated are cargo vehicles with operating costs of under $200 per 0.45 kilogram orbited by the year 2000, low-cost aerospace (ground to orbit) planes, orbital maneuvering vehicles, Space Station operation by 1994, permanently manned lunar outposts by 2005, an unmanned Mars sample return mission by 2005, a variable-gravity research facility, and fellowships in space science and engineering. Among the fifty-year goals listed in part 4 are a manned Mars outpost by 2015, full Moon manufacturing by 2017, full Mars manufacturing by 2027, and space-based factories. The benefits, *Pioneering the Space Frontier* avers, include achieving advances in science and technology, providing direct economic returns from new space-based enterprises, and opening new worlds, with resources that can free humanity's aspirations.

The NCOS also produced a thirty-minute videotape, for which they had been funded by Congress. It has been shown worldwide. The commission also provided copies to all the Teacher-in-Space candidates and made it available to all science teachers.

Because of problems with government schedules (caused, in part, by an international crisis concerning Libya), the NCOS could not officially deliver its report to either Congress or the President until July 22, 1986. On that day, both the House Committee on Science and Technology and the Senate Committee on Commerce, Science, and Transportation held hearings on the report. The committees agreed that the National Commission on Space had properly discharged its responsibilities. They complimented the Commission on its inspiring report but voiced concern at the proposed costs and benefits. Chairman Thomas Paine and Vice Chairman Laurel Wilkening presented the report to President Reagan later that day.

In August, 1986, as decreed by the sunset provision, the National Commission on Space disbanded. The many letters, videotapes, and reports resulting from the hearings and public forums are stored in the National Archives.

Context

The NCOS report, the most thorough study of the United States' space goals, was the first such report made by private citizens. It legitimized the ideas long studied by the space community and provided, in its "Declaration for Space," what may be the equivalent of the Declaration of Independence to future civilization.

Earlier efforts to establish long-term goals for the U.S. space program began in 1969, when Vice President Spiro Agnew chaired the Space Task Group. (Thomas Paine had been a member of this group.) After reviewing the findings of the group, President Nixon had decided on what seemed at first the least expensive option — proceeding with the building of a space shuttle only. Later, in 1975, President Ford and the Congress ignored NASA's report "Outlook for Space," which suggested broad goals except for a space station and a Mars return. Just before the publication of *Pioneering the Space Frontier*, the National Research Council's Space Science Board advocated the return to a fleet of both manned and unmanned launch vehicles, rather than sole reliance on shuttles. It saw no scientific need for a space station during the next two decades.

The January, 1986, the *Challenger* accident — in which the manned space shuttle exploded shortly after lift-off — overshadowed the NCOS. Preoccupied, NASA did not immediately respond to the NCOS report's call to suggest, by December 31, 1986, long-range implementation plans.

Former astronaut Michael Collins chaired the NASA Advisory Council Task Force on Space Program Goals, reporting in March, 1987. The task force's objective was to assess NASA's response to the NCOS report and advise on any plans, emphasizing certain shorter-term goals and broad policy issues. It recommended manned Mars missions, preceded by extensive research.

NASA also responded with the August, 1987, *Leadership and America's Future in Space: A Report to the Administrator* by astronaut Sally Ride, which coordinated various study findings. This report acknowledged the NCOS and *Challenger* reports. It listed initiatives around which to focus a space-goals discussion: the exploration of the solar system, an outpost on the Moon, and a manned mission to Mars. Ride was then the head of the Office of Exploration, formed in June,

1987, to develop technical options needed to realize space goals. Also in 1987, Congress approved NASA's Civil Space Technology Initiative to revitalize technology, enhance orbit access, and advance in-orbit science missions.

The Iran-Contra affair and other crises preoccupied the President, and Science Advisor William Graham stymied NASA's attempts to gain Reagan's support. Finally, on February 11, 1988, President Reagan unveiled his National Space Policy. It incorporated an interagency review of *Pioneering the Space Frontier*, an assessment of the implications of the 1986 *Challenger* accident, and a review of previous presidential decisions. It supported the goals of a strong commercial space program, international cooperation, obtaining benefits through space-related activities, and expanding humanity throughout the solar system.

Bibliography

Bova, Ben. *The High Road*. Boston: Houghton Mifflin Co., 1981. Bova, President of the National Space Society and former *Omni* magazine editor, discusses solving energy, environment, and nuclear-war problems with space resources. He shows how little extra energy is required to travel to the Moon and to asteroids to mine raw materials once a vehicle is in low Earth orbit.

David, Leonard. "America in Space: Where Next?" *Sky and Telescope* 74 (July, 1987): 23-29. David, the NCOS Director of Research, looks at the state of space in the United States and what the next goal should be. He compares space science obtained by astronauts versus robots and reviews the NCOS report.

McDonald, Frank B. "Space Research: At a Crossroads." *Science* 235 (February 13, 1987): 751-754. Assesses the current and future U.S. space science program in view of the *Challenger* disaster, NCOS, and the Gramm-Rudman-Hollings budget reduction act.

McDougall, Walter A. *The Heavens and the Earth: A Political History of the Space Age*. New York: Basic Books, 1985. This scholarly book traces the origins of the space age. McDougall considers whether the seemingly unending technocracy race against foreign competition will erode the basic values of Western civilization.

Mendell, Wendell W., ed. *Lunar Bases and Space Activities of the Twenty-first Century*. Houston: Lunar and Planetary Institute, 1985. Papers presented at the first (1984) Lunar Bases symposium. It is suggested that improved transportation technology will lead to routine payloads to and from the Moon by 2000. These papers consider lunar base concepts; transportation issues; lunar science, construction, materials, and processes; astronomy from the Moon; lunar oxygen production; life support and health maintenance; and societal issues. Several papers deal with Mars issues. Some illustrations.

————. "Solar System Bonanza: The Burden of Proof." *Space World* W-6-270 (June, 1986): 18-19. The NCOS held a public meeting in November, 1985, in San Francisco, California. This is a transcript of Commissioner Gerard O'Neill and Carl Sagan discussing commercializing the solar system.

Michaud, Michael A. G. *Reaching for the High Frontier: The American Pro-Space Movement, 1972-1984*. New York: Praeger Publishers, 1986. The pro-space movement (including more than fifty advocacy groups and more than 200,000 Americans) has developed since the end of the Moon landing program. Michaud traces key groups that have subtly influenced space policy and identifies their origins and goals.

National Commission on Space. *Pioneering the Space Frontier: The Report of the National Commission on Space*. New York: Bantam Books, 1986. This highly readable text resulted from the commission's thorough research. It sets forth goals and scenarios for space exploration into the twenty-first century. Illustrated.

O'Neill, Gerard. *The High Frontier: Human Colonies in Space*. New York: William Morrow and Co., 1976. An interesting look at concepts for space colonies, farms, and factories. O'Neill, an NCOS member, shows how they can be built from materials sent by "mass drivers" from lunar mines.

————. *2081: A Hopeful View of the Human Future*. New York: Simon & Schuster, 1981. Discusses predicting the future, shows the drivers of change — computers, increased automation, and space colonies — and provides scenarios of how life, including routine space travel, will be in the year 2081.

Reichhardt, Tony. "Predicting the Space Frontier: A Conversation with Laurel Wilkening." *Space World* W-10-274 (October, 1986): 8-13. An interview with Wilkening, the NCOS Vice Chairman, about the NCOS report and reactions to it.

Reichhardt, Tony, and I. Gilman. "Destination Mars: A Conversation with Michael Collins." *Space World* X-7-283 (July, 1987): 16-20. Former astronaut Collins chaired the Task Force on Space Programs Goals for the NASA Advisory Council. This article is an interview with him concerning the report.

White, Frank. *The Overview Effect: Space Exploration and Human Evolution*. Boston: Houghton Mifflin Co., 1987. White, who greatly influenced the commission's "Declaration for Space," interviewed astronauts and cosmonauts for this book. He reports on the "Copernican perspective," seeing Earth as part of the solar system, and the "universal insight," comprehending Earth's place in the universe. The space travelers share their unique visions.

Patricia Jackson

Navigation Satellites

Date: Beginning April 15, 1960
Type of satellite: Navigation

The United States and the former Soviet Union each began launching a series of two intrinsically different types of navigation satellites in 1960. These satellites are military spacecraft to which civilian interests have been allowed access. They have enabled ships, aircraft, and even land-based vehicles to pinpoint their positions on Earth, saving billions of dollars in fuel and lost time and helping to ensure safe travel for millions of people.

Principal personages

WILLIAM H. GUIER and

GEORGE C. WEIFFENBACH, physicists at The Johns Hopkins University who made use of the Doppler effect to track satellites in orbit

FRANK McGUIRE, a physicist at The Johns Hopkins University who conceived the original principles of the navigation satellite

Summary of the Satellites

Edward Everett Hale made the first-ever mention of a satellite in recorded literature in *The Atlantic Monthly* in 1869. Hale's idea, set in a fictional story, was to orbit a brick moon nearly sixty-five meters in diameter. The purpose of the brick moon was to provide navigators with a visual object by which to determine their positions. Thus, Hale must be credited with originating the notion of using artificial satellites for navigation.

Nearly a century later, the Soviet Union launched the first real satellite, Sputnik 1, an 83.3-kilogram object whose primary purpose was to beam repetitive signals to the Earth. The fact that a growing military power with an adversarial relationship to the United States could accomplish such a feat stirred much concern in the West. The immediate fear was that the Soviet Union could orbit nuclear weapons and drop them on any target in the United States at will. This fear gave rise to the need to track foreign satellites precisely so that their activities could be carefully monitored.

It quickly became clear that visual tracking alone did not provide experts with enough information. Two physicists at The Johns Hopkins University in Baltimore, Maryland, decided that a radio tracking system could provide them with the answer. Such a locating device had already been tested by scientists in West Germany. Eventually, William Guier and George Weiffenbach developed a system that measured the frequency shift of a satellite as it approached and again as it departed to determine its exact orbital characteristics — that is, making use of what is known as the Doppler effect.

A third physicist, Frank McGuire, had joined Guier and Weiffenbach. McGuire decided that if the satellite's exact position in orbit at any given time were known, the process could be reversed to determine the observer's position on Earth. The United States Navy seized upon the idea, and its scientists quickly worked to develop and launch the world's first navigation satellite, Transit 1B, on April 15, 1960. This satellite effectively reversed the science of determining spacecraft position using the Doppler principle.

This first satellite was to be one of a series of test satellites. As soon as the Navy had gathered data on them, it would build permanent ground stations and receivers that would be placed on U.S. naval vessels. The system developed after completion of all the tests would become known as the operational system. The Navy's primary purpose was to provide precise navigation for its submarines that carried nuclear-tipped ballistic missile warheads. In the event that a missile was launched at sea, any error of known position at launch would be magnified in proportion to the distance to the target. Before the Navy launched the first navigation satellite, it had had to rely on Earth-based radio navigation systems and sometimes celestial navigation. The Navy had also devised an on-board navigation system known as the Shipboard Inertial Navigation System (SINS, or INS, the latter including Inertial Navigation Systems employed on aircraft and land-based vehicles), which tracked the submarine's position based on recorded movements of the ship and worked with electronic navigation inputs. These systems were not precise enough for accurate targeting of sea-launched ballistic missiles.

The Transit satellites were launched into orbits that were called polar because they traveled over the Earth's poles at 1,000 kilometers in altitude. In this orbit, as the Earth rotated under them, every square kilometer of the Earth's surface would ultimately pass beneath them. As soon as the first Transit satellites were in orbit, scientists discovered several conditions that necessitated alterations to the satellite design. The original purpose of the Transit test series was to test the system designers' expectations regarding the satellites' behavior. It was discovered that orbital vehicles were subject to unexpected fluctuations, so that a more complex system design became necessary. The position of the Earth-based unit could not be determined with accuracy unless the precise position of the orbiting navigation satellite was known. Thus, these test orbits helped determine how the Earth-based system had to work so that the satellite's exact position could be continually monitored.

The experimental Transit system underwent almost four years of development, during which time the space-based navigation system and its Earth-based links were designed and built. During this entire period, the Transit system remained a carefully guarded military secret, a system tailored to the needs of the U.S. Department of Defense.

The Transit system became operational in late 1964, at first as a secret military program. It was not long, however, before private companies became aware of the Navy's new navigation system, and pressure began to be applied on the U.S. Congress to open the system for commercial use. Finally, in July of 1967, President Lyndon B. Johnson agreed to allow the Transit system to be used by civilian interests. Though businesses were given permission to build receivers to pick up and decipher the satellite's signals, the Transit system remained very much a military program. The Transit system and its nascent successor, the Global Positioning System, were both designed, financed, and maintained by the U.S. Department of Defense, even though civilians were allowed to use them. In this sense, then, all navigation are military satellites.

The U.S. Transit program has had many names over its lifetime. In 1968, for example, the acronym Navsat was the name of a series of naval navigation satellites. The designations Timation (from time navigation), Triad, Navigation Technology Satellite (NTS), Transit Improvement Program (TIP), and Nova were all used at one time or another to describe Transit-type satellites. Users of the system have often applied the term "Satnav" to the system as a whole.

The Transit-type satellites allowed an Earth-based observer to determine their position, receiving within two minutes the data transmitted from a single spacecraft. The ground-based observer need not transmit to the satellite. The ability to obtain a position (called a "fix") is determined by whether a satellite is within the observer's "receiving horizon," or a given distance above the receiver's actual horizon. Because of the altitude of the Transit satellite, a given satellite would be in a position to give an observer four fixes in a single day. Since normally there were five operating Transit satellites in orbit, an observer could count on twenty fixes per day, coming just over an hour apart, with an accuracy range of about 1.8 kilometers.

The Soviet Union entered the picture on November 23, 1967, launching its first navigation spacecraft from the Plesetsk cosmodrome. The satellite was designated Kosmos 192 and was placed in an orbit nearly identical to that of a U.S. Transit satellite.

The Soviets were secretive in regard to their space pro-

gram and even more so about their military space program. It is likely that the Soviets' air navigation needs were similar to those of the United States. It is a foregone conclusion that the first Soviet navigation satellites, like their American counterparts, were used exclusively for military purposes. It is also a fact that the Soviet military was much more closely associated with its civilian enterprises, especially in that its navy and its merchant marine often combined mission objectives. Typically, many if not all Soviet merchant vessels served in a surveillance role to some extent. Hence, it is presumed that many Soviet merchant vessels shared in the information provided by the military space-based navigation system.

In the West, knowledge of the early Soviet navigation satellites was gained by analyzing their orbits and signal characteristics. It was soon discovered that the Soviet system had much in common with the American Transit system. For example, both had polar orbits, their orbital altitude was nearly the same, and they broadcast on nearly the same frequencies, sharing a similar Doppler analysis scheme. Since the Soviets never acknowledged the existence of this system and a formal designation for it is not known, it will be referred to here as the "K-192-type system."

Government spokesmen in the Soviet Union made mention of a "civilian satellite navigation system" they call the Tsikada. These satellites were placed in the same orbits as the K-192-type spacecraft, operate on the same principles, and have a unique characteristic in that their orbital lifetime is twice that of a military K-192-type craft (the reasons for this are unknown). Tsikada satellites and their military counterparts are presumably both served by the same ground-based facilities.

While the American Transit and the Soviet K-192-type systems are still being utilized and replenished by both nations, new satellite navigation systems have been developed by both the United States and Russia. The U.S. version is called the Global Positioning System (GPS) and the Russian system, the Global Navigation System (GLONASS). The systems operate on very similar principles.

The GPS satellite is named Navstar. Orbits of these spacecraft are twenty times higher than those of Transit (20,000 kilometers), are not polar, have an orbital period of twelve hours, and are made up of a constellation of eighteen satellites. The first GPS satellite was launched on February 22, 1978, in the GPS test program. (The first ten Navstars were designated Block I and were used to test and validate the system.) The results of that test were spectacular. The GPS system, according to one published report, made it possible for a ground-based receiver to pinpoint positions to a reported accuracy of 2 meters. (The system's advertised accuracy is 10 meters.)

The GPS system uses between one and four satellites to "fix" a position. Using four satellites to formulate the fix enables multidimensional positioning, especially valuable to aviation users. The GPS system can fix position in latitude, longitude, altitude, velocity, and time. Hence, GPS has application to all forms of transportation from ships to aircraft to land-based vehicles.

Like Transit, the GPS system is designed to be used by both military and civilian users. The precision made available to civilian users, however, is less accurate. Civilian receivers' transmissions make use of a code termed the standard positioning system, which ensures an accuracy of 100 meters. The military uses a highly classified code, the precise positioning system, which permits the highest accuracy. GPS became fully operational in March, 1994, when the last of the NAVS-TAR II satellites was placed in orbit.

The Soviet GLONASS system was made available to the world in May, 1988, through details released to the International Civil Aviation Organization in Montreal. Signals similar to those of the U.S. standard and precise positioning systems show that the Soviet system is nearly identical in scope, design, and purpose to the U.S. system. As of August, 1995, nineteen of the planned twenty-four GLONASS satellites had been placed in orbit. Eventually, the GPS and GLONASS satellites will be combined.

Knowledge Gained

The Transit test series confirmed that Earth's gravitational attraction was not uniform over the planet, so that the spacecraft's altitude fluctuated slightly. Tests also confirmed that the density of the atmosphere was not uniform at the level of the satellites' orbits; this variation also affected their altitudes. This finding was important to the designers because in any given orbit the altitude of a spacecraft determines such things as its speed and exact position. Of more fundamental importance, however, was the simple fact that new and valuable knowledge of Earth's gravitational fluctuations and the behavior of the upper atmosphere had been gained.

Satellite navigation was enabled by making a reverse application of the knowledge and techniques gained from the first consistent measurements of spacecraft in orbit. Yet a method for determining an orbital body's precise orbit had to come first, even in an operational satellite navigation system. Once satellite navigation systems had been developed, the Earth-based leg of the system itself worked to refine the method of determining a spacecraft's precise orbital position, providing data for the very science responsible for its inception. This system also contributed to the knowledge needed to develop exact rules for Doppler positioning of other kinds of satellites.

These discoveries would be vitally important to the total body of astronautics. Precise determination of the positions of spacecraft by ground-based observers would become an integral component of manned spaceflights that require exact calculations of orbits for rendezvous. Such rendezvous would be vital for the manned lunar missions and later for linking with space stations such as Skylab.

Precise positioning was also essential in determining exact positions when igniting booster rockets of interplanetary spacecraft in parking orbits prior to departure for the Moon or distant planets. Though space-based navigation was designed to determine the position of Earth-based observers, the science itself in all of its reversible, interchangeable dimensions would be used for both Earth and space navigation.

Direct knowledge was gained regarding the design and use of space-based navigation systems, so that the accuracy of the GPS improved on that of Transit by more than a hundredfold. The ability to schedule, build, and launch a system to allow for uninterrupted use over decades has been refined and is reflected in modern systems that place satellite spares in orbit, lying dormant until called upon. This practice ensures service free of disruptions.

Microelectronics made possible the development of satellite navigation receivers whose cost was such that even recreational boaters could afford access to satellite signals. Systems now allow private and commercial access to the extremely accurate GPS. At least one company designed a system that will use both the American GPS and the Soviet GLONASS, selectable with the flip of a switch. Another uses a hybrid INS/GPS for ultra-precise positioning. These products have become available to anyone with a receiver, from commercial to military to recreational users.

Context

Transit 1B was the world's first navigation satellite. Until the inception of the Transit series of satellites, the art and science of navigation had relied on techniques that could not ensure accurate positioning of ships and aircraft. Most vessels could not be sure of their position to within about two kilometers at sea, regardless of the navigation technique employed.

With the Transit program and ultimately GPS, it became possible to fix one's position constantly, anywhere on Earth — on sea, land, or air — with an accuracy of 2 meters, a time standard accurate to 0.1 microsecond, and a velocity of plus or minus 0.1 meter per second.

The process of designing the system enabled Earth-based observers to track spacecraft in orbit precisely. It also provided a data base on the conditions of the upper atmosphere and their effect on orbiting spacecraft. The program and its antecedents also made the first fundamental discoveries of the fluctuations of Earth's gravitational field and their effect on orbital vehicles.

The science of space-based navigation originated as a result of reversing the knowledge gained in one of the first space sciences: precise tracking of orbital space vehicles. During the ensuing process of developing the space-based navigation network, it became necessary to track the navigation satellites meticulously, so that the system itself refined the progenitor science of space vehicle tracking.

Though navigation spacecraft were developed as military vehicles, they quickly became important to commercial users after the military allowed them access to the signals broadcast from the spacecraft. Their use has made possible the saving of billions of dollars in fuel and time since their inception by allowing precise tracking of ships and aircraft along their routes. This precise knowledge of position has also allowed for a greater margin of safety for ships and aircraft.

The Soviet Union pursued the same investigations as the United States, almost a decade behind in the early years of navigation, but nearly catching up with the inception of their GLONASS system. The GLONASS system was released by the Soviets in 1988, marking the first time one of its space systems was freely opened for world use. This change in Soviet policy came about the time Soviet General Secretary Mikhail Gorbachev instituted his policies of *perestroika* (social and economic restructuring) and *glasnost* (openness).

Use of the GPS and GLONASS systems promise to revolutionize navigation worldwide with the use of low-cost receivers on civilian ships, recreational boats, aircraft, and even automobiles.

Bibliography

Chamberland, Dennis. "Space Based Navigation." In *Proceedings of the United States Naval Institute*. Annapolis, Md.: Naval Institute Press, 1988. An article examining the unclassified military (primarily naval) use of the navigation satellite systems of the United States and the Soviet Union. This nontechnical article compares the latest system developments and naval applications. Includes photographs and historical charts.

Dahl, Bonnie. *The User's Guide to GPS: The Global Positioning System*. Evanston, Ill.: Richardsons' Marine Publishing, 1993. An explanation of GPS aimed at recreational boaters. Includes maps and illustrations.

Hassard, Roger. "A Layman's Guide to the Global Positioning System." *Navigator* July/August, 1985: 22- 25. This excellent article provides a concise and easy-to-understand description of the Global Positioning System (GPS). It makes comparisons with Transit-type satellites, examines the military versus civil uses, and briefly discusses civil receivers under development at the time of its publication.

Kaplan, Elliott D. *Understanding GPS: Principles and Applications*. Boston: Artech House, 1996. An overview of the Global Positioning System accessible to the general reader.

Logsdon, Tom. *The NAVSTAR Global Positioning System*. New York: Van Nostrand Reinhold, 1992. Overview of the NAVSTAR system. Includes illustrations.

————. Understanding the NAVSTAR: GPS, GIS and IVHS. New York: Van Nostrand Reinhold, 1995. Comprehensive explanation of the entire NAVSTAR system and its multiple functions.

Maloney, Elbert S. *Dutton's Navigation and Piloting*. 14th ed. Annapolis, Md.: U.S. Naval Institute Press, 1985. This text is considered one of the definitive works on navigation. A rather technical, stiff presentation of the science of navigation of all kinds. Nevertheless, it is a valuable reference for one interested in satellite navigation in that it presents several superb pages of narrative dealing with satellite navigation systems and detailing exactly how they are used. Enlightening for the more serious student of satellite navigation.

Paul, Günter. *The Satellite Spin-Off: The Achievements of Space Flight*. Translated by Alan Lacy and Barbara Lacy. Washington, D.C.: Robert B. Luce, 1975. This book describes the beneficial spin-offs of all kinds of satellite systems. The work lists in chapter-by-chapter fashion different kinds of satellites, from meteorological to navigation, gives a historical synopsis, and reviews the broader use scenario. Includes numerous drawings. Suitable for all readers interested in a general review of satellite systems up to the mid-1970's. An eminently readable translation from the German.

The Soviet Year in Space: 1987. Colorado Springs, Colo.: Teledyne Brown Engineering, 1988. This report in book format is the seventh annual report on Soviet space activities published by Teledyne Brown Engineering. It constitutes an intriguing look in surprising detail at what is otherwise considered one of the world's best kept secrets. Contains an entire section discussing navigation satellites, covering both the K-192-type system and Glonass. It is oriented to all readers in an easy-to-understand format with many photos and drawings.

Yenne, Bill. *The Encyclopedia of U.S. Spacecraft*. New York: Exeter Books, 1985. Reprint, 1988. This work is a pictorial reference, listing all unclassified satellites alphabetically, with photographs of many. It is a useful review of the Transit series, though the reader must keep in mind that the Transit-type spacecraft had many names over its operational lifetime. It is suitable for all readers.

Dennis Chamberland

NIMBUS METEOROLOGICAL SATELLITES

Date: Beginning August 28, 1964, to mid-1986
Type of satellite: Meteorological

The Nimbus satellites were used to develop new techniques for observing Earth, especially its atmosphere and oceans, by remote sensing from orbit. A number of methods that are now used routinely to gather data for weather forecasting, for example, were first tried on an experimental basis on one of the Nimbus satellites.

PRINCIPAL PERSONAGES

RONALD K. BROWNING and
CHARLES MCKENZIE, Nimbus Project Managers
    at Goddard Space Flight Center
WILLIAM R. BANDEEN,
MORRIS TEPPER, and
ALBERT J. FLEIG, Nimbus Project Scientists at
    Goddard Space Flight Center
I. S. HAAS, Earth Observatory Programs Manager,
    General Electric
WILLIAM FORNEY HOVIS, JR.,
H. JACOBWITZ,
JAMES MADISON RUSSELL III,
J. T. HOUGHTON,
MICHAEL PATRICK MCCORMICK,
DONALD HEATH,
P. GLOERSON, and
L. ALLISON, Principal Investigators, Nimbus 7

Summary of the Satellites

The Nimbus Earth observatory program consisted of seven satellite launches, separated by roughly two-year intervals, ranging from Nimbus 1 on August 28, 1964, to Nimbus 7 on October 24, 1978. In each case, the spacecraft's appearance and orbit were similar to the others'; the payloads, however, progressively increased in mass and sophistication as new experiments were developed and tested. Nimbus 1 weighed less than 400 kilograms; Nimbus 7, nearly 1,000 kilograms. All the satellites were built by the General Electric Corporation of Valley Forge, Pennsylvania. The dozens of experiments came from a wide range of institutions in the United States and England.

Nimbus has been described as a butterfly-shaped spacecraft. It is a cone three meters high and nearly two meters across near the base, with two large panels carrying solar cells to convert the Sun's energy into electricity, extending like wings at either side. The scientific instruments are inside, or hanging below, a doughnut-shaped "sensory ring" attached at the base of the cone. The satellite is stabilized by gyroscopes, so the base of the sensory ring faces toward Earth at all times as the satellite orbits the planet.

All the Nimbuses except the first were placed in nearly circular orbits about 1,000 kilometers above Earth's surface, and all were in orbits that passed close to the north and south poles of the planet, so that views of the whole of Earth's surface were possible. In addition, the orbits were Sun-synchronous, which means that a person at a particular place on the surface would see Nimbus go overhead at exactly the same times twice each day. A special type of Sun-synchronous orbit was used so that the Sun would be at noon when the satellite passed above an observer during the day. The second crossing would take place at local midnight. It has been estimated that, during their useful lives, Nimbus spacecraft have traveled about 7 billion kilometers in all.

Nimbus 1 was launched by a Thor-Agena B rocket from Vandenberg Air Force Base in California, the launch site that was to be used for the whole series. The launch was imperfect and resulted in an orbit that was too low and fairly eccentric. Furthermore, the spacecraft malfunctioned after twenty-four days, when the tracking system that kept the solar panels pointing at the Sun jammed, causing a serious loss of power. Nevertheless, many useful data and much experience were gained. The first attempt to launch Nimbus 3, in 1968, was a complete failure because of launch vehicle problems. A duplicate spacecraft was assembled and successfully launched in the following year. All the other launches in the series went well. By the time the relatively advanced and heavy Nimbus 7 was launched, the Thor first stage included nine Thiokol solid-fueled boosters to increase the thrust, and a Delta second stage was being used. Project management and mission operations were provided by the National Aeronautics and Space Administration (NASA) Goddard Space Flight Center in Greenbelt, Maryland.

The scientific payload of Nimbus 1 consisted of three sensors. The advanced vidicon camera system was a television camera that observed Earth with a resolution of about one kilometer. The automatic picture transmission system was an early version of the now-commonplace technique of transmitting pictures taken by a camera on a satellite directly to any user equipped with a low-cost receiving station. It was intended mainly for meteorologists, who need pictures of cloud formations as soon as they occur. Finally, the high-resolution infrared radiometer was an imaging system that could operate at night by making pictures from scanned maps of thermal infrared, or heat, emissions from the surface and clouds. Although Nimbus 1 was both in the wrong orbit and short-lived, all the experiments functioned successfully before the mission was terminated

Nimbus 2, launched May 15, 1966, repeated the Nimbus 1 experiments. All of them functioned well for thirty-three months, during which time more than 100,000 pictures were transmitted to Earth. In addition, this satellite carried the medium-resolution infrared radiometer to study Earth's heat balance, or the difference between the amount of heat absorbed from the Sun and the amount radiated from Earth as infrared waves. These two components must be the same for the whole planet, but they are quite different in different isolated regions, and these differences drive weather systems. One of the first really sophisticated weather satellites, Nimbus 2 provided data for, and stimulated the development of, computer models of the atmosphere that are the basis of accurate forecasting.

Nimbus 3 had a similarly long life, beginning on April 14, 1969, and it carried both high- and medium-resolution infrared sensors. It had an improved camera, called the image dissector camera system, which transmitted to ground-based stations. Most significantly, it had two particularly novel infrared sounders, the Infrared Interferometer Spectrometer (IRIS) and the Satellite Infrared Sounder (SIRS). The former was a fairly high-resolution spectrometer that scanned nearly all the infrared wavelengths at which Earth emits radiation, thereby sensing individual gases such as water vapor and ozone and allowing inferences about atmospheric composition and behavior to be made. SIRS used selected spectral channels specifically to measure the vertical profile of atmospheric temperature by monitoring atmospheric carbon dioxide emissions.

In addition to its meteorological experiments, Nimbus 3 carried three technology experiments: the IRLS (Interrogation, Recording, and Location Subsystem), the RMP (Rate Measuring Package), and the SNAP (System for Nuclear Auxiliary Power) 19. IRLS was used to collect data from surface stations, RMP was a test of a gas-bearing gyroscope, and SNAP was an early test of a nuclear isotope-driven power source for satellites. The plutonium in SNAP was valuable enough to be worth retrieving from 100 meters below the sea after the failed attempt to launch the first Nimbus 3 in 1968.

Nimbus 4 was launched April 8, 1970, on board a Thor-Agena D. In addition to repeats or improved versions of the image dissector camera, IRIS, SIRS, and IRLS from Nimbus 3, four new experiments were carried to test still more ways of monitoring the atmosphere by remote sensing. Three of these

were infrared spectrometers similar in principal and objectives to, although quite different in detail from, IRIS and SIRS. The most innovative sensor was the backscatter ultraviolet spectrometer, which allowed the amounts of ozone in the upper atmosphere to be inferred from maps of the solar ultraviolet radiation reflected back from the ozone layer. In addition, the first of four British experiments on successive Nimbus satellites used a special technique to measure temperature at greater altitudes than previous Nimbuses had (around 50 rather than 30 kilometers) and so was optimal for the study of the structure and dynamics of the stratosphere.

There were fewer but larger instruments on Nimbus 5, which was launched December 11, 1972. They included the first microwave sounders: the Nimbus E microwave sounder, or NEMS (Nimbus satellites were known by letter designations, A through G, before launch and were numbered after achieving orbit), and the Electrically Scanning Microwave Radiometer (ESMR). These instruments succeeded in their goal of proving microwave sounders' ability to penetrate clouds and measure the temperature of the surface beneath. The infrared temperature profile radiometer demonstrated how to eliminate cloud interference by making infrared observations with high spatial resolution and using ratios of adjacent soundings to eliminate images of rapidly varying features, on the usually reasonable assumption that these are clouds. Nimbus 5 also had an improved SCR and a surface composition mapping radiometer. At about the time of Nimbus 5, a series of satellites using the same basic configuration and called ERTS, for Earth Resources Technology Satellite, was instigated. They were to do for Earth's surface what Nimbus did for the atmosphere, and they carried some of the same instruments. The first ERTS was later named Landsat 1.

Nimbus 6, launched June 12, 1975, and Nimbus 7, launched October 24, 1978, completed the series. Nimbus 6 had an improved ESMR; a new Scanning Microwave Sounder, called SCAMS; a pressure modulator radiometer; and an improved Earth radiation budget experiment. The newest instruments were TWERLE, for Tropical Wind Energy and Reference Level Experiment, and LRIR, for Limb Radiance Inversion Radiometer. TWERLE was designed to obtain meteorological data by tracking balloons below. The LRIR was the first infrared limb-scanning instrument, which is to say it measured emissions while looking sideways at the atmosphere at the edge of Earth's disk, rather than downward. Most infrared sensors now use this technique, and Nimbus 7 had two such sensors: LIMS, for Limb Infrared Monitor of the stratosphere, an improved LRIR; and SAMS, for Stratospheric And Mesospheric Sounder, a limb-viewing version of the pressure modulator radiometer. Both of these devices measured temperature and a range of minor gas abundances in the stratosphere. Nimbus 7 also had an ocean monitoring device, the coastal zone color scanner, which could detect variations in plankton and other marine life, as well as an Earth radiation budget experiment.

Knowledge Gained

A large part of the new knowledge gained from the Nimbus program was the ability to apply new sensor technology, which, having been tested, went on to be widely used on operational weather, oceanographic, and Earth resources satellites. As a result, the oceans, surface, and especially the atmosphere were observed in unprecedented detail, and many significant scientific projects were conducted.

From as early as 1964, with the short-lived Nimbus 1, hurricanes Cleo and Gladys were observed from space in advanced videcon camera pictures and high-resolution infrared radiometer maps. The television images showed the character and evolution of these systems along with the rest of Earth's weather. Infrared maps added nighttime coverage and the vertical dimension, since cloud heights could be inferred from radiometric temperature readings. Nimbus images of the surface were also revolutionary; for example, a mountain in Antarctica was shown to be 75 kilometers from its presumed location, and a range shown on existing maps as double was clearly seen to be single.

The automatic picture transmission system on the early Nimbuses brought the lofty perspective of the Earth satellite to the ordinary user for any of a hundred applications, provided the user had a basic receiving station. One of these applications was the mapping of sea ice and the identification of channels in ice sheets, which previously had required a patrol using aircraft. The resolution from Nimbus 2's videcon camera was good enough to permit tracking of individual floes, allowing currents to be studied. Later versions could observe volcanic plumes, brush fires, and even dust storms. Nimbus 2 also produced temperature maps of the ocean and spearheaded the development of computerized data processing techniques for "cloud clearing," so that an extended region could be mapped without obscuration. These maps helped scientists to develop a first tentative understanding of the relation between ocean surface temperature and climatic behavior, including hurricane development.

The Nimbus series pioneered the study of Earth's ozone layer from space, including the monitoring of the flux of energetic ultraviolet radiation from the Sun, which produces ozone from oxygen. The brightness of the Sun varies significantly at the wavelengths of importance for ozone formation, and the gas itself is easily destroyed by a number of reactions. Starting with Nimbus 3, the satellites began an ongoing process of monitoring and studying the ozone layer. A correlation between stratospheric ozone and the pressure nearer the surface was discovered. Seasonal changes were also identified, including strong variability over the poles. On Nimbus 6 and 7, not only was ozone monitored with accuracy and higher vertical resolution through the use of the limb scanning technique, but measurements were made of the global distribution of other minor constituents that affect ozone

chemically. Among these are the oxides of nitrogen, produced in large part by human activities, which are thought to be making important changes in global ozone. Earth's atmospheric water budget has also been studied on a planetwide scale for the first time.

Predominant among the techniques pioneered by Nimbus is the measurement of atmospheric temperature structure, first in the lower atmosphere and then, through instruments such as the pressure modulator radiometer and SAMS, in the stratosphere and mesosphere. Many features of global temperature fields were discovered or first characterized fully by Nimbus instruments. Among these features are the various planetary wave modes that affect the general circulation, unexpected seasonal effects such as the warm winter mesopause, and dramatic phenomena such as the stratospheric sudden warmings, in which the air temperature can increase by 100 degrees Celsius in a few days.

Context

The Nimbus satellites provided new information about Earth's atmosphere, oceans, and surface on a scale never before contemplated. Even more important, however, Nimbus scientists pioneered, developed, and proved techniques and instrumentation that now form the backbone of the science of remote sensing. Sensors and subsystems developed from technologies first flown on Nimbus satellites now play major roles in data gathering for computerized weather forecasting, hurricane detection, climate research, agriculture, and even prospecting.

Earlier weather satellites — of which the most important were those known as TIROS, for television and infrared operations satellite — were much less capable of providing the observations that meteorologists require. These pioneering satellites were stabilized by their own spinning instead of by gyroscopes, were in orbits which did not view Earth continuously, did not cross the polar regions, and were too small, at about 300 kilograms, to carry instruments of the sophistication that Nimbus made possible. The eight original TIROS satellites, launched from 1960 to 1963, made important achievements; Nimbus, however, was the first significant platform in space for Earth observations.

Research into the various phenomena that make up the weather and routine operations such as forecasting require global data. Before satellites, these data came mainly from a network of stations that launched balloons carrying radio transmitters, which provided only fragmentary coverage. When the stabilized, polar-orbiting platform became available in the form of Nimbus 1, it rapidly stimulated the invention of new sensors that could map Earth in four dimensions — latitude, longitude, altitude, and time — with good coverage in all four. The availability of these data led in turn to the growth of computer processing methods that could make use of them, this advance being aided enormously by the rapid improvements in computer technology that were being made at the same time. From these processing methods has come the numerical weather forecast, in which the state of the atmosphere one day or a few days into the future is predicted by solving the physical equations that represent the fluid motions. To do this with sufficient accuracy and resolution requires large, fast computers and the rapid availability of extensive data about the "initial state," or today's weather. Therefore, although numerical forecasting was understood in principle as long ago as the 1930's, it was not feasible until the era of Nimbus. In the years to come, refinements of this technique may lead to reliable forecasts which extend, although with less detail, years or even decades ahead.

The overall state of Earth's atmosphere is subject to change, and the changes have profound effects on the suitability of the surface for the survival of life-forms, including humans. The great ice ages were examples of such changes, and even today there are smaller extremes of climate which lead to localized disasters: Missing rainy seasons lead to drought, and excessive monsoons lead to catastrophic flooding. There is currently much concern that Earth's climate is drifting permanently toward a more extreme, less comfortable state, possibly pushed by human activities such as the burning of fossil fuels or the release of chemicals harmful to the ozone layer. Satellite observations, most significantly those from Nimbus 7, have made it possible to commence an in-depth scientific study of some aspects of the ozone problem in particular and large-scale, long-term atmospheric behavior in general. These studies have helped to emphasize the seriousness of the global habitability situation, and, in time, they may suggest solutions.

Spacecraft derived from Nimbus will continue to gather meteorological data. The Nimbus design became obsolete for advanced scientific work at the end of the 1970's, however, and it has been replaced by the Upper Atmosphere Research Satellite (UARS), also managed by Goddard Space Flight Center and built by General Electric. This satellite is more than fives times as large as Nimbus 7 and outperforms the Nimbus satellites in factors such as stability, which controls a satellite's ability to obtain very high-resolution pictures or soundings, and data rate, which determines the number of observations that can be made in a given time. Planning called for UARS to be followed in the second half of the 1990's by an even larger and more powerful satellite called the Earth Observing System, or EOS.

Bibliography

Barrett, E. C. *Climatology from Satellites*. New York: Harper & Row, 1974. Includes a general discussion of meteorological satellite systems, including Nimbus. Contains particularly good discussions of the applications of satellite data to climatic aspects of interest to geography and agriculture students. Also casts light on physics and mathematics. For college-level readers.

Chen, H. S. *Space Remote Sensing System: An Introduction*. New York: Academic Press, 1985. This book describes satellite sensors from an engineer's point of view, but it is kept at a very basic level and so is suitable for the technically minded general reader. It is well illustrated and includes references.

Houghton, J. T., F. W. Taylor, and C. D. Rodgers. *Remote Sounding of Atmospheres*. New York: Cambridge University Press, 1984. Describes the various types of satellites, including Nimbus, that have been used to make atmospheric measurements. Gives details on the instruments and techniques used and the methods of analyzing the data. Intended for advanced college students, but early chapters are suitable for general readers. Contains many references to books and technical articles dealing with satellite observations of the atmosphere.

Hubert, Lester F., and Paul E. Lehr. *Weather Satellites*. Waltham, Mass.: Blaisdell, 1967. A good introduction for the non-specialist audience to the goals, methods, and achievements of weather satellites. Explains in general terms how satellite data is used for forecasting and research. Includes descriptions, with many illustrations of TIROS and early Nimbus satellites and their measurements.

Philosophical Transactions of the Royal Society. *Studies of the Middle Atmosphere*. London: Author, 1987. An up-to-date account of scientific work on the higher regions of the atmosphere, including the ozone layer, this work consists of individual review papers by specialists. Suitable mainly for those with a college-level background in atmospheric science.

Young, Louise B. *Earth's Aura: A Layman's Guide to the Atmosphere*. New York: Alfred A. Knopf, 1977. A particularly well-written account of the study of the atmosphere. Suitable for general readers.

F. W. Taylor

NUCLEAR DETECTION SATELLITES

Date: Beginning October 17, 1963
Type of satellite: Military

Nuclear detection is the spotting of secret nuclear explosions in space and within Earth's atmosphere in an effort to enforce the 1963 Nuclear Test-Ban Treaty. Nuclear detection was thus the first example of an arms-control treaty verified through satellites.

Summary of the Satellites

Nuclear detection satellites had their origins in talks during the late 1950's and early 1960's to ban nuclear tests in the atmosphere, under water, under ground, and in space. Policing a ban on nuclear tests in space posed difficulties. The verification system would have to detect explosions in the upper atmosphere, in near-Earth space, behind the Moon, and in deep space. Making this task more difficult was the natural background of solar radiation, cosmic rays, and the Van Allen radiation belts. One early proposal in 1958 involved placing five or six large nuclear detection satellites in 29,000-kilometer-high orbits. Additionally, 170 ground stations would be spread across the globe. Yet even this extensive network might not detect a weapon encased in radiation shielding.

To solve this verification problem, the Atomic Energy Commission and the U.S. Department of Defense jointly began Project Vela in the fall of 1959. (*Vela* is Spanish for "watchman.") One part of the project was called Vela Hotel. This part of the project envisioned the development of nuclear detection satellites able to spot an explosion as small as 10 kilotons and as far away as 160 million kilometers from Earth. The satellites would carry sensors to detect the X rays, neutrons, and gamma rays emitted by the fireball. The sensors were designed using data from satellites on the normal background radiation of space and five high-altitude nuclear tests. These measurements were made by the Explorer 4 satellite during the 1958 U.S. nuclear test series.

The Vela sensors were designed by the Lawrence Radiation Laboratory and built by the Los Alamos National Laboratory and the Sandia Corporation. The prime contractor for the satellite was TRW. The work went well, and on June 22, 1961, the Advanced Research Projects Agency gave approval for a five-launch program. Each launch was to carry a pair of Vela satellites. The first pair were to be launched in April, 1963, with the rest to follow at three-month intervals.

Over the next two years, as the Vela sensors were being developed, both the Soviet Union and the United States conducted high-altitude nuclear tests. The radiation from these tests was trapped by Earth's magnetic field and damaged several satellites, including the Telstar communications satellite. In addition to American and Soviet satellite measurements of these trapped radiation fields, Vela experiments were flown aboard the Discover 29 and 31 reconnaissance satellites.

It was not until August, 1963, that the United States, the Soviet Union, and Great Britain were finally able to reach an agreement on limiting nuclear tests. The treaty banned nuclear tests in the atmosphere, under water, and in space. Tests underground were permitted as long as no radiation escaped. The Nuclear Test-Ban Treaty went into effect on October 10, 1963.

On October 17, 1963, the first pair of Vela satellites were launched by an Atlas-Agena D from Cape Canaveral. The Agena placed the two Velas into 103,600-by-208-kilometer initial orbit. Eighteen hours after launch, a radio signal fired the first Vela's onboard engine, placing it into a 110,868-by-101,851-kilometer orbit. This was the highest orbit achieved by any military satellite and was equivalent to one-quarter of the distance to the moon.

On October 19, the second Vela's engine was fired, placing it into a similar orbit. The satellites were placed 180 degrees apart, so that their sensors could look into space in opposite directions. The detection package weighed 36 kilograms and was composed of twelve X-ray detectors and two diagnostic detectors on the outside of the satellite and four gamma-ray and neutron detectors inside. The Vela itself was a sphere made up of twenty triangular faces. By placing the X-ray detectors at the points where the triangles met, each detector had a field of view greater than a full hemisphere. Once in its final orbit, the Vela satellite weighed about 136 kilograms and was spin-stabilized at two revolutions per second.

The second pair of Velas were launched on July 17, 1964. A successful launch was marred when the Agena's engine continued to fire briefly after the two Velas had separated. The Agena bumped one of the satellites, damaging a few of the detectors and changing the spin axis slightly. The third pair went into orbit after a night launch on July 20, 1965. These two Velas carried improved instrument packages, which included an optical flash detector. These early Velas had a planned lifetime of only six months, yet they were still working after five years. Accordingly, the fourth and fifth pairs were not launched.

By this time, Vela's role had started to expand. In October, 1964, the Peoples Republic of China (PRC) exploded its first atomic bomb. As the PRC had not signed the test-ban treaty, the explosion was in the atmosphere. A Vela could, with optical flash detectors and other instruments, detect atmospheric nuclear tests and measure such things as the weapons yield. This, along with data from other sources, would indicate the status of the Chinese nuclear weapons program.

Accordingly, in March, 1965, TRW was awarded a contract to develop an advanced Vela. It would carry a larger instrument package — 63 kilograms. This package included a pair of optical flash detectors. Called Bhangmeters, they could indicate the size of a nuclear test by measuring the brightness of its flash. Two were used to cover a wide range of explosive yield: One Bhangmeter was more sensitive than the other. Because the Bhangmeters had to face Earth, the advanced Vela was stabilized by gas jets and gyros rather than by spinning. The sensors that made up the rest of the instrument package included eight X-ray detectors, an X-ray analyzer, four gamma-ray detectors, a neutron detector, two heavy particle detectors, an extreme ultraviolet detector, two Geiger counters, an electron-proton spectrometer, a solid-state spectrometer, and electromagnetic pulse detectors. The latter picked up the extremely strong burst of energy emitted by a nuclear fireball, which can damage electronic components such as transistors by generating very high currents and burning them out. The advanced Vela weighed about 227 kilograms and was a twenty-six-sided sphere.

The first pair of advanced Velas was launched by a Titan 3C on April 28, 1967, from Cape Kennedy. They were placed into orbits measuring 112,585 by 109,089 and 114,587 by 107,489 kilometers. A second pair followed on May 23, 1969, and the third was launched on April 8, 1970. The third pair carried an enlarged sensor payload weighing 74 kilograms. This was also the last Vela launch. Later, nuclear detection sensors would be carried piggyback on early-warning satellites. The advanced Velas performed as well as did the earlier ones — designed to operate for eighteen months, three of the six were still working ten years later.

It appears that this piggyback approach was used by the Soviets, since there were no Soviet satellites that can clearly be identified as dedicated to nuclear detection. The lone exceptions were Elektrons 1, 2, 3, and 4, which were launched in pairs on January 30 and July 10, 1964, from Tyuratam. One satellite went into a low orbit, while the other of the pair went into a highly elliptical orbit. It was thought at the time that they might have a nuclear detection role in addition to serving as orbiting geophysical observatories. Nevertheless, no subsequent launches were made using a similar profile. It must be assumed that any Soviet nuclear detection sensors were flown piggyback on other satellites rather than on specially designed satellites.

The shift by the United States to piggyback nuclear detection instruments began on June 12, 1973, with the launch from Cape Kennedy of a Program 647/Defense Support Program early-warning satellite. It used a Titan 3C booster and went into a 35,786-by-35,777-kilometer orbit. This satellite had proton counters and X-ray detectors on two of its solar panels. Optical flash detectors were fitted on the telescope housing.

Studies also began in early 1975 on fitting Bhangmeters and electromagnetic pulse sensors to Navstar navigation satellites. The advantage was increased coverage. Although the Navstars orbited at a lower altitude than did either the Velas or the early-warning satellites, their higher inclination (63 degrees versus 33 for the Velas and less than 10 for the early-warning satellites) would give increased coverage of the polar areas. Also, because a network of as many as eighteen satellites would be used, the probability was that several Navstars would observe any suspected nuclear tests. Ford Aerospace and Communications Corporation was selected to build the instruments in October, 1975. The first Navstar to carry the nuclear detection instruments was launched on April 26, 1980, by an Atlas F booster from Vandenberg Air Force Base. The instrument package weighed 27 kilograms. The Navstar was placed into a 20,288-by-170-kilometer orbit.

Knowledge Gained

Beyond the role they play as watchmen of the Nuclear Test-Ban Treaty, the nuclear detection satellites have also provided important scientific data. Vela 10 (one of the pair launched on May 23, 1969) provided measurements of the X-ray star Cygnus X-1 from May, 1969, until its X-ray detectors failed on June 19, 1979. From these data it was determined that Cygnus X-1 is a black hole orbiting a blue giant. The black hole is only about 8 kilometers across but has a mass between ten and fifteen times that of the Sun. This small size and high density mean that not even light can escape from it. As the black hole orbits the blue giant, it pulls hydrogen gas from the star's surface; the gas forms a ring around the black hole. As the hydrogen spirals into the black hole, it is heated to tens of millions of degrees and emits X rays, which can vary wildly in intensity over a period of a day.

Other objects of study are X-ray and gamma-ray bursts. The first is a normal star with a neutron star (the latter is the crushed core of a star after a supernova explosion). The neutron star pulls hydrogen gas from the normal star. This gas coats the neutron star's surface until pressure and temperatures become so high that a thermonuclear explosion takes place — a natural hydrogen bomb. The cycle repeats every few hours or days. The source of the gamma-ray burst is not yet clear. Two possibilities are a "star quake" on a neutron star and the impact of an asteroid on a neutron star. In any case, they are much more powerful than an X-ray burst and do not repeat. The bursts were first discovered in 1967 by Ian B. Strong from Vela data. Thus, the Vela can be considered the start of orbital X-ray astronomy. Before their launch, the only data available were from brief sounding-rocket flights.

Context

The success of nuclear detection satellites is the result of a number of factors. One is the directness of the test-ban treaty — it has none of the confusion or loopholes of other arms-control agreements. Another is data from earlier atmospheric nuclear tests. The American 1962 nuclear tests were planned with the belief that this could be the last opportunity for atmospheric explosions, and detailed information was recorded. A major reason, however, is the verification system in place; with not only Vela and the other satellites but also seismic networks and atmospheric fallout-sampling aircraft, the chance of making successful secret tests is too small for the data that could be gained.

Nuclear detection satellites provide the means to spot secret tests by would-be nuclear powers. Despite their capabilities, however, nuclear detection satellites are subject to false alarms. One such false alarm occurred on September 22, 1979, when a Vela satellite spotted what was thought to be the flash from a 2- or 4-kiloton explosion in an area of the South Atlantic. A year of analysis finally determined that it was a false alarm caused by a small bit of debris knocked off the Vela by a dust-sized meteoroid. The debris drifted in front of the Bhangmeters and was seen as a flash which mimicked that caused by a small nuclear explosion. Given both the variety of nature and the ingenuity of man, a small level of uncertainty is inevitable.

Bibliography

Hansen, Chuck. *U.S. Nuclear Weapons.* New York: Orion Books, 1988. Covers the physics of nuclear weapons, the history of American weapons programs, and the development of individual weapons themselves. Illustrated. Suitable for college readers (some knowledge of nuclear physics may be helpful).

Klass, Philip J. *Secret Sentries in Space.* New York: Random House, 1971. The first book to deal with military satellites. Covers the historical background and development of reconnaissance satellites. Includes a section on Vela.

Peebles, Curtis. *Guardians: Strategic Reconnaissance Satellites.* Novato, Calif.: Presidio Press, 1987. Covers the history and technology of reconnaissance satellites and their profound impact on international relations. Includes a full chapter on nuclear detection. Suitable for high school and college readers.

Richelson, Jeffrey. *American Espionage and the Soviet Target.* New York: William Morrow and Co., 1987. An overview of U.S. efforts to obtain information on Soviet military activities. Includes several mentions of fallout sampling and monitoring of nuclear testing.

Seaborg, Glenn T. *Kennedy, Krushchev, and the Test Ban.* Berkeley: University of California Press, 1981. Written by the man who was chairman of the Atomic Energy Commission at the time the test-ban treaty was negotiated. The book covers the events leading up to the treaty. Includes a chapter on the 1962 high-altitude nuclear tests.

Sullivan, Walter. *Assault on the Unknown.* New York: McGraw-Hill Book Co., 1961. A history of the International Geophysical Year by a science writer with *The New York Times.* Includes a chapter on the high-altitude nuclear tests of 1958.

Turnill, Reginald. *The Observer's Book of Unmanned Spaceflight.* New York: Frederick Warne, 1978. A brief overview of unmanned satellites. Includes a section on military satellites with a mention of nuclear detection.

Curtis Peebles

OCEAN SURVEILLANCE SATELLITES

Date: Beginning September 18, 1967
Type of satellite: Military

Ocean surveillance involves locating ships, identifying them, and determining their speed and course. As satellites pass over all Earth's oceans, they have a unique vantage point for this activity.

Summary of the Satellites

During the early 1960's, the Soviet navy underwent tremendous growth — from a coastal defense force to a fleet equipped with large vessels armed with long-range cruise missiles. To support this fleet, the Soviets also developed an ocean surveillance satellite. To be fully effective, the satellite would have to track ships under all weather conditions. The Soviets decided to use a radar system, a method which has certain advantages but also a major shortcoming. The radar transmitted 3 to 5 kilowatts of power. This meant that it could pick up ships up to 1,200 kilometers away through the heaviest clouds, but also that it would require a large amount of electrical power.

To supply this power, the Soviets developed the Topaz nuclear reactor. It contains 50 kilograms of enriched uranium 235, packed into seventy-nine multilayered tubes known as "electrogenerating channels." The uranium heats an outer tube of molybdenum to 1,500 degrees Celsius. At this temperature, molybdenum will lose electrons easily. In the Topaz, the electrons jump a gap a few microns wide to a cooler niobium collector, and so generate an electrical current. (In a commercial reactor, the heat from the nuclear material is used to turn water into steam, which spins a turbine to generate electrical power.)

The Topaz is relatively light in weight (105 kilograms) and powerful (5 to 10 kilowatts). Yet there are several problems with its design. It must be started on the pad before launch, because it is not self-regulating; as the Topaz becomes hotter, it also becomes more reactive. This in turn heats the reactor, which makes the Topaz more reactive, until, if left uncontrolled, it would suffer a meltdown. (A commercial reactor becomes less reactive as it gets hotter.) As the Topaz operates, it produces highly radioactive waste products — strontium 90, cesium 137, and xenon 135.

The first Soviet nuclear-powered ocean surveillance satellite was launched on December 27, 1967, from Tyuratam by an F-1m booster. Kosmos 198 was placed into a 281-by-265-kilometer orbit. The satellite was approximately 14.6 meters long and 2.4 meters wide and weighed about 4,785 kilograms. The radar antenna is believed to have been a flat panel 7.1 to 8.5 meters long and 1.4 meters wide.

The low orbit was necessary to prevent the radar echoes from being too weak. The inverse square rule applies here: If the altitude is doubled, the radar signal reaching Earth's surface is one-fourth as strong. Since the reflected echo (which is only a tiny fraction of the radar's transmitted signal) must also make the return trip, the final result is one-sixteenth the strength for an orbit twice as high. The low orbit, however, is unstable; the extremely thin traces of atmosphere slow down the satellite, lowering the orbit. After a few months at most, the satellite would reenter and scatter radioactive debris on Earth.

After two days in orbit, Kosmos 198 separated into three parts — the F-1m's third stage, the radar antenna, and the reactor unit. A booster rocket attached to the reactor fired and placed it into a final orbit of 952 by 894 kilometers. This orbit ensured that the reactor would not reenter for more than six hundred years. By that time, its radioactive waste products will have decayed.

Between 1967 and 1971, the Soviets launched a total of six nuclear ocean surveillance satellites, with poor results. One ended in a launch failure that scattered radioactive material in the upper atmosphere. Another, Kosmos 367, launched on October 3, 1970, failed almost immediately after reaching orbit, and the reactor had to be boosted. The best performance was by Kosmos 469 (launched December 25, 1971); it achieved ten days of operation.

During 1972 and 1973, Soviet fortunes improved somewhat. Despite another launch failure on April 25, 1973, the Soviets were able to operate one satellite for thirty-one days (Kosmos 516) and another for forty-four days (Kosmos 626).

It was not until 1974, seven years after test flights had begun, that the Soviets were able to shift to operational status. Kosmos 651 was launched on May 15, 1974, into a 276-by-256-kilometer orbit. Two days later, Kosmos 654 was placed in a nearly identical orbit. The two satellites worked together as a team, one following twenty to thirty minutes behind the other. This would provide multiple plots to indicate each ship's heading and speed. Kosmos 651 operated for seventy-one days, while Kosmos 654 lasted seventy-four days, nearly the full seventy-five-day lifetime that had been projected.

The year 1974 also saw the first launch of the second type of Soviet ocean surveillance satellite. On December 24, 1974, Kosmos 699 was launched by an F-1m booster from Tyuratam into a 454-by-436-kilometer orbit. Unlike the active radar on the nuclear-powered satellites, passive sensors were used on Kosmos 699. This Electronic Intelligence (ELINT) satellite listened for radar and radio transmissions from the target ships. As there was no power-hungry radar aboard, no nuclear reactor was carried.

The events of 1974 set the pattern for subsequent Soviet ocean surveillance activities. A pair of the nuclear-powered satellites were launched within a few days after Kosmos 699; a few months later, an ELINT satellite was orbited. After several months, another pair of nuclear satellites went into orbit. The performance of these later nuclear satellites remained erratic. The longest-lived was Kosmos 724, launched April 7, 1975; it operated for sixty-five days. The worst performance was by Kosmos 785, launched December 15, 1975; its reactor was boosted only sixteen hours after launch. Despite the experience Soviet designers had gained over four years of operations, the performance of these satellites did not improve but rather became worse.

During the mid-1960's, designers in the United States Navy explored the feasibility of ocean surveillance satellites. In 1969, development began on the Program 749 satellite. Like the Soviet nuclear-powered satellite, Program 749 would use active radar. Technical problems were too great, however, and the program was canceled in 1973.

To replace it, the Navy selected a three-step approach. The first was to borrow United States Air Force satellites to gain experience in orbital ocean surveillance. There has been speculation that this involved use of high-resolution photoreconnaissance satellites, launched by a Titan 3-B rocket and equipped with an infrared scanner. This scanner could pick up the hot gases from a ship's smokestack and the warm water discharged from a nuclear submarine's reactor. (ELINT satellites may also have been used.)

The next step was an advanced ELINT satellite, code-named White Cloud. The first was launched on April 30, 1975, by an Atlas F from Vandenberg Air Force Base. The payload went into a 1,128-by-1,092-kilometer orbit; then the three White Cloud satellites separated and went into parallel orbits, with the satellites a few kilometers apart. Each satellite was 2.4 meters long, 0.9 meter wide, and 0.3 meter thick. The upper side was covered with solar cells, while the underside carried the antennae to pick up transmissions from Soviet ships.

To locate ships, White Cloud uses a technique known as "interferometry." Radar and radio signals from a ship arrive at each satellite at a slightly different time. This difference is used to determine the ship's location. A similar procedure was suggested as the basis for a satellite navigation system. It was first tested on a December 14, 1971, launch. A long tank, thrust-augmented Thor-Agena D (LTTAT-Agena D) placed three satellites into an orbit similar to that used later by White Cloud. From a 1,126-kilometer orbit, a White Cloud cluster could pick up transmissions from ships as far away as 3,218 kilometers. This wide-area coverage, and the fact the cluster passed over the same spot on Earth twice a day, meant multiple position reports. To increase coverage further, a three-cluster network was used; each cluster of three satellites was positioned 120 degrees apart. The full network was not completed until March 2, 1980, when the third White Cloud was orbited.

In the meantime, the Soviet nuclear ocean surveillance program had suffered a major setback. Kosmos 954 was launched on September 18, 1977; ten weeks later, it suffered a malfunction that prevented the reactor from being boosted into the final parking orbit. Atmospheric drag lowered the orbit of the satellite until it reentered over northern Canada on January 24, 1978. The heat and stress of reentry broke up the reactor and scattered radioactive debris in an area between Yellowknife and Baker Lake. The clean-up lasted several months. Some of the debris was extremely dangerous — one metal slab, for example, emitted 200 roentgens per hour. Two hours' exposure could be fatal to humans. Fortunately, the search area was very sparsely populated (in 38,849 square kilometers, there were fewer than thirty-five people).

The third step in the United States Navy's ocean surveillance plan was the development of a radar-equipped satellite, code-named Clipper Bow. During 1978 and 1979, doubts arose regarding its cost, usefulness, and possible duplication of the data of other orbital radar reconnaissance projects. By mid-1980, Clipper Bow had been canceled.

At the time that the Clipper Bow program was coming into question, the Soviets had halted nuclear ocean surveillance operations, following the failure of Kosmos 954. Much time was put into a redesign effort. The lifetime of the nuclear satellite was increased to 140 days, and the reactor was modified so that the core could be separated. The latter change served two goals. Once the reactor was in its final orbit, separating the core would delay reentry for another fifty to several hundred years. If the boost maneuver failed, separating the core would ensure that both it and the reactor would burn upon reentry.

It took two years to make the necessary changes. The first of the new nuclear ocean surveillance satellites was Kosmos 1176, launched from Tyuratam on April 29, 1980. It operated for 135 days before the reactor was boosted and the core separated.

With the resumption of the program, certain changes were implemented. The nuclear satellites continued to operate in pairs, but their launches might be separated by as much as a month. The Soviets' ELINTs were now used in pairs as well. Their orbits were placed so that their ground tracks (the lines on Earth's surface directly under the satellites' orbital paths) overlapped. Starting in 1982, two pairs of ELINTs were used.

The resumption of nuclear ocean surveillance satellite

launches lasted for only two years. On December 28, 1982, the Soviets commanded Kosmos 1402's reactor to separate and boost into the higher, final orbit. The maneuver failed, and the reactor remained attached to the third stage. Soviet ground controllers immediately separated the core. The satellite body and reactor structure were viewed as the most dangerous, for its contact with the core had made it highly radioactive. It reentered and burned over the Indian Ocean on January 23, 1983. The core followed on February 7, reentering over the South Atlantic. In both cases, any debris that survived was lost in the deep ocean. Subsequently, launches of nuclear satellites were once again halted. For one and a half years the Soviets sent only ELINT satellites into orbit.

Nuclear satellite operations resumed with Kosmos 1579, launched on June 29, 1984. Again, design and operational changes had been made. Kosmos 1579 worked with an ELINT, Kosmos 1567. Their orbits were 150 degrees apart, so that their orbits crossed over the shipping lanes of the North Atlantic. The satellites made passes about forty-five minutes apart.

The major change during the next four years was the launch of the first heavy ocean surveillance satellite, Kosmos 1870, launched by a D-1 booster on July 25, 1987, from Tyuratam into a 249-by-237-kilometer orbit, carrying both radar and visual sensors. The design of Kosmos 1870 appears to have been based on that of the Salyut military space station. The electrical power needed to operate its radar is supplied by solar cells. Such heavy ocean surveillance satellites were probably designed as replacements for the nuclear ones.

In the meantime, launches of nuclear satellites continued. One of these was Kosmos 1900, which malfunctioned in mid-April, 1988. For the third time, a nuclear reactor was heading toward Earth. A command to separate the core was ineffectual. This meant that the radioactive material was shielded by the reactor structure. There was, however, a last-minute reprieve. The spacecraft carried a backup system to separate the reactor if the spacecraft lost attitude control, the reactor depressurized, or the reentry began to heat it. On October 1, 1988, the automatic system triggered, and the reactor separated and was boosted into a higher, final orbit.

Knowledge Gained

For both the United States and the former Soviet Union, control of the seas was critical. In the event of a war in Europe, North Atlantic Treaty Organization (NATO) strategy envisioned large convoys to resupply equipment and troops. If the Soviets were to stop these convoys, their ships and submarines had to know their positions. The stakes were high. In World Wars I and II, German U-boats were almost successful in starving England into submission. A battle in the Atlantic could easily decide the fate of Europe. If the United States Navy and other NATO navies were to ensure security, they had to know the location of attacking Soviet ships and submarines.

The North Atlantic convoy routes are not the West's only points of vulnerability. Most of the oil that fuels the West's economies comes through the Persian Gulf. Any interruption, such as a blockade, could spark inflation, unemployment, and social disruption.

Despite the critical role ocean surveillance satellites played for both sides, the United States and the former Soviet Union had very different policies. The Soviets were the most active. They deployed three different types of satellites and continued to fly their nuclear-powered satellites even after repeated failures. In contrast, the United States was relatively slow to fly ocean surveillance satellites. The first White Cloud was not orbited until nearly a decade after the first Soviet ocean surveillance satellites. Two different radar satellites were canceled by the United States before they were flown.

This difference in policy reflected the different situations facing the two navies. The Soviets, with few ice-free ports, had poor access to the open ocean. This fact made it easier for the American military to track Soviet naval movements. The major naval threat to the Soviets was American aircraft carriers. It was not until the 1970's that the Soviets themselves began operating such ships.

Context

The Soviet ocean surveillance program raised two major issues. The first is safety. Twice, nuclear reactors have fallen because of booster failures (and there was a near miss with Kosmos 1900). When Western governments requested information, the Soviet response was nearly useless; for the most part, Soviet leaders simply issued statements that the satellite was operating normally or that the reentry posed no danger. When reality proved otherwise, the Soviets blamed the West rather than accepting responsibility.

Questions have been raised about the safety of the orbital graveyard. Altitudes above 900 kilometers have the greatest amount of "space junk," and they are crossed by expended Soviet boosters; thus, collisions are possible. A reactor hit in a collision could be sent into an elliptical orbit and might reenter far sooner than expected. The reactor could also be fragmented. The fragments could hit other reactors, producing more debris, until there were clouds of shrapnel.

The other issue is Antisatellite weapons (ASATs). The American Program 437 Thor and Program 505 Nike Zeus ASATs were meant as counters to possible Soviet nuclear weapons in orbit. After both programs had been closed down, the danger was seen as coming from Soviet ocean surveillance satellites. The United States began development of ASATs that would be launched from F-15's. In the event of war, their primary targets would be ocean surveillance satellites. The F-15 ASAT ran into political opposition and was halted after only a single interception. The military danger posed by Soviet ocean surveillance satellites was largely ignored in debates regarding weapons in space.

Bibliography

Hart, Douglas. *The Encyclopedia of Soviet Spacecraft*. New York: Exeter Books, 1987. A heavily illustrated survey of Soviet space activities. Gives only brief attention to ocean surveillance.

Kennedy, William V. *Intelligence Warfare*. New York: Crescent Books, 1983. A survey of the various forms military intelligence can take, this volume includes a chapter on ocean surveillance. Suitable for high school and college audiences.

Oberg, James E. *Uncovering Soviet Disasters: Exploring the Limits of Glasnost*. New York: Random House, 1988. An overview of Soviet accidents on land and sea and in air and space. The book includes a chapter on Soviet nuclear reactors in orbit. (It was published only a few months before Kosmos 1900.) The author is an engineer with the space shuttle program.

Peebles, Curtis. *Guardians: Strategic Reconnaissance Satellites*. Novato, Calif.: Presidio Press, 1987. The book covers the history, technology, and impact on international relations of reconnaissance satellites. It includes chapters on United States and Soviet ocean surveillance programs, ending with an account of the resumption of launches after Kosmos 1402.

Richelson, Jeffrey. *American Espionage and the Soviet Target*. New York: William Morrow and Co., 1987. This book is a detailed survey of United States intelligence activities. Some material on ocean surveillance is included.

U.S. Congress. Senate. Committee on Aeronautical and Space Sciences. *Soviet Space Programs, 1971 – 1975*. Report prepared by Congressional Research Service, the Library of Congress. 94th Cong., 1976. This volume, which covers the years preceding the period addressed in the following entry, includes material on the Soviet ocean surveillance program (prior to Kosmos 954).

U.S. Congress. Senate. Committee on Commerce, Science, and Transportation. *Soviet Space Programs, 1976 – 1980*. Part 3, *Unmanned Space Activities*. Report prepared by Congressional Research Service, the Library of Congress. 99th Cong., 1st sess., 1985. Committee Print. Part of a multivolume series on Soviet space activities. The book includes an account of the ocean surveillance program, a table of launches, and drawings of the satellite and reactor.

Curtis Peebles

THE ORBITING ASTRONOMICAL OBSERVATORIES

Date: April 8, 1966, to August 21, 1972
Type of satellite: Scientific

The Orbiting Astronomical Observatories provided astronomers with an opportunity to conduct observations at specific wavelengths above Earth's atmosphere. These data significantly contributed to an understanding of certain astronomical objects and phenomena. The development of the observatories presented new challenges, the solutions to which made the satellites some of the most complex of their time.

PRINCIPAL PERSONAGES
FRED L. WHIPPLE,
ROBERT J. DAVIS, and
WILLIAM A. DEUTSCHMAN, advisers for
    the Celescope Project
ARTHUR D. CODE, Director, Washburn Observatory
LYMAN SPITZER, JR., Principal Investigator,
    OAO 3 32-inch ultraviolet telescope
JOHN E. ROGERSON, JR., Executive Director,
    OAO 3 32-inch ultraviolet telescope
ALBERT BOGGESS, Principal Investigator, OAO B

Summary of the Satellites

Astronomers have always been plagued by problems over which they have little or no control, including poor weather at observing sites and the atmosphere's absorption of certain wavelengths of light. These conditions often result in limited observations, which in turn can diminish the quality and quantity of collected data.

With the advent of the space age and satellite technology, it became possible to place observatory satellites in Earth orbit. Such a satellite project, called the Orbiting Astronomical Observatory (OAO), enabled astronomers to overcome the problem of atmospheric interference.

The National Aeronautics and Space Administration (NASA) chose Grumman Aerospace to be the major contractor for the OAOs, satellites that would allow the study of astronomical objects in the ultraviolet, X-ray, and gamma-ray ranges of the spectrum. The first OAO was scheduled to be launched in 1964, but several problems postponed the launch until 1966. These difficulties involved the OAO control and guidance system and the development and completion of one of the first OAO experiments, a project of the Smithsonian Astrophysical Observatory, designated "Celescope." Budget overruns and a lack of agreement on how such a program

should proceed were also problems. Many scientists and engineers believed that the technology for such an advanced satellite program would have to evolve slowly, for it did not exist at the time, and that several smaller, less complex and less expensive OAOs should first be constructed and launched.

The cost of the OAOs, as estimated by the NASA comptroller's office, was about $55,000 per satellite kilogram. When compared with the cost of other programs, however, that amount was not excessive. The Orbiting Geophysical Observatory program had cost about $70,000 per satellite kilogram, the Surveyor satellites had demanded $85,000 to $90,000 per kilogram, and the Lunar Orbiters had cost about $95,000 per kilogram. Moreover, manned missions cost hundreds of thousands of dollars per spacecraft kilogram.

The OAO's basic design included large paddlelike panels covered with solar cells that would furnish power. Specially designed star trackers were used to orient the satellite. The trackers operated by selecting several stars and locking onto their positions, thereby establishing a frame of reference for the OAO and allowing it to orient itself toward any desired object. The observatory's scientific equipment, along with the necessary satellite equipment, was installed in a framework that was the same for all OAO missions. The scientific equipment and experiments could be varied on each mission, depending on the equipment's availability and the mission's objectives.

A special network of ground-based control stations was built for the OAO program. It included stations at Quito, Ecuador; Rosman, North Carolina; and Santiago, Chile. These stations were linked via radio to Goddard Space Flight Center, in Greenbelt, Maryland.

The first OAO was launched by an Atlas-Agena D booster rocket; later OAOs were launched by the Atlas-Centaur. OAO 1 was launched from the Eastern Test Range in Florida on April 8, 1966, at 2:36 P.M. eastern standard time. The observatory satellite's experiments included a University of Wisconsin package, a Massachusetts Institute of Technology (MIT) experiment, and a Lockheed experiment.

The University of Wisconsin equipment, supplied by the Washburn Observatory, included four telescopes with 20.3-centimeter-diameter mirrors, one telescope with a 40.6-centimeter-diameter mirror, two scanning spectrometers, and some photoelectric detectors. The equipment was controlled by five hundred high-reliability, low-power, digital logic elements. At the focusing point of the 40.6-centimeter tele-scope, which weighed 33.6 kilograms, was a photoelectric photometer, a device that measures an object's brightness. The four 20.3-centimeter telescopes, each weighing 12.7 kilograms, were brought to focus at a diaphragm, which can limit the size of the field of view, and a special filter wheel with optical filters. The University of Wisconsin experiment was designed to measure the radiation of stars and nebulae (interstellar clouds of gas and dust).

The MIT experiment's purpose was to detect high-energy gamma rays using a device similar to that flown on the Explorer 11 spacecraft. On Explorer 11, the experiment had not worked well because of the spacecraft's proximity to the Van Allen radiation belts. The Lockheed experiment was designed to measure soft X rays, which had recently been detected by high-altitude sounding rockets, using a special counter. This instrument, similar to a Geiger counter, was many times more sensitive than those flown on the sounding rockets.

After a successful launch and orbital insertion in a 790-by-800-kilometer orbit, all appeared well. On the second day, however, a primary battery failed. The malfunction appeared to spread to the other battery systems, and the OAO 1 ceased to operate. The lessons learned from this failure were applied to future missions; equipment was redesigned, and procedures for activating satellite power were changed.

OAO 2 was successfully launched from the Eastern Test Range on December 7, 1968, at 3:40 A.M. eastern standard time. The observatory assumed an orbit that was 761 kilometers at its perigee (the closest approach to Earth) and 770 kilometers at its apogee (the farthest point from Earth). The satellite had eleven telescopes on board, representing experimental packages of the University of Wisconsin and the Smithsonian Astrophysical Observatory.

The Smithsonian Astrophysical Observatory's Celescope was built around a telescope with a 30.5-centimeter-diameter mirror. At the telescope's point of focus there were four television-type tubes, or uvicons, each of which operated at a different wavelength. A special calibration system, using an ultraviolet lamp source coupled with pinholes of various sizes, was used to align and calibrate the telescope and its associated equipment. Digital techniques were used to send images to the ground stations without delays. (The University of Wisconsin experiment stored data and later relayed it to a ground station.) The Celescope's main purpose was to record data for the construction of a star map in ultraviolet light.

OAO B was to orbit a single reflecting telescope with a 91.4-centimeter-diameter mirror, a project of the Laboratory for Optical Astronomy at NASA's Goddard Space Flight Center. The telescope, coupled with a spectrometer, was to have measured phenomena in the ultraviolet-to-blue regions of the spectrum. The objects to be observed included variable stars, emission and reflection nebulae, star clusters, stellar objects in the nearby Magellanic Clouds, and member galaxies of the Local Group, a cluster of twenty-eight known

galaxies of which the Milky Way is a member. Unfortunately, the launch of OAO B failed, and the satellite never reached its planned 746-kilometer-high orbit.

The final planned mission of the Orbiting Astronomical Observatory series, designated OAO C before launch and OAO 3 upon placement in Earth orbit, was named Copernicus, for the great astronomer. OAO 3 was successfully launched on August 21, 1972, at 6:28 A.M. eastern standard time. Its orbit had a perigee of 739 kilometers, an apogee of 751 kilometers, and a period of 99.7 minutes.

Overall, Copernicus was similar to OAO 2. Several changes had been made, however, including improvements to the star tracking system and the Sun baffles. There was also a new computer system. Two experiments were aboard: a Princeton University project and an experiment designed by the University College in London and included as a result of NASA's international cooperation policy.

The Princeton University project was based on an 81.3-centimeter-diameter reflecting telescope designed to observe objects in the ultraviolet region of the spectrum. The telescope's mirror was specially constructed by the Corning Glass Works. Most telescope mirrors of this size are solid cast, but the Copernicus mirror was made of thin plates of silica fused into an egg crate pattern. The egg crate structure was then sandwiched between thin front and rear disks of glass. This method of construction reduces a mirror's weight by a factor of three; the Copernicus mirror weighed about 44 kilograms. The mirror was also designed to perform in the varying temperatures of space. It had an aluminized surface coated with protective lithium fluoride.

The Copernicus telescope's tracking instrumentation was capable of unprecedented accuracy. One NASA official stated that the telescope could locate an object the size of a basketball at a distance of 640 kilometers.

Three small X-ray telescopes were the basis of the project designed by the University College in London. The telescopes were designed to observe already-identified X-ray sources.

Copernicus was the final mission in the OAO series. Two of the four satellites were highly successful. The OAO technology also served as a foundation for new generations of space observatories, including the High-Energy Astronomical Observatory and the Hubble Space Telescope. The OAO project produced the most massive and complex unmanned satellites ever launched at that time.

The OAOs' contributions to astronomy were significant, even though simpler satellites could have been built for less money and orbited sooner. Moreover, important technological knowledge was gained from designing and constructing the OAOs.

Knowledge Gained

The opportunity to observe objects without the interference of Earth's atmosphere began a new era in astronomical research. The two OAOs that completed their missions, OAO 2 and OAO 3, made significant contributions, especially to ultraviolet radiation observational astronomy.

OAO 2 had several observational goals, including the observation of young, hot stars. Such stars expend a large amount of energy in the ultraviolet region of the spectrum, and, because of Earth's atmosphere's absorption of ultraviolet light, they cannot be studied from ground-based observatories.

OAO 2 completed the first survey of space in ultraviolet light as part of the Smithsonian Astrophysical Observatory's program, using the SAO's experimental package. OAO 2 was the first satellite to detect ultraviolet radiation originating from the center of M31, the Andromeda galaxy. The Andromeda galaxy is a spiral galaxy much like the Milky Way and is part of the Local Group.

OAO 2 also made the first space-based measurements of a comet. It detected a large hydrogen cloud surrounding the comet Tago-Sato-Kosaka, discovered in 1969. This finding was significant because free hydrogen had never been seen associated with a comet. Additional cometary data indicated the presence of hydroxyl molecules. Other OAO 2 results included information on stellar X radiation, gamma radiation, and infrared radiation.

Some of the observations conducted by OAO 3, Copernicus, were continuations of OAO 2 experiments. Some unique and useful investigations, however, were also carried out. OAO 3 collected data on Cygnus X-1, an object that many astronomers believe is a black hole. (A black hole is formed by the gravitational collapse of a star with a mass three times that of the Sun; no matter or energy of any form may leave a black hole after a certain point in its collapse, called the Schwarzschild radius). Data returned by Copernicus suggested that Cygnus X-1, an X radiation source, is indeed a black hole.

Copernicus also collected data on some remnants of supernovae, stars that, near the end of their lives, throw off the outer layers of their atmospheres in cataclysmic explosions caused by the rapid conversion of elements and the resulting extreme rises in temperature and pressure. These studies of the remnants, or planetary nebulae, had results that varied from object to object. Copernicus provided evidence that some planetary nebulae produce X rays as a result of the supernova explosion and that some, such as the Crab nebula in the constellation of Taurus, produce radiation many times stronger than others and have a more discrete radiation source.

Copernicus measured the amounts of deuterium, a hydrogen isotope that is twice the mass of ordinary hydrogen, present in the universe. Scientists use this type of information when constructing models of the universe's size. Other investigations included observations of the flow of radiation from stars ("stellar wind") and additional hydrogen, X radiation, and gamma radiation studies.

Context

The two Orbiting Astronomical Observatories successfully launched by NASA, OAO 2 and Copernicus, began a new era, both in astronomical observations and in spacecraft design and engineering. The OAOs required entirely new instrumentation and designs for spacecraft operation and experimentation packages. The necessity of developing sophisticated new technology delayed the original launch target two years, from 1964 to 1966, and the spacecraft proved to be more expensive than was originally intended. Nevertheless, engineering problems and delays were overcome and one of the most successful and extensive astronomical studies of the time was conducted. Many of the special instrumentation packages and spacecraft designs would be adapted for use in future missions.

Observations conducted via the OAOs allowed astronomers to perform studies on radiation at the extreme ends of the spectrum: X rays, gamma rays, infrared rays, and ultraviolet rays. Ultraviolet radiation studies were especially useful and were the primary observations performed by OAO 2 and Copernicus, because of the atmosphere's absorption of ultraviolet radiation.

Studies of such phenomena allow astronomers to do more than collect information in specific wavelengths of the spectrum; they allow astronomers to act as historians. Many sources of radiation are so far away that their light may take billions of years to reach Earth. (Light travels at 3×10^8 meters per second; the light from the closest star to Earth besides the Sun, Proxima Centauri, takes more than four years to reach Earth.) Data collected from the two OAO missions allowed astronomers to construct particular theories regarding the cosmogony and cosmology, or the beginnings and development, of the Universe. The life of a star may last millions or billions of years, depending on several factors. Instrumentation carried by the OAOs allowed astronomers to measure the ages of many types of objects.

The OAOs collected data for several years. Some scientists — comparing the OAOs' lifespans with the lifespans of sounding rockets, which had performed similar experiments — said that one month of OAO data was worth fifteen years of research with sounding rockets.

Finally, the OAOs gave astronomers and engineers an opportunity to study the potential of, and problems with, Earth-orbiting astronomical observatories. It is appropriate that the final OAO was named Copernicus, for the astronomer Copernicus had opened the eyes of the astronomical world several hundred years earlier. The development of future observatory platforms will rely on the experience gained through the pioneering Orbiting Astronomical Observatories.

Bibliography

Braun, Wernher von, et al. *Space Travel: A History*. Rev. ed. New York: Thomas Y. Crowell, 1985. A complete and detailed history of space exploration. Covers the development of the rocket, Soviet and U.S. space programs, and the space programs of other countries. This text has been revised to keep up with technological advancements. Includes high-quality photographs. Suitable for general audiences.

De Waard, E. John, and Nancy De Waard. *History of NASA: America's Voyage to the Stars*. New York: Exeter Books, 1984. This book reviews U.S. achievements in space, with an emphasis on American manned space exploration. Presents a good overview of the programs and missions. Contains excellent color photographs and illustrations. Suitable for general audiences.

Newell, Homer E. B*eyond the Atmosphere: Early Years of Space Science*. NASA SP-4211. Washington, D.C.: Government Printing Office, 1980. An excellent overview of program and spacecraft development within NASA, this source offers a candid look at a variety of projects and programs, details some problems and experiences, and summarizes program results. A historical approach is taken. Suitable for readers with some background in space science.

"Observing a Comet from Space." *Sky and Telescope* 39 (March, 1970): 143. Discusses the OAO 2 observations of the comet Tago-Sato-Kosaka. A number of excellent articles outlining the OAO program, its development, and its results have been published in this popular astronomy and space sciences magazine.

"The Orbiting Astronomical Observatory." *Sky and Telescope* 24 (December, 1962): 339-340. Details the original concept of and plans for the Orbiting Astronomical Observatory program.

Watts, Raymond N., Jr. "Another Orbiting Astronomical Observatory." *Sky and Telescope* 40 (December, 1970): 349-350. This article details OAO B's planned experiments and equipment; it was published prior to the satellite's launch failure.

————. "An Astronomy Satellite Named Copernicus." *Sky and Telescope* 44 (October, 1972): 231- 232, 235. Discusses the final OAO of the series. Information on the satellite's launch, experiments, and equipment is presented.

————. "The Celescope Experiment: Ultraviolet Telescopes on Orbiting Astronomical Observatory." *Sky and Telescope* 36 (October, 1968): 228- 230. Outlines the Smithsonian Astrophysical Observatory experiment and equipment that was flown on OAO 2.

————. "More About the OAO." *Sky and Telescope* 28 (August, 1964): 78-79. An article about the University of Wisconsin experiment and equipment that was flown on OAOs 1 and 2.

————. "Orbiting Astronomical Observatory." *Sky and Telescope* 31 (May, 1966): 275- 276. Details the launch, in-orbit problems, and ultimate failure of OAO 1.

Mike D. Reynolds

THE ORBITING GEOPHYSICAL OBSERVATORIES

Date: September 5, 1964, to July 14, 1972
Type of satellite: Scientific

The six Orbiting Geophysical Observatory (OGO) spacecraft returned significant data on various geophysical phenomena, including the solar wind, Earth's atmosphere and magnetic field, the radiation belts surrounding Earth, and numerous properties of the Earth-Moon space environment.

PRINCIPAL PERSONAGES

WILFRED E. SCULL, Project Manager for all six of the
OGO satellites
G. H. LUDWIG, Project Scientist for OGOs 1 and 3
C. D. ASHWORTH, Program Manager for
OGOs 1, 2, 3, and 4
A. W. SCHARDT, Program Scientist for
OGOs 1, 3, and 5
N. W. SPENCER, Project Scientist for OGOs 2, 4, and 6
R. F. FELLOWS, Program Scientist for OGOs 2, 4, and 6
J. P. HEPPNER, Project Scientist for OGO 5
T. L. FISCHETTI, Program Manager for OGOs 5 and 6

Summary of the Satellites

The Orbiting Geophysical Observatory (OGO) program consisted of six missions, based upon a generation of spacecraft that were capable of operating a number of scientific instruments and experiments simultaneously and returning large amounts of data to Earth at rates considerably greater than had been possible with previous satellites. Although earlier spacecraft were responsible for the more spectacular discoveries in the near-Earth environment — such as the Van Allen belts (the bands of highly energetic, charged particles trapped in Earth's magnetic field), the solar wind (a flow of charged particles from the Sun), and the magnetosphere (the region around Earth where its magnetic field interacts with the solar wind) — the OGO missions provided more detailed data on the complexities of these (and other) phenomena and their interactions.

In the late 1950's and early 1960's, researchers at the Goddard Space Flight Center of the National Aeronautics and Space Administration (NASA) developed the concept of the OGO spacecraft, and on January 6, 1961, the TRW Systems Group of Redondo Beach, California, was selected as prime contractor for the program. Wilfred E. Scull, at Goddard Space Flight Center, was named project manager for

all six OGO missions; for OGO 1, G. H. Ludwig was project scientist, C. D. Ashworth program manager, and A. W. Schardt program scientist.

Beginning on September 5, 1964, with the launch of OGO 1, the six satellites were eventually placed (at the rate of one launch per year) alternately in two different types of orbits about Earth. The Eccentric Geophysical Observatory (EGO) satellites, OGOs 1, 3, and 5, used eccentric orbits with apogee (highest distance from Earth's surface) altitudes of approximately 150,000 kilometers and perigee (closest distance to Earth's surface) altitudes of approximately 300 kilometers, giving each spacecraft an orbital period (round-trip time about Earth) of approximately sixty-four hours. These orbits carried the satellites through the radiation belts and far enough from Earth to allow data collection in near interplanetary space. Closer to Earth, the three Polar Orbit Geophysical Observatory (POGO) satellites (OGOs 2, 4, and 6), had their apogees at approximately 1,000 kilometers and their perigees at approximately 400 kilometers, resulting in orbital periods of approximately 100 minutes. These orbits were oriented to carry the satellites close to Earth's polar regions, allowing studies of ionospheric and atmospheric phenomena involving the higher intensities of the magnetic field at those locations. All the eccentric orbit missions used Atlas-Agena launch vehicles and were launched from the Eastern Test Range (Cape Canaveral, Florida), while the polar missions employed thrust-augmented Thor-Agena vehicles and were launched from the Western Test Range (Vandenberg Air Force Base, California). Scheduling of the missions was such that, at times, as many as three satellites were returning data concurrently; while this imposed an additional burden on ground support, it afforded unprecedented opportunities for simultaneous measurements (for example, of variations in Earth's magnetic field intensity) at different points in space.

A nearly identical design was used for all six OGO spacecraft, following the program philosophy of demonstrating the feasibility of a standard observatory satellite that could house a relatively large payload (many scientific experiments and instruments). The main body of each OGO satellite was rectangular in shape, measuring 0.9 by 0.9 by 1.8 meters. One of the sides was kept pointing toward Earth; experiments could be mounted both on this side and on the opposite face. Instruments sensitive to magnetic fields generated by electrical

Orbiting Geophysical Observatory. (NASA)

equipment on the satellite itself were mounted on one of several booms extending from the vehicle; these included two 6.7-meter booms and four 1.8-meter booms. Electrical power was generated by solar arrays, that could provide up to 550 watts; these charged two 28-volt nickel-cadmium storage batteries, which were used for peak demands in power requirements by the experiments or during eclipses (shadowing of the spacecraft by Earth during part of each orbit). Attitude (orientation) control, important for solar power generation as well as for accurate pointing of the scientific instruments, was achieved through the use of horizon sensors (which determine the vehicle's orientation relative to Earth's visible edge), gas jets, and reaction wheels (electrically driven wheels that use the action-reaction principle to change the satellite's orientation). The OGO spacecraft were the first scientific satellites to be three-axis stabilized, with all three axes held relatively stable in space. For thermal control of each satellite's components, both active (requiring on-board power) and passive means were employed. Active control was limited to individual sites on the spacecraft's periphery. Louvers, radiating surfaces whose positions determine the rate of cooling of the

vehicle, provided a means of rejecting excess heat from the main body of the satellite. Alternatively, they could be adjusted to minimize heat radiation during times when the vehicle was eclipsed and in danger of cooling too much. Two 100-milliwatt transmitters and one 10-watt transmitter were used to aid in tracking each OGO satellite. Commands to the vehicle from the ground control station were received by two redundant receivers. Up to 86 million bits of data collected by the instruments could be stored on tape recorders for later transmission to Earth via the two redundant telemetry transmitters.

OGO 1 housed twenty experiments, including studies of solar cosmic rays, interplanetary dust, plasmas (gases consisting of charged particles), positrons (positively charged particles with mass equivalent to that of electrons), gamma rays (highly energetic photons), radio astronomical sources, radio propagation, the geomagnetic field, and ion composition. Following a failure to deploy two of its booms (thereby prohibiting operations of the three-axis stabilization system), OGO 1 was placed in a backup attitude control mode, in which it was spin stabilized (made to spin at a constant rate about a symmetry axis, relying on gyroscopic effects to keep that axis relatively fixed in space). Such spinning limited the data-gathering capabilities of the sensors, since their pointing directions were then constantly changing. This mission was generally considered quite successful, and it was decided to use the basic design of OGO 1 for the subsequent five missions. Operational support of OGO 1 was terminated on November 1, 1971.

OGO 2, carrying twenty experiments, was launched into a polar orbit on October 14, 1965. After exhausting all of its attitude control propellant ten days after launch, it went into the backup spin-stabilized mode, but the spin axis wobbled so much that the performance of the major subsystems degraded rapidly, requiring ground controllers to shut down the spacecraft approximately two years later; it was revived intermittently during the following four years. Operational support for OGO 2 was terminated on November 1, 1971.

OGO 3, carrying twenty-one experiments, was launched into an eccentric orbit on June 7, 1965. After forty-six days in orbit, its attitude control system failed, resulting in activation of the backup spin-stabilized mode. Operational support for this satellite was terminated on February 9, 1972.

OGO 4 carried twenty experiments into a polar orbit on July 28, 1967. It maintained nominal attitude control for eighteen months, until a failure in the data recorder system required that the attitude control system be shut down, resulting in activation of the spin-stabilized mode. The spinning motion was such that seven of the experiments no longer generated useful data, and the spacecraft was placed in standby mode, with only intermittent operations during the following two years. Operational support for OGO 4 was terminated on September 27, 1971.

OGO 5 carried twenty-five experiments into an eccentric orbit on March 4, 1968. After forty-one months its attitude control system failed, and the backup spinning mode was activated. Ground control revived the spacecraft from standby for a one-month period just prior to terminating operational support on July 14, 1972.

OGO 6, with twenty-six experiments, was launched into a polar orbit on June 5, 1969. After seventeen days of successful data collecting, solar panel failures led to reduced data storage capability and degraded operation of the experiments. After two years, the spacecraft was placed in the spin-stabilized mode, but it was returned to three-axis stabilized mode for two months just prior to the termination of operational support on July 14, 1972.

Knowledge Gained

Because of the long span of operations (September, 1964,

to July, 1972), the OGO satellites provided data on geophysical phenomena influenced by the Sun during one of the peak activity periods for sunspots (regions of intense magnetic disturbances on the solar surface), which occur approximately every eleven years. A large part of the data collected pertains to specialized areas in space physics; only those findings of a general interest are mentioned here.

OGO 1 conducted a thorough study of the outer magnetosphere and identified the existence of the plasmasphere (the definite outer boundary of Earth's ionosphere). OGO 2 helped map the geomagnetic field for the World Magnetic Survey (an international cooperative effort involving satellites from the United States and the Soviet Union). These measurements improved scientists' geomagnetic model immeasurably. OGO 2 also mapped the ion composition as a function of latitude, providing some of the first substantial data on this variation. OGO 4 performed the first global measurement of nitric oxide concentrations in the atmosphere.

Context

The OGO satellites served important roles, from both the scientific and the technological perspectives. Many of their scientific results, particularly the magnetic field studies, have been employed in scientific research, with the consequence that these data have appeared in numerous articles and textbooks. Many of the data were collected simultaneously with similar measurements from other spacecraft (such as the Interplanetary Monitoring Platforms, the Applications Technology Satellites, the Injun satellites, and the International Satellites for Ionospheric Studies). By correlating the data from these spacecraft with those from the OGOs, scientists were able to separate the temporal and spatial characteristics of such phenomena as cosmic rays, solar particles, charged particles trapped in Earth's magnetic field, and variations in the magnetic fields surrounding Earth and in interplanetary space.

Technologically, the OGOs served to demonstrate the feasibility of standard, general-purpose satellites into which experimental packages could be placed and operated successfully without the experiments' having to be integrated into the design of the overall vehicle. In addition, the OGOs demonstrated the feasibility of maintaining three-axis stabilization for this modular design, and thereby provided the necessary conditions for experiments with precise pointing requirements. The successful use of backup modes (resulting in extended spacecraft operational lifetimes) was demonstrated repeatedly with the spin-stabilized standby mode of attitude control.

Bibliography

Caprara, Giovanni. *The Complete Encyclopedia of Space Satellites*. Translated by John Gilbert and Valerie Palmer. New York: Portland House, 1986. Organized by type of satellite, this book is well illustrated and provides descriptions of "every civil and military satellite of the world" from 1957 to 1986. For a general audience.

Corliss, William R. *Scientific Satellites*. NASA SP-133. Washington, D.C.: Government Printing Office, 1967. Gives a history of scientific satellite missions from 1958 to 1967. Describes major subsystems common to all satellites, devoting significant coverage to scientific instrumentation for spacecraft use. Includes an appendix of all U.S. scientific missions (with descriptions of the satellites and their experiments) flown during the period covered. A rather technical work.

Jackson, John E., and James I. Vette. *The Observatory Generation of Satellites*. NASA SP-30. Washington, D.C.: Government Printing Office, 1963. A collection of six papers: Two pertain to OGO; the others describe similar observatory-type spacecraft. Written before the first OGO launch, these are useful overviews of the program and the engineering design of the spacecraft. For general audiences, although moderately technical.

———. *OGO Program Summary: The Orbiting Geophysical Observatories*. Springfield, Va.: National Technical Information Service, 1975. Provides an overview of the six OGO missions. Describes each satellite, including engineering systems and all scientific experiments on board (along with names of the associated principal investigators). Describes scientific findings from each experiment. For technical audiences.

Turnill, Reginald, ed. *Jane's Spaceflight Directory*. London: Jane's Publishing Co., 1987. This informative volume is a good reference for novices and space buffs alike. Written in clear language, it provides concise descriptions of the missions, programs, and spacecraft of the world. Illustrated. Contains an index.

Yenne, Bill. *The Encyclopedia of U.S. Spacecraft*. New York: Exeter Books, 1985. Reprint, 1988. Yenne has produced an indispensable volume for the student of U.S. space exploration. He provides spacecraft descriptions and mission summaries. Illustrated.

Robert G. Melton

The Orbiting Solar Observatories

Date: March 7, 1962, to September 26, 1978
Type of satellite: Scientific

The Orbiting Solar Observatories were designed to study the structure of the Sun and its outward flow of high-energy particles. They were used to study the influence of the Sun on Earth and to provide a basis for the study of more distant stars.

PRINCIPAL PERSONAGES
    L. T. HOGARTH, Program Manager
    JOHN C. LINDSAY AND
    JOHN M. THOLE, Project Managers
    WILLIAM E. BEHRING,
    STEPHEN P. MARAN, and
    ROGER J. THOMAS, Project Scientists
    MICHAEL E. MCDONALD, an Administrator at NASA
    G. K. OERTEL, Chief of Solar Physics, NASA
    STUART D. JORDAN, an OSO 8 Investigator,
        Goddard Space Flight Center
    NORMAN E. PETERSON, JR., an OSO
        Project Engineer

Summary of the Satellites

The Orbiting Solar Observatory (OSO) program was the first series of satellites to operate from the ground as orbiting observatories. The program was conceived to monitor the Sun continuously for long periods of time, particularly during one complete eleven-year solar sunspot cycle. (Sunspots are relatively cool regions on the Sun that appear dark in contrast with the hotter surrounding material. The occurrence of sunspots follows a semiregular cycle of eleven years.)

The study of the Sun from space began in the late 1940's, when V-2 rockets were launched to obtain ultraviolet solar spectrograms. (A spectrogram is a picture of the light from a radiant body separated into its component parts.) The major limitation of this type of research was that the rockets could only stay above the atmosphere of Earth for a few minutes. Observation above the atmosphere is essential to solar research, since the atmosphere filters out much of the lethal, but informative, radiation that comes to Earth from the Sun.

The OSOs were designed to carry solar experiments in a low orbit around Earth. Orbiting once every ninety minutes, they could view the Sun continuously for up to one hour before passing into the nightside of the orbit. The observatories were designed to last from six months to a year, and a series of satellites was planned to cover an entire solar cycle.

All the satellites in the OSO series were of the same basic design. Each satellite consisted of a nine-sided spinning base about 38 centimeters high, with a diameter of about 112 centimeters. Attached to it by a rod was a stationary sail-shaped platform about 58 centimeters high. The base rotated at a speed of about thirty to forty rotations per minute, allowing experiments housed there to scan the Sun every two seconds. The base also held the power, command, and communications systems. The sail section held experiments that could be continuously pointed at the Sun and the solar cell panels, which provided power to the spacecraft and experiments. As the wheel rotated, the different sensors in the wheel pointed first at the sky and then at the Sun to compare radiation coming from different directions. Nitrogen gas jets were used on both the wheel and sail sections in order to keep the experiments aligned correctly and to adjust the spin rate of the base.

The spacecraft contained sensors to measure X-ray, gamma-ray, and ultraviolet radiation. These ranges of radiation are of wavelengths shorter than those of visible light and are highly energetic. Experiments were also carried to study radio emissions, which are of wavelengths longer than those of visible light. The same basic satellite design could accommodate a variety of experiments as the program progressed. In many cases, the observations from the OSO satellites were coordinated with those of ground-based observatories to maximize the scientific usefulness of the data.

As the data were collected, they were recorded and stored on a continuous-loop magnetic tape. Upon command from the ground, the tape would deliver ninety minutes of observations in the five minutes it was within range of a receiving station. Tracking, data acquisition, and command generation were handled by NASA's Space Tracking and Data Acquisition Network (STADAN), later the Spaceflight Tracking and Data Network (STDN), with tracking stations located around the world. As the data were received at each of these stations, they were transmitted to the OSO control center at Goddard Space Flight Center and then to the various experimental laboratory locations. Among these were the Naval Research Laboratory, the University of California at San Diego, and the University of New Hampshire.

The life expectancies of the satellites ranged from six

months to one year. Several of the satellites performed for a period well over their expected lifetimes, providing useful information for up to three years.

OSO 1 was launched March 7, 1962, from Cape Canaveral. It was given the nickname "streetcar" because of its boxlike design and interchangeability of instrumentation. Although it was expected to perform for six months, a malfunction on May 22 prevented the solar cells from receiving power, and the satellite failed. During its short lifetime, the satellite was able to transmit more than one thousand hours' worth of useful data and record measurements on some 140 solar flares. (A solar flare is a large, sudden outburst of energy from the Sun. It is a phenomenon normally associated with sunspot groups.)

OSO 2 was launched February 3, 1965. It was originally scheduled to be launched in April of 1964 but was delayed because of a preflight accident in which the launch rocket ignited prematurely and exploded. Three technicians were killed, thirteen others were injured, and the spacecraft was severely damaged. OSO 2 incorporated many improvements over the first OSO, including the ability to scan across the entire solar disk rather than be confined to a single point on the Sun.

The third satellite in the series, OSO-C, was launched August 25, 1965, but failed to reach orbit because of a malfunction of the launch rocket. OSO 3 was designed as a replacement, but it was not launched until March of 1967. OSO 4 followed in October of the same year. This satellite focused on the structure and energy balance of the solar chromosphere, the layer of the Sun's atmosphere directly above the visible surface. It carried a device that could observe the Sun in the extreme ultraviolet portion of the light spectrum.

OSOs 5 and 6 were launched in 1969. Among their other experiments, these two spacecraft concentrated on the study of solar flares. The eventual goal in this study was to provide solar eruption predictions that could serve as an early warning to astronauts on manned spaceflights and extravehicular activity missions. With OSO 6 came an improvement in pointing accuracy and scanning capability. While previous satellites were able to point only at the center of the Sun or scan the entire disk, OSO 6 could aim at any desired place on the disk or in the corona and scan a select area. The space programs of Great Britain and Italy furnished two of the experiments flown aboard OSO 6.

A problem occurred in the launch of OSO 7 on September 29, 1971. A failure in the launch vehicle sent the satellite into an unplanned orbit. The spacecraft was spinning out of control and was unable to orient itself toward the Sun. Ground controllers were finally able to stabilize the satellite and position it properly, and all experiments began functioning normally. The primary target of OSO 7 was the solar corona, the outer layer of the Sun's atmosphere. This part of the Sun is normally only seen during a solar eclipse, when

the shadow of the Moon blocks the bright light from the disk of the Sun. A circular occulting disk on the spacecraft produced an artificial eclipse to allow the onboard cameras to observe the corona. OSO 7 was also used in conjunction with experiments aboard the orbiting manned space laboratory Skylab.

OSO 7 continued earlier experiments conducted by OSOs 5 and 6 to study solar flare prediction techniques for use during manned missions. It also studied the effect of solar flares on weather and communications on Earth.

The last satellite in the series, OSO 8, was launched June 21, 1975. While improvements had been made to each successive satellite in the series, OSO 8 was by far the most sophisticated. Its instrument package was larger, had more resolution, could be directed more accurately, and was able to cover a wider range of wavelengths than that of any OSO satellite before it. It observed fine details in the photosphere and explored the various ways energy is transported in the chromosphere and the corona.

OSO 8 also pointed its instruments away from the Sun in order to investigate other celestial X-ray sources. Of special interest was the X-ray background emission that comes from all directions in space, an emission that is now believed to be a remnant from the beginning of the universe. The French space program contributed to the experiment package on OSO 8, with an instrument designed to study the fine structure of the Sun's chromosphere. (The chromosphere is the layer of the Sun's atmosphere directly above the visible surface of the Sun.)

Knowledge Gained

The largest body of knowledge scientists have today about the Sun was gained by experiments flown aboard the Orbiting Solar Observatories. The first OSO alone gathered more than four thousand times the information previously known about the Sun. For the first time, scientists were able to study comprehensive data from the entire disk of the Sun in wavelengths and with resolution impossible from the surface of Earth.

Solar flares were one of the most exciting areas of discovery. Spectra taken showed some flares being heated to an excess of 30 million Kelvins. Fast periodic pulses of X rays were found to be emitted by flares, and the emission of gamma rays was also detected. The first images of the birth, buildup, and death of a flare were recorded. Flares seem to occur in two stages, showing two distinctly different bursts of X rays.

A new model of prominences emerged from OSO data. Prominences are bright, flamelike masses of gas that can be seen on the edge of the Sun. They appear to consist of a relatively cool material concentrated in threads and surrounded by a hot sheath that rises to the same temperature as the corona as it rises over a distance of a few thousand kilometers.

While the light and heat output of the Sun remains fairly

constant, large fluctuations were found in other wavelengths of energy. During active periods of the Sun (times with more sunspot and flare activity), it was found that ultraviolet and X radiation can be four times as intense as usual. An important result of one OSO experiment was the ability to generate a daily X-ray map of the Sun, along with a map of solar radio emissions and sunspots. This mapping allowed for a systematic approach to the study of the evolution of active regions on the Sun.

Many discoveries were made regarding the solar corona, usually visible only during a total solar eclipse. Evidence of gamma-ray emission was found. This indicates there are nuclear reactions occurring in solar flares themselves, rather than solely deep within the Sun, as was previously thought. OSO discovered "coronal holes," areas of reduced emission of radiation at high-energy wavelengths. In a coronal hole, the temperature is about 600,000 Kelvins cooler than in the surrounding corona, and the pressure of the gases in the area is about one-third lower. The solar wind, the flow of particles away from the Sun, seems to be increased in the area of a coronal hole.

Using an experiment designed by a team of French physicists, researchers found that there are huge periodic oscillations of the Sun occurring every fourteen minutes. The entire atmosphere pulses to a distance of about 1,300 kilometers. It is believed that sound waves are responsible for the pulsations.

Toward the end of the OSO program, resolution on the satellites had improved in such a way that the equipment could record the fine details of the solar surface, structures resembling granules. These granulations are constantly in motion and have a lifetime of only a few minutes. They contain clues about how energy is transferred in the Sun's atmosphere.

OSO discoveries were not restricted to the Sun. The satellites also studied Earth's atmosphere, in particular the zodiacal light, a band of faint light extending along the path of the Sun in the sky. This light is visible in the dark sky just after sunset and before sunrise. It was found that this light is not evenly distributed; it is much less apparent near the poles. Other atmospheric studies showed that lightning strikes more often over land than water and that certain land areas seem to experience more lightning than others. In other areas of the sky, OSO satellites searched for and mapped out other sources of X radiation, discovering many new sources and verifying the positions of other known sources.

Context

The Sun, while the most vital and central member of the solar system, is perhaps the least understood. With the advent of orbiting observatories such as the OSO series, the knowledge of the Sun and its effects on Earth has just begun.

Light and heat, which are the emissions from the Sun that scientists can most readily measure from the surface of Earth, are quite constant. Yet the Sun's energy varies considerably in other wavelengths, such as in the ultraviolet and radio regions. At apparently random intervals, the Sun will unleash a burst of cosmic rays (high-energy subatomic particles). Until researchers had the means to measure these emissions and variations, the data went unnoticed. With the invention of radio, however, these bursts were discovered, since they interfered with radio waves in the upper atmosphere and created a disturbance in radio transmission.

In comparison with instruments launched since the end of the OSO series, these satellites seem rather out-of-date. In the context of the times, however, these were the most sophisticated satellites that had been launched, the first of a new class of orbiting observatories. It was the first time an artificial satellite had been applied to astrophysical problems in a comprehensive way. Information that was obtained from these satellites was therefore revolutionary to the study of the Sun.

Since little was known of the Sun at that time, scientists were uncertain as to what might come of the information. One hope was that the study of the Sun might lead to its use as an unlimited and pollution-free power source for Earth. In fact, the understanding of solar physics is considered to be essential to the understanding of the physics of Earth and of the other planets. The study of the Sun is closely related to geology and biology, since the record of rocks and fossils on Earth confirms what scientists are discovering about the Sun. It is theorized that variations in the Sun's output of energy may have been responsible for the ice ages that have occurred at intervals throughout the history of Earth.

Besides its importance to Earth, the Sun provides astronomers with their only opportunity to study a star at close range. The next nearest star is 48 trillion kilometers farther away than the Sun.

With each OSO came improvements in the accuracy and capability of the spacecraft and in experiment capability. The OSO series can also be seen, then, as a test for the more sophisticated satellites that were to come later: the Orbiting Astronomical Observatories (OAOs), the International Ultraviolet Explorer (IUE), the Solar Maximum Mission (SMM), and the manned space station Skylab. In fact, OSO 7 was in orbit at the same time as Skylab and was used in conjunction with the space station.

The OSO program was also a test for spaceflight in general. Besides performing experiments that were directed at astronomical phenomena, satellites such as OSO 2 performed experiments to test the effects of solar radiation on different surface preparations and paint used on spacecraft. One of OSO's major missions was to find a method to predict solar flares, which would affect future missions, particularly the manned Apollo flights.

The OSO program can, therefore, be seen as an essential stepping stone in the space program, in terms of both scientific research and spaceflight.

Bibliography

Cornell, James, and Paul Gorenstein. *Astronomy from Space: Sputnik to Space Telescope*. Cambridge, Mass.: MIT Press, 1985. An overview of twenty-five years of astronomical research from space. Summarizes what has been learned in different areas of astronomy in the form of short articles written by experts in each field. Suitable for those with some science background.

Eddy, John A. *A New Sun: The Solar Results from Skylab*. NASA SP-402. Washington, D.C.: Government Printing Office, 1979. Describes the Skylab mission in detail and discusses what is known about the Sun, including the findings from Skylab. Written for the lay person with an interest in astronomy, although some background is helpful. Illustrated with color photographs, many taken aboard Skylab.

Fire of Life: The Smithsonian Book of the Sun. Washington, D.C.: Smithsonian Books, 1981. A collection of articles written by experts in the field, geared toward general audiences. Includes a history of beliefs about the Sun. Contains beautiful photography. Poetic as well as informative.

Gibson, Edward G. *The Quiet Sun*. NASA SP-303. Washington, D.C.: Government Printing Office, 1973. A fairly technical manual on current solar theory. Includes information obtained from the OSO program. Describes solar structure and processes in detail and includes those questions that remain unanswered.

Henbest, Nigel, and Michael Marten. *The New Astronomy*. New York: Cambridge University Press, 1986. Compares optical, infrared, ultraviolet, radio, and X-ray observations of well-known astronomical objects. Written specifically for general audiences. Illustrated.

Maran, S. P., and R. J. Thomas. "Last OSO Satellite: Orbiting Solar Observatory." *Sky and Telescope* 49 (June, 1975): 355-358. Describes the last OSO satellite and reports on the program's status and results of observations. Geared toward amateur astronomers with some background in astronomical and spaceflight principles. Illustrated.

"OSO-8 Program Keyed to Skylab Data: Orbiting Solar Observatory." *Aviation Week and Space Technology* 102 (June 30, 1975): 45. Describes the satellite and its discoveries. Reports on plans for satellites not yet launched. Written for professionals in the field of avionics. Thus, the language is technical. Illustrated.

Divonna Ogier

PIONEER MISSIONS 1–5

Date: October 11, 1958, to June 26, 1960
Type of mission: Unmanned lunar and deep space probes

These probes began the exploration of the solar system. Although Pioneer 5 was the only unqualified success, achieving its aim of a heliocentric orbit between the paths of Earth and Venus, all five Pioneers made important discoveries about the radiation belts around Earth.

PRINCIPAL PERSONAGES

LOUIS DUNN, President of Space Technology
  Laboratories (STL)
RUBEN F. METTLER, Executive Vice President of STL
ABE SILVERSTEIN, Director of Space Flight
  Development, NASA
T. KEITH GLENNAN, NASA Administrator
HUGH L. DRYDEN, Deputy Administrator, NASA
WILLIAM H. PICKERING, Director of the Jet
  Propulsion Laboratory (JPL)
JACK FROEHLICH, Project Manager for JPL
JAMES A. VAN ALLEN, scientist who helped design
  the radiation-counting equipment
KURT H. DEBUS, director of NASA's Cape Canaveral
  launch facilities
WERNHER VON BRAUN, Director of the Army
  Ballistic Missile Agency in Huntsville, Alabama

Summary of the Missions

The first Pioneer project, whose goal was to launch unmanned spacecraft toward the Moon for the International Geophysical Year (IGY), began in 1957 when the Advanced Research Projects Agency authorized the program. In March, 1958, Secretary of Defense Neil H. McElroy announced that the U.S. Air Force would launch three probes in an attempt to place a scientific payload in the vicinity of the Moon. When the National Aeronautics and Space Administration (NASA) was created in 1958, it assumed responsibility for the nation's space programs, including the Pioneer project. One of the purposes in forming NASA was to unify the diverse and competing space programs of the Navy, Army, and Air Force. At that time, NASA's intention was to compete with the Soviet Union in the conquest of space. Abe Silverstein, Director of Space Flight Development, made it clear that he hoped that the U.S. would leap ahead of the Soviet Union in lunar and planetary exploration with the Pioneer program.

Since the majority of the early Pioneers were failures, they never won much public respect, but they were interesting spacecraft that lived up to their name — they pioneered unmanned interplanetary travel. These probes were modest vehicles, weighing around 45 kilograms, designed to investigate the medium between the planets. The series started inauspiciously on August 17, 1958. An explosion in the first-stage engine ripped open the vehicle after it had traveled about 16 kilometers above Earth. Although "Pioneer" was a name applied by NASA only to successfully launched deep space vehicles, this initial launch was sometimes unofficially referred to as Pioneer 0.

On October 11, 1958, another attempt was made to "shoot the Moon." The Pioneer 1 probe was perched atop a three-stage experimental rocket configuration: The first or bottom stage was a Thor intermediate-range ballistic missile, the second stage was a modified Vanguard, and the third stage was an advanced version of the Vanguard. The overall objective of the mission was to circle the Moon and relay to Earth infrared images of its far side. Other objectives were to measure radiation, micrometeoritic density, magnetic fields, and temperatures in interplanetary space. A subsidiary but essential goal was to get all three main stages to fire in proper sequence.

Pioneer 1 was launched at 4:42 A.M. eastern daylight time, and all three stages fired properly. After the burnout of the second stage, eight small rockets mounted sideways imparted a spin of 120 revolutions per minute to the third and fourth stages to stabilize them in space. Unfortunately, an error in the Thor guidance system caused a speed slightly less than escape velocity (the speed necessary to break loose from Earth's gravitational pull) and a 3.5-degree inaccuracy in the angle at which the first stage was lofted into space. Later analysis revealed that the launch would still have been a success were it not for the third stage swerving sharply to one side as the rocket ignited. Pioneer 1 did reach a record distance from Earth of 115,000 kilometers, nearly a third of the distance to the Moon, before reentering Earth's atmosphere over the South Pacific forty-eight hours later.

According to Louis Dunn, the technical director of the entire project, Pioneer 1 was a success, despite its failure to reach the Moon, because this was the first time that the complex mechanisms — involving rockets, explosive bolts, and

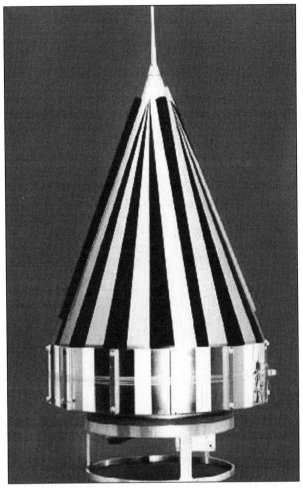

Pioneer 3 and 4 spacecraft. (NASA)

back to Earth. The instrument package burned up as the probe plunged into the atmosphere above east-central Africa. The flight had lasted less than 45 minutes.

The problems that plagued the first three Pioneers continued to plague Pioneer 3. This time, Pioneer's target would not be the Moon. Instead, it was to orbit around the Sun. The objectives of the mission were to measure radiation at extremely high altitudes and to test a moonlight-activated trigger mechanism for a photographic experiment. The overall vehicular engineering for Pioneer 3 was done jointly by the Army Ballistic Missile Agency, under the technical direction of Wernher von Braun, and by the Jet Propulsion Laboratory (JPL), a NASA facility. The first stage was a modified Jupiter intermediate-range ballistic missile. The second, third, and fourth stages were solid-fueled JPL rockets. The Army called its combination of rockets the Juno 2.

At 12:45 A.M. on December 6, 1958, the Juno 2 began to lift the 5.9 kilograms of delicate instruments from the launch-pad at Cape Canaveral. Shortly after launch, Thomas Keith Glennan disclosed that the course angle was 3 degrees too low and the rocket's speed was 1,600 kilometers an hour slower than the desired escape velocity. Von Braun blamed the first-stage Jupiter, which had stopped burning 3.7 seconds short of the planned burning time of 3 minutes. The shortened firing time was probably caused by improper adjustment of a valve designed to keep the rocket fuel under constant pressure. Initially, Jack Froehlich held out hope that the other stages could make up the deficit, but William H. Pickering later told a news conference that the payload would not reach the vicinity of the Moon. Pioneer 3 soared to an altitude of 107,313 kilometers before Earth's gravity pulled it back to a flaming end in the atmosphere over French West Africa at 5:15 P.M. eastern standard time on December 8, 1958. In one way, Pioneer 3 was a fruitful failure, because, as it fell back to Earth, it transmitted data on the newly discovered zone of radiation around the planet.

After the four failures, further disappointment was in store for the planners of the U.S. space program. On January 2, 1959, the Soviet Union sent a much heavier payload (361 kilograms) past the Moon and into orbit around the Sun, a feat the United States had been trying to accomplish with its Pioneer program. Thus, it was anticlimactic when Pioneer 4 was shot into orbit around the Sun a few months later. Pioneer 4 had the same program objectives as Pioneer 3, although an enhanced Geiger tube encased in additional lead shielding was added to improve the measuring of radiation.

The launching of the rocket was in the charge of an Army team from the Redstone Arsenal in Huntsville, Alabama, headed by Kurt H. Debus. The Juno 2 with its Pioneer 4 probe left the pad at Cape Canaveral at 12:10:30 A.M. on March 3, 1959. Hugh L. Dryden soon announced that the six-kilogram, gold-plated payload would pass about 56,350 kilometers from the Moon; unfortunately, this would not be close enough to trigger the probe's photoelectric sensor, sam-

precise timers — had worked in perfect sequence. Pioneer 1 had broken all altitude records, and it had sent back the first measurements of radiation above 4,000 kilometers. According to Dunn, the launch demonstrated the soundness of the design of this vehicle, which was capable of performing a wide variety of space missions.

Pioneer 2 had the same objectives as Pioneer 1, though some modifications had been made in the payload; for example, an image-scanning television system created by the Space Technology Laboratories (STL) had replaced the infrared device in Pioneer 1. Changes had also been made in the rockets to ensure cleaner separation of the stages, such as improved retro-rocket batteries better able to withstand cold temperatures. Despite these improvements, Pioneer 2 proved to be a disaster. After a successful launch at 2:30 A.M. on November 8, 1958, and a successful separation of the first and second stages, the third stage failed to ignite; Pioneer 2 climbed to an altitude of only 1,600 kilometers before falling ple the Moon's radiation, or measure its magnetic field.

Although Pioneer 4 became an artificial satellite around the Sun, it was a silent one, for its tiny batteries allowed intermittent transmissions for only a few months. Pioneer 5 became the first probe to test radio communications at interplanetary distances. This probe was mounted in the nose cone of a three-stage Thor-Able 4 rocket that went up from Cape Canaveral at 8:02 A.M. on March 12, 1960. Pioneer 5 was successfully placed in an 848.5-million-kilometer heliocentric orbit. Its radio was powered by solar energy through 4,800 photovoltaic cells covering the satellite's four extended vanes. The spacecraft radioed back information on magnetic fields, cosmic radiation, solar particles, micrometeorite collisions, and interplanetary temperatures. It continued to telemeter data to Earth from a distance of more than 36 million kilometers, a record for long-distance communication that went unchallenged for more than two and a half years.

Pioneer 5 was a case of *finis coronat opus*. It had been five months since the last successful American space project, and NASA anxiously desired a success. Silverstein saw Pioneer 5 as an important stepping-stone in the ten-year program of space exploration that he had helped to develop. Pioneer 5 had proved that a spacecraft could endure the rigors of interplanetary space and radio useful information back to Earth. The way was now open for unmanned exploration of Mars and Venus.

Knowledge Gained

The greatest scientific contribution of the early Pioneer probes was to clarify the ways in which various solar emissions interact with Earth's magnetic field. Explorer 1 had begun the discovery of Earth's radiation zones, called the Van Allen belts after James A. Van Allen, who had been in charge of the satellite and space-probe experiments that led to their discovery. The early Pioneers provided a more detailed understanding of the two doughnut-shaped radiation zones ringing the planet. Pioneer 1, for example, determined the radial extent of the radiation belt and first observed that it actually is a band. Pioneer 2 discovered that Earth's equatorial region has a higher energy radiation than previously thought.

Helped by these early findings, Pioneer 3 made the program's greatest discoveries in its two passes through these radiation zones. It found that there are two widely separated radiation belts around Earth: an inner band between 2,250 and 5,500 kilometers from the planet and an outer band extending from 13,000 to 19,000 kilometers. Geiger counters aboard Pioneer 3 showed that radiation grew more intense near the center of the Van Allen belts, whereas a thousand kilometers away the radiation dwindled to a low level.

The most striking result of Pioneer 4 was its discovery that an immensely greater quantity of trapped radiation existed in the outer zone on March 3, 1959 (Pioneer 4), than on December 6, 1958 (Pioneer 3). Pioneer 4 reported the same radiation intensity in the inner belt, but in the outer belt it

found conditions radically different from those reported by Pioneer 3. During the time between these probes, radiation had increased considerably and the belt had expanded in cross section. Van Allen suggested that these changes were the result of a magnetic storm that had occurred on the Sun, leading to five consecutive nights of strong auroral activity in the days immediately preceding Pioneer 4's launch. The period preceding Pioneer 3's flight had been geophysically quiet.

Besides its great accomplishment of long-range radio transmission, Pioneer 5 also investigated the space between the planets. It observed cosmic rays in space, completely free of any Earthly effects, showing the rays were truly galactic, not solar, phenomena. It detected particles accelerated by solar flares and found that the energetic electrons in Earth's outer radiation zone arose from an acceleration mechanism within its geomagnetic field rather than from direct injection of energetic electrons from the Sun. Pioneer 5 also found that Earth's magnetic field ended about 14 Earth radii from its daylight side, much greater than predicted earlier.

Context

The Pioneer program was begun prior to the creation of NASA and was a phenomenon of the earliest days of American space development. Despite the inevitable failures of a new program, sufficient successes occurred to convince several important scientists that similar interplanetary spacecraft were a necessity for future explorations of space. Indeed, the early Pioneers pointed to the need for further exploration, since they had raised more questions than they answered — questions about micrometeorites and magnetic fields, about the radiation milieu in outer space, and about the transient effects of solar flares.

The first group of Pioneer probes also encouraged progress in other areas. For example, to communicate with these flights, tracking hardware had to be developed. JPL was in the forefront of this development, and to support the Pioneer flights JPL built what ultimately became the Deep Space Network. Managed by JPL for NASA, this communications network began by tracking the Pioneer probes toward the Moon and into deep space. By late 1958, as the first Pioneers were launched, JPL had established tracking stations at Cape Canaveral, Puerto Rico, and Goldstone Lake, California. The basic functions performed by these stations were to track spacecraft positions with high precision, to issue commands to the spacecraft, and to acquire and process scientific data from the spacecraft. Because of JPL's experience with the early Pioneers, NASA assigned more responsibilities to the laboratory in its Ranger, Surveyor, and Mariner programs.

Besides illustrating the institutional context of the early U.S. space program, the history of the Pioneer 1 through 5 missions also reveals something of its political context. Although NASA was created to eliminate certain rivalries in the space program, the Air Force's failures with the first three

Pioneers and the eventual Army successes with the later Pioneers show that this rivalry continued even after NASA's creation. More important than this rivalry, however, was the competition between the United States and the Soviet Union. NASA wanted to surpass the Soviet Union by launching the first deep space probe, but because of the failure of Pioneer 3, Luna 1, later renamed Mechta, or "Dream," became the first artificial satellite orbiting the Sun. The Soviet Union had again bested its American rivals. These Soviet successes encouraged NASA officials to press for a more vigorous space program.

This political rivalry played a role in the creation of later Pioneers. The early Pioneers showed what could be done by modifying existing technologies, but they also revealed the need for new approaches. The later Pioneer spacecraft were therefore considerably more complex, with average weights of three and a half times the early Pioneers. While the new instruments still included a preponderance of particle and field sensors, more sophisticated imaging devices were used. Since JPL was already so heavily involved in lunar and planetary programs, the new Pioneers became a major responsibility of the Ames Research Center, with the Goddard Space Flight Center also contributing. Unlike the first group of Pioneers, which, except for Pioneer 5, were plagued by unreliable equipment, the second group, Pioneers 6 through 9, were extremely reliable. Though the early Pioneers failed in their ambitious overall objectives, they achieved many other goals and opened the path to the planets for later spacecraft.

Bibliography

Corliss, William R. *The Interplanetary Pioneers.* Vol. 1, *Summary.* NASA SP-278. Washington, D.C.: Government Printing Office, 1972. The first of three volumes on the Pioneer program and the most accessible for the general reader. Although he emphasizes the second group of Pioneer missions (Pioneers 6 through 9), Corliss, in his well-illustrated and extensively documented account, makes some interesting comparisons between the first and second group of probes.

Hirsch, Richard, and Joseph John Trento. *The National Aeronautics and Space Administration.* New York: Praeger Publishers, 1973. Hirsch was aerospace assistant to the National Aeronautics and Space Council. He died before completing this popular history of NASA. Using Hirsch's notes and files, Trento finished the book as he thought Hirsch himself would have done. Though the book does not avoid controversial issues, its emphasis is on NASA's remarkable accomplishments, including some of the flights of the early Pioneer spacecraft.

Koppes, Clayton R. *JPL and the American Space Program: A History of the Jet Propulsion Laboratory.* New Haven, Conn.: Yale University Press, 1982. JPL played an important part in some of the early Pioneer missions, and Koppes analyzes these in the context of the laboratory's evolution. This widely praised book relates the scientific and technical history of the American space program.

McDougall, Walter A. *The Heavens and the Earth: A Political History of the Space Age.* New York: Basic Books, 1985. McDougall is a professor of history at the University of California in Berkeley, and his book focuses on the race between the United States and the Soviet Union for technological supremacy in space. His book, which won the 1986 Pulitzer Prize for History, demonstrates the ways in which technology can bring about social and political change, and how politics can influence technology.

Newell, Homer E. *Beyond the Atmosphere: Early Years of Space Science.* NASA SP-4211. Washington, D.C.: Government Printing Office, 1980. This book, part of the NASA History Series, vividly captures the early years of space science. Newell tells a multifaceted tale, ranging from the technical to the political, the personal to the institutional, the national to the international. He presents space science as a continuation of traditional science.

Nicks, Oran W. *Far Travelers: The Exploring Machines.* NASA SP-480. Washington, D.C.: Government Printing Office, 1985. A popularly written account of the first twenty-five years of space exploration. The author's approach is thematic rather than chronological, and the Pioneer missions are not discussed until the end of the book, in an interesting chapter entitled "Spinners Last Forever." As a senior NASA official during most of the period about which he writes, Nicks played a major role in shaping and directing NASA's lunar and planetary programs, and his narrative makes excellent use of his personal experiences.

Shelton, William Roy. *American Space Exploration: The First Decade.* Boston: Little, Brown and Co., 1967. A popular and anecdotal history of the American adventure in space from 1957 to 1967. Shelton's emphasis is on the human element, and his account is clearly intended for a wide audience. A glossary of technical terms and an appendix are included.

Robert J. Paradowski

PIONEER MISSIONS 6-E

Date: Beginning December 16, 1965
Type of mission: Unmanned deep space probes

Pioneers 6 through E were the first spacecraft specifically prepared to obtain synoptic information on the effects in interplanetary space of solar activity that varies with the solar cycle, such as the incidence of solar flares.

PRINCIPAL PERSONAGES

GLENN A. REIFF, Program Manager, 1963 – 1970
CHARLES F. HALL, Project Manager
    through February, 1980
RICHARD O. FIMMEL, Project Manager
    beginning February, 1980
JOHN H. WOLFE, Project Scientist and Principal
    Investigator, plasma experiment
RALPH W. HOLTZCLAW, Spacecraft Systems Manager,
    Ames Research Center
NORMAN F. NESS,
CHARLES P. SONETT,
HERBERT S. BRIDGE,
JOHN A. SIMPSON,
KENNETH G. MCCRACKEN,
WILLIAM R. WEBBER,
VON R. ESHLEMAN,
FREDERICK L. SCARF, and
OLGA BERG, Principal Investigators

Summary of the Missions

In 1962, in response to a request from Edgar M. Cortright, Deputy Director of the Office of Space Science for the National Aeronautics and Space Administration (NASA), Space Technology Laboratories (STL) performed a two-and-a-half-month feasibility study for the Pioneer 6 through E spacecraft for the Ames Research Center. This Pioneer project was approved by NASA Associate Administrator Robert C. Seamans, Jr., on November 9, 1962. Following competitive bidding procedures, STL (later TRW) was selected to provide the five spacecraft; the definitive contract was approved by NASA headquarters on July 30, 1964.

In order to maximize scientific returns, the spacecraft were equipped with spin stabilization such that the spin axis could be oriented perpendicular to the ecliptic plane (the plane of Earth's orbit) and a high-gain transmitting antenna directed at Earth. This choice for the direction of the spin axis

meant that the scientific experiments could scan all azimuthal directions in the ecliptic plane. Also, the solar cells that provided electrical power to operate the spacecraft could all be oriented parallel to the spacecraft's spin axis, thereby minimizing their weight. Finally, a relatively simple transmitting antenna, called a modified Franklin array, provided a disk-shaped transmission pattern centered on the ecliptic plane, thus maximizing the distance from Earth at which data could still be returned. The 8-watt transmitters were relatively high powered for such small spacecraft. The orbits of the five spacecraft were to be essentially in the ecliptic plane. Long spacecraft lifetimes, at least six months, were another scientific requirement. The masses of the five spacecraft, including 16 to 19 kilograms of experiments, ranged from 62 kilograms for Pioneer 6 to 67 kilograms for Pioneer 9. The main body of these spacecraft was a right cylinder (a cylinder whose side is perpendicular to its base) 0.89 meter high and 0.95 meter in diameter.

The experiments for the five spacecraft were solicited from the scientific community and had to measure interplanetary plasma and the magnetic field or study cosmic ray gradients and radio propagation. Eighteen proposals were evaluated for experiments for Pioneers 6 and 7 and fifteen for Pioneers 8 and 9; seven experiments were selected for Pioneers 6 and 7 and eight for Pioneers 8 and 9. Since Pioneer E used spare parts from the earlier four spacecraft, it was decided that the Pioneer 8 and 9 experiments should be used. In addition to having the basic plasma, magnetic field, and cosmic ray experiments carried by Pioneers 6 and 7, Pioneers 8, 9, and E included additional cosmic dust and electric field wave experiments, at the expense of eliminating a second solar plasma experiment.

The five spacecraft, Pioneers 6 to 9 and Pioneer E, were launched from December 16, 1965, to August 27, 1969. The basic launch vehicle was the McDonnell Douglas Thrust-Augmented Improved Delta.

The trajectories of these spacecraft were heliocentric (centered on the Sun) to study phenomena such as cosmic ray gradients over a range of distances from the Sun. Pioneers 6 and 9 took trajectories closer to the Sun, with perihelia (the points in their orbits closest to the Sun) of 0.81 astronomical unit (fractions of the distance of Earth's orbit from the Sun) for Pioneer 6 and 0.75 astronomical unit for Pioneer 9.

Pioneers 7 and 8 took trajectories farther from the Sun with aphelia (the points in their orbits farthest from the Sun) of 1.1 astronomical unit. The two trajectories that passed beyond Earth's orbit also passed through Earth's geomagnetic wake, which is similar to a shadow in the plasma flowing from the Sun. Consequently, the experiments on those two spacecraft could collect data on this phenomenon.

Pioneer E was to have a special inward-outward trajectory that would keep it relatively near Earth for more than one-half year. The hydraulic pressure in the first stage of the launcher became too low while this stage was firing, however, and was lost altogether about seven seconds before that engine was to complete its firing. This caused the upper stages to move too far from their proper course, and their destruction was commanded about eight minutes after launch.

After each Delta third stage burned out and was separated from the spacecraft, three spacecraft booms unfolded (these included an orientation system nozzle, the magnetometer experiment sensor, and a wobble damper). Then, each spacecraft was automatically oriented, using the timed release of pulses of compressed nitrogen to move its spin axis so that it was perpendicular to the spacecraft-Sun direction. Finally, each spacecraft was moved in a similar way, but under ground control, so that it was perpendicular to the spacecraft-Earth direction, as well (permitting high data rates through the use of the high-gain spacecraft antenna), and the experiments were activated.

The plasma and magnetic field data collected as Pioneer 6 departed from Earth showed the magnetopause, or boundary of Earth's magnetic field in space, at an altitude of 11.8 Earth radii. Subsequently, the bow shock, which is "upstream" from Earth in the solar wind, was encountered at an altitude of 19.5 Earth radii. The Pioneer plasma experiments were improved over those used previously and thus were more detailed in their measurements of the properties of shocked plasma, and interplanetary plasma, and of the properties of the bow shock itself.

The Pioneer 6 experiments, soon after launch, were able to fulfill the plan to measure the effects of solar flares in interplanetary space. A flare on December 30, 1965, produced energetic solar protons, measured by the cosmic ray experiments. In this case, the effects were strong enough that the measurements were fairly detailed. After studying solar proton data from a number of events, combined with measurements of the interplanetary magnetic field, it was found that the energetic solar protons usually began to arrive first from the direction of the magnetic field. Also, for some events, fast solar plasma associated with the solar flares was observed. In addition, the relation between the speed of propagation and the time histories of the energetic solar proton fluxes, along with the longitudes of the solar flare sources, was examined.

Another contribution was an improved description of the discontinuous structure of the interplanetary plasma, using data from various experiments aboard Pioneers 6 and 7. Also,

a four-sector structure of the interplanetary magnetic field was found using magnetometer data from the first month after the launch of Pioneer 6. As the Sun rotates in space, the interplanetary magnetic field is first directed toward it and then away from it, forming the sector pattern. It is known that four such sectors are not always present; sometimes there are only two sectors. Such changes reflect the long-term variations of the solar magnetic field itself.

The Stanford University radio propagation experiment collected data on many solar plasma events while the interplanetary electron density was measured. Using a different technique, transient phenomena near the Sun were also detected using the spacecraft transmitter— at times when one of these Pioneer spacecraft passed behind the Sun, as seen from Earth.

In early August, 1972, a series of four large solar flares produced intense fluxes of energetic solar particles and large increases in the speed of the solar wind. Pioneer 9 was in a favorable position then in its orbit around the Sun to measure these effects. Increased solar plasma fluxes can produce important effects on Earth, such as geomagnetic storms, and these Pioneer 9 data, in combination with those from other spacecraft and from the ground, provided important information on the propagation speeds of solar plasma disturbances and of solar particles. As a result, the understanding of many geophysical effects was increased.

During the Apollo manned lunar excursions from 1969 through 1972, data from Pioneers 6 through 9 were used to give indications of solar particles to which the astronauts might be exposed. The Pioneer data were sent to a solar disturbance forecast center in Boulder, Colorado, for this purpose.

The Pioneer 7 and 8 spacecraft provided opportunities to explore the geomagnetic wake farther downstream from Earth. As concluded from the study of the plasma and magnetic field data, the geomagnetic wake is not ordered and regular inside the Moon's orbit as it is nearer Earth; instead, it is mixed with the solar wind, possibly in a turbulent or filamentary manner. Pioneer 7 initially traversed this wake region at a distance of about 1,000 Earth radii downstream from Earth, while Pioneer 8 initially did so at about 500 Earth radii. Following its heliocentric orbit, Pioneer 7 again passed through this wake region in 1977, at about 3,100 Earth radii, and plasma data were again collected showing the presence of the wake region. Again in 1987, Pioneer 7 passed through this region at a distance of about 3,000 Earth radii, but by then the decrease in power from the spacecraft's solar cells meant that only the University of Chicago cosmic ray experiment could be operated; thus, the wake region was not detected in the data. Similarly, Pioneer 8 in 1985 again passed through this region, at about 1,650 Earth radii downstream. The power available from this spacecraft was also limited and only the electric field wave experiment could be operated, but the wake region apparently was detected in the data. The structure of this wake has now been determined in more detail, out to a distance of about 220 Earth radii, by measurements made in 1983 by the International Sun-Earth Explorer 3 spacecraft.

The Pioneer 9 spacecraft was last tracked on May 18, 1983. Subsequent attempts to locate its transmissions were unsuccessful. The last of these attempts was in November, 1986, and it is not known exactly why the transmitted signal could no longer be located.

During 1986 while passing near the sun, Halley's comet also passed fairly close to Pioneer 7 (within 12 million kilometers of it, or only 8 percent of the distance from the Sun to Earth). Plasma and cosmic ray experiments were operated during several days of tracking, and there was an indication in the plasma data nearest the comet of solar wind helium being altered by the presence of cometary gas. The European Giotto spacecraft detected similar altered solar wind helium much closer to Halley's comet. The plasma data collected from Pioneer 7 had to be interpreted with a technique developed once the spacecraft's Sun sensor stopped operating in February of 1969.

Knowledge Gained

As the first spacecraft in the series, Pioneer 6 left Earth in December, 1965, and obtained new, more detailed data from plasma and magnetic field experiments concerning Earth's interaction with the solar wind. New features of Earth's bow shock and of the shocked plasma behind it were also measured, because the experiments carried were more sophisticated than earlier ones.

As time passed, data on the effects in interplanetary space of many solar flares were obtained from Pioneers 6 through 9. The evolution over time of energetic solar proton fluxes associated with flares was often found to progress from alignment close to the direction of the interplanetary magnetic field toward an isotropic distribution. The gradient of the solar protons was sometimes observed to have more intense fluctuations closer to the Sun. The effect of the flare's solar longitude relative to the Pioneer spacecraft was examined for some events, and it was concluded that a magnetic field line connection between the source and the spacecraft favored prompt, relatively intense effects. Fast solar plasma, accelerated in association with the flare, was observed for some events.

In August, 1972, particularly large solar flare events occurred, and solar wind plasma, magnetic field, and energetic particle data from Pioneer 9 proved valuable in combination with data from other spacecraft in determining the speeds and the longitudinal and radial extents of the associated accelerated interplanetary plasma and energetic solar particles.

The structure of the interplanetary medium was found to be characterized by discontinuous changes in the magnetic field and the solar wind plasma parameters by experiments on these Pioneer spacecraft. Also, experience was gained from the radio propagation experiments in the detection of solar plasma events and in the measurement of interplanetary electron densities.

For the first time, Earth's magnetic wake in the solar wind was observed at fairly large distances downstream by the Pioneer 7 and 8 spacecraft. The data from about 500 and 1,000 Earth radii downstream indicated that Earth's magnetic field, carried away by the solar wind, had generally lost its ordered near-Earth characteristics. At the greater distances, this wake may have become turbulent. In March of 1986, it was possible to collect interesting data from Pioneer 7 on the interaction of Halley's comet with the solar wind.

Context

Before the traversal by Pioneer 6 in 1965 of the magnetopause and bow shock boundaries of Earth's magnetic field, the general locations of these boundaries were known because of experiments aboard other spacecraft. The plasma data collected by Pioneer 6, however, provided actual measurements of the plasma flow vector, which previously had not been fully defined. The measurements of electrons in the solar wind provided by Pioneer 6 considerably improved upon those previously available.

Solar wind plasma, magnetic field, and energetic particle data collected by Pioneer 6, and by the later Pioneer spacecraft of this series, permitted scientists to prove the discontinuous nature of the solar wind near Earth's orbit. Hints of this characteristic had been present in earlier, less complete data.

Scientists had already understood that solar flares can generate energetic particles and had studied the transmission of these particles and the accelerated solar wind associated with them by the time that Pioneer 6 was launched. Serious research on these topics had been possible through measurements of geomagnetic activity and of effects associated with the energetic particles that could be measured from Earth, such as changes of the ionosphere, neutrons generated in the upper atmosphere, and the incidence of ionizing radiation measured with balloon-borne experiments. The Pioneer spacecraft, however, provided more complete, accurate, and sensitive data, unmodified by Earth's close proximity. This new data permitted important advances in the understanding of the physical effects in the interplanetary medium, and on Earth, of these solar phenomena.

Prior to the passage of Pioneer 7 in 1966 through Earth's magnetic wake, nothing was known about this phenomenon at those distances. The Pioneer 7 data indicated that this distant wake had a disordered nature, a characteristic that was confirmed with data from Pioneer 8 a little more than a year later. Pioneer 7 data were used to detect Earth's wake in the solar wind near 3,100 Earth radii downstream in 1977.

The Pioneer 7 solar wind plasma data obtained in the vicinity of Halley's comet during March, 1986, augmented the results obtained by other spacecraft by performing a relatively unsophisticated but fairly sensitive experiment at an extreme distance.

Bibliography

Corliss, William R. *The Interplanetary Pioneers*. 3 vols. NASA SP-278, 279, and 280. Washington, D.C.: Government Printing Office, 1972. An official history of these Pioneer spacecraft. Contains numerous photographs, diagrams, and illustrations.

———. *Scientific Satellites*. NASA SP-133. Washington, D.C.: Government Printing Office, 1967. An ambitious work that presents the rationales and engineering and experimental details for scientific spacecraft of the period. Includes a useful diagram of the electrostatic deflection arrangement for the Pioneer 6 and 7 Ames plasma experiments. Contains some college-level material. Provides numerous diagrams and photographs.

Gibson, Edward G. *The Quiet Sun*. NASA SP-303. Washington, D.C.: Government Printing Office, 1973. This volume was prepared by a Skylab scientist-astronaut to organize the available knowledge about the Sun, particularly in its undisturbed state. Contains considerable general information. Despite the title, a reasonable discussion of the solar flare phenomenon is included. Contains numerous photographs and illustrations.

Kennel, Charles F., et al., eds. *Solar System Plasma Physics*. 3 vols. Amsterdam: North-Holland Publishing Co., 1979. Volume 1 contains a discussion of solar flares that generally is at the graduate level, although some introductory material is also included. Volume 3 provides a discussion of some detrimental effects on Earth of solar activity. These three volumes contain photographs and diagrams.

Sturrock, P. A., et al., eds. *Physics of the Sun*. 3 vols. Dordrecht: D. Reidel Publishing Co., 1986. This advanced summary of scientific knowledge about the Sun contains considerable graduate-level material. Chapter 10 provides discussions of the solar wind and solar flares. Chapters 21 and 22 discuss the effects of solar wind and solar energetic particles on the terrestrial environment. References are given to more specialized treatments. Contains diagrams and photographs.

J. D. Mihalov

PIONEER 10

Date: Beginning March 2, 1972
Type of mission: Unmanned deep space probe

Pioneer 10 was the first spacecraft to provide close-up reconnaissance of Jupiter, the largest planet in the solar system, and to sample directly its magnetic and particle environment. It was also the first spacecraft to cross the asteroid belt between Mars and Jupiter and the first to escape the solar system, by passing beyond the orbits of Neptune and Pluto.

PRINCIPAL PERSONAGES
  CHARLES F. HALL, Pioneer Project Manager
  JOHN H. WOLFE, Project Scientist
  ROBERT R. NUNAMAKER, Flight Operations Manager
  NORMAN J. MARTIN, who succeeded Nunamaker as
      Flight Operations Manager
  RICHARD O. FIMMEL, science chief
  GILBERT SCHROEDER, spacecraft chief
  ALFRED SIEGMETH, Tracking and Data Systems
      Manager
  J. W. JOHNSON, launch operations, Pioneer Jupiter
      project

Summary of the Mission

The Pioneer 10 mission performed the first close flyby of the planet Jupiter. In 1967, as a part of the activities of the National Aeronautics and Space Administration's (NASA) Lunar and Planetary Missions Board, an outer planet panel chaired by Dr. James Van Allen of the University of Iowa (discoverer of Earth's radiation belts) recommended low-cost missions to the outer planets. In June of 1968, the Space Science Board of the National Academy of Sciences stated that Jupiter was probably the most interesting planet from a physical point of view, and that it was technically feasible to send probes to that planet. The board recommended that "Jupiter missions be given high priority, and that two exploratory probes in the Pioneer class be launched in 1972 or 1973."

A mission to Jupiter was officially approved by NASA Headquarters in February of 1969, and the program was assigned to the Program Office, Office of Space Science and Applications. The Pioneer Project Office at the NASA Ames Research Center was selected to manage the project, and TRW Systems Group, in Redondo Beach, California, was awarded a contract to design and fabricate two identical Pioneer spacecraft for the mission. The first feasible launch opportunity was in late February through early March of 1972, and the first spacecraft, "Pioneer F," was scheduled to make this launch window. After launch, the spacecraft was renamed Pioneer 10.

At the Ames Research Center, Charles Hall became manager of the Pioneer Project. The experiments on the spacecraft were the responsibility of Joseph Lepetich, and the spacecraft system was the responsibility of Ralph Holtzclaw. The Flight Operations Manager was Robert R. Nunamaker and later Norman J. Martin. Robert Hofstetter was launch vehicle and trajectory analysis coordinator, Richard Fimmel was science chief, and Gilbert Schroeder was spacecraft chief. The California Institute of Technology's Jet Propulsion Laboratory (JPL), in Pasadena, California, provided tracking and data systems support with Alfred Siegmeth as the Pioneer Tracking and Data Systems Manager, and NASA's Goddard Space Flight Center in Greenbelt, Maryland, provided worldwide communications to the various stations of the Deep Space Network. NASA's Lewis Research Center in Cleveland, Ohio, was responsible for the launch vehicle system. J. W. Johnson represented the Pioneer Jupiter project for launch operations at NASA's Kennedy Space Center, and Bernard O'Brien was manager of the project at TRW Systems Group.

The Pioneer 10 spacecraft was spin-stabilized so that it continually rotated, always pointing in the same direction, with the center of its dish-shaped communications antenna pointed toward Earth. The spacecraft was 2.9 meters long, from the end of its conical medium-gain antenna to the adapter ring that fastened it to the third stage of the launch vehicle. Its structure was centered around a 36-centimeter-deep flat equipment compartment. The top and bottom of this compartment consisted of regular hexagons with sides 71 centimeters long. Attached to one side of this compartment was a smaller compartment which contained most of the instruments for the scientific experiments. The main communications antenna, 2.74 meters in diameter and 46 centimeters deep, was attached to the front of the equipment compartment. Radioisotope thermoelectric generators were held about three meters from the spacecraft center. These provided the essential electrical power for the long flight time to Jupiter by converting heat from the radioactive decay of plutonium 238 into electricity. Solar cells would not be of much use in this mission, since, at the planet Jupiter, sunlight has only 3.7 percent of its power on Earth. A third single-rod boom, 120 degrees from the two trusses, projected from the experiment compartment to position a magnetometer sensor about 6.6 meters from the center of the spacecraft. Both trusses and the boom were stowed before launch and extended from the spacecraft after launch in order for everything to fit within the 3-meter shroud of the Atlas-Centaur launch vehicle.

Besides the magnetometer experiment (for which Edward J. Smith of JPL served as principal investigator), the other scientific experiments (listed with principal investigators in parentheses) included a plasma analyzer experiment (John H. Wolfe, Ames Research Center), charged particle composition experiment (John A. Simpson, University of Chicago), cosmic ray energy spectra experiment (Frank B. McDonald, Goddard Space Flight Center), Jovian charged particle experiment (James A. Van Allen, University of Iowa), Jovian trapped radiation experiment (R. Walker Filius, University of California, San Diego), asteroid-meteoroid astronomy experiment (Robert K. Soberman, General Electric Company and Drexel University), meteoroid detection experiment (William H. Kinard, Langley Research Center), ultraviolet photometry experiment (Darrell L. Judge, University of Southern California), imaging photopolarimetry experiment (Tom Gehrels, University of Arizona), and Jovian infrared thermal structure experiment (Guido Munch, California Institute of Technology). In addition, two experiments used information obtained from the radio communications link, the celestial mechanics experiment (John D. Anderson of JPL) and the occultation experiment (Arvydas J. Kliore, also of JPL).

Pioneer 10 could be launched between February 25 and March 20, 1972, arriving at Jupiter sometime between October, 1973, and July, 1974. The arrival had to be timed so that Jupiter and Pioneer 10 would not be too close to the Sun as seen from Earth. Initial electrical problems and high-wind conditions delayed the launch for several days. On March 2, 1972, at 8:49 P.M. eastern standard time, Pioneer 10 was launched from the Kennedy Space Center by an Atlas-Centaur that was equipped with an additional boost stage. After seventeen minutes of flight, Pioneer 10 was traveling 51,682 kilometers per hour, faster than any previous man-made object. After only eleven hours of flight, Pioneer 10 was beyond the orbit of the Moon. On March 7, a small correction was made to the spacecraft velocity to assure an arrival time at Jupiter that optimized the results of the scientific experiments. After only ten days, all the scientific experiments had been turned on, so that the health of each of the instruments could be monitored during the long flight time to Jupiter. Although there had been a few unexpected readings from the spacecraft, there were no serious problems, and all the experiments had survived the launch in operating condition. On July 15, Pioneer 10 became the first spacecraft to

enter the asteroid belt between the orbits of Mars and Jupiter. Also in July, a trajectory correction was made so that Pioneer 10 would be occulted by Jupiter's innermost main satellite, Io, near its closest approach to the planet.

During the portion of its trajectory between Earth and Jupiter, Pioneer 10 was active in collecting scientific measurements of the characteristics of charged particles, interplanetary plasma, and the zodiacal light — a faint glow in the plane of the zodiac that is visible in dark Earth skies after sunset and before sunrise. Pioneer also had the chance to observe the influence on the solar wind — the stream of plasma that moves outward from the Sun— of several unprecedented storms on the visible surface of the Sun. These measurements were compared with similar measurements from spacecraft still in orbit around the Sun in different parts of the sky, Pioneers 6, 7, 8, and 9.

Although scientists estimated that Pioneer 10 had a 90 percent chance of passing through the asteroid belt undamaged, there was still some concern. Encounter with any object as big as half a millimeter in diameter at speeds that could be as great as fifteen times the speed of a high-powered rifle bullet would do serious damage to the spacecraft. Pioneer 10 emerged from the asteroid belt in February of 1973, completing the seven-month asteroid belt passage without harm.

In early November of 1973, controllers readied Pioneer 10 for the most active phase of its mission, the encounter with Jupiter. Its trajectory was nearly equatorial with respect to Jupiter, and it would come within 140,000 kilometers of the Jovian cloudtops. By November 16, Pioneer 10 had crossed the bow shock where Jupiter's magnetic field affects the solar wind. This bow shock was farther from the planet than had been expected. On November 27, Pioneer crossed the magnetopause, the boundary between the shocked solar wind and the magnetic field of Jupiter itself. Pictures of Jupiter began coming in from the imaging photopolarimeter's narrow-angle telescope. Images were returned in two colors, red and blue, and a detailed color image of the planet was reconstructed later by mixing green with the image in proportions as seen from Earth-based observations.

Because it had no on-board computer, commands had to be transmitted to the spacecraft daily, increasing to four thousand commands per day as Pioneer 10 moved toward its closest approach to Jupiter. A series of contingency commands was periodically sent so that the spacecraft could be corrected in case of spurious commands generated by Jupiter's powerful radiation field. Some spacecraft systems showed the effects of the powerful radiation environment, with several particles and fields experiments reaching measurement saturation points. Although there was concern that some instruments might be destroyed in the deluge of radiation received from Jupiter at closest approach, the spacecraft continued to function well when it emerged from behind Jupiter just after periapsis — closest approach — on December 3.

Transmitting a series of crescent-shaped images of Jupiter,

Pioneer 10 left the Jovian environment on a trajectory that followed the heliotail — that part of the solar wind swept by the interstellar wind — gathering data about the solar wind at extreme distances from the Sun, crossed the orbit of Neptune, and is now being used to seek a possible tenth planet. Pioneer 11, a virtual twin, would reach Jupiter about a year after Pioneer 10 and gather more information about the planet, particularly its polar regions, eventually encountering the planet Saturn.

Knowledge Gained

Between the Earth and Jupiter, Pioneer 10 found that the solar wind temperature behaved differently from theory and had many nonuniformities from irregularities in the Sun's own atmosphere, the corona. In its mission past Jupiter, Pioneer 10 discovered that the heliopause (the boundary of the heliosphere, or expanding solar wind) was beyond the orbit of Neptune. Pioneer 10 experimenters found that the stream of uncharged hydrogen atoms which make up the gas between stars — the interstellar wind — enters the solar system along the path of Earth's orbit, but at a direction 60 degrees away from the direction the solar system is traveling through interstellar space. On the way to Jupiter, the number of small particles was not found to be particularly greater in the asteroid belt than anyplace else in the solar system.

Near Jupiter, Pioneer 10 mapped the intensity, direction, and structure of its magnetic field. The outer portion of the magnetic field was similar to the structure of that of Earth, but turned in the opposite direction. It was also tilted 11 degrees away from the rotational axis and offset from the center of the planet. Closer to the planet, between the cloudtops and a distance of three Jovian radii, the field was extremely complicated and extremely strong — at its cloudtop level more than ten times the strength of Earth's magnetic field at its own surface. The outer part of the field expanded and contracted under the influence of the solar wind and was blunted in the solar wind direction. Between 20 and 60 Jovian radii, the field was found to be extremely distorted by trapped plasma; there, ionized particles formed a sheet of electric current around Jupiter which produced its own magnetic field. Peak intensities of electrons in Jupiter's radiation belts were ten thousand times greater than the maximum in Earth's belts; protons were several thousand times more intense.

Imaging and infrared measurements of the atmosphere of Jupiter showed that dark regions were warmer than light regions. The Great Red Spot, a semipermanent oval in the southern hemisphere, was cool and had clouds higher than anywhere else on the planet. The temperature of the atmosphere reached a minimum of about −163 degrees Celsius, near a pressure of one-tenth that at Earth's surface, and warmed up to about −100 degrees Celsius at a pressure equal to that at Earth's surface. At lower pressures in Jupiter's stratosphere, the temperature warmed up again. Jupiter was found to emit about 1.7 times as much energy in the form of heat as it absorbed from sunlight. Jupiter's ionosphere was found to rise some four thousand kilometers above the visible cloudtops, and it has at least five distinct layers, similar to Earth's ionospheric layers.

Low-resolution images were made of Jupiter's second and third innermost large moons, Europa and Ganymede, respectively. Analysis of changes in the radio signal from the spacecraft as it went behind Io showed that the satellite had an ionosphere which extended seven hundred kilometers above its day side. Masses and sizes of the large satellites were observed with sufficient precision to determine that their densities decreased uniformly outward from Jupiter, indicating predominantly rocky compositions for the innermost moons and predominantly icy compositions for the outermost satellites.

Context

Prior to the Pioneer 10 mission, the only information about Jupiter came from Earth-based astronomy and theoretical investigations. Scientific interest in Jupiter is centered on the fact that it is the largest planet in the solar system and that its bulk density, like those of the other outer planets — Saturn, Uranus, and Neptune — indicates that it is primarily composed of gas, unlike the inner planets — Mercury, Venus, Earth, and Mars. Theory suggests that a large region of the interior of Jupiter is composed of metallic hydrogen. The measured behavior of the magnetic field conformed to this expectation by showing the generation of an immense magnetic field by a dynamo at the center of the planet with a complicated structure and gravitational measurements of the planet that confirmed that it is almost entirely fluid. Measurements of the radiation belts and electrical flow revealed the details of an electromagnetic environment that has, for many years, made Jupiter known as the largest source of radio emission in the sky.

Jupiter's atmosphere is mostly molecular hydrogen, but it also contains helium. The flow of heat from the interior, in excess of the energy received from the Sun, is consistent with the slow cooling off of such a large body from the time of its formation. Preliminary infrared measurements of Jupiter by aircraft had indicated that even more heat was emitted by the planet. Because of the flow of heat from the interior and the planet's rapid rotation (it has a 10-hour period), weather patterns appear quite different from those on Earth. The familiar pressure systems with clockwise and counterclockwise flow are stretched over many degrees of longitude; the major features are stretched all the way around the planet. Visible and infrared results are consistent with a picture in which the bright and cool bands are regions of upwelling atmosphere and the dark and warm areas are regions of subsidence. Rapid convection of the atmosphere causes incrementally higher temperatures toward the interior; the warm stratosphere is consistent with absorption of sunlight by methane and other hydrocarbon gases present in the atmosphere and by haze particles.

Pioneer 10 was the first spacecraft to visit Jupiter, and its scientific results are best viewed together with its "twin," Pioneer 11, which differed by the addition of another magnetometer experiment and a trajectory which took it closer to the planet and allowed the imaging and infrared experiments to view the polar regions. Pioneer results are often overlooked in the wealth of high-resolution imaging and detailed remote sensing spectral data obtained less than a decade later by the missions of Voyagers 1 and 2. For a relatively inexpensive mission, Pioneer 10 provided a treasure of information, some of which was not duplicated or improved by Voyager, such as the cloud photometry and polarimetry of the imaging system.

Besides being the first spacecraft to reach Jupiter, Pioneer 10 was the first spacecraft to penetrate the asteroid belt. It traversed and communicated from greater distances than ever before. It was also the first spacecraft to use all nuclear electrical power. It demonstrated the viability of spin-scan imaging, and its probing of the asteroid belt and the radiation environment of Jupiter paved the way for the later sophistication of the Voyager spacecraft. It also showed that a probe could survive penetration into the Jovian atmosphere and an orbiter could survive many months in the neighborhood of the planet.

Bibliography

Beatty, J. Kelly, et al., eds. *The New Solar System*. Cambridge, Mass.: Sky Publishing Corp., 1981. A general description of solar-system bodies that contains detailed articles pertinent to Jupiter and Pioneer discoveries. The book is readable at high school or college levels. In particular, chapters by James Van Allen, Edward J. Smith, Andrew P. Ingersoll, and J. W. Johnson discuss magnetospheres and the interplanetary medium, the Voyager mission (and pre-Voyager information, including a brief description of Pioneer results), the atmosphere of Jupiter, and the Galilean satellites (the four largest moons of Jupiter).

Belton, Michael, Gary Hunt, and Robert West, eds. *Time-Dependent Phenomena in the Jovian System*. Washington, D.C.: Government Printing Office, 1988. This book is a collection of articles that reviews time-dependent phenomena that are known or suspected in the Jovian magnetospheric environment in Jupiter's atmosphere, and on Jupiter's satellites (particularly Io's volcanic activity). The college and postgraduate-level material is intended to review what is known from existing data and from theoretical expectations.

Burns, Joseph A., and Mildred Shapely Matthews, eds. *Satellites*. Tucson: University of Arizona Press, 1986. A collection of review articles on the various satellites of the solar system authored primarily by American scientists. This work is a primary professional reference. Separate chapters specifically review the Galilean satellites Io, Europa, Ganymede, and Callisto. Other specific satellites are also reviewed, including Earth's Moon, the satellites of Mars, and the satellites of Saturn. In addition, general processes involved in the formation of satellites and their subsequent evolution are reviewed.

Fimmel, Richard O., William Swindel, and Eric Burgess. *Pioneer Odyssey*. NASA SP-396. Washington, D.C.: Government Printing Office, 1977. An authorized history of the Pioneer 10 and 11 missions to Jupiter. This work is a revision of the original NASA Special Publication with more Pioneer 11 results added, including several images of the Galilean satellites. It is suitable for high school and college levels. Appendices include a list of all images taken of Jupiter by Pioneers 10 and 11.

————. *Pioneer Odyssey: Encounter with a Giant*. NASA SP-349. Washington, D.C.: Government Printing Office, 1974. An authorized history of the Pioneer 10 mission to Jupiter with some early Pioneer 11 results included. It is suitable for high school and college levels and describes the mission from the earliest planning stages to its accomplishments by 1973. It contains many illustrations of the spacecraft, its instrumentation, the scientists involved with the experiments, and many of the images obtained of Jupiter by the imaging photopolarimeter. Appendices offer complete information on the details of image acquisition and the personnel involved with the mission.

Fimmel, Richard O., James A. Van Allen, and Eric Burgess. *Pioneer: First to Jupiter, Saturn, and Beyond*. NASA SP-446. Washington, D.C.: Government Printing Office, 1980. An authorized history of the Pioneer 10 and 11 missions to Jupiter and Saturn and their extended missions. This book is the most recent edition of the NASA series on the Pioneer 10 and 11 missions, and it includes all the information on the Pioneer 11 encounter with Saturn. It is suitable for readers with high school or college science backgrounds. Additions to the appendices include a list of Saturn images made by Pioneer 11.

Gehrels, Tom, ed. *Jupiter*. Tucson: University of Arizona Press, 1976. A collection of articles written shortly after the Pioneer 10 and 11 encounters with Jupiter. The authors are primarily from the United States, and final results of the Pioneer missions at Jupiter, as well as observations from Earth-based telescopes at the time, are described. This work is a primary professional reference, and it contains information suitable to readers with college or postgraduate science backgrounds.

National Aeronautics and Space Administration. *Voyager to Jupiter and Saturn*. NASA SP-420. Springfield, Va.: National Technical Information Service, 1977. Summary of the Voyager mission to explore planets of the outer solar system. The spacecraft, its trajectories, and its scientific instruments are described, along with relevant investigations. Color graphics, pictures of the spacecraft and its instruments, and the well-known Voyager images are included.

Yeates, C. M., et al. *Galileo: Exploration of Jupiter's System*. NASA SP-479. Washington, D.C.: Government Printing Office, 1985. This book reviews the Galileo mission to orbit Jupiter and send a probe directly into its atmosphere. Appropriate for high school and college levels, it reviews many of the results from Pioneers 10 and 11 as well as Voyagers 1 and 2. The details of the Galileo mission have been revised following the *Challenger* explosion. The general description of the scientific background and review of Galileo science instruments, however, remains the same. Separate chapters review current information on the atmosphere, satellites, rings, and magnetosphere of Jupiter.

Glenn S. Orton

PIONEER 11

Date: Beginning April 5, 1973
Type of mission: Unmanned deep space probe

Pioneer 11 collected critical data on the outer solar system, obtained images of Jupiter and Saturn, and continues to collect important data from the interplanetary space beyond Saturn.

PRINCIPAL PERSONAGES

CHARLES F. HALL, Pioneer 11 Project Manager
     until 1979
RICHARD O. FIMMEL, Pioneer 11 Project Manager
JOHN H. WOLFE and
PALMER DYAL, Project Scientists
TOM GEHRELS, Principal Investigator,
     photopolarimeter experiment
EDWARD J. SMITH,
MARIO F. ACUÑA,
JOHN A. SIMPSON,
FRANK B. McDONALD,
JAMES A. VAN ALLEN,
R. WALKER FILLIUS,
WILLIAM H. KINARD,
JOHN D. ANDERSON,
DARREL L. JUDGE, and
ANDREW P. INGERSOLL, other Principal Investigators

Summary of the Mission

The Pioneer 11 spacecraft mission was approved by the National Aeronautics and Space Administration (NASA) as a backup to the Pioneer 10 mission to Jupiter. Both spacecraft were managed by the Pioneer Project Office of NASA's Ames Research Center. Pioneer 11 was initially targeted to follow the path of Pioneer 10 through the asteroid belt and past Jupiter. If the asteroid belt or the radiation environment of Jupiter had proved hazardous to Pioneer 10, Pioneer 11 would have been retargeted to avoid the hazards.

The Pioneer 11 spacecraft, like other spacecraft designed to travel to the outer solar system, had to be very lightweight and extremely reliable. It also needed power sources other than solar cells, because the Sun's power is too faint at large distances. Finally, Pioneer 11 needed communications systems designed to operate over billions of miles.

The Pioneer 11 spacecraft was designed and built by TRW in Redondo Beach, California. The designers chose a spin-stabilized spacecraft with a large dish antenna for com-

munications and four radioisotope thermoelectric generators (RTGs), mounted on external booms, for power.

The communications dish antenna is 2.7 meters in diameter and 46 centimeters deep, and it is mounted on the main equipment compartment, which surrounds a hydrazine fuel tank that feeds the propulsion units. Ten propulsion rockets are located at the rim of the antenna and are used to control the spin rate, attitude, and velocity of the spacecraft. All electronic spacecraft components are housed in the main equipment compartment. A smaller hexagonal compartment attached to the side of the main compartment houses most of the twelve scientific instruments. A thermal insulation blanket surrounds both compartments to maintain the electronics at a comfortable temperature. On the bottom of the main equipment compartment are a number of thermally activated louvers, which open and close to expel excessive heat generated by the electronic equipment. Extending from the equipment compartment are two 3-meter booms holding the RTGs and one 6.6-meter boom holding the helium vector magnetometer instrument. The total weight of the spacecraft at launch was only 270 kilograms, including 30 kilograms of scientific instruments.

The scientific instruments were designed to obtain data on interplanetary space, Jupiter, and Saturn. The twelve instruments comprise two magnetometers, an infrared radiometer, a cosmic-ray detector, a charged particle instrument, a trapped radiation detector, an ultraviolet photometer, a Geiger tube telescope, an imaging photopolarimeter, a plasma analyzer, meteoroid detector panels, and an asteroid/meteoroid detector.

Pioneer 11 was launched at 9:11 P.M. eastern standard time, April 5, 1973, from Cape Canaveral aboard an Atlas-Centaur three-stage launch vehicle. The launch vehicle propelled Pioneer 11 to a speed of more than 51,000 kilometers per hour, fast enough to pass the orbit of the Moon in just eleven hours. At the time of Pioneer 11's launch, the sister spacecraft Pioneer 10 had been in flight for thirteen months and was still eight months from its encounter with Jupiter.

After the successful encounter of Pioneer 10 with Jupiter on December 3, 1973, at a closest approach of 132,252 kilometers above the cloud tops, the final decision was made to retarget Pioneer 11. On April 19, 1974, a pair of rockets on Pioneer 11 were fired to add an additional 63.7 meters per second to the spacecraft's velocity and to make it aim for an altitude of 43,000 kilometers above Jupiter's cloud tops. This very close approach caused the spacecraft to be accelerated by Jupiter's tremendous gravity to a velocity fifty-five times that of the muzzle velocity of a high-speed rifle bullet, or 173,000 kilometers per second, so that the spacecraft would intercept Saturn five years after passing Jupiter.

The Jupiter encounter trajectory was such that the spacecraft would approach Jupiter on the left side, as viewed from Earth, and hurtle almost straight up, thus obtaining the first polar view of Jupiter. This path had the added benefit of quickly crossing the planet's magnetic equator, where the most intense radiation belts are concentrated, thus minimizing the total radiation exposure of the spacecraft.

On November 2, 1974, the first Pioneer 11 images of Jupiter were made. The images were obtained by the imaging photopolarimeter instrument, designed by scientists at the University of Arizona. The instruments used the spacecraft's rotation (4.8 revolutions per minute) to scan across Jupiter with a small telescope. On one rotation of the spacecraft, a red filter was placed in front of the telescope and a digital image of a thin strip of the planet was transmitted to Earth. On the next rotation, a blue filter was used to obtain a digital image of the same strip of the planet. On the third rotation, the telescope was moved a tiny amount to a new viewing angle, and a red filter was again used. This sequence of filter changes and angle adjustments was repeated until the whole planet had been scanned. After ground equipment had converted the digital signals to a video signal, green was added to the combined red and blue signals to produce full-color pictures. As the image was created, strip by strip, the television screen in the Pioneer Mission Operations Center displayed a rather oddly shaped planet. Instead of having the familiar spherical shape, Jupiter appeared quite elongated, like a beach ball with an air leak. This curious shape was the result of the forward motion of the spacecraft, which had moved a considerable distance between transmissions of successive strips of the picture. Later, ground computers corrected for the spacecraft motion by placing the strips in the geometrically correct position.

Between November 2, 1974, and December 2, 1974, approximately twenty-five images per day were obtained, each one progressively filling more of the field of view as the spacecraft moved closer and closer to Jupiter. At 8:21 A.M. on December 2, 1974, the last full-disk picture was taken. Subsequent pictures would show only portions of the planet. At 4:00 P.M. that day, a two-hour scan produced a detailed image of the famous Red Spot, a puzzle to astronomers for centuries.

On December 2, 1974, at 9:02 P.M. Pacific standard time, twenty-two minutes before closest approach, Pioneer 11 passed behind Jupiter. Its on-board memory was busy recording data to be transmitted to Earth later, assuming the spacecraft's electronic equipment survived the fierce radiation. At 9:44 P.M., the spacecraft came out from behind Jupiter, but controllers on Earth had to wait an additional forty minutes to hear anything because of the time it took the signal to

travel the 720 million kilometers from Jupiter to Earth. Eleven seconds after 10:24 P.M. Pacific standard time, the tracking station at Canberra, Australia, picked up the first faint signal from the spacecraft and sent it on to the Pioneer Mission Operations Center at Ames Research Center. Pioneer 11 had survived the ordeal. The massive dose of radiation had caused minor irregularities in the functioning of the plasma analyzer, the infrared radiometer, the meteoroid detector, and the imaging system. Engineers quickly sent a string of 108 commands to the spacecraft to reconfigure the instruments so they could continue to provide data as the spacecraft hurtled away from Jupiter.

During the spring of 1978, engineers and scientists decided that the spacecraft should pass just outside Saturn's bright rings. The hydrazine rockets were used to adjust the spacecraft's velocity so that Pioneer 11 would swing under the ring plane at a distance of 35,400 kilometers and come within 21,400 kilometers of the cloud tops. Later, this trajectory would carry Pioneer 11 within 355,600 kilometers of Titan, the largest of Saturn's moons.

By August 2, 1979, Pioneer 11 had accelerated to a speed of 30,600 kilometers per hour under the influence of Saturn's enormous gravity. With increasing speed, the spacecraft continued its plunge toward Saturn, and at 7:36 A.M. Pacific daylight time on September 1, 1987, it crossed the plane of the rings. By this time its speed had increased to 112,000 kilometers per hour. Because of the large distance of Saturn from Earth (1.547 billion kilometers), project engineers had to wait eighty-six minutes for the signals to reach their computer screen at the Pioneer Mission Operations Center. The data confirmed that the spacecraft had not sustained any damage by particles in the ring plane, as had been feared. Images continued to arrive, showing beautiful close-ups of the rings around Saturn.

At 9:31 A.M., Pioneer 11 hurtled through its closest approach to Saturn, only 20,930 kilometers above the cloud tops, at 114,150 kilometers per hour. At 11:04 A.M. Pacific daylight time on September 2, 1979, Pioneer 11 made its closest approach to the giant Saturnian satellite, Titan. By September 3, 1979, the spacecraft was moving away from the ringed planet at 36,210 kilometers per hour.

Pioneer 11 and its sister ship, Pioneer 10, were the first man-made objects to travel through the asteroid belt and obtain images of Jupiter and Saturn. They continue to provide exciting scientific data as they travel toward the edge of the solar system. Sometime in 1991 or 1992 their signals will cease when their power sources run out, but they will continue to travel through the Galaxy.

Knowledge Gained

Pioneer 11's flyby of Jupiter revealed that the high-energy electron intensity in the radiation belts was as predicted, but that the proton flux was about one-tenth of what was predicted based on extrapolation from Pioneer 10 data. The particle

Pioneer 11 close-up view of Saturn's ring system showing details never before available. (NASA)

and field experiments verified that Jupiter's intense radiation belts are dangerous only at lower latitudes and pose relatively few hazards to spacecraft flying through them at high altitudes.

Images of the cloud tops at Jupiter's poles showed that they are significantly lower than at the equator and that they are covered by a thick, transparent atmosphere. Blue sky, caused by multiple molecular scattering of light by the gases in the atmosphere, was clearly visible at the poles. The Jupiter encounter also provided new details of the circulation and convection patterns in the clouds around the Red Spot.

After the Jupiter encounter, Pioneer 11 looped high across the ecliptic plane and across the solar system, providing new information about the magnetic field of the Sun. The magnetic field detectors on board the spacecraft showed that solar rotation produces a warped "current sheet" (a disk-shaped region separating the north and south magnetic fields around the Sun). The sheet wobbles up and down, as seen from Earth and the other planets, and this explains the reversals of the solar magnetic field that had puzzled scientists for years.

During the Saturn encounter, Pioneer 11 produced spectacular pictures of the ring plane from underneath, with the Sun shining through the gaps of the rings from above. The actual rings, which from Earth appear bright, were now seen as dark areas. The Sun's light, scattering off particles in the spaces between the rings, produced bright regions, which from Earth appear dark. A new ring, the F-ring, was discovered just outside the A-ring, which had been thought to be the outermost.

The close encounter also produced evidence — based on absorptions of charged particles — of additional Saturnian satellites. This was the first time that previously unknown satellites had been inferred from charged particles and magnetic field measurements.

Data from the infrared radiometer instrument showed that Saturn is still hot inside and that it emits more heat than it absorbs from the Sun. The magnetic field was shown to be almost exactly aligned with the axis of rotation, unlike the fields of Earth and Jupiter.

Images and data obtained during the Titan encounter showed Titan as a fuzzy ball with a slight orange tint and a suggestion of blue around the edge caused by its thick atmosphere. Radiometric data proved that Titan is in equilibrium with the solar radiation and has no internal heat source.

As Pioneer 11 speeds out of the solar system, it continues to provide particle and field measurements from regions where no other spacecraft has traveled. In addition, engineers are gaining valuable information about the reliability of spacecraft components on extended missions. Techniques developed to control and maneuver spacecraft at large distances from Earth have contributed to a valuable information base for future exploration of the universe.

Context

Pioneer 11 was the second spacecraft to traverse the asteroid belt and proceed on to Jupiter to take close-up pictures. Pioneer 10, launched in the spring of 1972, traveled through the asteroid belt in the summer of 1972 and made a flyby of Jupiter in December, 1973. The best Earth-based observations of Jupiter had previously established that the giant planet is a beautifully colored globe with bands of swirling clouds parallel to its equator. The Giant Red Spot, just south of the equator, had intrigued astronomers for centuries.

The Pioneer 10 flyby of Jupiter in 1973 had answered many questions about the intensity of the radiation belts and provided exquisite pictures of the cloud bands. Since it followed an almost equatorial trajectory, however, no information was obtained about the latitudinal distribution of the radiation belts, and no pictures of the polar regions were obtained. Pioneer 11 was the first spacecraft to fly by Jupiter on a near-polar trajectory and thus was able to complement Pioneer 10's data by providing measurements in the third dimension. Unexpectedly, Pioneer 11 discovered that the cloud tops are substantially lower near the poles than at the equator. Scientists were astounded to see a blue atmosphere at the poles similar to Earth's blue sky.

Before Pioneer 11 visited Saturn, the only information about this mysterious planet had come from Earth-based observations. As far back as 1675, the astronomer Jean-Dominique Cassini had studied the rings and identified a large gap, the Cassini Division, between the outer two rings. Before the Pioneer 11 flyby, astronomers had speculated that the rings consisted of ice or ice-coated rocks from several centimeters to several meters in size. Pioneer's perspective, not possible to achieve from Earth, allowed it to observe the rings illuminated from behind. The gaps between the rings were found to contain small particles that scattered the light, but a small amount of sunlight passed through. This allowed scientists to assess much more accurately the thickness of the ring material. From temperature and heat balance measurements of the rings and the ring shadow on the planet, scientists determined that the rings cannot be more than 4 kilometers thick. Sensitive measurements of the motion of the spacecraft as it was influenced by Saturn's gravity allowed scientists to estimate the mass of the rings to be less than one three-millionth of the mass of Saturn itself.

The infrared photopolarimeter and the ultraviolet spectrometer provided new information about the size and atmosphere of Saturn's moon Titan. The diameter was estimated to be approximately 5,600 to 5,800 kilometers. Analysis of the polarized light suggested a haze of methane particles high into the atmosphere. The discovery of a cloud of hydrogen atoms, extending at least 300,000 kilometers along the orbit, suggested that the methane in Titan's atmosphere is being broken down into hydrogen and carbon by solar radiation.

Pioneer 11 is historically significant because it was the first spacecraft to observe Saturn. The information it provided had a direct influence on the planning and design of the Voyager missions to Jupiter, Saturn, and beyond.

Bibliography

Alexander, Arthur Francis O'Donel. *The Planet Saturn: A History of Observation, Theory, and Discovery*. London: Faber and Faber, 1962. This is a comprehensive account of observations of Saturn from 650 B.C. to 1960. Suitable for advanced high school and college students, the book presents interesting accounts of the observations and discoveries made by most of the famous astronomers.

Briggs, Geoffrey, and Frederick Taylor. *The Cambridge Photographic Atlas of the Planets*. Cambridge: Cambridge University Press, 1982. This lavishly illustrated book contains detailed photographs of Jupiter, Saturn, and their moons, obtained mostly from the Voyager spacecraft. Particularly striking are the close-up photographs of Saturn's rings.

Elliot, James, and Richard Kerr. *Rings: Discoveries from Galileo to Voyager*. Cambridge, Mass.: MIT Press, 1984. This book describes planetary ring systems discovered by ground-based, airborne, and spacecraft telescopes. Suitable for high school and college students, the volume provides good insight into the characteristics and composition of planetary rings. The text is supported by numerous illustrations and diagrams.

Fimmel, Richard O., et al. *Pioneer: First to Jupiter, Saturn, and Beyond*. NASA SP-446. Washington, D.C.: Government Printing Office, 1980. Contains a good description of the Pioneer 10 and Pioneer 11 missions to Jupiter and Saturn. Suitable for high school and college students, the publication describes in detail the probes' launches, journeys through the asteroid belt, and first encounters with Jupiter and Saturn. Contains pictures transmitted from Jupiter and Saturn and illustrations and photographs of the spacecraft and their scientific instruments.

Gehrels, Tom, ed. *Jupiter*. Tucson: University of Arizona Press, 1976. A collection of scientific essays, this book is the authoritative reference on Jupiter. Its 212 contributors from around the world present scientific analyses of all aspects of Jupiter; many of the data are derived from the Pioneer missions. Suitable for college students, this book lists thousands of references to scientific papers in professional journals.

Washburn, Mark. *Distant Encounter: The Exploration of Jupiter and Saturn*. San Diego: Harcourt Brace Jovanovich, 1983. Describes the images obtained by the Pioneer and Voyager missions to Jupiter, Saturn, and their planets. Provides a good account of the activities and events occurring in the Pioneer Mission Operations Center at the time of the Pioneer 11 Saturn encounter. Suitable for high school and college students, this book contains numerous illustrations and photographs, many in exquisite color.

Manfred N. Wirth

PIONEER VENUS 1

Date: Beginning May 20, 1978
Type of mission: Unmanned Venus probe

Pioneer Venus 1 obtained important information on Venus's topography and atmosphere. Previous American missions to Venus were quick flybys, but the Pioneer Venus orbiter, over many months, mapped most of the planet's surface and investigated in detail its atmosphere and ionosphere.

PRINCIPAL PERSONAGES
    CHARLES F. HALL, Project Manager, Pioneer Venus
    RICHARD O. FIMMEL, Project Manager for the
        extended Pioneer missions
    RALPH W. HOLTZCLAW, Spacecraft Systems Manager
    BRUCE C. MURRAY, Director of the Jet Propulsion
        Laboratory
    GORDON H. PETTENGILL, leader, orbiter radar
        mapper team
    A. I. (IAN) STEWART, Principal Investigator, orbiter
        ultraviolet spectrometer experiment
    H. A. TAYLOR, JR., Principal Investigator, orbiter ion
        mass spectrometer experiment
    H. B. NIEMANN, Principal Investigator, orbiter neutral
        mass spectrometer experiment

Summary of the Mission

In 1967, shortly after the American Mariner 5 spacecraft flew by Venus and the Soviet Venera 4 spacecraft probed its atmosphere, the Pioneer Venus project began. Three scientists, Richard Goody of Harvard University, D. M. Hunten of Kitt Peak National Observatory, and N. W. Spencer of the Goddard Space Flight Center, formed a team to study the possibility of sending a probe to investigate Venus's atmosphere. Meeting at the Goddard Space Flight Center in Greenbelt, Maryland, these scientists considered several approaches for the Venus mission, including orbiters, atmospheric probes, and balloon probes. By 1969, the National Aeronautics and Space Administration (NASA) merged some of these ideas into what became known as the universal bus, a probe-orbiter combination that could be used to orbit the planet and send probes into its atmosphere. In 1970, twenty-one scientists of the Space Science Board and the Lunar and Planetary Missions Board of NASA analyzed possible missions to Venus and issued a report, *Venus: Strategy for Exploration*, informally referred to as the "Purple Book"

because of its cover's color. This report recommended that exploration of Venus be an important item on NASA's agenda for the 1970's and 1980's.

Congress was reluctant to fund new space programs in the early 1970's, and as a result NASA's Venus program fell behind its Soviet counterpart, whose Venera spacecraft were piling up successes. After the American program was transferred to the Ames Research Center in California, a study team was organized. This team worked closely with a science steering group, formed of interested scientists, to define payloads for the mission. In 1972, this group published a report, informally referred to as the "Orange Book," which became the American guide to Venus exploration by spacecraft. This Pioneer Venus report recommended that three missions to Venus be supported: a multiple probe in 1976, an orbiter in 1978, and a probe in 1980. Because of financial restrictions, NASA decided in August, 1972, to limit the project to a multiprobe in 1977 and an orbiter in 1978. During this same period, the European Space Research Organization expressed an interest in the orbiter mission and, after a period of study, a group of European and American scientists recommended that the spacecraft's mission should cover a period of one rotation of Venus, that is, 243 Earth days.

In 1974, NASA approved the Pioneer Venus missions and selected the scientific instruments to be developed for the orbiter and multiprobe. After competitive studies, NASA chose Hughes Aircraft Company's Space and Communications Group to build the spacecraft for the missions. Unfortunately, Congress continued to withhold funding for the Venus program, and, after much political maneuvering, NASA decided to delay the multiprobe until 1978, so that both the multiprobe and orbiter could use the same launch opportunity, thereby reducing operational costs. Nevertheless, political problems were not at an end; in 1975, the House cut $48 million intended for the Pioneer missions, causing many organizations to lobby intensely for the restoration of Pioneer Venus's funding. Eventually, funding was restored.

Meanwhile, scientists had been building and testing various scientific instruments for the mission. For example, they found ways to compress more than a thousand microcircuits and many intricate mechanical devices into a radar mapper that weighed only 11 kilograms. By February, 1978, all the scientific instruments for the orbiter had been shipped to

The Pioneer Venus Orbiter was inserted into an elliptical orbit around Venus on December 4, 1978. (NASA)

Hughes Aircraft Company in El Segundo, California, where, after further testing, the instruments were placed in the spacecraft. The orbiter was then shipped to the launch site at Kennedy Space Center in Florida, where it was mated to an Atlas-Centaur rocket, its launch vehicle.

The Pioneer Venus Orbiter began its 480-million-kilometer voyage to Venus at 9:13 A.M. on May 20, 1978. Pioneer Venus 1 traveled outside Earth's orbit for three months and then moved inside it for the next four months. NASA scientists chose this long, indirect route to minimize the accelerating influence of solar gravity and reduce the amount of fuel needed to go into orbit around Venus. Because cosmic rays had caused errors in the spacecraft's memory circuits, controllers were uneasy about the orbital insertion maneuver that would take place while the orbiter was hidden behind Venus. On December 3, technicians loaded the orbiter's memory circuits with the sequence of commands needed to fire the orbital insertion rockets on the following day. At 7:51 A.M. on December 4, the orbiter passed behind Venus and communications with Earth ceased. When the orbiter finally emerged, it sent a signal that took three minutes to travel the 56 mil-

lion kilometers to Earth. When that signal was received, controllers responded with shouts of joy.

The orbiter had been inserted into a highly eccentric, nearly polar Venusian orbit that would take it as close to Venus as 150 kilometers (perigee) and as far away from Venus as 66,900 kilometers (apogee). On December 5, the orbiter was maneuvered so that its antenna pointed to Earth, and several of its instruments were activated. By December 6, it beamed the first black-and-white images of Venus to Earth. During this early period of the mission, the radar mapper performed erratically and was deactivated for a month. Then, with redesigned operating procedures, it worked perfectly for the rest of the mission. Such was not the case with the infrared radiometer, which failed on the seventieth orbit. It was the only permanent instrument failure of the mission.

As the orbiter began its scientific tasks, not much was known about the upper atmosphere of Venus. As more and more data were transmitted to Earth, it became practical, through trial and error, to make orbital corrections. During the first phase of the orbiter's mission, the surface of Venus was mapped by radar. The radar mapping took 243 Earth days to accomplish (the time for Venus to rotate once on its axis). The radar mapper continued to function through 243 more orbits — a second Venus day — and with the completion of orbit 600 on July 27, 1980, NASA officials announced that phase 1 of the Pioneer Venus mission had been completed, though the radar mapper continued to return data through orbit 834, nearly eight months later. Other instruments were still active, however, and the orbiter will continue to investigate the interaction of solar radiation with Venus's ionosphere over an entire solar cycle (from sunspot minimum to maximum to minimum).

After orbiting Venus for more than fourteen years, the Pioneer Venus Orbiter spacecraft ceased operating on October 8, 1992. The spacecraft passed through the lowest part of its orbit and out of radio contact. A radio signal should have been received by ground stations when it emerged from behind the planet.

Knowledge Gained

Venus is perpetually shrouded by layers of pale yellow clouds, and most of the twelve scientific instruments aboard the cylindrical orbiter were developed to observe the planet's atmosphere and global weather patterns. Venus's surface had been totally hidden until Earth-based radar began to penetrate its veil of clouds. These earthbound efforts were crude, however, and it was not until the Pioneer Venus mission that the surface could be mapped. The orbiter's radar was able to map about 93 percent of the Venusian surface by sending radar pulses downward and measuring the time lag until the echo returned. In this way, the radar altimeter could detect features as small as 30 kilometers across and about 200 meters high.

The topographic map derived from the orbiter's radar data revealed that Venus is generally smoother than the other terrestrial planets (Mercury, Earth, and Mars), but that it is also a world of great mountains, expansive plateaus, enormous rift valleys, and shallow basins. The Venusian surface can be divided into three regions: rolling plains, lowlands, and highlands. The plains, which cover about 65 percent of the mapped surface, are pockmarked with circular features that may be the remains of impact craters. The lowlands, which comprise about 27 percent of the imaged surface area, are much less abundant than on Earth, where ocean basins cover about two-thirds of the globe. Venus's highlands, which comprise only 8 percent of the imaged surface area, are like Earth's continents. Venus has two major highland areas: Ishtar Terra and Aphrodite Terra. Venus's highest point, a mountain massif higher than Mount Everest on Earth, is known as Maxwell Montes. This huge area of uplifted terrain occupies the entire east end of Ishtar Terra, 11,800 meters above the average level and 9,000 meters above the adjoining Lakshmi Planum.

Besides the radar mapper, the orbiter carried other instruments for studying Venusian phenomena, including a tele-scope to observe the clouds in ultraviolet light, an infrared radiometer to measure heat radiated from the atmosphere, an ultraviolet spectrometer to track dark streaks in the clouds, and a variety of other sensors to examine the physical and chemical properties of the upper atmosphere. The orbiter found that the clouds that enshroud Venus horizontally also have an enormous vertical extent, about 50 kilometers. The orbiter's data also provide a more detailed picture of the temperature structure of Venus's atmosphere. Scientists have divided the atmosphere into two regions. In the lower region (called the troposphere), which extends from the surface to an altitude of 100 kilometers, the temperature decreases with height. Above the troposphere lies a thinner region (called the thermosphere), which is heated by ultraviolet solar radiation; thus, its temperature increases with height. One of the orbiter's most exciting discoveries was the enormous change in temperature between day and night in the upper atmosphere. Scientists have coined the term "cryosphere" (cold sphere) to describe the cold region of the upper atmosphere on the nightside of Venus. On Earth, the thermosphere is present day and night, whereas on Venus, the thermosphere disappears on the nightside.

Using such instruments as the photopolarimeter and ultraviolet spectrometer, the orbiter transmitted data to Earth about Venus's clouds that led to a detailed understanding of their morphology, composition, and motions. For example, ultraviolet photographs made on consecutive days showed that the dominant wind pattern on Venus drives cloud markings around the equatorial region in only four Earth days. The orbiter's instruments were also able to penetrate beneath the clouds to measure the high amount (96 percent) of carbon dioxide in Venus's atmosphere. Because Venus's dense carbon dioxide atmosphere inhibits absorbed sunlight from being reradiated into space as heat (the greenhouse effect), Venus has a higher surface temperature than any other planet.

Finally, the orbiter gathered considerable information about Venus's ionosphere, the region of its upper atmosphere characterized by a high density of electrically charged particles. The orbiter instruments also gathered data on how the ionosphere interacts with the solar wind, the outward flow of atomic particles from the Sun. Venus has no appreciable magnetic field, but the interaction of its ionosphere with corpuscular radiations from the Sun produces a well-developed bow shock. Since the solar wind travels faster than any atmospheric pressure wave that could divert its flow around Venus, a shock wave, or bow shock, forms in the solar wind in front of Venus analogous to the shock wave in front of a supersonic aircraft. The Venusian bow shock is in many ways similar to that of Earth, but there are differences. At Venus the ionosphere deflects the solar wind; at Earth the strong terrestrial magnetic field deflects it.

Context

Until the Pioneer Venus missions, Earth-based observations had contributed little detailed knowledge about the cloud-covered planet. NASA's flyby missions, Mariners 2, 5, and 10, along with the Soviet Union's Venera spacecraft, had shown that Venus's atmosphere is composed largely of carbon dioxide and that the surface temperature is hot enough to melt such metals as lead, tin, and zinc. These previous missions offered only fleeting glimpses of the planet, however, whereas the Pioneer Venus orbiter permitted repeated sensing of Venus's surface, lower and upper atmosphere, ionosphere, and solar wind interaction. The Pioneer Venus missions clearly marked a milestone in the American exploration of Venus.

Pioneer Venus 1 was the latest in a series of unmanned probes that had begun in the late 1950's. This series of probes sent back important information on magnetic fields and radiation in interplanetary space. Pioneers 10 and 11, which crossed the asteroid belt and studied Jupiter's atmosphere, conducted investigations of the interplanetary medium beyond Mars. Ames Research Center, which had taken over the Pioneer missions in 1962, viewed Pioneer Venus as a logical development within the tradition of small, relatively inexpensive Pioneer spacecraft.

From orbital insertion on December 4, 1978, through the following decade, Pioneer Venus Orbiter produced a wealth of scientific data on all aspects of the Venusian environment. Many of the instruments continued to be fully functional throughout this period. Even though the resolution of the orbiter's radar map of Venus is crude in comparison with that available for planets such as Mercury and Mars, this map does reveal enough about Venus to fuel speculations concerning its

geology and evolution. Certain features — for example, the vertical uplift of the Lakshmi Planum— suggest tectonic activity. Some scientists think that the ridges on Ishtar Terra may be the result of plate motion, though there is little evidence for integrated plate movements similar to continental drift on Earth.

In textbooks written before the Pioneer Venus missions, Venus was often depicted as Earth's twin because of their similar sizes and stable atmospheres. The data from the Pioneer orbiter has revealed that Venus is no twin of life-nurturing Earth but a strange inferno. This is not to say that there are no similarities between the planets. Like Earth, Venus appears to have experienced volcanism, impact cratering, and a complex geologic history, and some models used to understand the circulation of the atmosphere on Earth can be used to understand Venus's weather patterns. Nevertheless, the differences on Venus are even more striking: the poisonous atmosphere, the sulfuric-acid rain, and the furnacelike surface temperatures.

The Pioneer Venus Orbiter's mission has also raised a host of intriguing questions for future exploration. What caused the greenhouse effect on Venus? Was there ever a large amount of water on Venus, and if so, what happened to it? Is Venus now quiet geologically, or is it active? Are the Venusian highlands like Earth's continents, or are they gigantic volcanic piles? Are the rolling plains extremely old or relatively recent in origin? These questions, in addition to the Pioneer Venus mission's many scientific accomplishments, are the lasting legacy of the orbiter mission.

Bibliography

Beatty, J. Kelly, et al., eds. *The New Solar System*. Cambridge, Mass.: Sky Publishing, 1982. Summarized in this book are the chief results of the first decades of space exploration. Beautifully illustrated with photographs, paintings, and diagrams, this book explains for the general reader the most exciting discoveries of the planetary scientists. The Pioneer Venus missions are treated in the chapters on the surfaces and atmospheres of the terrestrial planets.

Briggs, Geoffrey, and Frederick Taylor. *The Cambridge Photographic Atlas of the Planets*. Cambridge, England: Cambridge University Press, 1982. This atlas of the solar system contains the official maps of the planets and their satellites as well as more than two hundred photographs. The chapter on Venus makes excellent use of the data and photographs from the Pioneer Venus missions. The approach is descriptive and nontechnical.

Burgess, Eric. *Venus: An Errant Twin*. New York: Columbia University Press, 1985. An informative, well-illustrated volume by a skilled science writer. He recounts, with perception, the history of the major discoveries about Venus, including those of Soviet and American exploration. He is willing to speak his mind, and his discussions of both the well-established findings and the controversial areas are entertaining as well as enlightening.

Fimmel, Richard O., Lawrence Colin, and Eric Burgess. *Pioneer Venus*. NASA SP-461. Washington, D.C.: Government Printing Office, 1983. All three authors have been deeply involved in the Pioneer Venus missions, Fimmel as project manager, Colin as project scientist, and Burgess as a science journalist. Their attractively illustrated book is both authoritative and clearly written. Though they do not shirk analyses of technical matters, they avoid advanced mathematical and physical concepts, and their approach is accessible to the layman.

Glass, Billy P. *Introduction to Planetary Geology*. New York: Cambridge University Press, 1962. Intended for college students, this book emphasizes an understanding of the geology of the planets in relation to the great discoveries made in the American and Soviet space programs. Because the author keeps scientific jargon to a minimum, his textbook is accessible to general audiences. The chapter on Mercury and Venus presents the findings of modern space probes, including the Pioneer missions. Glass also discusses these findings in the context of contemporary theories about the solar system.

Hunten, D. M., et al., eds. *Venus*. Tucson: University of Arizona Press, 1983. A scientific celebration of the exploration of Venus by spacecraft from Mariner 2 in 1962 to the Pioneer Venus and Venera missions of the late 1970's and early 1980's. Though the articles are technical, some of the more general ones are accessible to the layman, especially if use is made of the helpful glossary at the end of the book. Most of the articles, however, are intended for readers with a knowledge of physics, chemistry, and advanced mathematics.

Jones, Barrie William. *The Solar System*. Elmsford, N.Y.: Pergamon Press, 1984. A presentation of the contemporary picture of the solar system. Written at an introductory level, the book assumes no previous knowledge of planetary astronomy. Jones's chapter on Venus discusses its interior, surface, and atmosphere in the light of data from the Pioneer Venus 1 Orbiter and the Pioneer Venus 2 Multiprobe.

Muenger, Elizabeth A. *Searching the Horizon: A History of Ames Research Center, 1940–1976*. NASA SP-4304. Washington, D.C.: Government Printing Office, 1985. Muenger wrote this book for NASA's internal history program. Ames has made many important contributions to NASA space missions, especially the Pioneer projects. Although Muenger does not deal directly with the Pioneer Venus program (it lies outside her chronological limits), her account of Ames's technological achievements in the context of its managerial evolution sheds light on the early history of the Pioneer program and makes the later successes of the Pioneer Venus project more understandable.

Robert J. Paradowski

PIONEER VENUS 2

Date: August 8 to December 9, 1978
Type of mission: Unmanned Venus probe

Pioneer Venus 2, often called the multiprobe, was a cluster of five spacecraft designed to penetrate Venus's cloud cover and gather information about all levels of its atmosphere at widely separated locations.

PRINCIPAL PERSONAGES

CHARLES F. HALL, Pioneer Venus Project Manager
RICHARD O. FIMMEL, who succeeded Hall as
    Pioneer Venus Project Manager
LAWRENCE COLIN, Pioneer Venus Project Scientist
GORDON H. PETTENGILL, the team leader for the
    multiprobe's radio science experiment
H. A. TAYLOR, JR., Principal Investigator,
    multiprobe ion mass spectrometer experiment
A. SEIFF, Principal Investigator,
    atmospheric structure experiments
R. W. BOESE, Principal Investigator,
    large probe infrared radiometer experiment
J. H. HOFFMAN, Principal Investigator,
    large-probe mass spectrometer experiment

Summary of the Mission

During the 1960's, the Soviet Union and the United States used two different methods to explore Venus. The Soviets, with greater booster rockets, flew probe and lander missions as well as flybys, whereas the United States used flybys only. Conflicting information was sometimes obtained. For example, the Venera 4 lander recorded a surface temperature of 265 degrees Celsius, while in the same year the Mariner 5 flyby recorded a surface temperature of 526 degrees Celsius. Because of the many unanswered questions about Venus, a new approach was needed — a multifaceted mission to orbit Venus and probe its dense atmosphere. The impetus for this new approach came mainly from scientists.

By 1968, the Space Science Board of the National Academy of Sciences had completed a study, *Planetary Exploration, 1968–1975*, which concluded that planetary explorations should be undertaken as an integrated plan involving a wide range of scientific disciplines. A specific recommendation was for the National Aeronautics and Space Administration (NASA) to initiate a program to put Pioneer spinning spacecraft into orbit around Mars and Venus. In 1969, the Goddard Space Flight Center published a report, *A Venus Multiple-Entry-Probe, Direct-Impact Mission*, which advised the development of a program to send seven entry probes to Venus to measure its atmospheric temperature, pressure, and wind speeds. In 1970, twenty-one scientists of the Space Science Board and the Lunar and Planetary Missions Board of NASA studied the scientific potential of a mission to Venus and produced a report, *Venus: Strategy for Exploration*, which recommended two multiprobe missions for 1975 and two orbiter missions for 1976. Because of congressional unwillingness to provide funding, these missions never materialized.

NASA established a Pioneer Venus science steering group in 1972 to involve the scientific community in the Pioneer Venus missions. This group published a report known as the "Orange Book" that became a widely used guide to Venus exploration. Because of restricted budgets, NASA decided to limit the Pioneer Venus program to two flights, a multiprobe in 1977 and an orbiter in 1978. NASA then invited scientists to suggest experiments for the missions. A small number of instruments were made for these experiments, usually by the principal investigator. The instruments had to be miniaturized for inclusion in small, sealed shells. Unfortunately, Congress balked at the program, refusing to authorize mission starts, resulting in cancellations and a decision by NASA to launch both the multiprobe and the orbiter in 1978 to save costs. The biggest setback came in 1975 when the House of Representatives cut $48 million from NASA's appropriation for the Venus missions. By this time, however, many scientists had become ardent supporters of the program, and they successfully lobbied Congress to restore the funds.

Meanwhile, Hughes Aircraft Company, the primary contractor for the spacecraft, began assembling and testing the instruments at its plant in El Segundo, California. All the instruments were ready in time for the mission. During April, 1978, the multiprobe completed its preshipment review at Hughes. Then the large probe was shipped to Kennedy Space Flight Center in Florida separately from the three small probes and the bus, the spacecraft whose main function was to carry the probes to Venus and target their entries into the atmosphere. At Kennedy, the probes were placed on the bus and explosive bolts installed. The multiprobe was then transferred to the launchpad and mated with the launch vehicle, an Atlas SLV-3D booster and a Centaur D-1A second stage. The scheduled launch was August 6, 1978, but this was missed when a problem occurred in loading liquid helium into the Centaur.

Pioneer Venus 2 began its 350-million-kilometer voyage to Venus at 3:33 A.M. on August 8, 1978. During its trip, the multiprobe had to undergo several critical maneuvers. For example, on August 17, mission controllers, through a series of carefully timed rocket thrusts, effected a course change that put the spacecraft on target for encounter with Venus on December 9. Then on November 9, when the multiprobe was about 13 million kilometers from Venus, controllers oriented the bus so that the large probe would separate in the proper direction. Unfortunately, tracking stations on Earth encountered problems about the precise orientation of the bus. Some scientists thought that there might be a propellant leak; others believed that the problem arose from difficulties in observing a spacecraft racing away from a rotating Earth, which itself was traveling in orbit around the Sun and wobbling in concert with the Moon. Because of these unresolved problems, the project manager decided to delay the scheduled release of the large probe from the bus. After an all-night session, NASA scientists agreed on a compromise timer setting, and on November 16 the large probe was successfully released toward its planned entry point on Venus.

The next critical maneuver involved pointing the bus toward Venus so that the three small, identical probes could be correctly released. The spin rate of the bus had to be precise, 49.60 revolutions per minute, to ensure that the probes would separate along paths that would take them to their individual targets on Venus. On November 20, when the bus was twenty days away from the planet, the small probes were successfully released. On separation, the probes became silent, because they lacked the power to replenish their batteries, but their internal timers had been set to reactivate their instruments three hours before descent into the Venusian atmosphere.

On December 9, radio contact with the four probes resumed. The first radio signal came from the large probe; then, one by one and within minutes of one another, the three small probes were detected. All instruments on all probes were operating satisfactorily. Now the exciting entry phase of the mission began. The large probe, which was the first to enter, rapidly decelerated as it penetrated the ever-thickening atmosphere. Its speed was further slowed by the deployment of a Dacron parachute. At 45 kilometers above the surface, the parachute was jettisoned. The large probe hit the surface at 32 kilometers per hour, landing near the equator, some 55 minutes after first encountering the Venusian atmosphere. Its radio signals ceased immediately upon impact.

The three small probes, entering the atmosphere within 3 minutes of one another, were all quickly slowed, the dense gases retarding their fall without the use of parachutes. During their descents, windows opened on the probes and sensors began encountering the atmosphere and telemetering

data to Earth. Like the large probe, the small probes took about 55 minutes to reach the surface. One of the probes, called the north probe, landed in darkness near the north polar region. Another probe, called the day probe, landed in the southern hemisphere on the day side. The third, the night probe, landed in the southern hemisphere on the night side. Although radio signals from the north and night probes ended at impact, transmissions from the day probe continued for another 68 minutes before it became silent.

Meanwhile, the bus hurtled toward Venus close behind the probes. Its entry into the atmosphere occurred about 88 minutes after the last probe's entry. The bus plunged into the atmosphere on the day side. Unlike the small probes, the bus had no heatshield and it burned up within 2 minutes. During this time, however, radio transmissions poured back to Earth, carrying scientific data on the composition of the very high atmosphere of Venus. The entire entry phase, involving all five spacecraft, took only 1 hour, 38 minutes, but in that time the probes and bus generated data that would cause scientists to take a new look at the complex atmosphere of Earth's sister planet.

Knowledge Gained

Scientific instruments on the probes measured temperature, pressure, and density from the upper atmosphere through the clouds to the surface. These instruments located sources and sinks of solar and infrared radiation in the lower atmosphere. Because of the various locations of the probes' descents, they were able to explore the atmosphere under daytime, nighttime, low-latitude, and high-latitude conditions. Enough data were acquired from these probes that a general meteorological model of Venus could be developed and compared with the meteorologies of other planets.

One of the important questions the probes were designed to answer is why Venus's surface has such a high temperature. The ubiquitous cloud cover on Venus is so reflective that the planet actually absorbs less solar radiation than Earth does. Furthermore, the probes showed that only a small fraction of this absorbed radiation actually penetrates the clouds and the dense lower atmosphere to reach the surface. Nevertheless, Venus remains extremely hot because, like a gigantic greenhouse, its atmosphere allows passage of incoming solar radiation but greatly restricts the radiation of heat back into space. The probes also found water vapor in the lower atmosphere in sufficient amounts to trap infrared radiation even further.

Another discovery made by the probes is the enormous temperature difference between Venus's dayside and nightside upper atmospheres. Even with twice the incoming solar radiation, Venus somehow manages to keep a cooler upper-atmospheric temperature than Earth's, but the real surprise is the low temperature of the upper atmosphere on Venus's nightside. The basic difference between the atmosphere of Venus and that of Earth is that the atmosphere of Venus is hot at the bottom and cold at the top, whereas on Earth the reverse is true. In the lower Venusian atmosphere, the probes found that, below the clouds, there is very little thermal contrast between day and night or between the equatorial and polar regions.

The probes also detected several distinct layers in Venus' atmosphere. The dense cloud layer that enshrouds the planet begins at an altitude of about 70 kilometers and then disappears at about 50 kilometers. A lower haze layer extends down to 30 kilometers, and below that the atmosphere is remarkably clear. Data from a spectrometer aboard the large probe revealed that the Venusian clouds contain nearly pure sulfuric acid. Another probe discovered a large quantity of atmospheric sulfur dioxide that declined after a time. Some scientists interpret this data as indicating that this sulfur dioxide had been ejected by a volcano and then settled to lower levels.

The Pioneer probes also measured wind speeds. The winds of Venus are dominated by a global east-to-west circulation that reaches a maximum of about 150 meters per second at the altitude of the cloud tops. For the most part, wind speeds on Venus decline with decreasing altitude, and at the surface are only about 1 meter per second. Venus itself turns from east to west, but it takes 243 Earth days to complete a single rotation. When this slow rotation is compared with the rapid speed of Venus's high-altitude winds, it becomes clear why many scientists say that the atmosphere of Venus superrotates, for at the cloud tops the atmosphere moves more than 60 times as fast as the planet does. In contrast, Earth turns west to east, and its atmosphere rotates nearly synchronously with the solid planet below it. Less dramatic than this global circulation is the movement of atmospheric matter from the equator to the pole and back again. This gigantic circulation — in which heated air rises at the equator and cooled air descends at the poles, traveling more or less horizontally in between, poleward above the clouds and equatorward below — is called a Hadley cell, and on Venus these cells exist in a much simpler state than they do on Earth.

Finally, a surprising discovery resulted from the large probe's neutral mass spectrometer. Its data indicated that two isotopes of argon, argon 36 and argon 40, are present in Venus' atmosphere in equal amounts, whereas on Earth argon 40 is four hundred times more abundant than argon 36 is. This unexpected presence on Venus of argon 36 — relatively rare on Earth and Mars — might lead to a total revision of theories about the formation of the planets.

Context

The Pioneer Venus Multiprobe was an important part of the massive exploration of Venus by American and Soviet spacecraft in December, 1978. The multiprobe's success was particularly important for the United States, which had never before attempted to penetrate the clouds of Venus. Three Mariner spacecraft had reconnoitered the planet on flyby missions in 1962, 1967, and 1974, but it had been the Soviet Union that was the most significant explorer of Venus's atmosphere and surface. Following the missions of Pioneer Venus and of Veneras 11 and 12, Soviet and American scientists formally exchanged data and held several meetings at which results and interpretations were discussed. The Joint U.S./U.S.S.R. Working Group on Near-Earth Space, the Moon, and Planets fostered these cooperative efforts. These meetings ushered in a period of scientific détente between the two space superpowers, at least with regard to the exploration of Venus.

Soviet and American scientists agreed that the multiprobe mission resulted in a major increase in knowledge about Venus's gaseous environment. Unfortunately, space missions return ambiguous, even erroneous, data along with the unambiguous data and results. Thus, in addition to key scientific questions about Venus that are now answered, there remain several that are not. For example, the composition of many of the particles in Venusian clouds remains to be determined. More information about lightning on Venus is necessary before scientists can speak with certainty about its origin. Why a westward superrotation is the dominant circulation of the Venusian atmosphere remains a great mystery. Another intriguing quesion left unresolved by the multiprobe mission is the reason for the lack of water in Venus's atmosphere. Some scientists think that the high deuterium-to-hydrogen ratio in the water of Venus's atmosphere is an important clue toward solving this puzzle; other scientists are more skeptical.

The multiprobe mission was like much scientific research in that answering one question often raises many others. New mysteries of Venus were created by the multiprobe mission, and new space missions will be needed to resolve them. Some of this further exploration has occurred. For example, Veneras 13, 14, 15, and 16 built on the foundation established by Pioneer Venus 1 and 2. The multiprobe's success demonstrated the practicality of focused interplanetary missions, and it contributed to the success of such missions as the Voyager explorations of Saturn and Jupiter. Pioneer Venus became a model for exploring the surfaces and atmospheres of the other planets of the solar system.

Bibliography

Beatty, J. Kelly, et al., eds. *The New Solar System*. Cambridge, Mass.: Sky Publishing, 1982. Summarized in this book are the most important results from the first two decades of space exploration. Helpfully illustrated with photographs, paintings, and diagrams, this book is intended for the general reader. The editors have chosen authors who communicate their expertise and insights clearly. The Pioneer Venus missions are treated in the chapters on the surfaces and atmospheres of the terrestrial planets. The introduction is by Carl Sagan.

Burgess, Eric. *Venus: An Errant Twin*. New York: Columbia University Press, 1985. Burgess, a research scientist and journalist, presents a survey of modern knowledge about Venus for the nonscientist. His book emphasizes the modern explorations of Venus by spacecraft, but he also discusses previous attempts to understand the veiled planet by telescope and radar. He presents a knowledgeable discussion of the Pioneer multiprobe in his chapter on Pioneer Venus. His well-illustrated book also contains an adept examination of various theories about Venus's origin and evolution.

Fimmel, Richard O., Lawrence Colin, and Eric Burgess. *Pioneer Venus*. NASA SP-461. Washington, D.C.: Government Printing Office, 1983. This handsomely illustrated book was prepared at the Ames Research Center with the help of many of the scientists who were directly involved in the mission, as were Fimmel and Colin. The authors place the Pioneer Venus missions in context with previous explorations of the planet. There are also excellent chapters on the history of Pioneer Venus, the development of the orbiter and the multiprobe, the actual mission to the planet, and an analysis of the scientific results.

Glass, Billy P. *Introduction to Planetary Geology*. New York: Cambridge University Press, 1982. Intended for college students, this textbook presents modern scientists' understanding of the planets in the light of the great discoveries made through the American and Soviet space programs. Since the author keeps scientific discussions free of advanced mathematics and physics, the book is accessible to general audiences. The chapter on Mercury and Venus presents analyses of the discoveries of the Pioneer Venus missions in terms of contemporary theories of the solar system.

Hunten, D. M., et al., eds. *Venus*. Tucson: University of Arizona Press, 1983. Sixty-five authors have collaborated with the editors in analyzing the new data presented to scientists by the many spacecraft explorations of Venus. Topics discussed include the interior, surface, and atmosphere of Venus. Most of the authors are drawn from the Pioneer Venus and Venera scientific communities. The thirty chapters are almost evenly divided between major surveys and detailed technical discussions. The papers are all in English (Russian papers have been translated), but they require a knowledge of advanced mathematics, physics, and chemistry for a complete understanding.

Jones, Barrie William. *The Solar System*. Elmsford, N.Y.: Pergamon Press, 1984. A survey of contemporary knowledge about the solar system. The author presupposes no previous knowledge of planetary astronomy on the part of the reader. The chapter on Venus uses the Pioneer Venus data to discuss the interior, surface, and atmosphere of Venus. There is also a good discussion of why so little water exists in the Venusian atmosphere.

Muenger, Elizabeth A. *Searching the Horizon: A History of Ames Research Center, 1940 – 1976*. NASA SP-4304. Washington, D.C.: Government Printing Office, 1985. Muenger wrote this book for NASA's internal history program. Ames Research Center (ARC) has made many important contributions to several space missions for NASA and was intimately involved with the Pioneer program from Pioneer 6 onward. Although she briefly mentions the Pioneer Venus missions, her focus is on ARC's technological achievements in the context of its managerial development. This emphasis helps the reader to grasp the early history of the Pioneer program, which in turn makes the Pioneer Venus mission more comprehensible in terms of NASA's institutional evolution.

Robert J. Paradowski

PLUTO FLYBY MISSIONS

Date: Beginning early 2000's
Type of mission: Unmanned Pluto probe

The Pluto flyby missions, consisting of two probes to Pluto and Charon, will study the most distant, previously unprobed, and anomalous planet-moon pair, providing insights into the early solar system's development.

PRINCIPAL PERSONAGES

> ROB STAEHLE, preproject manager
> RICH TERRILE, preproject scientist
> LEON ALKALAI, preproject technologist
> STEVE MATOUSEK, Mission Design
> HOPPY PRICE, Flight System Design
> JOHN CARRAWAY, Mission Operations
> BRUCE CROW, Deep Space Net Operations
> BILL HUBBARD, Outer Planets Science
> Working Group chair
> JONATHAN LUNINE, Science Definition Team chair
> VIACHESLAV LINKIN, Pluto Science Definition Team
> ARNIE RUSKIN, Planning and Analysis
> KEVIN CLARK, Product Assurance

Summary of the Mission

The Pluto Express (formerly named the Pluto Fast Flyby mission) will constitute the first attempt to study Pluto and its moon Charon from within these bodies' proximity. Still in the planning stages, the mission must take many factors into consideration before launch.

One consideration in planning the mission is timing, which is critical. Many different launch dates have been suggested and planned, and just as many have passed. The earliest plan, employing the Mariner Mark 2 model, was set for launch in 1988. Another possible launch date was when Pluto was at perihelion (closest to the Sun) on September 29, 1989; the planet's proximity to Earth would have reduced costs and made the trip shorter. However, distance is but one of many criteria used to determine optimal launch dates. The "swingby effect"—the use of a planet's gravitational energy to "slingshot" a spacecraft farther into space—would reduce flight time to Pluto to below half; hence, another optimal launch date would be during a period when the planets are positioned to allow the spacecraft to take maximum advantage of this swingby effect. In one model, the planet Jupiter has been named as the "big benefactor"— the planet that could provide the greatest gravitational energy to boost the probe toward Pluto. Shortly before 1990, such an opportunity existed and was lost. The next time Jupiter will be optimally positioned is in 2001, which suggests that the program should be engaged with some urgency. At the same time, however, Pluto's slowing recessional velocity, combined with advances in technology with each year of delay, could argue for a later launch date. Alternate, possibly more complex trajectories and exploration agendas are possible. When considering the best launch date, therefore, planners must weigh the pros and cons of all these factors.

Another characteristic of a mission plan is its hardware checklist, which determines the spacecraft's weight, the amount of fuel required for the trip, and ultimately the cash and other resources needed to realize the program. In the Mariner Mark 2 model, the main spacecraft alone weighed 250 kilograms, with a payload of four flyby components weighing 85 kilograms. Adding the atmospheric probe and other equiprment, the total payload weighed 555 kilograms. Under tighter budgetary constraints, these loads and structures were reduced. In a later plan, two spacecraft are slated to be sent on the mission, a single year separating their independent final approaches toward Pluto. Each was planned to weigh 150 kilograms, have a hexagonal shape, be constructed of aluminum, and carry a high-gain antenna contributing approximately two meters to the probe's size. These size and weight dimensions still exceed budgetary constraints and therefore are being reduced in hopes of getting the project approved by the federal government.

Currently, the Pluto Express is planned to carry several hardware components. First is the Central Processing Unit (CPU) for the spacecraft's computer, which operates at 1.5 million instructions per second, its main chip using the modern, efficient Reduced-Instruction-Set-Computer (RISC) architecture. To support the processing of data, more than 400 megabits of memory will be available for the CPU to use. It can process 5 megabits of data per second, although data transmission to Earth can bottleneck this processing speed somewhat.

The spacecraft will gather information by means of two on-board spectrometers (both infrared and ultraviolet); a visible-range "camera" (applying charge-coupled device technology); and an occultation testing device. Radioisotopic

Thermal Generators (RTGs) will power the internal functions of the devices on this mission with 65 watts. RTG is a fundamental, standard technology that was applied in projects such as Galileo, launched in 1989.

Once a design has been approved and a launch date has been selected, the probe will be boosted into Earth orbit by either a Delta or a Russian Molniya booster. There, it will follow a curved path around the planet to achieve acceleration. The trajectory must be precise, for without achieving the best path, the spacecraft will be unable to take full advantage of Jupiter's reserves of gravitational energy, a need that will be magnified by the spacecraft's minimal propellant storage. After swinging by Jupiter, the probe will proceed rapidly toward Pluto.

At the point when the Pluto Express reaches its destination, a period of several years will have passed since the launch. Pluto and its moon Charon rotate approximately every six days. Both vehicles will enter the planetary system either within a few days of each other or, in another plan, within a year of each other. The spacecraft will first do a "flyby" of the planet, at a speed of roughly 20 kilometers per second. During both the approach and the flyby, the spacecraft will take sequences of photographs of Pluto and Charon using the 1024 rectangular grid model of the photosensitive circuitry in the camera. These pictures will be transmitted to Earth and should become clearer and display crisper planetary surface characteristics as the spacecraft enter flyby. Pluto and Charon have an interesting orbital relationship, one of relatively great size equity and proximity. The same part of Pluto faces the same part of Charon all the time, resulting in some distortion of the shapes of both bodies. To photograph all the surface area, therefore, the flyby trajectories must be carefully computed.

As the spacecraft moves nearer to Pluto, the atmospheric probe will disengage from the main craft, move along a path to Pluto, and continuously transmit data to the vehicle as it enters the atmosphere and hits the surface. This is a onetime entry and leaves little room for error. Despite a slight delay, the images observed in descent will be retransmitted to Earth, synchronized via computers.

The main spacecraft, in the meantime, will approach closer than 20,000 kilometers. High-resolution monochrome photographs of the planetary and lunar surfaces will be taken. Color images will also be recorded, although their resolution will be lower. (Color equipment must represent and process more information per "image-point," or pixel, and thus constraints in time, circuitry, and amount of information result in the lower resolution.) However, color images are useful in determining chemical composition and other characteristics of the terrain. Therefore, on the flyby craft, the infrared spectrometer and the ultraviolet spectrometer will cover their respective ranges of the spectrum and transmit additional data that will help in determining chemical composition.

While the foregoing itinerary describes the general mission as envisioned by many different plans, the actual mission may prove even more interesting. As Pluto Express mission plans continue to evolve, another, more advanced and technologically ambitious, plan is developing in its shadow, the Kuiper Express. The Kuiper Belt is an expanse of space outside Neptune's orbit. It is the realm of ice-rock bodies and comets and extends from 30 to 100 astronomical units away from the Sun (one astronomical unit equals the distance from Earth to the Sun, approximately 93 million miles or 150 million kilometers). The Kuiper Express will continue in the vicinity of Pluto but go beyond it both in mission scope and in distance, exploring the even more mysterious region for which it is named.

If adopted, the Kuiper Express will leap forward after its work is done at Pluto, using an alternate rocket model that is powered by xenon ion engines. This model is even less expensive and more elegant than the technology currently planned for the Pluto Express. Instead of using the RTG power supply, the Kuiper Express would use solar panels to collect energy, resulting in further weight reduction. While solar panels may not be as effective in the Kuiper Belt, where the Sun appears as not much more than a very bright star, solutions may be forthcoming. It is possible, for example, to envision a "local train" going from one ice body to the next: The need for extra power may be mitigated by the low masses of these ice-rock bodies, whose gravitational pulls are not as great as that of Pluto or Charon.

Knowledge Sought

In 1930, Pluto was discovered by Clyde Tombaugh. While only a tiny speck on photographs amid many stars of equal or lower magnitude, Pluto was detected because repeated photographs showed its motion across a background of stationary stars. Given its remote position and tiny size, it is still a cloaked enigma. For decades, few characteristics of this planet, except for its orbit, were deduced and verified from extensive telescopic observation. Charon's existence was discovered in 1978 by James W. Christy, a specialist in binary stars, observing what appeared to be elongations, minute wobblings detected by comparing several images of Pluto. Most of what is known about Pluto and Charon was determined and confirmed in the 1980's and 1990's, as a result of advances in ground-telescope design and their application in space. Another major breakthrough came with a powerful observational tool, the Hubble Space Telescope. In 1994, it provided an exceptionally clear picture of the Pluto-Charon system as two exactly defined "spherical" bodies.

Pluto is covered with methane ice and a tenuous methane atmosphere that consists of gases that cycle between gaseous and frozen states. Carbon monoxide, nitrogen, and other components are there, frozen as well. Water ice is below the surface, forming the mantle of Pluto. Rocks are present at the planet's core. Crater structures riddle the landscape. Pluto's moon Charon is large relative to Pluto when compared with other planets. They are also much closer together than any

Artist's rendering of the Pluto Flyby spacecraft, also known as the Pluto Express, in the vicinity of Pluto and its moon, Charon, in the background. (NASA/JPL)

other planet-moon pair in the solar system. Tidal effects are more pronounced. A warped bulge effect is caused by the intense mutual attraction of gravity. Consequently, only one hemisphere of Pluto faces Charon, and vice versa.

The high-resolution photographic images of surface and landscape features returned by the Pluto missions will prove invaluable to understanding the geomorphology and geology of Pluto and Charon. Images presently available from the Hubble Space Telescope do not offer the detailed resolution that will be available by means of probe and flyby. The statistical distributions of rocks and ices along surface structures will become possible. The nature of craters and their analysis will provide insight into these bodies' previous encounters.

Another goal is to determine the composition, density, and permanence of Pluto's tenuous atmosphere at several altitudes above the surface terrain. A surface landing probe would provide this cross-section perspective. Presently, our long-range spectroscopy has provided only a rough approximation of the blend.

The chemical composition of Pluto's surface is an independent issue. It is hypothesized that, as Pluto recedes, the Plutonian atmosphere may refreeze (within approximately thirty years of perihelion), another reason for urgency. If the probe can pass prior to the freezing process, other thin frozen gas layers, such as carbon dioxide ice and water ice, may be visible. The polar features may also be studied. Pluto's observation from close proximity may provide evidence of the pattern of flow in the refrozen surface. Spectrographic analysis of the polar regions and those areas leading to them may prove useful in analysis, possibly showing different compositions at different points along paths.

If joined or replaced by the Kuiper Express, by-products of the mission would include the first in-depth study of the Kuiper Belt. Kuiper Belt objects are relics of the initial nebula that evolved into our solar system. The Hubble Space Telescope has identified many Kuiper Belt objects, possibly more than thirty-five thousand with radii exceeding 50 kilometers. Ice, rocks, organic compounds, and other constituents have been detected in the Kuiper Belt. Some believe that Pluto and Charon themselves are large Kuiper Belt entities. The Kuiper Express's collection of data on the spatial distribution of Kuiper Belt objects would lead to a model of the structure and dynamics of the Belt (or possibly many distinct belts). Closer investigation would resolve questions of chemical composition. More important, the Kuiper Express might provide statistical data on Kuiper objects' distribution in the extended model of our solar system. By analyzing these data and observing patterns, scientists would gain insight into the origin of the planets.

Advances in technologies are expected to be another great by-product of this mission. The Kuiper Express is one example. The new cost efficiencies achieved will benefit all future expeditions, making it possible to explore the exterior region of the solar system.

Context

While exploring Pluto and its moon Charon is the primary goal of the Pluto Express, this mission represents a concentration of efforts directed toward advances in economy and "size of" technologies. In the 1990's, Congress began defunding the sciences with serious cutbacks, and missions have been forced to become more cost-effective. Whereas the Voyager missions cost a billion dollars each in the 1970's and 1980's, the original Pluto Fast Flyby plan was rejected as too expensive, even at half a billion dollars. Extensive hardware integration efforts resulted in the emergence of the Pluto Express, which proposed to use smaller probes with a correspondingly smaller bill of $300 million.

This mission is intended to serve as a paragon for the next century's explorations by reducing costs through the application of technological advances, alternate approaches to management, and innovative operational, design, and developmental methodologies. Recycling of components from one mission to the next will constitute another cost-saving method. While the design concepts developed for the Voyagers and previous missions are applicable in principle, the Pluto missions will require higher levels of compression (machine, component, and circuit size and integration) to

deliver more function in a smaller, lighter, and less expensive package. Advances in material science are one source of these reductions and refinements.

Another important measure of technology is reliability. From a remote site, problem resolution is possible only via computer and machine. Any subsystem failures must be repaired without direct human contact or must not be critical to the mission. Simulations and testing on models help in detecting weak points. With the hard tradeoff between reliability and cost reduction, one solution is redundancy of large-scale components. To minimize risk and cost, for example, the model of the Pluto missions that includes two independent crafts was designed so that, if one fails, it will not be necessary to wait decades for an optimal launch date because a second probe must be constructed from scratch.

Cooperation and funding from the private sector and industries is expected to play a greater role. Stronger ties with educational institutions are developing as well. College students are getting involved in projects such as the Art Center College of Design's partnership with the Jet Propulsion Laboratory in Pasadena, California; other students are gaining access to ground telescopes such as the one at Mount Wilson Observatory in Southern California.

Greater international cooperation is also being encouraged in the space program, in contrast to the competitiveness of the Cold War era. In the Pluto missions, subsystems such as the atmospheric probe will be Russian; launch vehicles will be both U.S. (Delta) and Russian (Molniya). Germany may also join in this mission.

Perhaps a century from now, interstellar travel will be possible. The Voyager and Pluto probes may be recovered as artifacts. Progress is measured in small steps, and the Pluto missions target the last region of our solar system. Rather than Pluto and the Kuiper Belt being the end of our solar system, they may be the gateway to the Oort Cloud. This next step in exploration requires new technologies and solutions; radically different methods of propulsion and space travel are essential if space exploration is to continue beyond the solar system. The Pluto missions could provide the opportunity for beginning the development of those new technologies.

Bibliography

Asker, James R. "Pluto Fast Flyby Slated for 2006." *Aviation Week & Space Technology* 138, no. 7 (February 15, 1993): 46–51. Overview of the Pluto Fast Flyby with emphasis on the vehicle design and craft technology as it was formulated up to 1992. Comparisons of the different crafts of previous missions (with illustrations) with the Pluto mission plans demonstrate the reductions in size and cost.

Burgess, Eric. *Uranus and Neptune.* New York: Columbia University Press, 1988. While several monographs exist on other planets, there are relatively few on Pluto at the level of those texts. This study of Uranus and Neptune devotes one chapter to Pluto and the search for Planet X. Gives a scientific overview of the outer planets, their orbits, moons, and other characteristics. Illustrated with photographs from the Voyager missions.

Dyson, Freeman J. "21st Century: Pluto Fast Fly By and the Kuiper Express." *Scientific American* 273 (September 1995): 114–116. Excellent assessment of the space program's Pluto projects, cost reductions, and the nature of future explorations to Pluto and beyond. Concisely contrasts the different plans that have emerged for exploring Pluto and provides an exposition of the technologies to achieve those goals. Inspiringly focused on future progress and scientific possibilities.

Littmann, Mark. *Planets Beyond: Discovering the Outer Solar System.* New York: John Wiley & Sons, 1988. Compelling coverage of the outer planets (Uranus, Neptune, and Pluto), including the historical and personal side of their discovery, observation, and exploration by the Voyager probes. The author captures the drama and excitement of the process of science, not just facts. Coupled with his inspiring perpective are abundant illustrations and plates.

Miller, Ron, and William K. Hartmann. *The Grand Tour: A Traveller's Guide to the Solar System.* New York: Workman Publishing, 1981. An aesthetically pleasing review of the planets and moons in the solar system. Artists' renditions are commonplace but motivating. It devotes only a few pages to Pluto based on its relative "importance" and amount of available information.

Tombaugh, Clyde W., and Patrick Moore. *Out of the Darkness: The Planet Pluto.* Harrisburg, Pa.: Stackpole Books, 1980. Good background material, readably presented by the original discoverer of Pluto. A foreword by James W. Christy, the discoverer of Charon, complements this exposition. Strong coverage of the historical aspects of Pluto's observation.

John Panos Najarian

PRIVATE INDUSTRY
AND SPACE EXPLORATION

Date: Beginning January 1, 1955
Type of issue: Private industry and space

Since the mid-1950's, utilization of launch vehicles, artificial satellites, and space shuttles by private companies that produce or consume space-related technologies and services has grown to multibillion-dollar proportions, creating new challenges as well as opportunities.

PRINCIPAL PERSONAGES
    RONALD REAGAN, the fortieth President
        of the United States
    ROBERT TRUAX, an initial developer of
        private launch vehicles
    GARY HUDSON, a pioneer of private launch vehicles
    HOWARD HUGHES, the corporate initiator of the
        Syncom satellite series
    JAMES ABRAHAMSON and
    JEROME D. ROSENBERG, directors of NASA
    JOSEPH CHARTYK, a president of
        Communications Satellite Corporation
    LEW ALLEN, Director of the Jet Propulsion Laboratory
    GERARD K. O'NEILL, head of Geostar Corporation

Summary of the Issue

After the mid-1950's, with the development of increasingly sophisticated rocketry, missiles, space shuttles, and space laboratories — and their support technologies — it became clear that space afforded economic opportunities for individual and corporate private enterprises.

Commercial and industrial interest in space exploitation have been affected by shifts in the national policy-making climate. Ever since the eighteen-month International Geophysical Year (July, 1957, through December, 1958), heavy emphasis has been placed on peaceful, cooperative use of space, and every American president since Dwight D. Eisenhower has reaffirmed those principles; nevertheless, by the 1960's space had become one more arena for U.S.-Soviet competition.

For percipient business interests, however, there were positive signs. Created in 1958, the National Aeronautics and Space Administration (NASA) was charged with maintaining U.S. superiority in space science and technology and with contributing to the peaceful exploitation of space. By 1962, with its communications satellite program under way, NASA was asking private enterprises to enter with it into joint pub-

lic/private ventures. For private interests this meant a sharing of investment, a relative abundance of money, and a reduction of risk. There was general recognition, too, that the United States enjoyed the technological edge in space, despite the Soviet Union's launching of Sputnik, the first satellite, in 1957. It had more experience; its launch vehicles lifted the heaviest payloads; not only did it possess the world's largest and richest scientific establishment, but it also was then the leader in computerization and communications; and until the mid-1980's, the aerospace-defense corporations — Litton, EG&G, and Rohr Industries, among others — enjoyed close working relationships with NASA and the Department of Defense (DOD) and also were comfortably placed in the most profitable sector of the U.S. economy.

Added incentives arose from the entrepreneurial sense of adventure, the desire to serve the national interest, the prospect of making hundreds of millions or even billions of dollars, and the awareness that competition elsewhere was building rapidly. By the early 1970's, Western European nations, realizing aspirations that they had harbored for twenty years, had launched the European Space Agency (ESA); Japan, China, India, and Brazil were advancing claims to the limited resources of inner space; and the Soviet challenge, whose strength could not be accurately gauged, continued.

NASA's public/private ventures were effective in encouraging private enterprise in space. By the early 1980's, Ford Aerospace, McDonnell Douglas, American Telephone and Telegraph (AT&T), Martin Marietta, Boeing, and forty other major corporations — some well-known, such as General Electric and General Dynamics, and some not, such as Loral, Rockwell International, Morton Thiokol, and United Technologies — had moved partially under NASA's umbrella. NASA, with annual budgets exceeding $5 billion and with more than twenty thousand employees, had created an economic and scientific imperium of its own. By 1983, it included thirteen flight, space, and research centers (the Goddard Space Flight Center, the Kennedy Space Center, the Johnson Space Center, and the Ames Research Center, for example); the Jet Propulsion Laboratory in Pasadena, California, run for NASA by the California Institute of Technology; the Slidell Computer Complex; and the Landsat (land satellite) and Seasat (sea satellite) programs.

Changes in this NASA conglomerate began in 1983. First,

James Abrahamson, NASA's director, decided to begin using space shuttles to launch payloads instead of using the Titan, Atlas-Centaur, and Delta missiles, known generically as Expendable Launch Vehicles (ELVs). Giants of aerospace such as Martin Marietta, McDonnell Douglas, and General Dynamics, which produced these missiles, were outraged, and they pointed to the missiles' excellent performance record; their protests, however, were to no avail. This initial step toward opening the space transportation business to competition was followed in May, 1983, by President Ronald Reagan's executive decision to sell NASA's spare rocket parts and lease its launchpads to private enterprises. The field of space transportation, even in 1983 a billion-dollar market, thus was rendered more competitive.

Almost immediately, more than twenty transportation enterprises began to produce ELVs, minishuttles, or shuttles, among them Starstruck, Pacific American Launch Services (PALS), Space Services Incorporated of America (SSIA), and Earth Space Transportation System. Many, like Robert Truax's Truax Engineering, had been trying since the 1960's to develop space products — in Truax's case, a "Volksrocket." Gary Hudson, one of the pioneers of space launchers and a founder of PALS, had been looking for opportunities since the early 1970's. SSIA, a Houston firm, swiftly raised $6 million from sixty investors, bought equipment from NASA, and soon succeeded in launching its Conestoga successfully and joining with Space America in exploring the economic potential of the remote sensors that have been part of many satellite missions. Investors William and Klaus Heiss, heads of Space Transportation Company, were able to offer NASA a one-billion-dollar bid to buy a space orbiter from NASA and lease it back to the agency if NASA would allow STC to market the shuttle fleet to private firms. A doubling of this bid by a company seeking to buy a fifth shuttle from NASA was indicative of the competition to tap the ostensibly lucrative space transportation market.

Space-based communications were placed on a business footing somewhat earlier than space transportation. A Christmas message from President Dwight D. Eisenhower was broadcast in 1958 from a U.S. military satellite, and two years later NASA's Echo 1 balloon relayed signals over intercontinental distances. AT&T soon persuaded NASA to launch its Telstar 1 satellite, the first active satellite with a transmitter, so that it might be tested and so that the United States could be assured of superiority in space applications. Telstar carried the first live transatlantic television broadcast, underscoring its immense commercial potential. Since the Telstars were limited by being line-of-sight transmitters, Hughes Aircraft developed its Syncoms (Synchronous Communications Satellites), which provided immediate and inclusive communications. The development of more powerful rocketry by 1964 allowed the placing of communications satellites in high orbit 35,680 kilometers above the equator, where they could move with the speed of Earth (in a geostationary position) and provide con-tinuous worldwide transmissions. Meanwhile, the Communications Satellite Act of 1963 made Communications Satellite Corporation (COMSAT) the first private enterprise in space, with NASA providing launch services and funds for research and development but obedient to a largely private board of directors that sold stock publicly. Under American auspices, the internationalization of space communications, directly and indirectly involving more than one hundred American firms alone, produced an economic phenomenon: By 1988, 7.2 percent of the world's communications assets were space-based — $13 billion of a $165 billion market.

Domestically, the Federal Communications Commission (FCC) authorized use of the first communications satellites, thereby creating business opportunities independent of federal funding. Satellites were used to provide long-distance phone service and, later, live and cable television. By 1988, more than eighty satellites were in service and the FCC had been licensing as many as twenty new satellite systems annually. Hughes Aircraft, Ford Aerospace, RCA, TRW, and General Electric were major contractors in the production of communications satellites. Moreover, the complexity of the satellites necessitated subcontracting to hundreds of firms, not counting subcontracts for launch services, ground control facilities, production of transponders, and production of other specialized equipment for data transmission.

Technology underlying the profitable deployment of space communications satellites was essentially the same as technology for the development of remote-sensing satellites. These are satellites that monitor Earth from remote points in space, either by photographing objects or by using infrared, radar, radio-wave, or multispectral scanners. Since 1960, the United States' Television Infrared Observations Satellite (TIROS) spacecraft have been photographing Earth. NASA's Landsat and Seasat series of satellites have provided immense quantities of data concerning crop yields, the state of forests, fish populations, weather prediction, the availability of water, ocean currents, air pollution, water quality, topography, and the locations of mineral resources. Although some of these data are in the public domain, others have come close to commercial viability. Of Landsat's $2.94 million in revenues in the mid-1980's, for example, 38 percent came from American academic and industrial users. Still, NASA was not expected to break even on Landsat until 1989, and, as a consequence, General Electric Space Systems, RCA, Bendix, Lockheed, and Computer Science Corporation were pressing NASA to privatize the series.

Because NASA can manage cargoes on its space shuttle ranging in size from 90 to 27,000 kilograms, its Materials Processing Space Division has been selling cargo space since 1978. To stimulate sales, it started the Get-Away Special Experiments, which in only a few years brought more than three hundred users into the program, among them Coors Brewing Company, Dow Chemical, Dupont, *Columbia Pictures*, Corning Ware, and Ford Motor Company.

To market space-manufactured biological products, which include pharmaceuticals — which alone are expected to constitute a twenty-billion-dollar market — NASA has entered into an agreement with Johnson and Johnson and McDonnell Douglas for exclusive use of its continuous flow electrophoresis (CFE) process. The unique gravity-free environment of space permits CFE, and variations of the process, to produce purer products. The Johnson and Johnson CFE experiments in the mid-1980's, moreover, yielded quantities five hundred times greater than would have been possible on Earth. Similarly, a Florida firm, Microgravity Research Associates (MGR), has contracted with NASA for the installation of a 13,500-kilogram factory aboard a space shuttle to grow crystals. Gerard K. O'Neill, an advocate of solar power satellites, estimates that satellite conversion of sunlight to supply world energy needs represents a sixty-billion-dollar market.

Such private industrial demands and visions prompted President Ronald Reagan in 1984 to direct NASA to have a shuttle orbiting within a decade and to press plans for a U.S. space laboratory. Meanwhile, between 1984 and 1988, the United States orbited sixty-nine satellites, nearly all of them engaged in commerce-related missions. By the mid-1980's, the predominantly private aerospace industries and related companies represented the nation's leading growth industry. McDonnell Douglas Astronautics reckoned that commercial uses of space and of facilities in space offered markets of $4 billion to $6 billion for reusable orbital transportation; $10 to $20 billion for communications; $1 billion for remote sensing; $1 billion for leasing space platforms; $2 to $3 billion for assembly, deployment, retrieval, servicing, and repair of space vehicles; and $20 to $40 billion for materials processing and manufacturing in space. More than 370 private companies, employing about 1,350,000 people, were seeking to exploit these market opportunities by 1988.

Knowledge Gained

The private, or semiprivate, commercialization of space was largely a dream in the 1950's. In the next decade, under the auspices of NASA, communications satellites were being orbited and the inner space frontier was being exploited, the technological lead firmly in American hands. The late 1980's found a broadened aerospace market for an increasingly wide range of ventures, products, facilities, and services. For private business, the period from the 1950's to the 1980's involved much trial and error, ingenious negotiations between private companies and government, and the continuous accumulation of an immense body of hard-earned experience — technological, financial, political, and legal. Few directly applicable precedents existed in any of these areas to which private enterprises could turn for guidance. Astrobusiness was new and risky. A failed launch vehicle, an unsatisfactory orbit, or malfunctioning transponders or solar energy cells spelled disaster: years of planning, and often tens of millions of dollars, lost.

Since American private enterprises were going to space

for profit, they first learned important lessons about the financial characteristics of their novel undertakings. Commonly requiring between thirty and sixty million dollars for development, and sometimes several hundred million dollars, space ventures are capital-intensive. Pioneering firms may enjoy some protection from competition, but those that develop later face lesser financial rewards. Moreover, space investments are long-term propositions, and they are perceived by both institutional and individual investors as entailing very high risks. Investors cannot be certain that the sophisticated, often experimental technology will work and that market conditions will be favorable, and they cannot know how they may be affected by domestic and international political changes. The almost exclusive dependence on high technology has increased the risks, and many potential investors or entrepreneurs have chosen to remain outside aerospace ventures until more years of experience have been acquired. Those who have ventured, however, have been able to draw on an outstanding pool of talent, and management quality is now viewed as critical to success.

NASA is a repository of immense quantities of data, information, experience, and intelligence essential to novel enterprises. As it was intended to do, NASA has directly and indirectly shared expenditures the better to curtail risks and encourage space ventures. Similarly, risks have been shared by groups of private corporations and by private companies and the federal government. COMSAT, a company formed in 1963 under the Communications Satellite Act, represents this type of hybridization; that is, it was created by the federal government, but it functions as a private, profit-making company. Indeed, when the Federal Communications Commission sought in 1980 to regulate COMSAT's soaring profits, COMSAT, though formed with a monopoly on U.S. satellite communications with other countries, promptly joined with RCA, AT&T, IBM, and Xerox, among others, to enter the domestic communications satellite business.

Other means to reduce risks have been discovered. In the early days of space ventures only one major American insurance company, Associated Aviation Underwriters, would insure spacecraft. By 1972, however, RCA was able to purchase insurance for its Satcom satellite, and the way was paved for a sharing of losses in the $200 million range. A recognition of the benefits of risk sharing, which occurred early with communications satellites under Intelsat (International Telecommunications Satellite Corporation), dawned on other U.S. enterprises, so that many private firms began sharing financing, design, construction, allocation of payloads, and losses and profits with foreign companies and governments, as was the case with Skylab, use of the space shuttles, and the launch vehicles of the ESA.

Context

With the federal government's use of NASA as a handmaiden to U.S. astrobusiness, two overriding precepts have established a basic framework for American enterprise in space. The first precept holds that the United States must maintain world leadership in the peaceful and military use of space. The second precept is that inner and outer space be opened as widely as possible to U.S. private enterprise. Seven space age presidencies, fluctuating domestic economic conditions, a changing international scene, and startling technological advances have affected the application of these precepts to the United States' private space efforts.

The word "private" invites qualification. Without U.S. military experience with rocket launch vehicles during the decade after World War II and, later, the military's classified communications, navigation, weather, and spy satellites, it is doubtful that much private investment in space activities would have been forthcoming. Public moneys originally paved and paid the way for the costly experimentation essential to opening private prospects in space. Just as had been the case with Columbus's voyages, with the founding of many of the American colonies, and with the construction of the nation's great railway system in the nineteenth century, stimulus, seed money, and security came from government. In return, space enterprises have added enormously to the economy at large. Initial investments from the public sector have been more than repaid by the generation of tax revenues, ideas, and technology, and most of the original publicly borne risks were perceived to have been undertaken in the national interest.

Although every national administration has been eager to broaden private participation in space programs, the Reagan Administration in particular vigorously sought to transfer an increasing measure of responsibility for them to the private sector. It continued leasing ELVs, launch sites, and NASA equipment to private firms. It encouraged private companies to undertake the construction of launch vehicles and space shuttles. It took the position that any well-conceived space project can be privately financed without government subsidies, guaranteed loans, or any other kinds of direct or indirect investment. Nevertheless, the aerospace industry still considered certain governmental support desirable, even essential, for fledgling projects. Consequently, many aerospace companies called for government to help underwrite demonstration projects, to provide cheap payload integration services, and to provide free, or very low-cost, shuttle flights. They wanted speedier access to government information on new technologies, the removal of what they believed to be discriminatory taxation, and eligibility for tax credits for investment and research. They also wanted government to assist them with legislation or regulations making it easy for them — as it was for the aircraft industry — to compete or, as occasion demands, cooperate with foreign interests in the development of space stations.

Although the federal government has consistently espoused privatization, the stronger push in this direction has come from the aerospace industries. Generally, they have supported legislative efforts to deregulate space telecommunications and to place licensing and regulation of private launchings in the hands of the Department of Commerce or the Department of Transportation. Such views imply that NASA should function principally as a research and development resource that would ensure the United States' primacy in space. Some have advocated that NASA concentrate on long-term space projects, such as development of permanent space research, mining, and manufacturing stations; placement of large space structures such as space lasers and antennae-farm satellites in orbit; and establishment of a lunar colony that would mine lunar ores and serve as a scientific outpost.

Bibliography

Aviation Week and Space Technology 120 (June 25, 1984). This entire volume is devoted to prospects for the commercialization of space. The articles are readily understandable by the layperson. A few footnotes are included with the articles, but there are no appendices and no index. Despite the lack of these aids, the articles are germane and worth reading.

Bainum, Peter M., and Friedrich von Bun, eds. *Europe/United States Space Activities with a Space Propulsion Supplement: 23d Goddard Memorial Symposium, 19th European Space Symposium.* San Diego: American Astronautical Society, 1985. These Goddard symposia draw upon experts in many space-related fields. Their papers are intended for intelligent, but not necessarily specialist, audiences. Twelve of the twenty papers presented relate to this topic. Profusely illustrated with drawings, charts, graphs, and photographs. Reference endnotes are appended to papers, but there is no overall bibliography. Author index only.

Finch, Edward Riley, Jr., and Amanda Lee Moore. *Astrobusiness: A Guide to the Commerce and Law of Outer Space.* New York: Praeger, 1984. A concise and readable review of the development and future of American space commerce. Only twenty pages deal with national and international space law; attention is focused on a condensed overview of space commerce. The text is augmented by many pictures and schematics. There is no overall bibliography, but there are four very informative appendices and a useful, if minimal, index.

Goldman, Nathan C. *Space Commerce: Free Enterprise on the High Frontier.* Cambridge, Mass.: Ballinger, 1985. Written in authoritative style for nonspecialist readers, this is a very useful review of American commercial space enterprises. It is both synthetic and specific. Charts, tables, and a few photographs augment the text. End-of-chapter notes partially compensate for lack of a bibliography. The six appendices are informative and the index is sufficient.

Gump, David, ed. *Space Processing, Products, and Profits, 1983–1990.* Arlington, Va.: Pasha, 1983. Contains specific descriptions and analyses of the commercial potentials offered by space's low-gravity, high-vacuum environment and the opportunities for electrophoretic biological separation, containerless liquid processing, and crystal manufacture for semiconductors and solar cells. Aimed at a narrow business audience, this is still an interesting, readable study. There are adequate notes, a bibliography, and an index.

McLucas, Charles, and Charles Sheffield, eds. *Commercial Operations in Space, 1980–2000: 18th Goddard Memorial Symposium*. San Diego: American Astronautical Society, 1981. The editors are representative of the oldest profitable private American enterprise in space: communications satellites. Each of the eighteen papers, or addresses, is authoritative and relevant; the papers cover topics such as manufacturing in space, space power systems and technology, Earth resources location, commercial launch operations, and international commercial opportunities. Papers include numerous charts, graphs, and photographs. There is no bibliography and only an inadequate index.

O'Neill, Gerard K. *The High Frontier*. New York: William Morrow, 1977. The bright, imaginative, and eminently readable author was head of Geostar and a member of the Space Studies Institute at Princeton University. He was especially interested in lunar and asteroid mining. The book is delightful and is intended for a general readership, despite O'Neill's scholarly credentials. There is an adequate bibliography and index.

Schwarz, Michiel, and Paul Stares, eds. *The Exploitation of Space: Policy Trends in the Military and Commercial Uses of Outer Space*. London: Butterworth and Co., 1986. Taken from a special issue of *Futures* that was devoted to space, these are eminently readable and highly informed articles by business or scholarly specialists, intended for nonspecialist readers. Tables, charts, and graphs abound throughout. Notes and references conclude each article. There is no index or overall bibliography, nor are there appendices. Nevertheless, it is an excellent series.

Clifton K. Yearley
Kerrie L. MacPherson

THE RANGER PROGRAM

Date: February 23, 1961, to March 24, 1965
Type of program: Unmanned lunar probes

The Ranger program, the first part of NASA's three-stage plan leading to the manned exploration of the Moon, provided some of the earliest detailed information about the lunar surface. The images sent back by Rangers 7, 8, and 9 helped scientists to determine landing sites for the Apollo missions.

PRINCIPAL PERSONAGES

> T. KEITH GLENNAN, NASA Administrator
> JAMES E. WEBB, his successor
> ABE SILVERSTEIN, Director of Space Flight
>   Programs, NASA
> HOMER E. NEWELL, Assistant Director of Space Flight
>   Programs, NASA
> WILLIAM H. PICKERING, Director of the
>   Jet Propulsion Laboratory
> JAMES D. BURKE and
> HARRIS SCHURMEIER, project managers
> GERARD P. KUIPER, Principal Investigator

Summary of the Program

The National Aeronautics and Space Administration (NASA) first considered the Ranger program as early as 1959; authorities had thought that the early lunar missions could act as an appropriate answer in part to the successful Earth-orbiting missions launched by the Soviet Union. Since the best Earth-based telescopes could achieve only a limited resolution, it became obvious that a reconnaissance of the lunar surface at much higher resolution was needed for the benefit of later lunar-landing missions. It was important, for example, to know whether the flat parts of the lunar surface, the so-called seas, were littered with rocks and other debris. Also, scientists needed to know the number of smaller craters, which were invisible to the telescope, and how extensively they covered the floor of the seas. Finally, there was the possibility that the lunar surface was covered with a dust layer, perhaps several meters deep; if the layer were present, it would present considerable dangers for a landing spacecraft.

The Ranger program was expected to help supply answers to these questions. The spacecraft, directed toward selected points of the lunar hemisphere, were not supposed to land softly on the Moon; before crashing, however, they would carry out various studies of the surface and transmit the results back to Earth. A particularly important part of the program involved taking television pictures of the surface with increasing resolution as the spacecraft approached the Moon. Such a reconnaissance mission gained particular favor after May, 1961, when President John F. Kennedy called for the country to put a man on the Moon and bring him back safely before the end of 1969; the Ranger program was a necessary preliminary for this more ambitious project, which was later known as the Apollo program.

As originally conceived, the Ranger program was to carry out some other types of observations in addition to the visual imaging experiment. After some reverses in the early attempts, however, designers agreed in mid-1961 that only a restricted version of the program, called the Block 3 project, would be carried out. This project — realized later by Rangers 6 through 9 — was limited to imaging experiments using television cameras. There was, nevertheless, some doubt remaining whether a mere picture-taking experiment would provide all the information needed for planning the Apollo missions. An even more technically demanding, soft-landing project (the Surveyor program) was initiated, as was a project for high-resolution mapping of the whole Lunar surface (the Lunar Orbiter program).

The spacecraft of the Block 3 project were launched with Atlas-Agena rockets from Cape Canaveral, Florida. Each craft weighed about 370 kilograms and was equipped with solar energy panels, a high-gain directional antenna for telecommunications, and an instrument system consisting of a battery of television cameras. The contractor for the imaging system was RCA. The cameras' focal lengths were 25 millimeters or 76 millimeters; cameras A and B had wide fields while cameras P1 through P4 had restricted fields. The last complete pictures in the successful missions were taken with the A and B cameras between two and five seconds before impact, with the P cameras less than one second before impact. The resulting maximum resolution was about one meter with Rangers 7 and 8; somewhat better, about one-half meter, with Ranger 9.

The picture-scanning rate was generally slower than in commercial systems; this made possible the use of a narrow electrical bandwidth, greatly simplifying the telecommunications process. The signals from the spacecraft were received with two 26-meter antennae near Goldstone, California. Commands from the mission center at the Jet Propulsion

The Ranger spacecraft shown in its flight position to the surface of the Moon (left); Ranger 9 picture of Crater Alphonsus (right).   (NASA)

Laboratory (JPL) were also sent through the Goldstone antennae.

Ranger 6, the first mission of the restricted program (which was to provide images only), was launched on January 30, 1964. It proved unsuccessful: After its flawless flight, the craft's camera systems failed, and no pictures were transmitted before the probe's impact on the Moon. The remaining missions, Rangers 7, 8, and 9, however, were spectacularly successful. Ranger 7, launched on July 28, 1964, crashed as planned in the Sea of Clouds, one of the lunar seas. These seas, also known as maria, were apparently flat, roughly circular areas with diameters ranging from a few hundred to a thousand kilometers. The first close-up images of the floor of the Sea of Clouds, showing a wealth of detail, were greeted with worldwide enthusiasm; the region around the impact site was renamed the Sea of Knowledge. These first photographs of the lunar surface clearly suggested that the seafloors of the Moon would not prove as dangerous as expected to landing missions.

After Ranger 7 tested the Sea of Clouds, Ranger 8, launched on February 17, 1965, was sent to another lunar sea; NASA wished to discover whether the seafloors were similar to one another. The target chosen for Ranger 8, the Sea of Tranquillity, indeed revealed similarities to the Sea of Clouds. Later, Apollo 11's lunar module, carrying the first humans to set foot on the Moon, would also land in the Sea of Tranquillity.

Since its predecessors had already gathered important prac-

tical information about the nature of the lunar lowland, Ranger 9 was sent on a purely scientific mission. Crater Alphonsus, a large crater in the central highland of the Moon that showed patches of unusual coloration and a central peak where reports suggested possible volcanic activity, was to be investigated. Ranger 9 made impact only a few miles from its assigned target point, close to the central peak of the crater, and also achieved a somewhat higher resolution by taking its last pictures even closer to the surface prior to impact.

Knowledge Gained

In total, the Ranger 7, 8, and 9 missions sent back more than seventeen thousand pictures, showing the lunar surface with resolutions more than one thousand times greater than that available to Earth-based telescopes. Images taken by Rangers 7 and 8 showed the floors of the seas to be flat, slightly undulating lowlands, mostly free of large debris. Craters of various sizes, down to about one meter and probably smaller in diameter were numerous, but the sizable ones, those which could offset an attempted landing, occupied no more than 1 to 2 percent of the surface. The depth of a possible dust layer proved more difficult to judge, but experiments tentatively decided that this layer was probably less than one millimeter thick.

The floors of the two seas studied by Rangers 7 and 8 proved to be basically similar in nature; all lunar lowlands probably share the same features. Ranger 9's target was a different type of object: the large crater Alphonsus. No confirmation of any recent or ongoing volcanism was found, but the study of a different type of lunar feature proved highly rewarding. The crater floor itself resembled the seafloors studied in the earlier missions.

In addition, two previously unknown features emerged from the study of the Ranger images. Along the conspicuous "rays," the long straight streaks of brighter surface material emanating from several of the large craters (known for centuries from telescopic studies), the pictures show an occasional dense clustering of smaller craters, ranging in size from one hundred to two hundred meters. These small craters are obviously of secondary origin, that is, they were formed by swarms of rocks or other debris hurled out when the large primary crater was formed. The second discovery was a new and unexpected type of lunar crater, the dimple crater: These are shallow depressions without sharp edges and are located on the floors of seas and craters. In size they range from a few meters up to about one hundred meters. Scientists have not been able to agree on the nature of these craters. If they are collapse features (caused by surface matter slumping into a cavity beneath), they might indicate the lunar surface's inability to bear heavy weights.

Context
The Ranger Missions were intended to prepare the way

for the manned lunar missions of the Apollo program by gathering knowledge of the small-scale structure of the lunar surface. The most important result of the program was the discovery that the lunar lowlands, the seafloors, presented no serious surface hazards to landing spacecraft. The presence of a deep dust layer could not be disproved with absolute certainty, but the Ranger missions did render it rather improbable. Nevertheless, the bearing strength of the surface, even without a significant dust layer, remained problematic; the discovery of the dimple craters emphasized this problem. This question was the only one not resolved by the otherwise successful Ranger program. As a spokesman for the program reported in May, 1966, "the bearing strength of the lunar surface cannot be determined by photographs, even at this high resolution."

Fortunately, the subsequent soft-lander missions of the Surveyor project, as well as their Soviet counterparts, the Luna missions, clearly demonstrated the lunar surface's ability to bear up under massive structures such as the Apollo lunar module. The first soft-landings on the Moon (Luna 9 and Surveyor 1) followed within fifteen months of the flight of Ranger 9. With the results of the lunar-landing studies of the soil, even the spectacular Ranger pictures were quickly replaced by new knowledge; they became mere documentation of the progress toward the first Apollo landing in July, 1969.

Bibliography

French, Bevan M. *The Moon Book*. New York: Penguin Books, 1977. This non-technical overview focuses on the information gained during the 1960's and portrays the excitement as well as the facts of the search. Includes descriptions of the various lunar spacecraft and their experiments. Very well illustrated.

Hall, R. Cargill. *Lunar Impact: A History of Project Ranger*. NASA SP-4210. Washington, D.C.: Government Printing Office, 1977. A detailed history of the Ranger program, this highly readable volume is well documented and brings out the human story behind the technical developments.

Koppes, Clayton R. *JPL and the American Space Program: A History of the Jet Propulsion Laboratory*. New Haven, Conn.: Yale University Press, 1982. This text provides virtually the entire history of the Jet Propulsion Laboratory; it also contains considerable information on the pre-Apollo lunar probes, including the various Rangers.

Kuiper, Gerard P. "Lunar Results from Rangers 7 to 9." *Sky and Telescope* 29 (May, 1965): 293 -308. This article summarizes the results achieved by the program. Written for the reader with only a minimal background in science.

Shoemaker, Eugene M. "The Moon Close Up." *National Geographic* 126 (November, 1964): 690-707. This well-written summary provides a general overview of the results of the Ranger program. Includes a large selection of photographs.

Smith, Gerald M., et al. *Ranger VII, Photographs of the Moon: Part I, Camera A Series*. NASA SP-61. Washington, D.C.: Government Printing Office, 1964.

———. *Ranger VII, Photographs of the Moon: Part II, Camera B Series*. NASA SP-62. Washington, D.C.: Government Printing Office, 1965.

———. *Ranger VII, Photographs of the Moon: Part III, Camera P Series*. NASA SP-63. Washington, D.C.: Government Printing Office, 1965.

———. *Ranger VIII, Photographs of the Moon: Cameras A, B, and P*. NASA SP-111. Washington, D.C.: Government Printing Office, 1966.

———. *Ranger IX, Photographs of the Moon: Cameras A, B, and P*. NASA SP-112. Washington, D.C.: Government Printing Office, 1966. These volumes, filled with spectacular images taken by the various successful Ranger missions, are enjoyable for anyone interested in astronomy. Each book contains a short technical description of the spacecraft and instrumentation.

T. J. Herczeg

SATURN LAUNCH VEHICLES

Date: April, 1957, to July, 1975
Type of technology: Launch vehicles, expendable

The Saturn launch vehicle, which evolved through three phases, made possible the placement in orbit of very large payloads. These payloads would later include the technology to allow the launching of three astronauts and their subsequent safe return to Earth.

PRINCIPAL PERSONAGES

WERNHER VON BRAUN, Director of
Marshall Space Flight Center
EDMUND F. O'CONNOR, Director of Industrial
Operations, Marshall Space Flight Center
KURT H. DEBUS, Director of Kennedy Space Center
who developed the Saturn launch facilities
ROCCO A. PETRONE, Director of Launch Operations,
Kennedy Space Center

Summary of the Technology

In April of 1957, a scientific group headed by Dr. Wernher von Braun began study of a vehicle with a booster that could launch payloads of between 9,074 and 18,149 kilograms in orbital missions, or between 2,722 and 5,445 kilograms for "escape" (that is, missions designed to place a spacecraft outside Earth's gravitational field). In December, 1957, the von Braun group, together with the Army Ballistic Missile Agency (ABMA), presented a proposal to the U.S. Department of Defense (DOD) for the creation of a vehicle capable of 337,079 newtons of thrust. In August, 1958, the Advance Research Projects Agency (ARPA) initiated the Saturn program, which provided for the clustering of rocket engines already in existence. Various configurations were explored, including the use of the Atlas-Titan and Centaur for second and third stages. The S-3D engine used on both the Thor and Jupiter missiles could be modified to 42,247 newtons of thrust (becoming the H-1), and the Redstone and Jupiter tools and fixtures could be used with little modification. The Redstone and Jupiter liquid oxygen and fuel tanks could also be adapted to become the tanks of the proposed booster. The idea behind these adaptations and modifications was to save time and resources in the creation of what was at first called Juno 5. In January, 1959, a contract was awarded to the Rocketdyne Division of North American Aviation to develop and test a single engine capable of 337,079 newtons of thrust, using liquid oxygen and RP-1, a kerosene-type fuel. The program name was formally changed to Saturn on February 3, 1959.

Three distinct Saturn vehicles were developed during the course of the program, with different types of engines and boosters used. The testing procedures were complicated and extremely thorough; the first stage of Saturn 1B was tested more than five thousand times as a single engine, then checked seventy-two more times in vehicle tests. The various elements were built as far away as the Douglas Aircraft facilities in California, then taken for testing to the Mississippi Test Facility (MTF) or to Kennedy Space Center (KSC), then returned for adaptation or repair. The launchings took place either at the MTF or at KSC. The logistics of moving the giant boosters were solved with a huge, adapted aircraft, called the "Pregnant Guppy," and with barges constructed especially for this purpose. A canal was even dredged for easier access to the MTF.

The first Saturn vehicle series was designated Saturn 1. The successive ten flights were labeled SA-1 through SA-10, and were launched from KSC. The first stage, S-1, was powered by eight H-1 engines and fueled with liquid oxygen and RP-1. The second, S-4, was propelled by six RL-10 engines, which used liquid oxygen and liquid hydrogen.

SA-1 was launched on October 27, 1961. It stood 49.4 meters high and weighed 417,422.9 kilograms at lift-off. It flight-tested the cluster of H-1 (adapted Thor-Jupiter) engines, the S-1 (the first stage) clustered propellant tank structure, the S-1 control system, and other structural functions.

SA-2, "Project Highwater," was launched April 25, 1962. It carried almost 86,208 kilograms of water, which it delivered at an altitude of 106.6 kilometers. Scientists were able to observe conditions as the ionosphere regained equilibrium following the water dump.

SA-3, launched on November 16, 1962, was also a research and development flight, and, like SA-2, delivered a payload of water. Flight SA-4, launched on March 28, 1963, had much the same characteristics and goals as SA-3. In addition, a planned early cut-off of one of the eight engines proved that a mission could continue in spite of an engine failure.

On November 27, 1963, the first extended-duration firing of the J-2 engine occurred. This engine powered the upper stages of both the Saturn 1B and 5 vehicles.

Another significant engine test took place in December, 1963 — the F-1 initial firing at Marshall Space Flight Center

(MSFC). The F-1 was later used in the first stage (S-1C) of the Saturn 5 vehicle.

The next flight, SA-5, launched on January 29, 1964, was the first in a series of more sophisticated trials, with instrumentation for on-board guidance and a live second stage. The first and second stages were separated in flight, and eight cameras were carried, seven of which were recovered.

Flight SA-6, launched on May 28, 1964, contained an active guidance system that was able to correct a deviation from the planned trajectory, caused by a premature shutdown of one engine. A boilerplate (simulation) was placed in orbit, where it stayed for 3.3 days, for a total of fifty orbits. Twelve thousand performance measurements were telemetered to ground stations, until battery power was depleted during the fourth orbital pass.

SA-7 was significant in that the Saturn 1 launch vehicle was declared operational following the flight, a full three missions earlier than planned. The orbit attained by the spent second stage and the payload was similar to that of a three-man lunar mission. The payload closely resembled that of a lunar mission as well, containing boilerplate Apollo spacecraft command and service modules and instrumentation.

The last three Saturn I flights carried Pegasus, a meteoroid detection satellite designed by Fairchild Hiller Corporation. Once released from its protective Apollo boilerplate, Pegasus spread its wings to more than 29 meters. It carried solar panels to supply its power to collect data, sort information, and transmit. This satellite helped determine how thick the walls of spacecraft needed to be to provide protection from meteoroids during spaceflight.

The second series of vehicles flown was the Saturn 1B series. The first stage, S-1B, was powered by eight H-1 engines, fueled by liquid oxygen and RP-1, and the second stage, S-4B, by one J-2 engine, which burned liquid oxygen and liquid hydrogen. It was 43 meters long, its weight at lift-off was 588,475 kilograms, and its Earth-orbit payload was 18,149 kilograms. There were only five Saturn 1B vehicles launched. One of the prime objectives of Apollo-Saturn (AS) 201, the first flight, was to test the ablative heatshield during reentry. It was launched on February 26, 1966, from KSC.

One of the primary objectives of AS-203, launched July 5, 1966, was to perform an engineering study of the behavior of liquid hydrogen during orbit. The unmanned satellite placed in orbit was the heaviest to date. The flight of AS-202, on August 25, 1966, provided valuable inputs into the behavior of the S-4B stage fuel dynamics. This stage was to be used as the third stage of the Saturn 5 on lunar missions.

Tragedy struck during a manned launchpad test on what was initially AS-204, which was to have been the first manned lunar flight. On January 27, 1967, a flash fire in the command module killed the three astronauts: Virgil I. Grissom, Edward H. White, and Roger B. Chaffee. In honor of the crew, this mission was renamed Apollo 1.

The fourth successful Saturn 1B launch came nearly a year later. A second AS-204, launched on January 22, 1968, was also known as Apollo 5. It was the first flight of a lunar module. It carried instrumentation, an adapter, the lunar module, and a nose cone. Its objective was to check the ascent and descent propulsion systems and the abort staging function for future manned flights. The final Saturn 1B flight was the first manned mission in the Apollo program, Apollo 7, an orbital mission. It was launched on October 11, 1968.

Twelve flights were made in the third series of Saturn launch vehicles, the Saturn 5. It had three stages, with a total vehicle length of almost 922 meters. The first stage, S-1C, was powered by five F-1 engines, for a total thrust of 1,710,112 newtons, and burned liquid oxygen and RP-1. Stage two, S-l, was propelled by five J-2's, with a total thrust of 258,427 newtons; it was fueled by liquid oxygen and liquid hydrogen. The third stage, the S-4B, which used only one J-2 engine, was propelled by liquid oxygen and liquid hydrogen. The Saturn 5 weight at lift-off was 2,903,211 kilograms. Its translunar payload capability was 45,372 kilograms; the Earth-orbit capability was 129,310 kilograms.

Few major changes were made from flight to flight. The only alterations made were performed to reduce weight, to improve safety or reliability, or to increase the possibilities of a larger payload.

On May 14, 1973, Skylab 1 was put into orbit. The body of Skylab 1 was a Saturn 4B booster structure. A two-stage Saturn 5 was used to launch the workshop, and Saturn 1Bs launched the crews.

The Saturn launch vehicle continues to be used in configurations with other vehicles, particularly in commercial launch enterprises. It has proved to be not only effective but also reliable. The Saturn program was an engineering and managerial marvel.

Knowledge Gained

The most significant knowledge acquired from the Saturn program was the technological know-how needed to produce a rocket booster capable of launching large payloads of more than 5,000 kilograms to destinations as far away as the Moon. The development of engines of enormous thrust, combined with the development of a range of boosters, was a tremendous breakthrough in the space program.

Testing facilities and procedures were also greatly advanced to meet the demands of the huge boosters. New types of static and gymbal test facilities were used, and the transporter with crawler treads was developed to facilitate the Saturn program. Transportation problems were resolved with adaptations to aircraft and watercraft, and test areas were further developed to accommodate these craft.

In the entire Saturn program, only one vehicle failed to fly. On one unmanned Saturn 5 flight, two of the second-stage engines shut down early, and the third stage failed to fire. What is more significant is that no vehicle failure prevented the accomplishment of a mission. The Saturn launch vehicle was remarkably successful and extremely reliable.

The Saturn vehicle made possible the retrieval of a tremendous amount of data from the many varied missions for which it has been used, beginning with Project Highwater in April of 1962.

Context

The American space program started on the heels of the Soviet program. Yuri Gagarin was successfully launched into space only three weeks ahead of Alan Shepard. The Soviet Union also launched the first multiple crew into space, as well as the first woman. The Saturn launch vehicle thrust the United States into the lead for the "race for the Moon," placing the American public firmly behind the future of the space program. The Soviet Union was supposedly working on the "G" booster at the same time, but no test-firings were ever reported. A Russian cosmonaut has never walked on the Moon. The development of the Saturn launch vehicle, capable of delivering immense payloads, made it possible for the United States to achieve such goals.

The Saturn launch vehicle invigorated the American public's support of the development of space technology. Following the Apollo 1 tragedy involving the command module, eighteen months elapsed before another manned flight was attempted. The course of space exploration was considerably slowed as a result. (Sadly, a similar tragedy occurred in the Russian space program only three months later: Vladimir Komarov was killed upon impact when the parachute lines of the Soyuz 1 spacecraft were tangled. The Soviet program would also take eighteen months to recuperate before another cosmonaut could fly in space.)

By the end of the Saturn launchings, however, the American public had begun to lose interest in space exploration. The final three launchings were canceled. The Saturn program had been extraordinarily expensive, and the launch vehicles were not reusable (hence the name "expendable launch vehicle"). The public seemed to wonder whether the returns were worth the billions of dollars spent. The American space program began to move in the direction of a spacecraft which could be re-used, in the belief that costs could thereby be cut. In addition, space technology seemed to be directed toward the development of a space station, the ultimate goal being to reach other terrestrial planets via an inhabitable way station. Many payloads would be required to launch and outfit such a facility, and it was thought that a space shuttle, functioning as an airbus, would serve that end. The research and development spent on the Saturn program helped prepare the way for this next stage of the space program.

Bibliography

Baker, David. *The Rocket: The History and Development of Rocket and Missile Technology*. New York: Crown Publishers, 1978. A highly detailed, well-illustrated volume telling the story of rocketry from the invention of gunpowder to the landing of the first man on the Moon. Suitable for a general audience.

Bilstein, Roger E. *Stages to Saturn: A Technological History of the Apollo/Saturn Launch Vehicles*. NASA SP-4206. Washington, D.C.: Government Printing Office, 1980. A thorough treatment of the Saturn program, including the personnel and managerial aspects as well as the technology. A good resource for in-depth understanding. For technical audiences, but of value to the general reader as well.

Historical Office Management Services Office, Marshall Space Flight Center. *Saturn Illustrated Chronology*. Huntsville, Ala.: Author, 1968. A detailed presentation of all stages of development of the Saturn program. Many unretouched photographs give a realistic account of the launch sites and hardware. College-level material.

Ley, Willy. *Rockets, Missiles, and Men in Space*. New York: Viking Press, 1967. A complete overview of rocketry through 1967. Begins with mankind's earliest dreams of flight and continues through the Saturn program. Technical appendices provide more technical data. Illustrated. College level.

National Aeronautics and Space Administration. *Apollo Program Summary Report*. Houston: Johnson Space Center, 1975. This clear, succinct narrative of the Apollo program includes a treatment of the role of the Saturn launch vehicle in that program. College level.

———. *Countdown! NASA Launch Vehicles and Facilities*. Washington, D.C.: Government Printing Office, 1978. A collection of short articles giving data on various NASA launch vehicles, both active and inactive, and a description of NASA facilities. Suitable for general audiences.

Ellen F. Mitchum

Search and Rescue Satellites

Date: Beginning September 1, 1982
Type of satellite: Search and rescue

Search and rescue satellites are used to detect emergency beacons of downed aircraft, capsized boats, and individuals involved in exploration. In its first few years, the COSPAS/SARSAT program helped to save eleven hundred lives.

PRINCIPAL PERSONAGES

JAMES T. BAILEY, head of SARSAT

Y. ZURABOV, Vice President, COSPAS

J. R. F. HODGSON, Executive Director, National Search and Rescue Secretariat

I. REVAH, Director of Programs, Centre National d'Études Spatiales

Summary of the Satellites

COSPAS/SARSAT is an international program using satellites to aid in search and rescue operations. SARSAT stands for Search and Rescue Satellite-Aided Tracking System. COSPAS is a Russian acronym for Space System for Search of Vessels in Distress.

The idea for a satellite-aided search and rescue program arose at almost the same time that artificial satellites were first placed in Earth orbit. It was not until the 1970's, however, that the National Aeronautics and Space Administration (NASA) began to experiment with the Doppler effect using the Nimbus satellites. The Doppler effect is the apparent change in frequency (the number of wave crests passing a point per unit of time) of an electromagnetic wave when an object moves toward or away from the source of the wave. A satellite in low Earth orbit would observe an emergency beacon to have a higher frequency as it approached the beacon and a lower frequency as it receded from the beacon. This shift in frequency allows the satellite to calculate the location of the beacon's source. The Nimbus satellites succeeded in locating weather buoys, drifting balloons, and other remote sensors, and a search and rescue system was proved to be feasible.

The first operational Doppler data collection system was the French ARGOS, which was carried on the U.S. National Oceanic and Atmospheric Administration's TIROS (Television Infrared Observations Satellite). Later systems evolved from ARGOS.

In 1976, the COSPAS/SARSAT program became an international effort when the United States, France, and Canada signed agreements to test a satellite-aided search and rescue system. The United States would contribute its advanced TIROS-N weather satellites, Canada would supply the electronic repeaters (devices that receive a radio signal, amplify it, and relay it to a ground station), and France would supply the electronic processors for the collection of identification data transmitted by the system's users. In 1980, the Soviet Union joined the program by agreeing to place the same type of electronic equipment in its Kosmos satellites. The system became operational on September 1, 1982.

Other countries began participating in the system, although their participation was limited to ground operation of a local user terminal and/or a mission control center. Local user terminals receive the distress signals relayed from the repeaters in the satellites. Mission control centers collect the information and send it to the appropriate rescue control center. In 1981, Norway and Sweden began operating local user terminals. The United Kingdom joined in 1983; Finland, in 1984. Norway and the United Kingdom also operate mission control centers.

A number of other countries either use the system, operate a user terminal, or are discussing possible operation of user terminals or control centers. These countries include Chile, Denmark, India, Italy, Japan, Pakistan, Sweden, Switzerland, and Venezuela.

COSPAS/SARSAT works in the following manner. An emergency transmitter— either an Emergency Locator-Transmitter (ELT) or an Emergency Pointing/Indicating Radio Beacon (EPIRB) — is activated when an emergency occurs. ELTs are carried on general aviation aircraft, and EPIRBs are carried on marine vessels that venture into open ocean. The signal from an ELT or EPIRB has a frequency of 121.5 megahertz or 243 megahertz. The signal is received by a COSPAS/SARSAT satellite and relayed to a local user terminal at a frequency of 1544.5 megahertz. Some more advanced ELTs can transmit a 406-megahertz signal that provides data on the sender's identification and nationality as well as the transmission's time and location. If necessary, the information can be stored until the satellite is in range of a ground station. After the information is processed by the local user terminal, it is relayed to the appropriate mission control center, which may be in another country. The mission control center then passes the information on to a rescue control center, which is then responsible for the rescue.

The type of rescue effort varies with geography, national boundaries, and available resources. In the United States, the Air Force, sometimes with help from the Civil Air Patrol, coordinates inland airplane rescue attempts. The Coast Guard is called on for rescue attempts in the ocean to a distance of 320 kilometers from the U.S. coastline; farther out, the Air Force is responsible for searches. COSPAS/SARSAT's pinpointing the signal sender's location helps reduce the amount of money spent on the search and, more important, the time spent in finding the downed aircraft or ship in distress. Studies show that the survival rate for airplane crash victims is 50 percent if they are rescued within eight hours. If the rescue takes place more than two days after the disaster, however, the chance of survival drops to 10 percent.

Knowledge Gained

In the 1970's, using satellites in search and rescue was proved feasible. In the early 1980's, enough satellites were placed in orbit and enough ground stations were set up to allow the system to operate.

The first COSPAS/SARSAT rescue took place in September, 1982. A young couple's plane went down in British *Columbia* in July, 1982, and the Canadian government launched an extensive search that ultimately cost nearly two million dollars. The couple was not found, and the search was stopped. The young man's father, along with a pilot, and a friend decided to embark on their own search and rescue mission, which continued into September, 1982.

During this search, however, their plane crashed in a mountainous, tree-covered area. They were injured, but alive. An ELT had been activated by the crash, and as Soviet COSPAS 1 passed over, it relayed the information to the Trenton, Ontario, Air Rescue Station. A rescue control center was contacted, and the three victims became the first persons to be rescued with the aid of a satellite.

The two disasters serve as contrasting examples of search and rescue operations. The location of the first crash was not known, and, consequently, much time and money were expended in searching. The second crash location was known to within 22.5 kilometers, so only a relatively small area had to be searched; costs were much lower, and the victims were found while they were still alive.

In July, 1985, the COSPAS/SARSAT steering committee, which includes experts from Canada, France, the Soviet Union, and the United States, adopted the 406-megahertz frequency for signals broadcast by ELTs. Since the 406-megahertz signal allows for more precise location calculations than does the 121-megahertz signal, it makes faster rescues possible. Moreover, such a signal can carry identification data.

One problem that plagues the COSPAS/SARSAT system is the large number of false alarms. In 1988, the false alarm rate was 98 percent for ELTs and more than 50 percent for EPIRBs. False alarms are caused by unintentional activation of an ELT or EPIRB beacon resulting from improper handling, equipment failure, or improper shipment or testing. Since each distress signal must be tracked down, much time and effort are wasted.

False alarms probably can be reduced by better enforcement of laws governing the use of emergency frequencies, redesign of equipment so that it is harder to trigger a false alarm, and better education for ELT and EPIRB users.

Context

Before the COSPAS/SARSAT program, search and rescue operations were conducted on a hit-or-miss basis. During the 1970's, NASA showed that it was possible to locate radio signal sources on Earth by making use of the Doppler effect, and this technology spurred the development of COSPAS/SARSAT. New knowledge has been gained in the area of locating points on Earth's surface, and, as a result, rescue teams can locate disaster victims much more quickly.

COSPAS/SARSAT saves money as well as lives. When the area to be searched is a few square kilometers instead of hundreds or thousands of square kilometers, fewer searches are needed. More efficient searches also mean that rescue planes or ships can be used for shorter times. COSPAS/SARSAT has effected savings of $20 million per year, according to estimations by Owen Heeter, commander of the Air Force's Aerospace Rescue and Recovery Service.

COSPAS/SARSAT is important for one other reason: international cooperation. The United States and the Soviet Union were partners in this program almost from its beginning. For nearly six years, it functioned only on a memorandum of understanding, but on July 1, 1988, an international agreement requiring nations to cooperate in COSPAS/SARSAT was signed by the United States, Canada, France, and the Soviet Union. William E. Evans, Undersecretary of Commerce for Ocean and Atmosphere, signed for the United States. This agreement is binding for the first ten years, with automatic renewal every five years thereafter. Countries and organizations such as the United Nations International Maritime Organization can require that ships and/or aircraft have the emergency 406-megahertz beacon on board. In the late 1980's, NASA was trying to develop low-cost 406-megahertz ELTs.

By 1988, more than eleven hundred persons had been rescued with the aid of COSPAS/SARSAT. NASA and others continue to seek new methods to cut costs and time in search and rescue operations while saving even more lives.

Bibliography

Bailey, James T. "Satellites Answer SOS." *Mariner's Weather Log* 32 (Spring, 1988): 8-11. A nontechnical description of search and rescue satellites, this article requires minimal technical knowledge. Contains several photographs and one illustration showing how the COSPAS/SARSAT system works and features stories about searches that demonstrate the usefulness and cost-effectiveness of the COSPAS/SARSAT system. Emphasizes international cooperation.

Haggerty, James J. *Spinoff 1984*. NASA TM-85596. Washington, D.C.: Government Printing Office, 1984. This multi-color pamphlet contains a summary of NASA's major accomplishments in space and the ways in which those accomplishments have benefited mankind. One page is devoted to a brief description of how the COSPAS/SARSAT system operates. Illustrated. For general audiences.

Kachmar, Michael. "SARSAT/COSPAS: Saving Lives Through Cooperation In Space." *Microwaves & RF* 23 (October, 1984): 33-35. This journal contains a news section in which uses of microwaves and radio frequency waves are described. Includes photographs. Technical language is used mostly without explanation, but some terms are defined in the text.

McElroy, John H., and James T. Bailey. "Saving Lives at Sea… Via Satellite." *Sea Technology* 26 (August, 1985): 30-34. A nontechnical description of the COSPAS/SARSAT international satellite network. Includes illustrations of a rescue at sea and a discussion of searches and their costs. Suitable for general audiences.

National Aeronautics and Space Administration. *COSPAS/SARSAT*. Greenbelt, Md.: Author, 1986. Describes the COSPAS/SARSAT system and gives a brief history of how the system was developed. Outlines the operation's successes and the ways in which it will expand in the future. Contains color diagrams. Suitable for general audiences.

National Oceanic and Atmospheric Administration. *COSPAS-SARSAT Search and Rescue Satellite System*. Washington, D.C.: U.S. Department of Commerce. This brochure gives a very brief overall view of the COSPAS/SARSAT system. It contains several drawings showing how COSPAS/SARSAT works, the locations of local user terminals, an advanced TIROS-N satellite, and emergency beacon devices. Suitable for general audiences.

Ola, Per, and Emily D'Aulaire. "The Starduster's Last Flight." *Reader's Digest* 131 (July, 1987): 75-80. This is a nontechnical account of a successful search and rescue operation. It is part of the "Drama in Real Life" series of articles that appears in *Reader's Digest*. Written as a historical account of an air disaster, with a description of the efforts of the Air Force Rescue Coordinator Center to help in the search. Suitable for general audiences.

U.S. Coast Guard. *National Search and Rescue Manual*. Washington, D.C.: Author, 1986. This two-volume work gives detailed information on the procedures for performing search and rescue operations. These manuals could be used in training exercises. Satellite-aided search and rescue operations are discussed. For general audiences.

U.S. Department of Commerce. *NESDIS Programs, NOAA Satellite Operations*. Washington, D.C.: Author, 1985. This publication provides a very brief description of the COSPAS/SARSAT system and a diagram of how the system operates. Most of this publication covers other topics, such as the weather satellites of the NESDIS program. For general audiences.

U.S. Department of Transportation. *1986 SAR Statistics*. Washington, D.C.: Author. This publication offers tables, photographs, and diagrams detailing search and rescue statistics for 1986. Does not distinguish between satellite-aided and non-satellite-aided rescues. For general audiences.

T. Parker Bishop

Seasat

Date: June 26 to October 6, 1978
Type of satellite: Scientific

Seasat was an artificial satellite designed to perform a variety of experiments, including photographic and remote-sensing procedures, above the world's oceans. Though short-lived, it provided invaluable data about ocean currents, wave action, water temperature, wind direction, ice floes, and other marine phenomena.

Summary of the Satellite

In the post-Sputnik era, the use of satellites for communication, reconnaissance, and remote sensing of Earth's surface became commonplace. Early experiments with ground-controlled orbital spacecraft demonstrated the usefulness of such technology for cartography, natural resource identification, and the monitoring of agricultural production. It was only a matter of time before scientists would begin to explore the uses of satellite technology for the study of the world's oceans.

On January 9, 1975, the National Aeronautics and Space Administration (NASA) announced a satellite program to monitor the world's oceans and provide continuous data on sea conditions and ocean weather. The satellite, dubbed Seasat-A, was scheduled for launch in 1978. Its purpose was to provide evidence that an artificial scientific satellite could function in this role. Thus, it was not an operational system, although indications were that such a system might be launched if Seasat accomplished its mission successfully. Instrumentation aboard the satellite would measure wave height, current direction, surface wind direction, and the temperature of the surface of the ocean for a period of one year.

It was hoped that the information gathered by Seasat could be used to predict the weather, help ships avoid storms and icebergs, provide coastal disaster warnings, and locate currents that could be used to advantage by ships. The satellite was also designed to provide data on ocean circulation, interaction between the atmosphere and the sea, and the movement of heat and nutrients by ocean currents.

At the time of the announcement, the cost of Seasat was projected to be $58.2 million. In November of 1975, a $20-million contract for the design and construction of Seasat was awarded to Lockheed Missiles and Space Company of Sunnyvale, California, by the Jet Propulsion Laboratory (JPL).

As the launch date approached, NASA began to seek potential users of the data that it expected would be forthcoming from Seasat. Early in 1977, representatives of a group of European nations — including Denmark, Finland, France, West Germany, Great Britain, and Spain — came forward to express an interest in the project. These potential data users suggested two common goals: that Seasat would be useful in forecasting floods in the North Sea region, and that the information it provided might be useful in designing ships and offshore installations such as oil-drilling platforms and in-harbor structures.

Among U.S. organizations interested in the program were the National Oceanic and Atmospheric Administration (NOAA), the National Science Foundation, the Office of Naval Operations of the U.S. Coast Guard, and the U.S. Geological Survey. Early studies done by NASA suggested that an operational oceanographic satellite data-gathering system could produce returns valued at up to $2 billion by the end of the twentieth century. The semicommercial nature of the venture was important in overall strategic planning at a time when NASA was being encouraged by the U.S. government to find markets for its programs.

Seasat was launched successfully on June 26, 1978, from Vandenberg Air Force Base in California. The satellite itself, an Agena rocket stage with sensor module attached, was the second stage of the launch vehicle. The first stage was an Atlas-F rocket. Seasat's overall length was 12 meters, and it weighed 2,273 kilograms. It was placed in an orbit 800 kilometers above Earth, and made 14 $\frac{1}{3}$ orbits of Earth each day. Each orbit took 100 minutes to complete, and during each thirty-six-hour period, 95 percent of Earth's surface was covered. After an initial adjustment, the orbit was to be stabilized and devoid of variation for three months at a time, in an effort to create the best operating environment for one of the onboard instruments.

Seasat was fitted with five remote-sensing instruments, designed and programmed by various NASA agencies. These devices included a radar scatterometer, a Synthetic Aperture Radar (SAR) system, a scanning multispectral microwave radiometer, a radar altimeter, and a visual and infrared radiometer.

The microwave wind scatterometer, built under the direction of the Langley Research Center, measured surface wind speeds over the oceans in a range of between 4 and 28 meters per second.

The SAR system measured the direction and size of ocean waves, day and night. It was able to detect ice, oil spills, and ocean current configurations, even through fog and moderate precipitation. It was designed by JPL.

The five-channel multispectral scatterometer, also designed by JPL, was able to detect the surface temperature of the ocean to within 2 Kelvins. It also detected wind speed to 50 meters per second and mapped ice fields. It measured atmospheric moisture content as a housekeeping chore, allowing controllers to correct altimeter and scatterometer readings. It also measured radiation on five wavelengths.

The radar altimeter measured the distance between the ocean surface and the satellite, with an accuracy of +/−10 centimeters, and could detect currents, tides, and weather-related storm surges, of importance to shipping and coastal population areas. It was designed by the Applied Physics Laboratory at The Johns Hopkins University for NASA's Wallops Flight Center.

The scanning visible and infrared radiometer observed surface features during clear weather only. It was designed to help in the identification and location of phenomena such as intense currents, storm systems, and ice floes.

Soon after Seasat was launched, users began to extol the high quality of the heavy volume of data being generated by its sensors and transmitted to ground stations around the world. The mission was widely proclaimed a success, despite equipment malfunctions that cut it short after only four months of operation. Each day of its life, Seasat generated more data than had accumulated over decades using land-based techniques and aircraft, with their relatively narrow fields of vision. Seasat had met the expectations of NASA and led to the development and eventual deployment of advanced sensing instrumentation for Earth resources satellites. Soon, NASA and other space agencies began to plan the construction and deployment of similar satellites, as the space era moved into the 1980's.

Knowledge Gained

Before the advent of satellites, oceanographers had been able to obtain information about the seas, but the efficiency of operation and quantity of data were limited by the relatively narrow field of vision of traditional sensing technology and the expense of deployment and operation. For example, ocean vessels and aircraft were able to contribute to the body of knowledge about the weather, currents, ice fields, and the like by relaying visual-sighting information to federal agencies responsible for monitoring such activity. Sensing equipment had also been mounted on automatic transmission buoys on the surfaces of oceans around the world. Using radio signals, these buoys could detect and transmit information about ocean surface temperatures, wave action, wind direction, and salinity.

When satellites did appear on the scene, these buoys were able to receive data for relay to control sites around the world. Seasat, on the other hand, was able to acquire and transmit a much higher volume of significantly more sophisticated data in real time, twenty-four hours a day.

Until Seasat was developed, most satellite systems contained only passive sensors. Among their functions were the measurement of radiation emission, infrared and microwave, and reflected solar radiation. Such passive systems are useful in nighttime detection of icebergs and ice floes. Sea ice emits a different rate of microwave radiation than does seawater, a fact that is key to the process of oceanographic sensing of the delineation between them. Furthermore, old sea ice emits radiation at a different rate than new sea ice, giving scientists another dimension to the picture: the relative age of the ice field and of its individual components.

In addition to monitoring radiation, passive systems could collect visual images of clouds, ice fields, and ocean surfaces during daylight. They could also determine the amount of water vapor in the atmosphere and thus constituted a useful tool in the creation of global precipitation maps. Yet passive systems had certain limitations.

Seasat carried three different types of radar, two of which were active, including a new and relatively untried sensor, a scatterometer. The scatterometer measured back-scattered radiation from the ocean surface at a specific wavelength. The theory behind its design was that the intensity of such radiation and its direction related to the amount of wind at the surface and the direction of that wind. Empirical analysis of the data received from the scatterometer aboard Seasat suggested that it was extremely accurate in its findings. The other active system aboard Seasat was the SAR system, which provided radar images of the ocean surface. Scientists were surprised to discover that it also detected waves below the surface and bottom topography, even in relatively deep water. For the first time, scientists were able to witness the effect the ocean floor has on the ocean surface under varying weather and wave conditions.

Seasat was also able to detect time variations of intermediate-sized eddies in the Atlantic Ocean to the east of the North and South American continents, a development that has led researchers to conclude that weather forecasters will soon be able to include eddy maps in their portfolios of atmospheric maps, also generated from satellite data.

The one major limitation of active sensing radar is the fact that seawater is opaque to infrared and microwave radiation, so that this kind of sensing is useful only in monitoring the surface. Once the radio waves slip beneath the surface, they are reflected by objects such as the ocean floor and do not show transitory characteristics of ocean movements. Thus, their usefulness in other than relatively shallow waters is quite limited.

Seasat set the standard for the application of satellite technology to oceanography. This experimental program delivered on its promise of advancements in the study of the world's oceans. Its function was twofold: to prove that Seasat satellite technology would work, and to provide evidence that commercial applications exist for such technology. In both cases the evidence was overwhelming that an operational satellite system would be extremely useful to oceanographers.

Context

The Seasat program was developed largely at the request of a number of academic, scientific, and governmental entities. Satellite technology in the mid-1970's had come to represent a viable, dependable, economical alternative to existing techniques for the gathering and analysis of oceanographic data. In addition, international cooperation in the development of space technology was at an all-time high.

Satellites already satellites had become important to the maritime industry for a number of reasons. The fishing industry, for example, had used satellite photography to track such periodic sea phenomena as red tides and abnormal currents, which have a significant effect on commercial fishing. In some fishing grounds, commercially important species of sea life tend to gather near the outer edges of major currents and, under certain weather conditions, along coastal areas, where large concentrations of nutrients congregate. Satellite imagery was useful to commercial fishing boats because it provided detailed, up-to-the-minute information on the location of these fish, allowing significant savings on the cost of fuel and the amount of time necessary to obtain such data. During 1975, the year in which the Seasat program was announced, Exxon, the giant U.S. oil company, saved more than $350 thousand in fuel for fifteen oil tankers by using satellite imagery to track the Gulf Stream and the Gulf Loop current. The tankers rode the currents when going in the same direction and avoided them on the return trip.

Seasat overlapped other remote-sensing satellite programs in many ways. Among these other programs was Landsat, the U.S. Earth resources satellite series that aided in the location and identification of natural resources, such as mineral and oil deposits. These satellites were also used to monitor urban growth, track crop yields, monitor snowmelt, and map shallow lakes and oceans. In fact, Landsat sensors were able to penetrate clear water up to a depth of 20 meters, allowing them to chart underwater features in most lakes and along coastal areas, including much of the shallow Caribbean basin. Other satellites of a similar configuration were the ITOS, TIROS, and other meteorological satellites that monitored weather conditions and developing storm centers.

Seasat was the first satellite with remote-sensing capabilities devoted exclusively to the study of oceans. The success of its instrumentation resulted in modifications to existing Landsat equipment, making it more reliable and sophisticated. Many other satellites have been designed to engage in multitask service, monitoring the earth and sea, yet simultaneously providing communications and navigation capabilities to a variety of users. In addition, other nations and international consortiums began to develop satellite programs similar to Seasat.

Oceanographers continue to propose and develop Earth-based systems to support and, in many cases, refine findings derived from data acquired by Seasat and other scientific satellites. Among these are buoys equipped with sensing instrumentation and telemetry for relaying data to satellites, ships, aircraft, or radio receiving stations. Others include specially equipped aircraft that can be sent to the area of a developing storm, tidal surge, or ice field for direct observation.

Satellites can also be used to police the oceans. Seasat went beyond its mission by focusing attention on man-made as well as natural phenomena. It detected oil spills and air pollution and was capable of detecting ocean dumping from ships at sea. It was a multidimensional satellite that set the agenda for oceanographic research for many years after its launch.

Bibliography

Braun, Wernher von, et al. *History of Rocketry and Space Travel*. 3d ed. New York: Thomas Y. Crowell Co., 1975. An excellent account of the evolution of rocketry and satellite technology, beginning with the very early days, long before the Sputnik era. This book contains a thorough account of the accomplishments of the U.S. space program during the 1970's and includes an assessment of the most successful of the U.S. applications satellites that were deployed just prior to the launch of Seasat, which was a proof-of-concept mission rather than an operational system. Includes photographs and an index.

Brun, Nancy L., and Eleanor H. Ritchie. *Astronautics and Aeronautics, 1975*. Washington, D.C.: National Aeronautics and Space Administration, 1979. This annual represents a collection of press accounts and NASA publicity releases documenting national and international events related to space exploration during 1975. Includes accounts of various aspects of the planning stages of the Seasat program. Contains an index.

Corliss, William R. *Scientific Satellites*. NASA SP-133. Washington, D.C.: Government Printing Office, 1967. Most of the scientific satellites launched during the first decade of the space age by various countries are described in highly technical accounts. Also included is a chronology of international satellite launches. Contains a chapter describing geophysical instruments and experiments, preceded by an excellent overview of space science. Includes photographs and illustrations. Indexed.

Gatland, Kenneth. *The Illustrated Encyclopedia of Space Technology: A Comprehensive History of Space Exploration*. New York: Crown Publishers, 1981. Contains numerous chronologies of Soviet, American, and international space programs. Earth observation systems are discussed in detail. Specific examples of successful applications of space technology in this area are cited, and a very good description of the Seasat program is given, with technical parameters of the satellite and its instrumentation. Well illustrated. Indexed.

Ritchie, Eleanor H. *Astronautics and Aeronautics, 1976: A Chronology*. NASA SP-4021. Washington, D.C.: National Aeronautics and Space Administration, 1984. Abstracts of press accounts of space activity during 1976 are compiled here in chronological order. These accounts were gleaned from popular publications and space agency memoranda made available to the general public. Early activity in the Seasat program is documented in trade publications and NASA releases. Indexed.

Wells, Neil. *The Atmosphere and Ocean: A Physical Introduction*. London: Taylor and Francis, 1986. A primer on the relationship between the oceans and the physical elements of the atmosphere, this book contains an excellent account of the accomplishments of the Seasat project in terms of its oceanographic research techniques, the instrumentation aboard the satellite, and the significance of the data acquired from that instrumentation. Highly technical. Indexed.

Michael S. Ameigh

SHUTTLE AMATEUR RADIO EXPERIMENT

Date: Beginning November, 1983
Type of program: Amateur radio communications

Amateur radio operators (hams) have communicated with space shuttle astronauts orbiting the Earth since 1983. The National Aeronautics and Space Administration's (NASA) intention in making astronauts available for the Shuttle Amateur Radio Experiment (SAREX) is educational in nature as it seeks to involve a large number of people in technology and the space program.

PRINCIPAL PERSONAGES

ROY NEAL, K6DUE, Chairman of the SAREX committee

FRANK BAUER, KA3HDO, Radio Amateur Satellite Corporation (AMSAT), SAREX committee member

ROSALIE WHITE, WA1STO, American Radio Relay League (ARRL), SAREX committee member

LOU MCFADIN, W5DID, NASA Johnson Space Center, SAREX committee member

OWEN K. GARRIOTT, W5LFL, NASA astronaut and STS-9 Mission Specialist, the first amateur to operate from space

Summary of the Program

The SAREX project is a continuing program of an educational nature that combines the missions of the space shuttle with amateur radio. It is sponsored by NASA, AMSAT and the ARRL, with the approval of the Federal Communications Commission (FCC). During a SAREX mission, the astronauts typically make several types of amateur radio contacts, including scheduled radio contacts with schools, random contacts with the ham community, and personal contacts with the astronauts' families. The contacts are made during breaks in the astronauts' work schedule. The costs of this ongoing mission are borne by radio amateurs who donate their time and equipment on the shuttle and on the ground.

In 1983, the first SAREX mission was carried out when Owen Garriott, a mission specialist aboard the shuttle *Columbia*, was the first to carry his amateur radio equipment into orbit during STS-9. Before any communications could take place, however, there were many complex problems that had to be overcome by ground stations.

The problems of communicating between an earth station and a satellite (in this case, the shuttle in orbit) present the

amateur radio operator with a new set of challenges. Ordinary communications between fixed earth stations is dependent on frequency, time of day, and season of the year and is all related to the cycle of sunspots recurring over a course of eleven years. While these factors are well understood, and they allow a long or short communications link depending on the station the amateur wishes to contact, the resulting communication (or QSO, in amateur parlance) will be of a quality that varies greatly: Interference, both human-induced and natural, as well as fading, will affect the signal's readability.

Communicating with an orbiting satellite, however, is both simpler in terms of frequency choices involved and more complex in that it requires that the amateur know considerably more about the station with which he or she is communicating. When successful, the signals exchanged will be eminently readable. For example, two earth stations wishing a communications link find that signals are subject to the whims of the ionosphere, the chief cause, along with overcrowding, of variability in the communications in the most common ham bands (1.8 MHz to 30 MHz or the short-wave frequencies). For satellite work, however, the frequencies chosen are more limited: Direct, line of sight communications make sense when the distance involved is 150 or 200 miles away, such as the shuttle, as opposed to thousands of miles distant, such as normal earth communications. As a result, the frequencies of choice for SAREX missions and other satellite communications are what are known as VHF, or Very High Frequency. These frequencies have the virtue of traveling short distances but doing so in a straight (line of sight) path. This is normally a relative disadvantage on earth, but in space communications it is eminently predictable. However, where does one point the antenna for a line of sight communication with an orbiting body? Not only does the antenna on earth need to be oriented in azimuth (a north-south-east-west manner), but the elevation above the horizon must be set, too.

Fortunately, computers aid in determining precisely where the satellite (in this case, the shuttle) is at any given time. To do that job, two things are necessary: a computer with appropriate software and Keplerian elements. Keplerian data are orbital or tracking elements used to pinpoint the location of a satellite. They provide the computer and its software with a "snapshot" of the orbital track the shuttle is following, and they allow users to determine where in space the shuttle is located at any given moment. This enables the observer to know precisely when the shuttle will appear above his or her horizon (the window of time that the shuttle's signal can be heard may last only eight minutes). These data are updated frequently and available via amateur radio, on the Internet (using techniques such as FTP, File Transfer Protocol), and via the World Wide Web.

Armed with the data yielded by a computer program and the Keplerian elements, the amateur next readies his station. Typically, a SAREX earth station consists of a 2-meter FM transmitter and receiver (one that operates on the VHF frequency band of 144-148 megahertz) capable of an average output power of 50 to 100 watts, plus an antenna. The antenna is crucial to successful communications with the shuttle because the shuttle protocol allows the amateur operators on board primarily hand-held equipment with antennae that are limited in scope and, therefore, gain. Indeed, the amateur operators on board the shuttle can place an antenna only directly facing the windows on the flight deck. On earth, the ham station will use a circularly polarized (not directional) antenna of the crossed-Yagi type. Successful communications have been established with sufficient power from the transmitter using only vertical (straight up and down) antennae, but more modest stations will need antennae that are capable of being adjusted in two ways. First, the azimuth must be altered as the need arises. Second, the elevation or degrees above the horizon must be equally variable.

Finally, there is one additional issue the operator may wish to consider: the Doppler effect. As the shuttle moves through space at 17,000 miles per hour, its position is constantly changing with reference to a ground station. For part of the time, the shuttle will approach that station. As a result, the signal it transmits will appear to be slightly higher in frequency than its nominal signal. Conversely, as it flies away from the station, its signal appears to be somewhat lower. A station on earth, then, may choose to set its local transmitter frequency approximately several kilohertz above the frequency set as nominal and its local receiver frequency correspondingly lower. This situation is reversed when the shuttle is moving out of range. While not critical in achieving basic voice communications for the average station, the effects of Doppler shift on packet communications may be greater.

While Garriott's communications equipment aboard STS-9 consisted of little more than a hand-held 2-meter (144 -megahertz) transceiver and an antenna mounted in the window on the shuttle's flight deck, the shuttle's amateur station has grown somewhat over the years. In more recent shuttle operations, it has consisted of approximately sixty pounds of equipment, a not inconsiderable weight when one must account for every spare pound due to orbital constraints. Typically, this equipment is a hand-held transceiver or one similar in nature, operating on the 2-meter amateur radio band; a module to permit a means of interconnecting SAREX equipment with the standard crew headset and

microphone; an equipment assembly cabinet, which holds a converter for SSTV or Slow-Scan Television (the converter takes normal TV from a camera or the shuttle TV distribution system and makes still TV pictures from it so that they may be transmitted on a voice frequency without taking up too much bandwidth), packet radio terminal node controller (to allow digital reception and transmission using an amateur-written protocol), and power supplies, displays, and connectors; a television camera to televise via slow-scan technology scenes within the Orbiter and external scenes viewable from the windows; a TV monitor for viewing the output of the SSTV converter; an antenna; and a payload general support computer, which acts as a data terminal for the packet radio portion of the experiment.

There are a number of different configurations of SAREX equipment, and the choice of configuration for a specific flight depends on the number of amateur operators among the astronauts (some crews, notably of STS-37, STS-56, and STS-74, consisted entirely of hams) and the amount of space available for that particular operation. The configurations, designated by a letter of the alphabet starting with A and running through E (with a separate configuration for approaches to the space station Mir designated Configuration M), vary from the simple station brought aboard by Garriott, to a Configuration D, which includes fast-scan television. Configuration A includes the modes of transmission mentioned but only permits slow-scan television; Configuration C does not include facilities for television but does include the hand -held transceiver and the digital packet radio.

Recently, the shuttle has been equipped with an experimental antenna mounted on the outside of the Orbiter in addition to the usual window antenna on the flight deck. Amateur radio operators were helpful in testing this antenna when it was used on STS-63.

Knowledge Gained

Consistent with its stated goal of introducing a large number of people to the technology of the space program, SAREX has been employed to bring space to locations such as the classroom. For each mission, a school completes a SAREX application and writes an educational proposal. These proposals are collected by the ARRL and forwarded to the SAREX Working Group for final selection in collaboration with the astronauts involved. The educational proposal is an important element of an application, because it describes how the SAREX activity will be incorporated into the school's curriculum. For example, a school might sponsor an essay or poster contest or encourage letter writing to secure more information about an experiment planned for a given shuttle flight. The proposal also specifies the capabilities of the school organizer to set up the necessary equipment, usually with the aid of experienced hams, who may or may not be associated with the school. Even schools whose applications are not chosen can participate in the SAREX activity, as can all stations with at least a receiver and an antenna.

It should be emphasized that communications between a student and a given amateur radio operator aboard the shuttle are not at all rehearsed. Questions about geography, about if a bird can fly in space, about how astronauts sleep in space or move about the shuttle, or about what it feels like to return to earth are generally easy to answer, but they provide the young student with the unmatched realism that comes from someone on the spot.

Of course, the astronaut-hams have not forgotten the typical, random, contact that amateurs make. Garriott's method of contacting earth-bound amateur operators was relatively simple. When operating, he transmitted on even minutes and listened on odd minutes; a station wishing to contact him would send his or her call sign a few times during odd minutes (time was synchronized via WWV, the National Bureau of Standards atomic clock in Colorado) and then listen. This was about the extent of his QSO's (two-way communications) then, but much has changed over the years; nevertheless, many stations established communications using that mode. On subsequent flights, slow-scan television and packet radio via a robot station have been added. In addition, FSTV, or Fast Scan Television, has been added to the mix so that amateurs can receive live broadcasts from the shuttle on amateur radio.

Context

The SAREX goal is educational in nature and, as such, it is a never-ending quest. The questions it answers are not crucial to the flight in question but may awaken a certain interest in a young student. It is likely that shuttle amateur radio will be with us for the foreseeable future, and contacts on voice, packet radio, or television (depending on which configuration the shuttle takes into orbit) will continue.

Indeed, contacting human beings in space is international in flavor. With information made available weekly in a newsletter, amateur operators the world over can contact more than the orbiting space shuttle. The space station Mir, a Russian project, accommodates amateur radio, too, making it possible to talk to space on a regular basis. Using approximately the same frequencies as SAREX missions do, the Mir space station passes within reach of most North American locations at least six to eight times each day for up to ten minutes each pass. The equipment used by the shuttle to establish contact with Mir before docking is the same equipment the shuttle astronauts use for SAREX missions.

SAREX is invaluable for a number of reasons. First, it serves the purposes that created it: Namely, it awakens interest in technology among the youngest as they talk to an astronaut about real matters. Second, it encourages those students interested to become amateur radio operators themselves. Third, it provides countless other amateurs and short-wave radio listeners with an insight into the astronauts' daily living conditions in space. Finally, it serves to make all this public. It is indeed a successful experiment.

Bibliography

Davidoff, Martin, et. al. *ARRL Satellite Anthology.* 3d ed. Newington, Conn.: ARRL, Inc., 1994. Contains recent QST satellite articles on all amateur spacecraft from OSCAR (Orbital Satellite Carrying Amateur Radio) 10 through OSCAR 27. Includes how to work long distances via OSCAR's 10 and 13, and how to get on the "Pacsats" and the Russian "Easysats." Gives information regarding the future of advanced amateur satellites.

Davidoff, Martin. *The Satellite Experimenter Handbook.* 2d ed. Newington, Conn.: ARRL, Inc., 1990. Basic information on communications via the large number of satellites of interest to amateur radio operators. Contains advice for those who would like to receive weather, amateur radio, or TV broadcasts from spacecraft. Does not cover SAREX expressly, but the information may be applied toward receiving SAREX transmissions.

"STS-9 and Amateur Radio." *NASA Educational Briefs.* EB-83-9, Washington, D. C.: NASA, 1983. Gives a cursory overview of SAREX and Owen Garriott's mission; details how one can get a QSL card (a card confirming reception or a two-way communications) with SAREX.

The Shuttle Amateur Radio Experiment. Washington, D.C.: NASA, 1994. A collage of five pictures of astronauts in the shuttle making communications with school children on the ground during flights STS-37 and STS-47 make up one side of this sheet. The reverse side gives an explanation
oœœ

The Skylab Program

Date: August 6, 1965, to February 8, 1974
Type of program: Manned Earth-orbiting space station

Skylab, the first American space station, saw three crews spend a total of 171 days in space. The astronauts accumulated volumes of scientific data, establishing the value of humans in space research.

PRINCIPAL PERSONAGES

JAMES E. WEBB, NASA Administrator, 1961 – 1968

THOMAS O. PAINE, NASA Administrator, 1968 – 1971

GEORGE E. MUELLER, Associate Administrator for Manned Space Flight, 1963 – 1969

DALE D. MYERS, Associate Administrator for Manned Space Flight, 1970 – 1974

WILLIAM C. SCHNEIDER, Skylab Program Director

LELAND F. BELEW, Skylab Project Manager, Marshall Space Flight Center

ROBERT F. THOMPSON, Skylab Project Manager, Manned Spacecraft Center, 1967 – 1970

KENNETH S. KLEINKNECHT, Skylab Project Manager, Manned Spacecraft Center, 1970 – 1974

Summary of the Program

The project that evolved into Skylab originated in the summer of 1965, when researchers at the National Aeronautics and Space Administration (NASA) were developing projects to follow the manned lunar landing. Engineers at Marshall Space Flight Center (MSFC) proposed a "spent-stage experiment," in which astronauts would enter the empty upper stage (the S-4B stage) of a Saturn 1B launch vehicle and perform experiments in weightlessness. Almost nothing was known about how easily men could work in zero gravity, and the hydrogen tank of the S-4B, 6.5 meters in diameter and nearly 9 meters in length, could provide a protected environment to test mobility aids and restraint devices and experiment with various tools and techniques.

The spent-stage experiment, also called the orbital workshop (or the "wet workshop," because the fuel tank was to be filled with fuel at launch), was assigned to the Apollo Applications Program (AAP), a program established in August, 1965, in the Office of Manned Space Flight (OMSF). AAP developed ambitious plans to use Apollo spacecraft and launch vehicles for frequent flights in low Earth orbit, synchronous orbit, and lunar orbit, as well as long-duration lunar surface missions (up to fourteen days). Its major objectives were to use the Apollo hardware to produce useful data and to keep the Apollo team together while a major post-Apollo goal was defined and developed.

As defined in early 1967, the Orbital Workshop (OWS) was a semipermanent orbital station adaptable to many types of research. Besides the workshop itself, the station comprised an air lock, through which astronauts could leave the workshop for extravehicular activity, and a Multiple Docking Adapter (MDA) carrying extra docking ports to permit additional spacecraft, experiment modules, and supply vehicles to be attached. In 1966, AAP acquired a major scientific project, the Apollo Telescope Mount (ATM), which consisted of six high-resolution telescopes designed for systematic study of the Sun. Missions of twenty-eight and fifty-six days were planned, and longer stays were projected.

Technical problems multiplied as the OWS missions were planned in detail, and following the fatal fire in Apollo 1 (January 27, 1967) the OWS project was seriously underfunded. Until the Apollo program was successfully completed, AAP garnered little support. Between 1966 and 1969 most of the projected AAP missions had to be deleted; only the OWS remained. An alternative plan, long considered as an advanced project to follow the wet workshop, was to equip the S-4B stage as a laboratory before flight (a "dry workshop"). Such a laboratory, however, could only be launched on the much larger Saturn 5, and no Saturn 5 could be made available until the success of the lunar landing was assured. By early 1969, however, the dry workshop appeared to be the only solution to the many problems of the OWS missions. On July 22, 1969, the change to a dry workshop was implemented, and the development of Skylab began.

The redefined OWS (given its new name, Skylab, in February, 1970) comprised four modules. The workshop itself was an S-4B stage outfitted on two levels as a laboratory and living space. On the lower level, separate compartments were provided for scientific work, dining, sleeping, and personal hygiene. The upper level was mostly open; storage lockers for film, food, clothing, and scientific equipment lined the walls, and two small air locks permitted extension of scientific instruments into space. Atop the workshop at launch sat the air-lock module, basically a tunnel capable of being sealed at each end and equipped with a hatch opening to space. On its outer structure the air lock carried tanks of oxygen and nitrogen for the OWS atmosphere and the electrical power system for the cluster (batteries, chargers, and regulators connected to two large arrays of solar cells, one on the ATM and one on the OWS). The upper end of the air lock opened into the MDA, where the control console for the ATM was located. Topping off the launch stack was the ATM itself, mounted on a truss structure hinged to permit deployment of the telescope canister at right angles to the long axis of the cluster. The air lock, MDA, and ATM were enclosed at launch in a shroud that was jettisoned in orbit. The entire cluster measured about 30 meters in length and weighed nearly 100,000 kilograms at launch.

Major experiment programs planned for three manned missions involved medical investigations, to determine the effects of long-term weightlessness, and solar physics investigations. In 1970, a third set of experiments was added — a group of Earth-pointed sensors designed to explore the utility of human-tended instruments in studying Earth. Adding this package required changing the planned inclination of the orbit from 30 degrees to 50 degrees so the OWS could fly over the entire United States (as well as most of the populated area of the world).

Skylab progressed through 1970 toward a launch date in July, 1972. By 1971, however, delays in development pushed that launch date forward to April, 1973. While the cluster modules were built and flight operations plans developed, launch facilities at Kennedy Space Center (KSC) were modified for the project. Launch Complex 39 was built for the Saturn 5, and a 39-meter pedestal was built to raise the smaller Saturn 1B (to be used for crew launches) to mate properly with ground-support equipment. Skylab cluster modules began arriving at KSC in July, 1972, for the most extensive preflight test program ever conducted for a manned mission. Testing soon required one more postponement of the launch; it was rescheduled for May 14, 1973. By late April, both launchpads at Launch Complex 39 were occupied simultaneously for the first time ever: the workshop on its Saturn 5 at Pad 39A and the first crew's Apollo command and service module on its Saturn 1B at Pad 39B.

Skylab 1 was launched from KSC at 1:30 P.M. eastern daylight time, May 14, 1973, on the last Saturn 5 ever to be launched. The orbital cluster went into its 435-kilometer orbit as planned, but before the first revolution was completed the mission was in serious trouble. Temperature readings on the skin of the OWS were off the high end of the scale (above 82 degrees Celsius), and neither of the two solar panel arrays on the workshop was producing power. Flight controllers soon deduced that the workshop had lost its micrometeoroid shield during launch. This shield, a thin metal girdle surrounding the OWS (designed to prevent micrometeoroids from striking the OWS), had been tightly secured to the workshop at launch but in orbit became detached, standing 13 centimeters off the outer surface of the OWS. (Thermal engineers later incorporated a passive thermal control system into the shield, using

coatings on either side to regulate the flow of heat from the shield to the workshop.) When the shield was ripped away in orbit, the workshop surface absorbed much of the solar heat to which it was exposed. In the week after launch, the temperature inside the workshop approached 50 degrees Celsius, far above tolerable limits, and engineers worked frantically to design a temporary shield that could be taken up and installed by the first crew. A backup workshop had been built, but to abandon the first one and postpone the mission until the backup could be prepared for launch would require some $250 million, which the project could not afford.

In the eleven days following the launch of Skylab 1, two designs for a temporary shield were produced and tested: a parasol that could be extended through one of the air locks in the OWS and a "twin-pole sail," which could be stretched over two long poles by astronauts working outside the OWS. The first crew (Skylab 2, launched May 25, 1973) carried the parasol and successfully installed it; the temperature inside the OWS soon dropped to a tolerable level. The Skylab 2 crew discovered that one of the workshop's two solar arrays had been torn off and that the second was held down by a fragment of the micrometeoroid shield. After freeing it, they were able to complete their twenty-eight-day mission without incident.

A postmission investigation concluded that a design flaw, which had gone undetected throughout development, had allowed air pressure to lift the micrometeoroid shield away from the OWS. At maximum dynamic pressure, 63 seconds after lift-off, the shield had become detached. Pieces of the shield had ripped the latches that held down the solar arrays; one had been torn by the blast from the retro-rocket that separated the second stage from the OWS, and a strap of metal from the shield had wrapped around the second array, holding it against the OWS.

Crew members of Skylab 3 installed the twin-pole sail, further protecting the OWS against solar heating; their fifty-nine-day flight and the eighty-four-day mission of the third crew were completely successful. After the third mission ended on February 8, 1974, the OWS was shut down and left in a 435-kilometer orbit, where it was expected to stay for nine years. Tentative plans were made to visit the derelict workshop on an early flight of the shuttle orbiter so that a propulsion unit could be attached to boost the OWS into a much higher orbit. Too big to burn up when it reentered the atmosphere, the Skylab workshop was a potential hazard.

Yet events overtook plans; orbiter development was delayed, and atmospheric resistance pulled the OWS down much faster than had been predicted. In mid-1978, a team of engineers reestablished contact with the workshop and took control of its attitude in orbit. In this way it was possible to control its descent to some extent, since atmospheric drag was lower in certain attitudes. For ten months, flight controllers maneuvered the OWS by telemetry, trying to make sure that its final dive into the atmosphere would take place over an unpopulated area. On July 11, 1979, they successfully guided it to reentry over the southeastern Indian Ocean. Some pieces fell on the sparsely populated ranch country of southwestern Australia, but no injury or damage was reported.

Knowledge Gained

As intended, the Skylab missions accumulated a vast amount of scientific data, principally on the effects of long-term weightlessness on the human body and on the processes that take place on the Sun. Data from the three manned missions showed that no permanent ill effects resulted from nearly three months of weightlessness and that astronauts readapted rapidly to Earth's environment. Solar scientists were rewarded with more data than they could readily interpret; eventually, the ATM data would lead to a better understanding of how energy is produced on the Sun and how solar radiation is transmitted into space and eventually interacts with Earth's atmosphere. Earth observations showed that instruments and astronauts could gather information useful in oceanography, pollution control, and exploration for mineral resources. Finally, experiments in zero-gravity manufacturing suggested that the environment in orbit might enable advances in the production of semiconductor crystals and the casting of alloys that could not be made on Earth.

Context

Skylab had been conceived at a critical time for manned spaceflight. In 1965, the first lunar landing was at least four years in the future, and little effort could be spared for post-Apollo planning; at the same time, failure to get started on a new venture could lead to a serious hiatus in manned space-flight operations after Apollo. Support for expensive space projects was waning, and the nation faced internal and external problems that demanded attention. Like other federal programs, manned spaceflight felt the political pressure to reduce expenditures in the face of an expanding war in Vietnam and the Johnson Administration's costly social programs. In these circumstances, and lacking any mandate for a program comparable to Apollo, OMSF director George Mueller proposed to use the tremendous capacity developed for Apollo to produce useful results. His proposal attracted little support, even within NASA. Administrator James E. Webb was fully committed to the lunar landing but reluctant to propose any large subsequent project until he could feel sure of support from the White House and Congress. This support came only grudgingly, with the result that the AAP muddled along with an inadequate budget until Apollo succeeded and NASA settled on the space shuttle as its next major manned project. Only then did the OWS appear to bridge a gap between Saturn/Apollo and the shuttle, providing information that would be essential in further planning.

When Skylab became the only manned project available for several years, managers, eager to exploit this opportunity, tended to overload the missions with experiments. The result was increased difficulty in flight operations planning and crew training. The crews' workdays were crowded and poorly organized until ground-based personnel realized the difficulties of working in orbit, especially on the last mission. Despite all of its problems, however, Skylab was the most productive manned project ever flown in terms of its advancement of scientific knowledge and its proof of the utility of man in orbital operations.

Bibliography

Compton, W. David, and Charles D. Benson. *Living and Working in Space.* Washington, D.C.: Government Printing Office, 1983. An official overview of the Skylab program and all of its results, this volume details the origin of the program and all four of the manned missions. The launch accident and the repairs performed on subsequent missions are covered. Copiously illustrated, including several excellent shots of the ATM instrumentation and the Earth-resources cameras.

Cooper, Henry S. F., Jr. *A House in Space.* New York: Holt, Rinehart and Winston, 1976. Containing essays originally published in *The New Yorker,* this slim volume offers a readable account of the Skylab flights. Its journalistic tone is a refreshing change from other texts concerning spaceflight.

Cromie, William J. *Skylab.* New York: David McKay, 1976. A full description of the Skylab spacecraft, mission aims, and data collected on all the Skylab flights, this book is written for the beginning student of space exploration.

Ertel, Ivan D., and Roland W. Newkirk. *Skylab: A Chronology.* Washington, D.C.: Government Printing Office, 1977. The development of Skylab is traced from its conception to the uses of the data it brought back. Includes appendices on flight information and summaries of the missions.

W. David Compton

SKYLAB 2

Date: May 25 to June 22, 1973
Type of mission: Manned Earth-orbiting space station

The crew of Skylab 2 installed a makeshift sunshade to protect the orbital workshop from overheating, released the panel of solar power cells that had been jammed in the launch accident of Skylab 1, and went on to conduct a twenty-eight-day program of medical, astronomical, and Earth resources experiments, establishing a record for the longest manned Earth-orbiting flight.

PRINCIPAL PERSONAGES

CHARLES "PETE" CONRAD, the Mission Commander
PAUL J. WEITZ, Pilot
JOSEPH P. KERWIN, physician and scientist

Summary of the Mission

The crew of Skylab 2 was launched on a Saturn 1B rocket from Launch Complex 39B at Kennedy Space Center at 9 A.M. eastern daylight time on May 25, 1973, following eleven days of hectic activity. The launch had been postponed while engineers at the Manned Spacecraft Center and Marshall Space Flight Center sought ways to repair the overheated and underpowered orbital workshop. The crew carried with them three improvised sunshields: a "parasol" that could be extended through an air lock located in the center of the area exposed to the Sun, a fabric "sail" that astronauts might attach to the workshop while remaining in the command module, and a "twin-pole sail" to be stretched over two long poles by crewmen working from inside the workshop. They also carried several metal-cutting tools and other devices that engineers hoped would enable the astronauts to free the jammed solar panels.

Arriving at the workshop seven and one-half hours after launch, Charles "Pete" Conrad and his crew described the damaged Skylab vehicle and sent back television pictures to Mission Control. They confirmed that one array of solar panels was missing and that the other was held down by a metal strap, apparently debris from the lost meteoroid shield. No other damage was obvious, and the area to be covered was clear of debris. Conrad then soft-docked the command module to the workshop, using three of the twelve docking latches. After discussing strategy with flight controllers, the astronauts began an attempt to free the solar array.

After the astronauts had put on their pressure suits, Conrad maneuvered the command module nearer the workshop while Paul Weitz stood up in the open hatch and grap-

pled for the solar array with a long "shepherd's crook." In the absence of gravity, Weitz could not exert enough force to break the restraining strap, and after nearly an hour of frustrating labor Conrad abandoned the attempt and returned the command module to the docking port to secure it for the night. Three attempts at docking failed; the crewmen were finally forced to don their suits, depressurize the command module, and disassemble the docking probe to wire around an apparent electrical fault. It was nearly midnight (eastern daylight time) when they successfully docked, ending a twenty-two-hour working day.

Shortly after noon the next day, the crew entered the workshop, finding a dry, desertlike heat approaching 47 degrees Celsius but tolerable for short periods. Over the next four hours they performed several tasks to start up the workshop systems. Late in the afternoon they attached the canister containing the parasol to the solar air lock and pushed the parasol out some 6 meters, using extension rods. The parasol opened, and its four supporting arms were extended, spreading the aluminum-coated nylon-and-Mylar fabric shield to cover an area 6.7 by 7.3 meters. The external temperature of the workshop immediately began dropping; internal temperature followed more slowly, falling below 35 degrees Celsius within twenty-four hours and eventually stabilizing at 24 degrees Celsius.

Meanwhile, ground-based technicians were constructing a mock-up of the workshop solar array, using descriptions and television pictures provided by the crew, to work out procedures for freeing it. After experimenting in the large water tank at the Marshall Space Flight Center, where the buoyancy of water provided an approximation of weightlessness, technicians concluded that the job could be done without danger to the astronauts. Tests continued, for the power shortage in the workshop was not yet critical.

In the cooling workshop, Conrad, Kerwin, and Weitz began their program of experiments on May 28, taking blood samples and performing some initial medical evaluations and checking out the solar instruments in the Apollo Telescope Mount. At midday they conferred with Mission Control, reporting that they were in good physical shape and were adapting easily to weightlessness. Physicians in Houston welcomed this news, for they had feared that the freedom of motion in the cavernous workshop might trigger nausea and

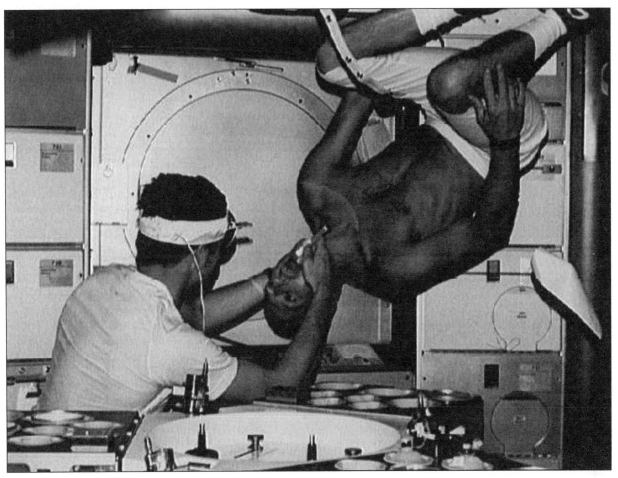

Skylab 2 Commander Charles Conrad is seen undergoing a dental examination. (NASA)

vomiting while the astronauts' inner ears adjusted to weightlessness. One out of every three Apollo crewmen had suffered from this malaise, and it was not at all understood.

On May 29, the Skylab 2 crew began operations, even though power was limited. Kerwin made four series of solar observations, and his crewmates prepared the Earth-observing instruments for use the following day. Conrad reported difficulty in running one of the major medical experiments, a bicycle ergometer that measured his metabolism while he performed physical work. In the absence of gravity, it proved difficult to ride the bicycle; the harness designed to hold the subject in position did not work as expected, and in the heat the prescribed exercises could not be completed at the expected levels of exertion. A week would elapse before the crew could use the ergometer as it was intended.

The first use of the Earth-sensing instruments on May 30 made clear the importance of freeing the solar panel. The workshop's batteries were depleted to about half their capacity, whereupon they were cut out of the circuit — a feature

designed into the system to protect the batteries from damage by excessive discharge. One battery failed to reconnect to the system after recharging, leaving the workshop with only 4,200 watts of power. Of this supply, 3,600 watts were required for operation of the workshop's essential systems, leaving only 600 watts for experiments and seriously curtailing the science program. On June 7, engineers on the ground held a long conference with the crew, discussing procedures they had worked out for releasing the solar panels. Conrad was only cautiously optimistic about success, since they had not trained for this exercise, and estimated their chances at about 50 percent.

The next day, Conrad and Kerwin suited up and went outside to attempt the repair. After considerable difficulty caused by the lack of adequate body restraints, Kerwin used a cutting tool (a pair of tree-trimming shears attached to an 8.3-meter pole) to snip the 1-centimeter metal strap holding the array. With the aid of a hook attached to a long nylon rope, Conrad and Kerwin were able to break a frozen hydraulic cylinder, allowing the array to come free. In a few hours, the solar panels extended fully, adding 3,000 watts to the power supply and clearing the way for the planned science program. The astronauts completed their three-and-one-half-hour excursion by changing a film magazine on one of the solar telescopes and freeing a jammed door on another.

After the two major repairs, Skylab 2 could proceed more or less as planned. Investigators were eager to make up for lost time, however, and the crew occasionally complained that flight controllers were rushing them through the experiments and not allowing time for the difficulties of working in weightlessness. Soon, however, they settled into a routine of experiments, housekeeping chores, and occasional recreation that seemed humdrum after the excitement of the first two weeks.

On June 19, Conrad and Weitz made their last excursion outside the workshop to retrieve exposed film from the solar telescopes. Before going back inside, Conrad, following instructions from Houston, sharply rapped one of the malfunctioning battery modules with a hammer; technicians on the ground had decided that the problem was a stuck relay that could be released by mechanical shock. Their diagnosis proved correct. The battery immediately began recharging and soon became fully operational.

Three days later, on their twenty-ninth day in space, the crew of Skylab 2 left the workshop. In spite of their difficulties they had completed nearly all of the pre-planned medical experiments and 80 percent of the scheduled solar observations. Completed Earth observations, however, came to only 60 percent of what had been expected, largely because most of them had been scheduled during the first half of the mission, when the power shortage was most acute.

The Skylab 2 command module landed in the Pacific Ocean 1,340 kilometers southwest of San Diego, California, at 9:50 A.M. eastern daylight time on June 22, 1973. Half an hour later, crew and spacecraft were taken aboard the USS *Ticonderoga*, having set a new world endurance record for manned flight in Earth orbit (28 days, 49 minutes, and 49 seconds).

Knowledge Gained

Skylab 2 made its major contributions in space physiology and solar physics. The astronauts were the subjects of three major medical studies of the human body's adjustment to weightlessness: effects on the muscular and skeletal systems, effects on the cardiovascular system, and effects on the vestibular system (the inner ear, which controls balance). Earlier spaceflights had indicated that there was a possibility of serious physiological harm from long-duration missions. Skylab's medical program was designed to study physiological changes during weightlessness over progressively longer periods to see whether such changes continued.

Skylab 2 showed that a loss of bone and muscle tissue continued throughout the mission, an effect attributed to the absence of compressive forces on weight-bearing bones (similar results are seen in patients confined to bed for long periods). Absolute losses were not large enough to constitute a

problem, but the fact that losses continued for twenty-eight days implied some risk for missions lasting several months.

The cardiovascular system was markedly affected by nearly a month of weightlessness. In the absence of gravity, the heart and major blood vessels function at a slower rate. This deconditioning was measured in flight by applying a partial vacuum to the lower body, simulating the effect of gravity by pulling blood into the legs; the resulting stress on the heart was measured by changes in blood pressure and pulse rate. On Skylab 2, the crew had difficulty completing this exercise, developing near-fainting symptoms. At one point, Kerwin strongly urged discontinuing the experiment, but investigators persisted in order to determine whether adaptation would occur. For the first Skylab crew it did not, and they did not regain their normal preflight responses for nearly three weeks after returning to Earth.

The bicycle ergometer and metabolic analyzer showed that the body's tolerance for exercise did not diminish during the mission. When they returned to normal gravity, however, the crewmen could no longer perform this exercise at preflight levels of efficiency, which indicated that some adaptation to weightlessness had occurred. Again, readjustment to Earth's gravity after the mission required almost three weeks.

Response of the human vestibular system to zero gravity was tested during flight by seating the crewman in a rotating chair and having him make rapid up-and-down head motions. The Skylab 2 crew showed no tendency to become ill under the most severe conditions tested, whereas on the ground all the crew members could be brought to the verge of nausea. Nor did they succumb to motion sickness on entering orbit, as many Apollo astronauts had.

The Apollo Telescope Mount produced the largest return of scientific data from Skylab 2. The astronauts spent 117 hours at the control console and took more than twenty-eight thousand photographs. Skylab's solar telescopes had been designed to allow simultaneous observation of different layers of the Sun: its thin, hot corona (outer atmosphere), visible on Earth only during solar eclipses; the chromosphere (outermost surface layer), which radiates most of the Sun's visible light; and the underlying photosphere. The instruments took advantage of Skylab's orbital altitude to study those wavelengths that do not penetrate Earth's atmosphere — ultraviolet and X-Rays. A white-light coronagraph, which blocks off the Sun's disk so that the faint corona can be seen, produced more data than had been obtained in all Earth-based studies preceding the mission.

Although thorough analysis of the data would require years, astronomers immediately recognized the immense value of the Skylab data, which will ultimately contribute to an understanding of the processes taking place in the Sun and their effects on Earth's atmosphere. The coronagraph, for example, along with the X-ray telescopes, confirmed the existence of "coronal holes," huge voids in the corona through which subatomic particles flow outward; these "holes" will eventually cause magnetic storms and atmospheric disturbances on Earth.

Skylab revealed a much more active Sun than scientists had anticipated and facilitated the investigation of phenomena for which little information had previously existed.

The Earth-observing instruments, added late in the development of the Skylab science program, were in large part designed to assess the value of orbiting sensors for the study of Earth. Among the methods evaluated were those used for the determination of changing land utilization, the search for undetected mineral resources, geodesy (measurement of the detailed shape of Earth), detection of water pollution, and crop surveys.

Besides the major experiment projects, Skylab 2 conducted some experiments relating to manufacturing in space. Using a specially designed furnace, the crew melted and resolidified samples of metals and semiconductors, experimented with welding and brazing metals, and attempted to cast spherical specimens of metals in zero gravity. Such evaluations provided basic data on the possible use of a zero-gravity facility (for example, a space station) to carry out processes not easily effected on Earth.

Context

Skylab 2 was the National Aeronautics and Space Administration's first step in a program which, manned space-flight enthusiasts believed, would establish the value of man in scientific research in space. To that end, the mission made its intended contribution, but its more immediately important contribution was the demonstration of man's adaptability to unexpected situations. The repair in orbit of the fatally damaged workshop literally saved the entire Skylab program. Although a backup workshop had been built and was available for launch, the estimated $250 million it would cost might have been an insurmountable barrier to continuing the program. In a time of shrinking space budgets, declining public interest in space, and growing public antipathy toward large technological projects, the resourcefulness of the Skylab team revived something of the spirit of earlier times in the space program.

Since it was only the first of three missions, Skylab 2 can hardly be considered in isolation. Still, on the first long-term manned flight, the easy adaptation of the crew to the weightless environment and their general good health and spirits throughout the mission boded well for subsequent missions and for future Earth-orbital projects. After an initial period of adjustment, crew and flight planners worked out a system for scheduling their daily activities that was improved further on the later missions.

Bibliography

Belew, Leland F., ed. *Skylab: Our First Space Station*. NASA SP-400. Washington, D.C.: Government Printing Office, 1977. Description of the entire program (evolution, plans, spacecraft, and missions); for high school and college students.

Compton, W. David, and Charles D. Benson. *Living and Working in Space: A History of Skylab*. Washington, D.C.: Government Printing Office, 1983. The officially sponsored history of the Skylab project, this volume exhaustively details the development of Skylab from earliest concepts to the missions and their results. Two chapters discuss the launch accident and the first manned mission. The book is intended for serious students of the space program but is not beyond the general reader. Numerous illustrations include several spectacular photographs from the Apollo Telescope Mount instruments and the Earth resources cameras.

Cooper, Henry S. F., Jr. *A House in Space*. New York: Holt, Rinehart and Winston, 1976. This slim volume is a collection of the author's articles first published in *The New Yorker* magazine — a chatty journalistic account of all three Skylab flights, based on transcripts of the air-to-ground communications and interviews with participants.

Cromie, William J. *Skylab*. New York: David McKay Co., 1976. Written mainly for high school students, this book describes the spacecraft, missions, and experiments on all three Skylab missions.

Ertel, Ivan D., and Roland W. Newkirk. *Skylab: A Chronology*. Washington, D.C.: Government Printing Office, 1977. A chronological tabulation of important events in the development of Skylab (and some other early space station concepts) from 1960 to 1975. Appendices contain summaries of flight data and other factual information.

Ise, Rein, ed. *A New Sun: The Solar Results from Skylab*. NASA SP-402. Washington, D.C.: Government Printing Office, 1979. A discussion of the Apollo Telescope Mount solar telescopes, the data they yielded, and their importance to solar physics. Profusely illustrated.

Johnston, Richard S., Lawrence F. Dietlein, and Charles A. Berry, eds. *Biomedical Results from Skylab*. NASA SP-377. Washington, D.C.: Government Printing Office, 1977. This volume is directed toward the biomedical professional and summarizes the medical findings from all three Skylab missions.

Lundquist, Charles A., ed. *Skylab's Astronomy and Space Sciences*. NASA SP-404. Washington, D.C.: Government Printing Office, 1979. A discussion of the corollary experiments on Skylab: stellar and galactic studies and Earth's atmosphere. For high school and college students.

National Aeronautics and Space Administration. *Skylab Explores the Earth*. NASA SP-380. Washington, D.C.: Government Printing Office, 1977. A technical summary of Skylab's Earth resources experiments and their results, for a technically literate audience. Includes many photographs illustrating the scope of the Earth-sensing instruments and the crew's observations of surface phenomena.

Summerlin, Lee B., ed. *Skylab, Classroom in Space*. NASA SP-401. Washington, D.C.: Government Printing Office, 1977. A discussion of the experiments devised by high school students to be performed on Skylab and the science demonstrations conducted on the last two missions. Contains many excellent photographs.

W. David Compton

SKYLAB 3

Date: July 28 to September 25, 1973
Type of mission: Manned Earth-orbiting space station

Skylab 3 continued the occupation and operation of the Skylab orbital workshop for fifty-nine days. After an initial period of motion sickness, the crew adapted well to weightlessness and exceeded the objectives of most of their experiments.

PRINCIPAL PERSONAGES
    ALAN L. BEAN, Mission Commander
    JACK R. LOUSMA, Pilot
    OWEN K. GARRIOTT, the scientist-pilot

Summary of the Mission

Skylab 3 began with the launch of the crew from Launch Complex 39B, Kennedy Space Center, at 7:10 A.M. eastern daylight time on July 28, 1973. Their command module docked with the orbiting Skylab workshop eight and one-half hours later, and the crew began activating the laboratory's systems shortly thereafter. All three crewmen experienced motion sickness on entry into orbit, the first time an entire crew had been thus affected on any manned mission. For the first three days, Alan L. Bean, Jack R. Lousma, and Owen K. Garriott were partially incapacitated by nausea. Flight controllers adjusted the astronauts' work load to allow for prolonged rest periods during each day. Only on the sixth day in orbit were they able to handle a normal work schedule.

A serious anomaly appeared on launch day when one of the command module's attitude control thrusters began to leak; four such units were used to orient the spacecraft and served as the backup system to bring it out of orbit in case the main propulsion system failed. After some analysis, engineers concluded that the mission could be completed with only three thrusters operating, but on the sixth mission day a second unit developed an apparently similar leak. Fearing that the entire attitude control system might fail, in which case the crew could not return to Earth, mission officials ordered a rescue mission to be made ready. Launch teams at Kennedy Space Center began round-the-clock preparation of the launch vehicle and spacecraft that had been scheduled to be used on the last Skylab mission. The command module was to be modified to carry five men, so that a crew of two could fly to the workshop and bring back the three Skylab astronauts. Preparations continued until engineers satisfied them-

selves that no generic problem existed and that the Skylab 3 command module could return safely using modified procedures. The problem was later traced to loose fittings in the lines carrying oxidizer to the thrusters.

Once they had recovered from nausea, the Skylab 3 crew set out to make up for lost time. On August 1, program officials extended their stay from the planned fifty-six days to fifty-nine. Experience on Skylab 2 indicated that recovery closer to a seaport was desirable, to minimize the time spent aboard ship after recovery, and the new mission duration did allow recovery closer to a port. On August 3 the astronauts made the first of thirty-nine Earth-observing passes; four days later Garriott activated the Apollo Telescope Mount (ATM) for solar observations. Medical experiments occupied a part of every mission day. This crew never found their days to be "humdrum," as Joseph P. Kerwin had described the last two weeks of Skylab 2. They felt no need for more than six hours of sleep, and during the second week commander Bean told Mission Control that they were not working as hard as they had in preflight training. Flight planners then increased their workday from eight hours to twelve. This made Skylab 3 highly productive, but it set a pace that would create problems for the crew that followed.

Eleven days into the mission, Garriott and Lousma went outside the workshop to rig the twin-pole sail, or sunshade. Assembling two 16.8-meter poles from sections of tubing, Garriott secured them to the ATM and extended them over the parasol deployed by the first crew. He and Lousma then hoisted the 3.6-by-7.3-meter sunshade by ropes until it covered the exposed area. The difficulty of working in zero gravity stretched the job from an anticipated two hours to four. Lousma then retrieved and replaced film cartridges in four ATM instruments, removed a balky door latch from another, and set out three new experiments.

Mechanical problems arose in mid-August. A dehumidifier in the environmental control system sprang a leak. An electrical short circuit destroyed one of two video display tubes on the ATM control console, and the cooling system for the workshop electronics appeared to be leaking. A continuing annoyance to Mission Control was the erratic operation of the rate gyroscopes, which sensed rotation of the workshop and provided input to the attitude control system computer. These had gone awry after the launch accident, and the

Skylab 3 crew had brought replacements. On their second trip outside the workshop, the crew connected the new gyroscopes into the system, giving ground controllers better ability to regulate the workshop's maneuvers. Finally, the mechanism that drove a scanning antenna on one of the Earth resources experiments failed. This and the other defects would be left for repair by the last crew.

While the major experiment programs occupied most of their time, the Skylab 3 crew found opportunities to conduct several small experiments and demonstrations. Owen Garriott brought some live minnows with him, along with fertilized minnow eggs and two common spiders, to see how other living organisms might behave in weightlessness. He observed that the full-grown minnows seemed unable to orient themselves; they swam in tight circles, perhaps searching for the gravity force that would determine a downward direction. The hatchlings from the minnow eggs, however, seemed to have no difficulty, and the spiders, after an initial period of apparent confusion, soon learned to spin normal webs.

Among the experiments important to future spaceflight was a series of tests of astronaut maneuvering aids, which the Skylab 3 crew tested in the large upper compartment of the workshop. The most complex device, a gyroscope-stabilized, thruster-propelled manned maneuvering unit, proved extremely useful and easy to operate in spite of being large and cumbersome (it resembled a large chair into which the astronaut strapped himself). Its ease of operation enabled Garriott to learn to use it in a short period of on-the-job training. A hand-held gas pistol much like the one used on early Gemini flights was less precise and controllable, and a foot-controlled unit (called "jet shoes") was less useful still. The large unit was later to be developed for use on space shuttle flights.

Weekly meetings of program officials evaluated the crew's performance during Skylab 3 and assessed the state of the workshop systems. Had any serious anomaly developed, the flight could have been terminated early. Yet no problems arose to justify ending the flight prematurely, and the crew continued to collect scientific data for the full fifty-nine days. In early September, Bean asked for an extension of the mission, but his request was turned down because medical investigators wanted to evaluate the fifty-nine-day information. On September 25, the Skylab 3 astronauts deactivated the workshop, powered up their command module, and returned to Earth. They were picked up by the recovery ship USS *New Orleans* late that afternoon, about 400 kilometers southwest of San Diego.

Knowledge Gained

Skylab 3 added substantially to the store of information about the Sun and Earth. The crew returned nearly twenty-five thousand photographs of the Sun, almost seventeen thousand photographs of Earth, and more than twenty-five thousand meters of magnetic tape containing Earth resources data. In all experiment categories they had exceeded preflight plans, in several cases by more than 50 percent. They had observed several rare occurrences on the Sun, among them coronal transients (outbursts of energy and matter from the Sun that distort the corona as they pass through it).

Of more importance to manned spaceflight, Skylab 3 showed that over long periods the human body slowly acclimates to weightlessness. Skylab 2 astronauts had experienced deterioration of their cardiovascular systems throughout their flight; the Skylab 3 crew returned in better condition. Physicians concluded that adaptation began between thirty and forty days into the mission. After their disabling nausea at the outset, the Skylab 3 astronauts experienced no further trouble with motion sickness. Nevertheless, the episode worried officials who were planning missions for the space shuttle; they were contemplating flights lasting no more than seven days, and the prospect of illness among the crew for half that time was not encouraging.

Context

Skylab 3, lacking the urgency and drama of the first manned mission, was more indicative of what humans might accomplish on long-term Earth-orbital missions. This, indeed, was the theme that the National Aeronautics and Space Administration (NASA) had emphasized in justifying the first post-Apollo project: the use of Apollo-developed capability to produce results that would be useful on Earth. Both the Skylab 3 crew and their flight planners strove to maximize the time spent on experiments, and in the context of a normal mission they developed considerable facility in using the available time to best effect.

The major science projects on Skylab meshed well with national concerns in 1973. A growing shortage of energy (to be exacerbated by an oil embargo imposed by Arab states later in the year), concern for environmental pollution, and a growing fear of depletion of natural resources all heightened concern for problems that Skylab's Earth resources experiments might help to solve. Although the contributions of these experiments were actually minimal, NASA officials and influential newspapers continually stressed the practical aspects of the manned space program.

Bibliography

Belew, Leland F., ed. *Skylab: Our First Space Station.* NASA SP-400. Washington, D.C.: Government Printing Office, 1977. This publication describes the Skylab program in detail (its evolution, plans, spacecraft, and missions). Appropriate for high school and college students.

Compton, W. David, and Charles D. Benson. *Living and Working in Space: A History of Skylab*. Washington, D.C.: Government Printing Office, 1983. The official history of Skylab, this volume details the development of Skylab from earliest concepts to the missions and their results. Includes discussion of the launch accident and the first manned mission. Intended for serious students of the space program but is not beyond the general reader. Numerous illustrations include several spectacular photographs from the Apollo Telescope Mount instruments and the Earth resources cameras.

Cooper, Henry S. F., Jr. *A House in Space*. New York: Holt, Rinehart and Winston, 1976. This slim volume, a collection of the author's articles first published in *The New Yorker* magazine, provides an informal, journalistic account of all three Skylab flights. Based on transcripts of the air-to-ground communications and interviews with participants.

Cromie, William J. *Skylab*. New York: David McKay Co., 1976. Written mainly for high school students, this book describes the spacecraft, missions, and experiments on all three Skylab missions.

Ertel, Ivan D., and Roland W. Newkirk. *Skylab: A Chronology*. Washington, D.C.: Government Printing Office, 1977. A chronological tabulation of important events in the development of Skylab (and some other early space station concepts) from 1960 to 1975. Includes appendices with summaries of flight data.

Ise, Rein, ed. *A New Sun: The Solar Results from Skylab*. NASA SP-402. Washington, D.C.: Government Printing Office, 1979. Included in this volume is a discussion of the Apollo Telescope Mount solar telescopes, the data they yielded, and their importance to solar physics. Profusely illustrated.

Johnston, Richard S., Lawrence F. Dietlein, and Charles A. Berry, eds. *Biomedical Results from Skylab*. NASA SP-377. Washington, D.C.: Government Printing Office, 1977. Written for the biomedical professional, this volume summarizes the medical findings from all three Skylab missions.

Lundquist, Charles A., ed. *Skylab's Astronomy and Space Sciences*. NASA SP-404. Washington, D.C.: Government Printing Office, 1979. Includes a discussion of the corollary experiments on Skylab: stellar and galactic studies and Earth's atmosphere. For high school and college students.

National Aeronautics and Space Administration. *Skylab Explores the Earth*. NASA SP-380. Washington, D.C.: Government Printing Office, 1977. This technical summary of Skylab's Earth resources experiments and their results is written for a technically literate audience. Includes many photographs illustrating the scope of the Earth-sensing instruments and the crew's observations of surface phenomena.

Summerlin, Lee B., ed. *Skylab, Classroom in Space*. NASA SP-401. Washington, D.C.: Government Printing Office, 1977. A discussion of the experiments devised by high school students to be performed on Skylab and the science demonstrations conducted on the last two missions. Contains many excellent photographs.

W. David Compton

SKYLAB 4

Date: November 16, 1973, to February 8, 1974
Type of mission: Manned Earth-orbiting space station

*Skylab 4 was the last flight to the United States' first space station.
The three astronauts on Skylab 4 spent a total of eighty-four days in
Earth orbit, making this mission the longest manned U.S. spaceflight
up to that time.*

PRINCIPAL PERSONAGES
    GERALD P. CARR, Mission Commander
    EDWARD G. GIBSON AND
    WILLIAM R. POGUE, astronauts

Summary of the Mission

The Skylab 4 mission was launched from the Kennedy
Space Center on November 16, 1973. Gerald Carr, Edward
Gibson, and William Pogue were ferried to the Skylab space
station via an Apollo spacecraft, which was boosted into orbit
by a Saturn 1B launcher. Each of these astronauts was
embarking on his first flight into space.

The Skylab space station had been used prior to the
Skylab 4 flight by two other crews. Skylab 2 astronauts
Charles Conrad, Joseph Kerwin, and Paul Weitz spent a total
of twenty-eight days aboard Skylab. Skylab 3 astronauts Alan
Bean, Owen Garriott, and Jack Lousma more than doubled
their predecessors' stay by remaining aboard Skylab for fifty-
nine days.

The launch of Skylab 4 did not occur as originally scheduled. Hairline cracks in the stabilizing fins of the Saturn 1B
launch vehicle were discovered, and the launch was delayed
nine days in order for engineers to replace the fins. Once
orbit was reached, it took three attempts for the crew to
complete a successful docking with Skylab. The three astronauts orbited at an altitude of approximately 435 kilometers.

Many major scientific experiments were planned for the
Skylab 4 mission. This flight would be longer than the flights
of the previous two missions. Thus, the experiments aboard
Skylab 4 could be increased in number and sophistication.
The crew of Skylab 2 had spent a major portion of its twenty-eight days in space repairing faulty equipment, and space
officials hoped that the Skylab 4 mission would not be similarly compromised.

Yet the Skylab 4 crew also had problems with equipment.
Three control-moment gyroscopes, mounted on three sides
of the Apollo Telescope Mount, were used to control the

position of the spacecraft. (The Apollo Telescope Mount was
used for observations of the Sun, utilizing a variety of ultraviolet and X-ray instrumentation.) Shortly after the crew's
arrival, one of the three gyroscopes failed, and the performance of a second gyroscope began to deteriorate. This malfunction required the use of the station's attitude control system, normally reserved for large attitude changes. Aside from
the gyroscope failures, a few smaller repair tasks also demanded the crew's attention.

The National Aeronautics and Space Administration
(NASA) had proposed four major objectives for Skylab. The
Skylab crews were to make Earth resources observations, aid
scientists in their study of materials processing in weightlessness, advance astronomical knowledge about objects in the
universe, and contribute to a better understanding of manned
spaceflight capabilities and basic biological processes. All four
of these broad goals were met during the Skylab 4 mission.

To prepare for the Earth resources observations, Carr,
Gibson, and Pogue had spent twenty hours in lectures given
by nineteen scientists in that field. This background provided
the crew with an insight into the significance of particular
phenomena and features on Earth, the type of data desired,
and procedures for conducting the observations. Some 165
features and phenomena were identified prior to the mission
as those with which scientists were principally concerned.

The equipment used by the Skylab 4 crew included
binoculars and two handheld cameras, a 70-millimeter
Hasselblad and a 35-millimeter Nikon. A variety of camera
lenses were used to provide narrow to wide fields of view.
Color Ektachrome slide film was used for most of the photography, but some photographs were taken with Ektachrome
infrared film. Thousands of photographs were taken by the
crew. In addition, more than 850 verbal descriptions were
made.

A variety of geologic features were observed during the
mission. These included desert regions, global tectonic features, meteoritic impact craters, and volcanic regions. Studies
of crops and their relationship to the environment were performed. Crew members also observed land-sea interaction
and certain oceanic events. In addition, meteorological observations were made as time and conditions allowed.

Carr, Gibson, and Pogue observed major duststorms while
passing over Africa. In addition, extensive foliage burning was

observed. Observations of other Earth features provided data important for understanding desert formation. Studies of the southwestern United States and the adjacent area of Mexico contributed information about major fault zones in Baja California, southeastern California, and Sonora, Mexico. Other tectonic studies focused on the Alpine fault in New Zealand, the African rift zone, the Zagros and Asian Caucasus mountains, Guatemalan fault zones, and the Atacama fault in Chile.

The Manicouagan impact area in Quebec, Canada, one of many meteoritic impact areas on Earth, was observed from Skylab 4. Carr, Gibson, and Pogue also had a rare opportunity to observe and photograph an erupting volcano, that of Sakurajima in Japan. The Skylab 4 crew took photographs of this feature, which included a set of stereophotographs, the first such images produced in space. Glacial development in the Gulf of Saint Lawrence, eddies in ocean currents, upwelling areas, and many other oceanic phenomena were observed and photographed. Meteorological experiments included height determinations of multilayered clouds; analyses of cloud structure, cloud movement, and cloud patterns; and tropical storm studies.

Materials processing experiments conducted during Skylab 4 were continued from previous Skylab studies. In Earth orbit, the extremely low gravity level — or microgravity — allows for the production of certain materials that cannot be processed on Earth. Skylab 4 experimentation with materials processing confirmed that space factories do indeed have a place in the future of space exploration.

The approach of the comet Kohoutek presented an unusual opportunity to add to the scientific knowledge of astronomical phenomena. The proximity of the comet, discovered earlier in the year, made it possible for the crew to study it without interference from Earth's blanket of air. Observations of Kohoutek included those made by Carr and Pogue during a seven-hour spacewalk on Christmas Day, 1973, and those made during a spacewalk four days later. Observations of the comet were also made from the space station itself.

One of the primary interests of scientists involved with Skylab was a study of the Sun. The Skylab 4 crew was able to observe the birth, development, and death of a medium-sized solar flare. It was the first time such an event was observed from space. Much information about the Sun was collected. More than seventy-three thousand photographs of the Sun were taken during the eighty-four-day mission. Other solar observations included a study of the corona in white light and a study of the corona in extreme ultraviolet light.

Most of the solar observations were made from the Apollo Telescope Mount. This instrument, controlled from the multiple docking adapter, weighed about 10,000 kilograms on Earth. Results from the observations were in either photographic or spectrographic format. (Spectrographic data reveal the components of light emitted by the Sun.)

Many biological experiments were performed on this final Skylab mission. The most obvious gain in this area would be determining the reaction of the human body to long-term weightlessness. The Skylab 4 crew recovered from the effects of microgravity more quickly than had any previous long-endurance flight crew. During the mission Carr, Gibson, and Pogue actually grew taller by at least a few centimeters as a result of the stretching of the spinal column and the shift in body fluids from the lower to upper body extremities. Yet the astronauts returned to their preflight height after their return. The crew spent much time on the station's bicycle ergometer and portable treadmill. A strict exercise regimen in space probably contributed to the astronauts' excellent condition upon their return to Earth.

Additional biological and medical experimentation included studies of body fluids, red blood cells, bones, human vestibular function, metabolic activity, and mineral balance.

Student projects were also a part of Skylab 4. The projects, selected by NASA and the National Science Teachers' Association, utilized the space environment to study the effects of weightlessness on a variety of activities.

The flight came to an end on February 8, 1974, with the crew's Apollo spacecraft landing in the Pacific Ocean southwest of San Diego, California. Skylab 4 was the longest manned flight up to that time. The mission lasted 2,017 hours and 16 minutes. It included four spacewalks totaling 22 hours and 21 minutes. The final mission of Skylab had provided scientists with important information about Earth, the Sun, materials processing in space, and the human body and its reaction to weightlessness.

Knowledge Gained

When the Skylab program was first proposed, space planners had specific objectives in mind. The flights of the three manned Skylab missions, however, went beyond the anticipated goals of the project.

The potential of studying Earth from orbit was realized early in the space program. Yet most spaceflights were not long enough to provide a sustained overview. The longest U.S. spaceflight before Skylab was the fourteen-day Gemini 7 mission in 1965. The Skylab missions provided an opportunity for sustained scientific study of Earth from space, and the flight of Skylab 4 was a prime example of this capability.

A better understanding of Earth's geologic features resulted from studies made by the Skylab 4 crew. Data collected provided geologists and other scientists with information on desert conditions and formations, volcanic formation and eruptions, meteoritic crater formation, faults, and mountains.

The study of the eruption of the Sakurajima volcano provided geologists with spectacular stereoscopic imagery. Data collected by the Skylab 4 crew contributed important information about the height of the eruption plume.

Crops were studied from orbit. Changes in vegetation patterns and the effects of weather on crops were observed.

Astronaut Gerald Carr floats in forward dome area (top); Astronaut Edward Gibson stands at Apollo Telescope Mount in Skylab (bottom). (NASA)

Meteorological events were also discernible from Skylab. The crew could determine where snow had fallen and what its general depth was.

Materials processing experiments made clear to scientists the advantages of microgravity. The weightless environment allows for a faster rate of materials processing and results in a purer product; it also allows for the production of materials that simply cannot be made on Earth. Many believe that space factories, of which Skylab is a forerunner, will be an important aspect of space exploration in the future.

Earth's atmosphere, while essential for sustaining life on the planet, prevents astronomers from making accurate observations. Space-based telescopes and observatories such as those on Skylab provide astronomers with more accurate data on astronomical phenomena. The data collected by the Skylab 4 crew led to a better understanding of the Sun and comets. An improved understanding of solar activities leads to an improved understanding of stars in general.

In the past, people thought that the space environment would be too harsh for human beings. It was known that humans could be protected from the extremes of heat and cold, the excessive radiation, and the vacuum of space. Yet many believed that the weightless environment might be the one factor that would prevent humans from becoming space travelers. The Skylab 4 mission demonstrated that humans could survive a long stay in space. Additional experiments provided more information on the effects of the weightless environment on other living things.

Context

If humans are to work in space on a full-time basis, then several questions must first be answered. These questions concern the importance of such a plan, the types of equipment and spacecraft necessary, and the reaction of the human body to the space environment.

Since the early days of the space program, many have questioned the benefit of space exploration. Early satellites and manned spacecraft have since provided information and services that have proved their usefulness. Yet some argued that the problems of long-duration manned spaceflight might outweigh the benefits. Skylab 4 proved that the gains are worth the risks. Important information about Earth as a planet was returned by the crew. This information contributed to a better understanding of the geological, meteorological, agricultural, and hydrological sciences.

The production and processing of certain materials in space would prove advantageous to future space stations. The crew of Skylab 4, along with the crews of two previous Skylab missions, demonstrated that materials processing can be performed successfully and that there are some important benefits from processing materials in space.

Astronomical observations made from Skylab provided important data about the Sun. These data will help astronomers and scientists better predict sunspot activity and solar flare eruptions. Such solar events directly affect life on Earth in various ways, from determining general weather cycles to interfering with radio transmissions.

Comets are thought to be very old phenomena. An understanding of these objects, once called omens, may lead to a better understanding of the history of the solar system. The Skylab 4 astronauts' opportunity to study the comet Kohoutek provided astronomers with new cometary data.

Even though NASA officials experienced significant problems with Skylab, they proved that a space station could be constructed and that such a station could be used for worthwhile experimentation during long-duration flights. Possibly more important, NASA proved that man could repair satellites and spacecraft in orbit. While the first Skylab crew, that of Skylab 2, did the major repair work on the space station, the Skylab 4 crew effected several repairs also.

If humans are to inhabit space, then the effects of the weightless environment must be understood. The Skylab 4 mission proved that with proper in-orbit exercise and diet, the effects of space are not detrimental. The future of long-term flights depended on this one important observation.

Even though many Soviet flights have surpassed the U.S. record set by Skylab 4, the mission of Skylab 4 will be remembered as a milestone. It proved that humans could adapt to the space environment and that this environment had much to offer those who could make use of it.

Bibliography

Baker, David. *Conquest: A History of Space Achievements from Science Fiction to the Shuttle.* Topsfield, Mass.: Salem House, 1985. A general overview of space exploration. The text summarizes significant advances and programs with appropriate illustrations. Suitable for general audiences.

Belew, Leland F., and Ernst Stuhlinger. *Skylab: A Guidebook.* NASA EP-107. Washington, D.C.: Government Printing Office, 1974. This text provides the reader with details of the Skylab project. It outlines the history, missions, operation and design, and experimental proposals of the Skylab program.

Braun, Wernher von, et al. *Space Travel: A History.* Rev. ed. New York: Thomas Y. Crowell, 1985. A complete and detailed history of space exploration. Covers the development of the rocket and the evolution of the Soviet and U.S. space programs. Well illustrated.

De Waard, E. John, and Nancy De Waard. *History of NASA: America's Voyage to the Stars.* New York: Exeter Books, 1984. This book reviews the United States' achievements in space, with an emphasis on manned exploration. A good overview of programs and missions is presented. Contains excellent color photographs and other illustrations.

Gatland, Kenneth. *The Illustrated Encyclopedia of Space Technology: A Comprehensive History of Space Exploration.* New York: Crown Publishers, 1981. This encyclopedia is a collection of chapters written by well-known experts in the space sciences. Each of the twenty-one chapters presents a different stage in the history of space exploration. Excellent references at the end of each chapter.

———. *Manned Spacecraft.* Rev. ed. New York: Macmillan, 1976. An excellent reference for both Soviet and American manned space exploration. Text provides complete details on spacecraft design and lists mission highlights. Fully one-quarter of the book is illustrations, including illustrations of Soviet spacecraft. A useful text for anyone desiring additional detailed information.

Lyndon B. Johnson Space Center. *Skylab Explores the Earth.* NASA SP-380. Washington, D.C.: Government Printing Office, 1977. The text is a detailed log of the many geological, meteorological, and hydrological observations made from Skylab 4. The chapters are actually papers given by prominent scientists. Illustrated with photographs returned from the Skylab 4 mission.

Mike D. Reynolds

SMS and GOES
Meteorological Satellites

Date: Beginning May 17, 1974
Type of satellite: Meteorological

Synchronous Meteorological Satellites (SMS) were prototypes placed in orbits over the equator at parameters that made them appear stationary in relation to Earth's surface. They provided continuous coverage of weather conditions such as temperature, wind speed and direction, and cloud cover. Geostationary Operational Environmental Satellites (GOES) eventually replaced the SMS, engaging in the same type of reconnaissance as part of an operational system under the direction of the National Aeronautics and Space Administration.

Summary of the Satellites

During the l960's, launch capabilities in the United States increased to the point at which craft could be placed in orbits high enough to allow them to become synchronous with the turning of Earth on its axis. In order to accomplish this synchronicity, they had to reach altitudes of around 35,880 kilometers directly above the equator and move with the turning of Earth at a speed that made them appear to stand still when viewed from Earth's surface. This concept is precisely what makes possible satellite communication with permanently positioned satellite receiving antennae such as backyard dishes. Such geosynchronous, or geostationary, orbits are useful for continuous, uninterrupted photoreconnaissance of a specific region of Earth's surface.

In the mid-1960's, the National Aeronautics and Space Administration (NASA) began to apply geosynchronous technology with a series of Applications Technology Satellites (ATS) that carried high-resolution cameras for monitoring the atmosphere. This series was experimental but did achieve success, leading NASA to launch two Synchronous Meteorological Satellites as precursors to an operational system, under the aegis of the Global Atmospheric Research Program of the United States Department of Commerce.

SMS 1 was the first of the two satellites designed for the Goddard Space Flight Center, under the direction of NASA's Office of Applications. It was launched from Cape Canaveral aboard a Thor-Delta launch vehicle on May 17, 1974, and placed at 45° west longitude. After being placed in orbit and tested, it was turned over to the National Oceanic and Atmospheric Administration (NOAA).

SMS 1 was a spin-stabilized cylinder 190.5 centimeters in diameter and 254 centimeters long. It was covered with solar cells and weighed 628 kilograms. The satellite carried a visible infrared radiometer, a meteorological data collection system, and a space environment monitoring system.

SMS 1 was equipped to provide day and night cloudcover imagery, take radiance temperatures of Earth's atmosphere, and measure proton, electron, and solar X-ray fluxes and magnetic fields. It could also transmit processed information from NOAA control facilities to regional stations and pick up and transmit data from thousands of manned and unmanned weather-monitoring sites. High-resolution photographs of the Western Hemisphere were transmitted every 30 minutes. The primary instrument aboard SMS 1 was a visible infrared spin-scan radiometer.

SMS 2 was launched from the Eastern Test Range on February 6, 1975, and placed at 115° west longitude.

The Geostationary Operational Environmental Satellite (GOES) program followed SMS. NASA had contracted with the Department of Commerce to build and operate SMS 1 and SMS 2 as well as the first of the GOES series, SMS-C, which became GOES 1 after it was launched and tested. Five additional GOES satellites were developed to give the program full global coverage and backup. The entire program was eventually taken over by NOAA.

GOES 1 was launched from the Eastern Test Range aboard a Thor-Delta rocket on October 16, 1975. It was a spin-stabilized cylinder, 344 centimeters long and 190 centimeters in diameter; it weighed 243 kilograms. Like the SMS series, it contained a radiometer and data collection and transmission instrumentation. It also contained a telescope. It worked in concert with SMS 2 to provide twenty-four-hour-a-day coverage of the Western Hemisphere.

GOES 2 was launched on June 16, 1977, GOES 3 in June, 1978, and GOES 4, GOES 5, and GOES 6 during the 1980's. The latter three satellites were equipped with advanced sensing technology, including a visible infrared spin-scan radiometer and atmospheric sounder that provides improved imagery. This sophisticated technology allows scientists to determine the three-dimensional configuration of Earth's atmosphere at any given time.

The SMS/GOES series was also used to relay meteorological data that had a bearing on navigation. During 1976, a program was established by the U.S. Coast Guard to use a channel of GOES 1 to relay radar imagery of ice fields from aircraft to control centers, which then distributed the analyzed data to ships for navigation purposes. This program, called Project Icewarn, was conducted over the Great Lakes during the winter and spring seasons in an effort to keep shipping lanes open longer. The SMS's were also used to monitor forest conditions in remote regions as part of an early warning system against forest fires.

These satellites were also equipped with maneuvering rockets, which allowed them to be moved from one location to another by ground controllers. As more GOES spacecraft were deployed, the older SMS and early GOES craft were moved into positions that would permit them to serve as backups for newer, more sophisticated equipment. Working together, they provided early warning of hurricanes, typhoons, and tropical storms. This information was of major importance to coastal areas and island locations in which weather information had not been readily available by other means.

The instrumentation aboard all these satellites also gave controllers the opportunity to monitor phenomena in space such as solar flares, which might not have a direct impact on the weather but which appear to play a role in the formation of magnetic storms.

The SMS/GOES project was incorporated into an international program known as the Global Atmospheric Research Program, a model of international cooperation. After the launch of GOES 3, GOES 1 was positioned over the Indian Ocean to collect and transmit data to a ground station in Spain operated by the European Space Agency (ESA), where it was analyzed and from which it was distributed. Control functions were turned over to the Darmstadt site in West Germany.

Knowledge Gained

While the ATS series proved the concept that weather satellites could function in geostationary orbit, SMS 1 and SMS 2 proved the reliability of satellite technology for providing continuous, real-time coverage of developing weather conditions from the same orbital location above the equator. The more sophisticated equipment aboard these satellites was the first to provide continuous high-quality color reconnaissance of Earth's surface. The photographs from these satellites were processed and looped together to create a visual progression of weather fronts from which meteorologists were able to track and predict the progress of a weather system. This imagery also gave scientists the opportunity to study the formations in an effort to understand more fully what caused them and what conditions caused them to develop in one way or another.

SMS/GOES also acted as part of an early warning system operated by the U.S. government's National Environmental Satellite Service. By 1976, several hundred sensors for forecasting natural disasters — floods, tidal waves, earthquakes, and forest fires — had been installed in various remote regions of the country. It was projected that eventually several thousand such instruments would be deployed. In the winter of 1976, two such sensors placed in a watershed in the state of Washington warned of impending flood conditions upstream from Deming, Washington. At the time, GOES 1 was stationed over the West Coast of the United States, while SMS 2 was stationed over the East Coast. Both satellites picked up signals from the sensors and relayed them to NOAA facilities in Maryland, where the information was analyzed and an advisory was transmitted to Deming area authorities. The same technology has been employed to keep track of ocean vessels, aircraft, and highway traffic.

During that same month, SMS 2 was also employed to transmit data from twenty-three ground stations in an area of one thousand square miles in northern California. Famous for its fragile redwood forests, the area was susceptible to forest fires. Sensors in each of these ground stations kept track of air temperature, humidity, wind speed and direction, solar radiation, rainfall, air pollution, and the moisture content of the forest floor, which was blanketed by dry pine needles and grasses. Data from each of these sensors were sent to the satellite for relay to ground stations.

The principal advancement achieved with the SMS/GOES series was the full-time, day and night coverage of the Western Hemisphere in real time. Earlier programs had established the dependability of satellites as data gatherers, but the nonsynchronous orbits of earlier systems had meant relatively long delays between sweeps of the same area.

The earliest meteorological satellites — Television Infrared Observations Satellites (TIROS), Environmental Science Services Administration Satellites (ESSA), and Improved TIROS Operational System Satellites (ITOS) — were placed in polar Sun-synchronous orbits. From that configuration, the more advanced ITOS satellites were designed to cover 95 percent of Earth's surface in 36 hours, which was not fast enough to track a developing tropical storm that might turn into a hurricane, threatening island and coastal regions. Nor were many of these early satellites able to provide nighttime imagery. The infrared sensors aboard SMS/GOES, developed to complement or replace less sophisticated technology aboard the earlier satellites, allowed meteorologists to update information on developing weather systems much more frequently.

Another major advancement was the establishment of a satellite system that could be maneuvered at will to serve a multitude of purposes, as described above. Semiretired ATS's were moved into position to relay television coverage of the 1968 Summer Olympics in Mexico City after an international communications satellite failed at launch, thus providing a dependable backup that later served in similar capacities. Since their launches, both SMS's and several of the GOES's have been maneuvered from one location to another along the equatorial plane to accomplish a variety of tasks.

The geostationary equatorial position of the SMS/GOES system is ideal for the United States and much of Europe and

Asia. The Soviet Union, on the other hand, uses a combination of geostationary and polar orbital configurations, because a large portion of its land mass is in extreme northern latitudes that cannot be seen adequately from a position above the equator.

Context

As the space era entered the 1980's, the space shuttle program demonstrated its ability to transport satellites into space, and even to repair them while in orbit. Thus the era of the serviceable, reusable satellite was at hand, offering operators the opportunity to upgrade onboard systems and make repairs rather than replace entire spacecraft.

In the 1960's, when the ATS program was launched, scientists had begun to look for ways to make satellites capable of performing various tasks over long periods of time. Early satellites had been in orbit for short periods, often only a few days or weeks. The need had been established for long-term, dependable, multifunction satellites.

A major problem in those early days had been that rocket technology had not attained the capability of boosting satellites or other spacecraft into the orbits required for geosynchronization. By the time the ATS series was ready for launch, however, more powerful rockets and second- and third-stage jump motor technology were available to accomplish the task.

The SMS and GOES series, along with communications satellites, were among the first to take advantage of the geosynchronous capability. From their stationary vantage points,

they were able to provide high-quality imagery of Earth's surface and simultaneously to monitor extraterrestrial phenomena such as solar flares and interplanetary radiation. They were also equipped with communications circuits that served dependably when unexpected demand outstripped supply of space channels for international communications traffic, including transoceanic broadcasts. They also served as relay platforms for a variety of Earth-based sensing instruments that provided critical monitoring of remote areas and natural-disaster-prone regions of the Western Hemisphere and its oceans.

As the GOES series continued to provide reliable data for a multitude of users, international cooperation continued to grow, allowing the program to be tied in with those of other nations to assure complete coverage of the globe in real time. The three-dimensional character of the imagery available from the advanced GOES 4, GOES 5, and GOES 6 has also enhanced the capabilities of other programs, including Landsat, which monitors Earth resources such as oil, minerals, and water. Scientists are still looking for ways to use the deluge of data from these satellites to enhance such endeavors as urban planning and environmental protection, perhaps putting the variable of climatology into the mix with infrastructure and available water resources.

A major concern in the scientific community over the decades following the launch of SMS 1 was the reported depletion of the ozone layer of Earth's atmosphere. SMS/GOES spacecraft were among the first to relay reliable information regarding the state of worldwide atmospheric ozone.

Bibliography

Brun, Nancy L., and Eleanor H. Ritchie. *Astronautics and Aeronautics, 1975.* Washington, D.C.: National Aeronautics and Space Administration, 1979. A chronologically arranged compilation of press releases and accounts from popular publications regarding key events in the space program during 1975. The reports cover not only NASA activities but also the projects and discoveries of other nations' programs. Includes material regarding SMS and GOES spacecraft. Indexed.

Corliss, William R. *Scientific Satellites.* NASA SP-133. Washington, D.C.: Government Printing Office, 1967. Highly technical descriptions of most of the early scientific satellites designed by NASA are offered, as well as of those launched by other nations. Includes a chronological list of launches, with comments and technical data. A chapter is devoted to geophysical instruments and experiments; there is also a helpful overview of space science in general. Surveys the technical nature of the instrumentation installed in many of the satellites that preceded the SMS and GOES series. Contains photographs, other illustrations, and an index.

Gatland, Kenneth. *The Illustrated Encyclopedia of Space Technology: A Comprehensive History of Space Exploration.* New York: Crown Publishers, 1981. Gatland has assembled a number of chronologies of American, Soviet, and international space programs. This well-illustrated volume includes a chapter that discusses in detail weather observation by satellites and specifically summarizes the SMS/GOES program and its accomplishments. Specific examples of successful applications of space technology in this area are cited. Indexed.

Hirsch, Richard, and Joseph John Trento. *The National Aeronautics and Space Administration.* New York: Praeger Publishers, 1973. A well-executed chronology of the accomplishments of NASA during the first decade of space exploration. Discusses early plans for NASA's weather observation satellite program, which eventually led to SMS, GOES, and the NOAA operational systems. Photographs supplement the text. Includes an index.

Ritchie, Eleanor H. *Astronautics and Aeronautics, 1976: A Chronology.* NASA SP-4021. Washington, D.C.: National Aeronautics and Space Administration, 1984. In chronological order, offers abstracts of press accounts of space activity during the extremely active year of 1976. These accounts include the names of hundreds of individuals associated with scores of international space programs. Perhaps most useful are summaries of updates on programs such as SMS/GOES, which were ongoing at that time.

Taylor, L. B., Jr. *For All Mankind: America's Space Programs of the 1970's and Beyond.* New York: E. P. Dutton and Co., 1974. Taylor examines the commercial applications and social benefits from the research associated with the space program and the then-anticipated priorities of that program. The discussion of weather observation by satellites is general, but the visionary nature of the author's approach to the subject is worth attention.

Michael S. Ameigh

THE SOLAR MAXIMUM MISSION

Date: Beginning February 14, 1980
Type of satellite: Scientific

The Solar Maximum Mission was designed to study the Sun during the 1980 peak of the eleven-year solar cycle. In addition, the satellite was the world's first to be retrieved, repaired, and redeployed in space.

PRINCIPAL PERSONAGES
  PETER T. BURR and
  FRANK J. CEPOLLINA, Project Managers
  KENNETH J. FROST and
  DAVID SPEICH, Project Scientists
  DAVID RUST, Science Coordinator for the
    International Solar Maximum Year
  RICHARD C. WILLSON,
  ROBERT M. MacQUEEN,
  EDWARD CHUPP,
  BRIAN R. DENNIS,
  CORNELIS DE JAGER,
  EINAR ANDREAS TANDBERG-HANSSEN, and
  LEONARD CULHANE, Principal Investigators

Summary of the Satellite

The Solar Maximum Mission (Solar Max) was initiated to continue the investigation of the Sun begun by Skylab's Apollo Telescope Mount and by the Orbiting Solar Observatory satellites. Solar Max's mission was to study activity on the Sun during the 1980 peak of solar activity, simultaneously recording information in a broad spectrum of wavelengths. (The spectrum refers to the electromagnetic spectrum, the entire range of wavelengths of electromagnetic energy, from gamma rays to radio waves and including visible light.) A primary focus of the mission was to study the mechanism behind flare activity. (A solar flare is a tremendous eruption on a small area of the Sun's surface that generates highly energetic charged particles and light.) The original estimated lifetime of the Solar Max satellite was one to two years.

Solar Max was not a solitary venture but was a key component in the International Solar Maximum Year's program to study the Sun. More than fifty observatories around the world participated in the International Solar Maximum Year, coordinating their observations of the Sun during the time of its maximum activity. Observations from orbiting instruments such as Solar Max and the International Sun-Earth Explorers were supported by simultaneous observations from ground-based observatories. From a control room at Goddard Space Flight Center, current images of the Sun could be obtained from several observatories. All the active regions of the Sun were monitored and predictions made as to likely targets for Solar Max. Amateur observers were also involved, photographing and sketching those flare events that were intense in visible light.

The Solar Max spacecraft was the first satellite to use a new system designed specifically to take advantage of the capabilities of the space shuttle. This "Multi-Mission Modular Spacecraft" (MMS) was a standardized base used for a variety of satellites. Multiple missions could be achieved using a system of modules designed to be replaced in space by shuttle crews. Such replacements would require the removal of only two bolts and could be accomplished with a special tool designed for use in zero gravity. Previously, expensive satellites could be completely disabled by problems as trivial as a blown fuse. With the space shuttle and this new modular design, satellites could be repaired in space and restored to usefulness for a fraction of the cost of replacing them. Originally, the National Aeronautics and Space Administration (NASA) had planned to retrieve Solar Max with the space shuttle in 1984 and return it to Earth before the spacecraft reentered Earth's atmosphere.

The instruments aboard Solar Max were the most diverse and sophisticated yet orbited to study solar activity. There were six major experiments to record information in wavelengths ranging from visible light to gamma rays. A group of Dutch scientists built the first orbiting instrument with the ability to photograph the Sun in "hard," or penetrating, X rays. Another instrument was designed to record gamma-ray emissions in solar flares, and other instruments included hard and soft X-ray detectors, a device to record the ultraviolet spectrum of the Sun, and one to produce images of the Sun's corona and follow the effects of flares into interplanetary space. (The corona is the part of the Sun's atmosphere above the visible surface and is observable only during a total solar eclipse.)

A seventh experiment, called the Active Cavity Radiometer Irradiance Monitor (ACRIM), was intended to observe variations in the solar constant in wavelengths from the far ultraviolet to the far infrared. (The solar constant is the

The Solar Maximum Mission (top); view of the damaged Solar Maximum Mission Satellite from STS 41-C (bottom). (NASA)

amount of solar energy falling on a unit area of Earth every second.) ACRIM produces one solar constant measurement every two minutes. All the instruments were designed to look at the same area of the Sun at the same time.

Solar Max was launched on February 14, 1980, into a circular orbit 570 kilometers above Earth. Instruments aboard began recording flares almost immediately. The spacecraft was to operate virtually automatically, monitoring preselected regions on the disk of the Sun for signs of impending activity. The areas were to be chosen from daily activity forecasts prepared by a solar observatory at NASA's Marshall Space Flight Center in Alabama. Solar data gathered there would reveal

areas where solar activity was most likely to occur.

As an area brightened rapidly, the X-ray detector aboard the satellite would electronically alert the other instruments, which then would focus on the region and conduct a series of preplanned observations. In this way, activity on the Sun could be studied from the "preflare" phase through the first major release of energy and then through the entire process.

Solar Max functioned perfectly for about ten months, returning useful information on flare and sunspot activity. Then in November, 1980, the spacecraft suffered a malfunction. The fuses began to blow in its primary attitude control system (the system that controls the satellite's orientation toward its targeted areas of study). In the vacuum of space, the small fuses had degraded to the point at which they could no longer carry their load of current. By mid-December, three of four attitude controls had been lost, and the spacecraft was disabled. Although the mission ended before its estimated lifetime was over, the wealth of information returned by the satellite made the mission a successful one.

In the spring of 1981, a team of engineers from NASA's Johnson Space Center in Houston began studying the possibility of an in-orbit repair of the satellite. Such a repair would require Extravehicular Activity (EVA) by a shuttle astronaut, who would first stop the satellite's motion by using a special grappling device. The shuttle's remote manipulator arm would then take over, moving the satellite into the shuttle's payload bay, where the failed modules would be replaced. Brought back to full function, the satellite was expected to be useful for two years or more.

Besides restoring an expensive and useful satellite, the repair project was seen as an opportunity to demonstrate the capability of the space shuttle's crew to retrieve and repair satellites in orbit. This demonstration would be important to civil and military space planners and would be a crucial factor in planning upcoming space missions. In early 1982, however, it appeared that the repair mission would be canceled because of funding priorities, but the mission was considered important enough to transfer funding from other projects. After two years of planning and rehearsal, the mission was scheduled for April of 1984; *Challenger's* crew would make the attempt.

Although Solar Max was successfully approached by the space shuttle, a first attempt to grapple the spacecraft failed. Since there was a danger that the depleted batteries would destroy sensitive instruments, a second attempt was made, two days later. On April 10, 1984, Solar Max was successfully secured in the shuttle's cargo bay. The repairs were completed, and the satellite was returned to orbit on April 12. Within a month, all the spacecraft's systems were again operational.

In the interval between its failure and repair, improvements were made to the Solar Maximum Mission. NASA's Tracking and Data-Relay Satellite (TDRS) was launched in 1983, and it provided a much more efficient way to communicate with Solar Max. Astronomers could receive images from Solar Max immediately and formulate new observing sequences based on the solar activity in progress. The more than three years that Solar Max was dysfunctional provided scientists with the opportunity to analyze the data they had already received. Perhaps even more important, the time allowed scientists to alter some of the experiments aboard the satellite.

Because the solar cycle was approaching a minimum after the repair, Solar Max studied the different types of activity that appeared during a decline in solar activity. The satellite was also expected to last until the 1991 solar maximum. Solar Max also studied other objects in the sky. Beginning in late 1986, investigators attempted to measure X- and gamma-ray spectra from Cygnus X-1, an unusual object that is suspected to harbor a black hole. Solar Max observed Comet Halley in 1986 when the comet was too close to the Sun for ground-based observation, and the satellite is now being used to study the supernova that appeared in January, 1987, in the Large Magellanic Cloud. The gamma-ray instrument aboard Solar Max is the only orbiting instrument sensitive enough to detect emissions coming from the supernova.

As the mission stretched toward the beginning of a new solar maximum, it continued to provide solar data not only at its maximum activity but also during the entire solar cycle. Solar Max also extended the reach of its sophisticated instruments beyond the Sun to study the processes in other stars. From a spectacular beginning and a disappointing end to an unexpected revival, Solar Max challenged the astro-physical community with information that is creating a new understanding of the Sun. Solar Max recorded its final data in November, 1989.

Knowledge Gained

A "new" Sun is emerging from the results obtained from Solar Max, along with studies from previous orbiting solar satellites. Together, these solar satellites have contributed a major percentage of scientists' knowledge of solar processes. The primary area of interest has been the study of solar flares. The 1980 solar maximum was especially active, and Solar Max recorded information on thousands of flares. Solar flares are usually associated with sunspots, the regions on the Sun's surface that appear dark and are associated with intense magnetic fields. Solar Max provided the first direct evidence that the interaction of solar magnetic fields can give rise to a flare.

At the onset of a flare, there are bursts of high energy X-ray and ultraviolet radiation. The timing of the bursts gives information about where they take place. It was found that the bursts occur simultaneously, and that they occur at the locations where magnetic fields are anchored in the lower solar atmosphere. Surprisingly, the largest flares do not necessarily produce the largest amount of radiation. Some flares have been known to produce ten times the gamma rays that are produced by flares more than twice their size.

Researchers have found that flare mechanisms can produce higher temperatures than were thought to exist on the Sun. Curiously, the solar material does not expand when heated but increases in density, indicating that something is pressurizing it. This is an important finding, because thermonuclear fusion researchers are trying to accomplish the same effect on Earth.

It was previously believed that the chemical composition of the Sun was uniform and constant. Solar Max observations, however, have shown evidence of variations in the abundance of calcium from flare to flare and in the proportions of other chemicals in different levels of the Sun's atmosphere.

Solar Max has also made important observations on the nature of coronal transients. These are huge bubbles of plasma (superheated gases) often containing 9 billion kilograms of matter, which are thrown away from the corona at speeds up to 1.6 million kilometers per hour. It seems from the data that these transients may precede the flare itself and not be caused by it, as was previously theorized. Thus, flares may not cause the many phenomena associated with them but may be merely one step in the entire process.

It has been observed that as a group of sunspots crosses the face of the Sun there is a slight drop in the Sun's total output of energy. It is not yet known whether this is a true drop in energy or whether the energy is being redirected in some way. In fact, the variability of the solar constant is one of the most notable discoveries made by Solar Max. ACRIM data show a downward energy trend of 0.019 percent per year between 1980 and 1985. As new data are obtained throughout the next solar cycle, researchers will be looking for this trend to reverse, possibly indicating a variation linked with the solar activity cycle.

Context

The Solar Maximum Mission came at a time when solar physicists had had several years to study the data returned by previous solar studies, particularly from the series of Orbiting Solar Observatories and from the Apollo Telescope Mount aboard Skylab. As much as those studies advanced knowledge of solar processes, they raised a new set of questions requiring a new, more sophisticated generation of instrumentation. Solar Max was designed in direct response to these needs.

Solar processes are felt to a great extent on Earth. During a large flare, the ejected particles, X rays, and gamma rays disrupt Earth's atmosphere, in turn disturbing radio communications and electrical power. Familiar manifestations of this influence are the auroral displays seen in polar regions. Radiation during the largest flares could also be lethal to astronauts who are above the protective layer of Earth's atmosphere.

Weather patterns and climate are ultimately determined by the interaction between Earth and radiation from the Sun. Theoretically, a change in the Sun's output by as little as 0.5 percent per century could produce profound changes in Earth's climate. A decrease of 6 percent in the Sun's radiation would cover the planet with ice. It is only by monitoring the solar constant and its effects that scientists will be able to determine this effect and predict future changes in climate.

Scientists now have data showing that the solar constant does indeed vary. By monitoring this variability over time, it may be possible to predict long-term climatic cycles and changes. With Solar Max able to record the Sun's activity for more than a complete solar cycle, an important record is being made to establish a baseline on the Sun's activity.

The Sun is the most important factor influencing life on Earth. Its study, therefore, is of utmost importance. Besides being of vital concern to life, the Sun is also a conveniently nearby star and thus provides an excellent laboratory for the study of stars. Moreover, since the Sun provides the energy needs of Earth, there may come a time when the Sun's power can be harnessed to power human technology. Perhaps with further study of the Sun, more secrets of its generation of power will be revealed.

In addition to the vast harvest of scientific data, however, Solar Max was an important milestone in space science. The reusable design of the satellite and its in-orbit repair can be seen as a first step toward humanity's increased ability to live and work in space.

Bibliography

Chaikin, Andrew. "Solar Max: Back from the Edge." *Sky and Telescope* 67 (June, 1984): 494-497. This article, one of several published by *Sky and Telescope* about the Solar Maximum Mission, discusses the status and results of the mission. Describes the satellite and includes illustrations of the spacecraft and images taken by its instruments. For the amateur astronomer.

Cornell, James, and Paul Gorenstein, eds. *Astronomy from Space: Sputnik to Space Telescope*. Cambridge, Mass.: MIT Press, 1985. An overview of the previous twenty-five years of astronomical research from space. Written by experts in the fields covered, these pieces are intended for those with some scientific background.

Eddy, John A. *A New Sun: The Solar Results from Skylab*. NASA SP-402. Washington, D.C.: Government Printing Office, 1979. Describes the Skylab mission in detail and discusses what is known about the Sun. Written for the layperson with an interest in astronomy.

Fire of Life: The Smithsonian Book of the Sun. Washington, D.C.: Smithsonian Exposition Books, 1981. A collection of articles geared for general audiences. Includes cultural history as well as projections of how the Sun might be used once a better understanding of its secrets is gained. Beautifully illustrated.

Gregory, William H. "Aftermath of a Rescue." *Aviation Week and Space Technology* 120 (April 23, 1985): 13. This article is part of a series of articles on the space shuttle mission to repair Solar Max. Written for professionals in the field of avionics, but readable for advanced college students. Illustrated.

Divonna Ogier

SPACE CENTERS AND LAUNCH SITES

Date: Beginning 1945
Type of facility: Space centers and launch sites

With the advent of the German V-2 rocket in World War II, areas separated from population centers for testing this new weapon became a necessity. As the rockets' size grew and as their purpose changed from carrying warheads to launching satellites, more and larger testing facilities were needed.

PRINCIPAL PERSONAGES

FORREST S. MCCARTNEY, Director, Kennedy Space Center

LAWRENCE GOOCH, Commander, Eastern Space and Missile Center, Patrick Air Force Base

WAYNE PENLEY, Commander, Cape Canaveral Air Force Station

THOMAS J. P. JONES, Commanding General, White Sands Missile Range

THOMAS W. HONEYWILL, Commander, Space and Missile Test Organization, Vandenberg Air Force Base

ARLEN D. JAMESON, Commander, First Strategic Aerospace Division, Vandenberg Air Force Base

ORLANDO C. SEVERO, JR., Commander, Western Space and Missile Center, Vandenberg Air Force Base

WARREN KELLER, Director, Suborbital Missions, Wallops Flight Facility

Summary of the Facilities

The primary launch sites for space vehicles in the United States are Kennedy Space Center, just outside Cocoa Beach, Florida, and Vandenberg Air Force Base, just outside Lompoc, California. Each lies on the coastline of an ocean, permitting relatively safe launches over large, unpopulated areas. Two other key launch sites are the U.S. Army's White Sands Missile Range, between Las Cruces and Alamogordo, New Mexico, and Wallops Flight Facility, of the National Aeronautics and Space Administration's Goddard Space Flight Center, in Virginia on the Chesapeake Bay's Eastern Shore. White Sands and Wallops are used primarily for sounding rockets, small rockets — usually from surplus military inventories — that carry scientific experiments on brief, suborbital flights. Wallops also manages the National Scientific Balloon Facility (NSBF) in Palestine, Texas, where many experiments are

launched on huge balloons. Wallops supports sounding rocket and balloon campaigns worldwide, as well, using mobile facilities in locations from Andoya, Norway, to Alice Springs, Australia.

Established in the early 1960's to serve as a launch site for the Apollo lunar landing missions, Kennedy Space Center is 240 kilometers south of Jacksonville and 80 kilometers east of Orlando. Kennedy and the Cape Canaveral Air Force Station form a complex that stretches for 55 kilometers and varies in width from 8 to 16 kilometers. The total land and water area occupied by the installation measures 568 square kilometers. When fully manned, the center has a work force of approximately two thousand National Aeronautics and Space Administration (NASA) civil servant employees and between twelve and thirteen thousand personnel who work for private companies under contract to NASA.

Since the completion of the Apollo lunar landing program, Kennedy Space Center has concentrated on supporting the United States' newest launch vehicle, the space shuttle. The first space shuttle (known as STS 1, for Space Transportation System 1) was launched from Launch Complex 39A on April 12, 1981. Many other launches from the Florida site take place at Cape Canaveral Air Force Station, which is adjacent to Kennedy. All the launches using the Delta rocket, for example, are from the air force station. Kennedy Space Center also operates a Vandenberg launch site resident office, which supports NASA activities at Vandenberg Air Force Base. Vandenberg was formed when an operational site was needed to train Air Force crews to launch intercontinental and intermediate range ballistic missiles. The Department of Defense decided to use an old Army facility, known as Camp Cooke, that occupied a strip of land on the Pacific Coast of California halfway between Los Angeles and San Francisco. The site was named Cooke Air Force Base in June, 1957, and the name was changed to Vandenberg Air Force Base in October, 1958, in honor of the late Hoyt S. Vandenberg, the second Air Force chief of staff.

The first missile was launched from Vandenberg in December, 1958. In February, 1959, Discoverer 1, the first satellite to be placed in a polar orbit, was launched from the base. Discoverer was a precursor to the first successful U.S. reconnaissance satellite, Discoverer 13, which was launched from Vandenberg in August, 1960. By 1988, more than sixteen hundred missiles and booster rockets of forty-eight different types had been launched from Vandenberg. Known as "America's western spaceport," Vandenberg is the only U.S. military installation that launches both land-based intercontinental ballistic missiles and space boosters.

Covering more than 396 square kilometers, Vandenberg is the third-largest U.S. Air Force installation. The facilities are linked by 837 kilometers of roads, 27 kilometers of railroad tracks, 129 kilometers of gas lines, 476 kilometers of water mains, and 727 kilometers of electrical power lines.

The launch site's economic impact on the region is significant. In the fiscal year ending September 30, 1987, the base generated revenues of more than $500 million. Military and civilian employees earned $300 million, and another $181 million was paid to aerospace contractors. An estimated forty-three hundred secondary jobs were created in the area because of the base.

In the 1980's, the Air Force started development of a space shuttle launch complex at Vandenberg, spending about $3 billion in facility costs and another $400 million annually in operating costs. The facility was named Space Launch Complex 6 (SLC 6). Because of a number of problems at the site, however, the Air Force postponed further work on SLC 6.

The original plans called for shuttle operations to be conducted both at North Vandenberg and at South Vandenberg. The runway, orbiter maintenance and checkout facility, orbiter lifting facility, thermal protection facility, supply warehouses, and most of the support personnel are at North Vandenberg. SLC 6, which includes the launch control center, payload preparation room, payload changeout room, shuttle assembly building, access tower, launch mount, mobile service tower, and three exhaust ducts, is at South Vandenberg.

The Vandenberg approach to vehicle assembly differs from the procedures used at Kennedy Space Center. At Kennedy, the shuttle's components — the orbiter, external tank, and solid-fueled rocket boosters — are assembled on a mobile launch platform in the vehicle assembly building and then moved to the launch area. At Vandenberg, the components would be assembled on the launchpad.

The Army's White Sands Missile Range (WSMR), in the Tularosa Basin of southern New Mexico, is 160 kilometers long by 65 kilometers wide — larger than the states of Delaware and Rhode Island and the District of Columbia combined. The range stretches more than half the distance from El Paso, Texas, to Albuquerque, New Mexico. In area, it is the largest military reservation in the United States.

WSMR was established on July 9, 1945, as the White Sands Proving Ground; the name was changed in 1958. The first missile fired at the range was a Tiny Tim sounding rocket, in September, 1945. The facility supports missile development and test programs for the Army, Navy, and Air Force and for NASA and other government agencies.

Since 1960, WSMR has had part-time use of a 4,225-square-kilometer area adjoining the range's northern bound-

ary. Two other areas, adjacent to the western boundary, also have been used to extend the range. Used many times a year, these areas — which total approximately 5,150 square kilometers — permit testing of longer-range missiles. When firings are scheduled in these areas, residents leave their homes, usually for a maximum of twelve hours. The ranch families are paid for the use of their land and for the hours they must spend away from home during these evacuations.

WSMR has served as an impact area for Army Sergeant and Pershing missiles launched from sites in Utah as far as 643 kilometers away. In 1982, the range added a launch complex near Mountain Home, Idaho, thereby acquiring the capability of firing the Pershing 2 missile and other test missiles with ranges of more than 1,200 kilometers.

On the northern portion of the range, on July 16, 1945, the world's first atomic device was detonated. The spot, known as Trinity Site, is in a missile impact area and is open to the public only once a year.

Several tenant organizations share the White Sands facilities. Among those organizations are the Naval Ordnance Missile Test Facility, the Atmospheric Sciences Laboratory of the Army's Electronics Research and Development Command, the Army's Communications Command Agency, NASA, and the Air Force Range Operations Office.

The range has more than one thousand precisely surveyed instrumentation sites and approximately seven hundred sophisticated optical and electronics instrument systems. Another of its assets is the National Laser Test Facility. This installation will be used for laser lethality testing by the Army, Navy, and Air Force and by government agencies.

The Wallops Flight Facility is managed by Goddard Space Flight Center. Wallops is 64 kilometers southeast of Salisbury, Maryland, and approximately 241 kilometers southeast of Goddard, which is in Greenbelt, Maryland. Composed of a main base, the Wallops Island launch site, and the Wallops mainland, the site covers more than 25 square kilometers and has eighty-four major facilities valued at about $105 million. Approximately 380 civil service workers and 560 contractor employees staff the installation, which has an annual payroll of more than $30 million.

Development of Wallops was started in 1945 when the National Advisory Committee for Aeronautics (NACA) authorized the Langley Research Center to establish a site for research on rocket-propelled vehicles. Since then, Wallops has been used as a launching site for scientific missions as well as a facility for aeronautical research.

The Navy decided in 1958 to close its Chincoteague Naval Air Station, about eleven kilometers northwest of Wallops Island. A year later, Wallops took over those facilities, which included buildings, utilities, hangars, and an excellent airport.

Wallops is at the center of NASA's suborbital programs. Sounding rockets, balloons, and aircraft are used in NASA programs concerned with space science, applications, advanced technology, and aeronautical research.

Sounding rockets carry scientific payloads to altitudes ranging from 48 to 965 kilometers. Experiment time above the atmosphere is usually about fifteen minutes. Scientific data are collected and returned to Earth by telemetry links, and the payloads are parachuted to Earth for refurbishment and reuse. Scientific balloons carry payloads to altitudes of up to 48 kilometers and normally remain aloft for from one to sixty hours. The balloon program supports approximately forty-five launches each year. Between 1975 and 1986, 493 balloons were launched, with an overall success rate of 85 percent. In addition, more than twenty satellites have been launched from Wallops since 1961, including Explorer 9, the first satellite to be launched by an all-solid-fueled rocket.

Along with its headquarters in Washington, D.C., NASA operates nine major space centers in the United States. Ames Research Center, Mountain Home, California, was founded in 1940 and is located on 1.7 square kilometers of land adjacent to the U.S. Naval Air Station at Moffett Field. It employs twenty-two hundred civil service employees and approximately twenty-one hundred contractor employees at the main center and at a subsidiary installation, Dryden Flight Research Center, which merged with Ames in 1981.

Goddard Space Flight Center, in Greenbelt, Maryland, was established in 1959. Goddard is 16 kilometers northeast of Washington, D.C. It has 4.45 square kilometers of land and employs approximately twelve thousand civil service and contractor employees at Greenbelt and at its subsidiary, Wallops Flight Facility.

The Jet Propulsion Laboratory, in Pasadena, California, is a government-owned facility operated by the California Institute of Technology under a NASA contract. It is 32 kilometers northeast of Los Angeles.

Johnson Space Center is about 32 kilometers southeast of Houston. The center was established in 1961 as NASA's primary center for manned spaceflight. Most of the one hundred buildings on the 6.6-square-kilometer site house offices and laboratories, some of which are dedicated to astronaut training and mission operations.

Kennedy Space Center is responsible for the assembly, testing, and launch of space shuttles and their payloads, for shuttle landing operations, and for the servicing of space shuttle orbiters between missions. Kennedy also operates a Vandenberg launch site resident office.

Langley Research Center is in the Tidewater area of Virginia between Norfolk and Williamsburg. The center occupies 3 square kilometers, one of which is used under permit from the U.S. Air Force. An additional 13.3 square kilometers of marshland near Langley are used as a model-drop zone.

Lewis Research Center is adjacent to the Cleveland Hopkins International Airport, approximately 32 kilometers southwest of Cleveland. Established in 1941, it is staffed by twenty-seven hundred civil service employees and thirteen hundred contractor personnel.

Marshall Space Flight Center, Huntsville, Alabama, is situated on 7.3 square kilometers inside the Army's Redstone Arsenal. The center has thirty-five hundred civil service employees, of whom 58 percent are scientists and engineers. Formed in July, 1960, Marshall has been most often identified as NASA's launch vehicle development center. Marshall manages two other sites: the Michoud Assembly Facility, in New Orleans, where space shuttle external tanks are made; and the Slidell Computer Complex, in Slidell, Louisiana, which provides computer service support to Michoud.

Stennis Space Center, Bay St. Louis, Mississippi, is located on the East Pearl River, which provides deep-water access to oversize cargo. Stennis occupies an area of 562 square kilometers, 54.6 of which are covered by the operational base. The center conducts static test firings of space shuttles' main engines.

Context

At the end of World War II, the United States was operating five small missile sites: one in West Virginia, three in California, and one in Texas. The maximum range of the missiles being tested at those facilities was less than 15 kilometers. To meet the greater distance requirements of the postwar era, three facilities were created in 1945. The first of these was Wallops Flight Facility, which was established by the National Advisory Committee on Aeronautics (NACA) with an 80-kilometer range. The second was the 161-kilometer-wide White Sands Missile Range, set up adjacent to the already-established Hueco Range at Fort Bliss. The third was the Naval Air Facility, established at Point Mugu, California, with a 100-kilometer range.

Hardly were these installations established, however, before officials recognized that they were not large enough. With the development of intermediate and intercontinental ballistic missiles already under way, the need for sites with ranges of thousands of kilometers became obvious. Moreover, further expansion of the existing ranges would be extremely difficult. As a result, in 1947, a War Department research committee was established to find a suitable range for testing medium- and long-range weapons.

The committee selected three possible sites. In order of preference, they were El Centro Naval Station, in California; Cape Canaveral, in Florida; and a third site, in Washington. The selections were made based on favorable weather and the location of island chains downrange on which tracking stations could be established. El Centro was found unsuitable when Mexico refused rights to some of its islands, and the weather in the Aleutians proved a setback for the Washington site. Great Britain was more cooperative with its Bahamas, and Cape Canaveral was chosen.

In October, 1949, President Harry S Truman established the Joint Long Range Proving Ground (later known as the Air Force Eastern Test Range), a huge overwater range extending 8,000 kilometers across the Atlantic from Cape Canaveral to Ascension Island. The first launch from the Cape was conducted by a military-civilian team on July 24, 1950. The rocket was a modified German V-2 with an upper stage, and it attained an altitude of 16 kilometers. In the early 1950's, the focus turned from missile tests to satellite launch-es. On January 31, 1958, the United States' first satellite, Explorer 1, was launched from the Cape by a military-civilian team led by Kurt Debus, a key member of the famed Wernher von Braun rocket team. Thereafter, Cape Canaveral became the launch site for the Mercury missions and, in the 1960's, for the Gemini and Apollo missions.

Bibliography

Hartman, Edwin P. *Adventures in Research: A History of Ames Research Center, 1940 – 1965.* NASA SP-4302. Washington, D.C.: Government Printing Office, 1970. Published in celebration of Ames Research Center's twenty-fifth anniversary, this work captures the excitement of the persons who staffed the center in its early years. Includes appendices and references.

John F. Kennedy Space Center. *America's Spaceport.* Washington, D.C.: National Aeronautics and Space Administration, 1987. Published in celebration of the center's twenty-fifth anniversary, this comprehensive overview of the history and facilities of Kennedy Space Center is illustrated with works from the NASA art program.

Koppes, Clayton R. *JPL and the American Space Program: A History of the Jet Propulsion Laboratory.* New Haven, Conn.: Yale University Press, 1982. Discusses research at JPL during World War II and the years that followed. Describes the relationship between the California Institute of Technology and JPL and between JPL and NASA. More than half the volume is devoted to JPL's space projects.

National Aeronautics and Space Administration. *Marshall Space Flight Center 1960 – 1985: Twenty-fifth Anniversary Report.* Washington, D.C.: Government Printing Office, 1985. A booklet on Marshall's history and role in the U.S. space program. Includes photographs of Marshall at work, contrasting early activities with later ones, and a timeline showing the dates of major projects.

Rosenthal, Alfred. *The Early Years, Goddard Space Flight Center: Historical Origins and Activities Through December 1962.* Washington, D.C.: Government Printing Office, 1964. This commemorative manual provides a comprehensive look at the founding of Goddard Space Flight Center.

Sloan, Aubrey B. "Vandenberg Planning for the Space Transportation System." *Astronautics and Aeronautics* 19 (November, 1981): 44-50. A detailed description of the original plan for the Vandenberg shuttle complex. Provides a good history of the decision-making process involved in bringing the shuttle complex to Vandenberg. Includes illustrations, diagrams, and a useful bibliography for further study. Full of technical information but still readable.

Wallops Flight Facility, Office of Public Affairs. *Wallops: A Guide to the Facility.* Greenbelt, Md.: Goddard Space Flight Center, 1988. This guide, prepared by the Office of Public Affairs, provides a thorough explanation of the facilities and programs at Wallops.

James C. Elliott

THE SPACE SHUTTLE

Date: Beginning January 5, 1972
Type of program: Manned spaceflight

The Space Transportation System, popularly known as the space shuttle, was established to develop an economic and reusable system that could transport humans, satellites, and equipment to and from Earth orbit on a regular basis. It would also provide support for a wide range of other activities.

PRINCIPAL PERSONAGES

JAMES BEGGS and
JAMES C. FLETCHER, NASA Administrators
ROBERT A. SCHMITZ, Program Manager, NASA
   Headquarters
JOHN S. THEON, Program Scientist, NASA
   Headquarters
DONALD K. "DEKE" SLAYTON, Manager, Orbital Flight
   Test Program
MAXIME A. FAGET, Director of Engineering and
   Development, Johnson Space Center
ROBERT F. THOMPSON, Program Manager of the
   Space Shuttle Office, Johnson Space Center
JAMES W. BILODEAU, the chief of crew procedures for
   shuttle missions

Summary of the Program

The concept of a reusable launch vehicle and spacecraft that could return to Earth was developed early in the U.S. rocket research program. The National Advisory Committee for Aeronautics (NACA), the predecessor of the National Aeronautics and Space Administration (NASA), and NASA itself cooperated with the U.S. Air Force on early studies in the X-15 rocket research program and the Dyna-Soar hypersonic vehicle program in the late 1950's and 1960's.

In 1963, NASA and the Air Force began to work on a design for a manned vehicle that would be launched into orbit and then return to Earth. The craft was called an aerospaceplane and could take off and land horizontally in the manner of a conventional aircraft. In addition, joint tests of wingless lifting bodies, such as the M2 series, HL-10, and eventually the X-24, laid the foundations for a future craft that could safely reenter the atmosphere.

At about the same time, NASA scientists began to talk about a craft that would serve as a cargo transport to other vehicles or space stations in Earth orbit or on the Moon.

Experts at NASA's Marshall Space Flight Center engaged in research on the recovery and reuse of the Saturn 5 launch vehicle. The Department of Defense and private industry also became involved in the development of a reusable transport spacecraft during the late 1960's.

In its 1967 budget briefing, NASA referred to an advanced studies program for a "ferry and logistics vehicle." The president's Science Advisory Committee agreed that developing a reusable, economic space transportation system was necessary.

Between 1969 and 1971, NASA awarded contracts to major aerospace industries to study and define such a system. After considering these findings, NASA decided in 1972 to develop the Space Transportation System (STS). Eventually, this program would be known simply as the space shuttle. On January 5, 1972, President Richard M. Nixon officially announced the inauguration of the space shuttle program. NASA's ambitious schedule called for suborbital tests by 1977 and the first orbital tests by 1979. The shuttle was scheduled to begin regular launchings by 1980.

NASA's goal was to establish a national space transportation system that was capable of substantially reducing the cost of space operations by providing support for a wide range of scientific, military, and commercial applications. Space officials hoped that the space shuttle would operate at a fraction of the cost of the expendable rockets that were in use at the time.

Designed to be a true aerospace vehicle, the space shuttle would be able to take off like a rocket, maneuver in space and attain an Earth orbit, and return to Earth and land like an airplane under a pilot's control. The four-part vehicle would include the reusable orbiter (it would be able to make one hundred launches) mounted atop the expendable External Tank (ET) and two solid-fueled rocket boosters. The boosters, too, would be recoverable and reusable.

The reusable space shuttle would make spaceflight and cargo transport routine and cost-effective, thereby encouraging and enhancing the commercial use of space. The system could be used as a base from which to deploy payloads, repair and service satellites, and launch or retrieve satellites. It could also serve as a platform for scientific research and the manufacture of certain materials requiring a zero-gravity environment.

Launch of STS-66 Space Shuttle Atlantis *(top left); STS-69 EVA view (top right); STS-56* Discovery, OV-103, *with drag chute deployed lands at Kennedy Space Center (bottom).* (NASA)

Additional advantages would be the shuttle's ability to carry large payloads, such as the Hubble Space Telescope, into orbit and to provide a vantage point from which to observe astronomical events, weather disturbances, and environmental changes on Earth. Very important, the shuttle could carry the component parts of a space station into orbit and then act as the transportation system between the station and Earth.

About the same length and weight as a commercial DC-9, the wedged-shaped orbiter could carry a crew of three to seven persons. The crew would consist of a commander, pilot, and one or more mission specialists, all astronauts. There would also be room for one to four payload specialists, nonastronauts responsible for conducting specific experiments and nominated for flight by the payload sponsor. Certified by NASA for flight, they could be chosen from the civilian population. The fact that "ordinary" people could travel on the shuttle added to the program's appeal.

A typical mission would last seven to thirty days. The crew would live and work in a shirtsleeve environment, without cumbersome spacesuits or breathing apparatuses. Maneuvering in the weightlessness of the microgravity environment would prove the biggest challenge to crew members.

In January, 1977, unmanned and manned tests of the shuttle's approach and landing capabilities began at Dryden Flight Research Center, Edwards Air Force Base, California. An orbiter prototype, *Enterprise*, was the first shuttle-type vehicle tested, and it flew successfully.

On April 12, 1981, two years behind schedule, the first of four manned orbital test flights was launched from Kennedy Space Center (KSC) at Cape Canaveral, Florida. The flight was designated Space Transportation System 1 (STS-1), and the orbiter was named "Columbia." The two-day mission demonstrated the spacecraft's ability to reach orbit and return safely. The mission provided data on temperatures and pressures at various points on the orbiter. There were also tests of the cargo bay doors, attitude control systems, and orbital maneuvering system. STS-1 landed safely at Dryden Flight Research Center.

STS-2 was launched in November of 1981. The orbiter *Columbia* was used again. For the first time, the Remote Manipulator System (RMS) was tested. STS-2 also marked the first time that a spacecraft had ever flown twice.

STS-3, launched in March, 1982, was the longest of the initial test flights. The spacecraft stayed in space eight days, and activities included a special test of the RMS and experiments in materials processing. This mission was the first to use the secondary landing site at White Sands Missile Range in New Mexico. (Rain at Edwards Air Force Base had prevented a normal landing at that site.)

The final test flight of this series, conducted in the summer of 1982, featured another test of the RMS, further materials processing experiments, and the launch of the first Department of Defense payload. Once these tests were completed, the space shuttle was declared operational.

From 1981 to 1983, NASA launched a total of nine STS missions. All were launched from KSC. The space shuttles *Columbia* and *Challenger* were used for the missions, and they carried a variety of payloads and experiments. Highlights included the launching of commercial and government communications satellites (the first Tracking and Data-Relay Satellite, or TDRS, was launched), the first retrieval of an object from orbit, the first shuttle-based extravehicular activity, and the launching of the first Spacelab — a portable science laboratory carried in a shuttle's cargo bay.

Beginning with the first shuttle launch of 1984, NASA began to use a new numbering system to identify the missions. The first number would refer to the year, the second number would refer to the launch site (1 for KSC, 2 for Vandenberg Air Force Base), and a letter would refer to the order of assignment in that year. The first 1984 mission, known as STS 41-B, used the orbiter *Challenger* and marked the first landing on a concrete runway specifically designed for shuttle landings at KSC. The shuttle's tenth flight featured the introduction of the Manned Maneuvering Unit (MMU), a backpack propulsion unit that allowed astronauts to maneuver in space independent of the orbiter. The next mission, 41C, was important because it demonstrated the ability of the orbiter and crew to retrieve, repair, and redeploy a malfunctioning satellite.

Throughout 1984 and 1985, the shuttle program continued. With each mission, payloads became more sophisticated, and the orbiters continued to meet NASA's expectations. Spacelab was flown two more times, and several missions were devoted to Department of Defense programs. Two new orbiters, *Discovery* and *Atlantis*, were added to the shuttle fleet.

Shuttle flights were considered so safe and routine that civilian scientists and even a U.S. senator traveled on the spacecraft. In 1984, at the suggestion of President Ronald Reagan, a national search for a candidate to be the first schoolteacher in space was started. Christa McAuliffe, a New Hampshire high-school teacher was chosen; she was scheduled to fly on STS 51-L. The prospect of a civilian schoolteacher riding aboard a shuttle captured the imagination of the nation. The attention of the world, especially the world's children, was focused on the launchpad at KSC the cold morning of January 28, 1986. Seventy-four seconds after the launch, the space shuttle *Challenger* exploded, destroying the spacecraft and killing all seven crew members, including McAuliffe. The accident had been viewed by millions on television. Once the initial shock and grief passed, another kind of anguish replaced it.

President Reagan immediately appointed an independent board of inquiry headed by former Secretary of State William Rogers. Shuttle astronaut Sally Ride and former astronaut Neil Armstrong also served on the commission. NASA conducted its own inquiry, and the U.S. Congress oversaw NASA's investigation. As the committees began to probe, it became obvious that NASA was in a difficult position. It was revealed that warnings about faulty equipment had been ignored because of scheduling pressures and that there was a general lack of communication among NASA, its contractors, and the government. NASA's director, James Beggs, resigned and was replaced by a former NASA director, James C. Fletcher. The old controversy over a manned program versus an unmanned program reemerged, and the United States' thirty-year-old love affair with space exploration seemed to end. The space shuttle was put on hold, and it appeared that it might never recover from the devastating blow. In addition to the serious social and moral ramifications, the scientific and economic costs were high. A $100 million orbiter, one-fourth of the space shuttle fleet, had been destroyed, and the deployment of several expensive and important scientific projects would have to be delayed indefinitely.

In June of 1986, the official findings of the Rogers Commission were released. Failure of the O-rings, critical connecting seals in the Solid-fueled Rocket Boosters (SRBs), was identified as the primary cause of the catastrophe. The report also revealed that there had been warnings that the rings might fail in the abnormal cold of January 28, 1986, warnings that had been ignored so that the shuttle could fly on schedule. In the wake of the report, harsh criticism fell on NASA management.

NASA began to reorganize the space shuttle program from its very foundations. Every aspect of the system was scrutinized, from the SRBs to the orbiter. More than forty major changes were made to the SRBs, at a cost of $450 million. Thirty-nine changes were made to the liquid-fueled main engines, and sixty-eight modifications were made to the shuttle itself. Perhaps the most significant changes, however, concerned the way NASA conducted business, especially with respect to safety. In July, 1986, a new safety program was instituted. NASA officials believed that their new system would safeguard against any poor or uninformed decision that could endanger the lives of shuttle crew members.

On September 29, 1988, approximately two and one-half years after the *Challenger* explosion, the space shuttle *Discovery* (STS-26) was successfully launched from KSC. On October 3, after sixty-four orbits and 2.7 million kilometers of successful spaceflight, it safely touched down on the runway at Edwards Air Force Base. The improved solid-fueled rocket boosters and modified main engines performed perfectly. NASA's goal had been reached: The space shuttle program had been reborn.

Knowledge Gained

The space shuttle is unique in its design and function. Thus, the development of the spacecraft has contributed much to the fields of aerodynamics and rocket research. Early test flights confirmed the capabilities and versatility of the vehicle that acts as rocket, spacecraft, and airplane in the course of a single mission. In addition, equipment and systems developed for the shuttle — such as the remote manipulator system, which allows for the deployment and retrieval of payloads — have raised the level of space technology and built a foundation for further advances. In fact, thousands of new products and techniques in such diverse fields as medicine and archaeology have been developed indirectly from advanced shuttle technology. The knowledge gained from the missions themselves has also covered a broad spectrum of disciplines.

The shuttle's early flights demonstrated and tested the capabilities of the craft and its systems. In addition, STS-2 carried a payload for NASA's Office of Space and Terrestrial Applications; experiments concerning land resources, environmental quality, ocean conditions, and meteorological phenomena were performed. A precedent was set for scientific study.

STS-3 carried the first Get-Away Special payloads, small, self-contained, low-cost experiments that are packaged in canisters. These payloads are available to educational organizations, industries, and governments. STS-3 also carried the first student project, an experiment designed to collect data on the effects of flight motion on insects. The first materials processing experiment was also conducted on STS-3, with the monodisperse latex reactor. The experiment produced microsized latex particles of uniform diameter for commercial use in laboratories. It marked the beginning of the shuttle program's commercial materials production capabilities.

As shuttle flights became more routine, the number and sophistication of experiments conducted in or from the orbiter increased. In 1983, STS-9 carried the first Spacelab. Spacelab 1 was an orbital laboratory and observation platform designed to remain inside the shuttle's cargo bay. When the bay doors were open, it was directly exposed to space. Spacelab 1 was funded by the European Space Agency as a major contribution to the STS program. It had the capability of performing numerous experiments in the areas of plasma physics, astronomy, solar physics, material sciences, life sciences, and Earth resources. The vast amounts of data collected in this and the subsequent Spacelab missions would take decades to analyze. The essential assumptions of the Spacelab program were proved: that non-NASA astronauts could perform as trained payload specialists working closely with a scientific command center on Earth, that the Tracking and Data-Relay Satellite System could return the information to ground stations, and that Spacelab could support complex experiments in space.

Shuttle missions also included many experiments on the effects of spaceflight, and especially of zero gravity, on human beings. It was discovered that zero gravity affects every component of the motor control system. The heart and other large muscles decrease in size because of the absence of gravity's pull. Astronauts also experienced a decrease in bone tissue and red blood cells. Certain parts of the endocrine system may be affected, too. Finally, space sickness, a type of motion sickness, is not uncommon in microgravity.

These medical experiments and tests provided important data on zero gravity's effects and the ways in which those effects could be controlled. Medical experiments also raised questions about how humans would respond to long-term missions in space — missions undertaken on a space station, for example.

Over the course of the space shuttle program, hundreds of experiments and observations in almost every scientific discipline — from computers to agronomy — have been conducted. As the program moves forward and technology improves, the data base will continue to grow. The impact this scientific study will have on mankind will be overwhelming. Already, the amount of information collected is so great that it will take many years and thousands of man-hours to evaluate it all.

Context

Human technology has often been spurred by the human imagination. Nowhere has this been illustrated better than in the space shuttle program. The ability to launch manned vehicles into space, maneuver them around the universe, and return them home safely was once the exclusive province of science-fiction characters such as Buck Rogers. The space shuttle changed all of that.

Based on years of research in the fields of rocketry and aerodynamics, the flights of *Enterprise* and the later orbiters placed mankind on the threshold of a new era. The ability to conduct routine space missions had arrived. The space shuttle was the first vehicle to travel into space and return, to be used again. The ramifications of that capability have been far-reaching.

Satellites have been deployed, repaired, and retrieved by shuttle astronauts. Thousands of scientific and medical experiments and hundreds of hours of solar, Earth, atmospheric, and cosmic observations have been conducted aboard the shuttle. Critical communications satellites and Department of Defense payloads have been launched from the shuttle, and the potential for the manufacture of commercial alloys, crystals, and other materials is enormous.

NASA's ambitious plans for an international space station in low Earth orbit by the end of the twentieth century depend on the shuttle's viability. After the success of STS-26, the first shuttle launched after the *Challenger* explosion, many of the more vocal critics of the shuttle program became silent. It appeared that NASA officials had learned their lesson, and the program took on a new vitality.

It is apparent that the shuttle itself will eventually become simply another step in mankind's space efforts. Already, NASA has accelerated studies on a new manned spacecraft that will replace the shuttle early in the twenty-first century. Two design concepts have been reviewed. One calls for extensive modifications to the current shuttle. The other is a design for a vehicle significantly different from the existing orbiter. Each study will be funded at $1 to $2 million. NASA has begun to define what a new U.S. manned spaceflight capability should involve so that a development program can begin.

Bibliography

Gore, Rick. "When the Space Shuttle Finally Flies." *National Geographic* 159 (March, 1981): 316-347. The article provides an in-depth look at the space shuttle program before it became operational. Written for a general audience, it includes diagrams, drawings, and color photographs.

Joels, Kerry M., and Gregory P. Kennedy. *The Space Shuttle Operator's Manual*. Rev. ed. New York: Ballantine Books, 1987. A complete guide to the space shuttle. Written from the pilot's point of view, it contains information on space shuttle systems and flight procedures. It includes checklists, diagrams, charts, and photographs. For general audiences.

National Aeronautics and Space Administration. *NASA: The First Twenty-five Years, 1958 – 1983*. NASA EP-182. Washington, D.C.: Government Printing Office, 1983. A chronological history of NASA and its programs. The text is designed for teachers to use in the classroom. It contains many illustrations.

Ride, Sally, and Susan Okie. *To Space and Back*. New York: Lothrop, Lee, and Shepard Books, 1986. This book offers an astronaut's insight into the living and working conditions aboard the shuttle. Contains color photographs.

Turnill, Reginald, ed. *Jane's Spaceflight Directory*. London: Jane's Publishing Co., 1987. Updated annually, this resource is invaluable for a quick overview of progress made in space exploration. Capsule summaries of manned and unmanned missions are provided, and the text is heavily illustrated with diagrams and black-and-white photographs. A helpful index is also included.

Lulynne Streeter

ANCESTORS OF
THE SPACE SHUTTLE

Date: 1928 to 1971
Type of technology: Humans in space

Ancestors of the space shuttle helped aerospace engineers learn how to build a reusable vehicle that can withstand the rigors of outer space and reentry into the atmosphere, and some experimental craft gave pilots vital experience in flying at supersonic and hypersonic speeds.

PRINCIPLE PERSONAGES

WERNHER VON BRAUN, German-American rocket
engineer and space exploration visionary
EUGEN SÄNGER, German aerospace scientist
FRITZ VON OPEL, German industrialist and
early rocketeer
WALTER DORNBERGER,
KRAFFT EHRICKE and
DARRELL C. ROMICK, rocket plane designers
MAXIME FAGET, space shuttle designer
FRIEDRICH STAMER, early German rocket plane
test pilot
CHARLES E. "CHUCK" YEAGER,
A. SCOTT CROSSFIELD and
JOSEPH A. WALKER, American rocket plane test pilots

Summary of the Missions

From the outset, space travel visionaries proposed reusable vehicles for carrying people into Earth orbit and beyond. Spacecraft designers themselves preferred reusable craft, since they would be much less expensive than the use-once rockets and capsules of the early American and Soviet manned space programs. When in the late 1960's American aerospace engineers had a chance to build a reusable craft for the Space Transportation System, they had two decades of technological development to build on. Following World War II, the United States military and National Advisory Committee for Aeronautics (NACA) cooperated in testing two types of craft: rocket airplanes and lifting bodies. They are the ancestors of the space shuttle.

A forebear of the shuttle in spirit, if not in direct line of development, the first rocket plane was a glider with solid rockets strapped to it. Sponsored by a German glider club, the rocket plane carried pilot Friedrich Stamer more than two hundred meters in 1928. The next year, German car manufacturer Fritz von Opel flew an improved design for about ten minutes and at nearly 120 kilometers an hour. These hybrids

promised no quick improvement on the internal combustion engine then used in airplanes, and so, despite quasi-military support, experimentation continued only sporadically.

In 1935 German engineers, including Wernher von Braun, designed and tested rocket engines for military aircraft. Shortly before and during World War II, the Germans brought at least two models to the prototype test stage, but interest of the military turned to jet engines late in the war, and no rocket planes saw combat. The Japanese, however, used a rocket-propelled suicide bomber, the *Baka*, late in the war. The most notable contribution from the Axis powers to the history of rocket planes came from a secret German study by Eugen Sänger. He proposed a winged rocket as a long-range bomber. Carrying eighty metric tons of bombs, this "antipodal bomber" (antipodes are two points on opposite sides of the Earth) would take off from a runway and boost to about forty-two kilometers altitude, above the densest layers of atmosphere. It would save fuel on the way to its target by skipping repeatedly off the atmosphere much as a flat stone, thrown at a shallow angle, skips off the surface of still water. Finally, the bomber would land under its own power. The Snger report, which was not widely distributed among Nazi scientists, did not prompt developmental efforts until long after the war. A similar proposal in 1949 by California Institute of Technology professor Hsue-shen Tsien envisioned a winged rocket for passenger service.

Meanwhile, the American military had been interested in advanced propulsion systems since early in the war. In 1944, Congress approved a research program to explore both jet- and rocket-powered airplanes. NACA (later the National Aeronautics and Space Administration, NASA), the Navy, and the Army Air Force were to share in the design and testing. The program produced the X-1, America's first rocket plane.

Designed to climb to 24,000 meters and reach 2,800 kilometers per hour (kph), the X-1 was astonishingly successful for a first research craft. Cigar-shaped with straight wings and tail fins, the ten-meter-long craft was launched in air from a B-29 bomber. In 1947, Air Force Captain Charles E. "Chuck" Yeager accelerated it to 1,266 kph, becoming the first person to fly faster than Mach 1, the speed of sound. In 1953, NACA test pilot A. Scott Crossfield flew a Navy-sponsored rocket plane, the D-558-II, past Mach 2.

Despite several crashes, further designs of the X-2 espe-

Bell X-1 aircraft in flight (top left); Douglas D-558-1 aircraft in flight (top right); X-15 just after launch from B-52 mothership (bottom left); 3 Lifting Bodies on lakebed (bottom right). (NASA)

cially steadily increased speed and altitude records until the X-plane research program culminated in the X-15. After launch from a B-52 bomber, the X-15 was intended to soar to the edge of outer space at speeds up to Mach 10. Engineers faced serious design problems to meet these goals. The airframe had to withstand temperatures up 650 degrees Celsius, acceleration of seven gravities (that is, the change in velocity increased weight seven times), and turbulence at hypersonic speeds (greater than Mach 5). Since it would climb out of the atmosphere into near vacuum, the pilot had to steer by firing small control rockets, a new technique. The final design, shaped something like a ball-point pen with short triangular, or delta, wings and three tail fins, eventually climbed to more than 111 kilometers altitude in 1963 and a speed of nearly Mach 6. The NASA test pilot who flew to the record altitude, Joseph A. Walker, became the first rocket plane pilot to achieve the status of astronaut, awarded to all who fly above eighty-three kilometers (fifty miles). Among the other pilots in the X-15 program was Neil A. Armstrong, later an Apollo astronaut and the first man to step on the Moon. Proponents of rocket planes wanted to mount the X-15 on a big booster and lift it to Earth orbit, but NASA official wor-

ried the vehicle could not survive reentry. Besides, the Mercury program had already successfully put Alan B. Shepard, Jr., and John H. Glenn, Jr., into space. NASA administrators decided to devote American efforts to the space capsule programs and discontinued the X-15. On October 24, 1968, the 199th and final test of the X-15 was flown.

Even without the X-15, the idea of a reusable space plane had steady and powerful support. First of all, it held an important place in von Braun's grand vision for space exploration and colonization, and von Braun had long guided American space policy. Beginning in 1952, he articulated the vision in popular books and articles. He wanted a space station built in orbit that would serve as an outpost for travel to the planets, Mars first. The station would require regular ferry service from Earth. For that, his solution was a massive three-stage vehicle, two boosters and a winged spaceship that could rendezvous with the station and then land under its own power on a runway. Darrell C. Romick independently presented a proposal for a three-staged winged passenger vehicle, the Meteor, at the American Rocket Society annual convention in 1954. At the same time, engineers Walter Dornberger and Krafft Ehricke, drawing upon the ideas of Snger, developed plans for a reusable, manned booster to start a Bomber-Missile space plane (BOMI) on its way between continents; the booster would return to base on its own.

These proposals impressed the United States Air Force. Beginning in 1957, it started development of the X-20, or Dynamic Soaring program, which had the unfortunate acronym Dyna-Soar. NASA joined the project in 1958. The X-20 had triangular wings with a 6-meter span and was 10.6 meters long. A one-man glider with a rounded ceramic nose, it was to be boosted into space by a Titan IIIC and an orbit-transfer stage. Retrorockets and gas jets would slow it for reentry and permit it to maneuver to a landing strip. Planners intended the program to identify and solve problems encountered in reentering the atmosphere and test new technology involved in the solutions. Although manned orbital flights were the goal, rising costs, delays, and uncertainty about the program's benefit to the Apollo program caused cancellation of Dyna-Soar in 1963. However, by then scientists had conducted wind-tunnel tests on models, built full-scale mock-ups, started a pilot-training program with simulators, and created and tested much new hardware. In 1963, just before Dyna-Soar was closed down, a related program, Spacecraft Technology Reentry Tests (START), launched a scaled-down version of the glider to measure acceleration, vibrations, temperature, and structural pressure as it dived back to Earth.

START thereafter began a new phase as the Precision Recovery Including Maneuvering Entry (PRIME) project. PRIME was to test lifting bodies, especially their maneuvering capabilities upon reentry from space and landing. A lifting body is a wedge-shaped vehicle whose entire fuselage acts as a wing to support it in the air. The project tested three unmanned X-23A vehicles, launched on Atlas boosters, in 1966 and 1967. A manned version, the M2-F1, had already been used to practice low-speed landings. Dropped from B-52 bombers, manned supersonic lifting bodies the M2-F2, M2-F3, HL-10, and X24A and X24B tested the aerodynamics and maneuvering properties of the basic design from 1966 to 1975.

Both the testing programs and the design ideas of space pioneers influenced the first space shuttle designs in the late 1960's. Several configurations called for a rocket plane or lifting body piggybacked on a manned, winged booster that could itself land and be reused, similar to Romick's Meteor. Other designs bracketed a lifting body or X-15-like vehicle with boosters or required a single engine that could operate as a jet in the atmosphere and as a rocket in space, a versatility beyond the technology of the 1960's. Maxime Faget, principal designer for the shuttle, initially favored a rocket plane with straight wings, much like the X-1 and X-2. However, because the Air Force demanded a large payload bay and cross-orbital maneuvering capabilities, Faget changed to a delta-wing configuration, which was adopted into the final design in 1971.

Knowledge Gained
Although most were far too grandiose for the technology of

their times, early rocket plane proposals, such as those by von Braun and Snger, furnished important ideas for the Space Transportation System and other programs. They introduced the delta-wing configuration that the space shuttle uses. They suggested lowering costs by launching spacecraft on a reusable booster, which the space shuttle program partially adopted in the form of two recoverable solid rocket boosters. Most significant, they implanted the idea that a spacecraft could be amphibious — an airplane in the atmosphere and a rocket in space.

More practically, rocket-plane and lifting-body experiments taught essential lessons in piloting and vehicle design. Many scientists believed that the speed of sound amounted to a "brick wall in the sky," as rocket pioneer Theodore von Kármán facetiously put it. The fears were not entirely unreasonable. They worried that supersonic speeds would compress air along the leading edges of aircraft and create turbulence. Aircraft might then turn unstable and uncontrollable. The X-1 was designed to settle the stability question, and it proved quickly and finally that no insurmountable sonic barrier exists. Moreover, the X-15 demonstrated that hypersonic speeds are also manageable. Pilots of these craft developed skills to handle the difficulties of extreme velocities in the atmosphere, especially attitude control, and to wrest back control if the craft began to spin. Data from the X-15 flights suggested piloted vehicles are superior to unmanned craft because thoroughly trained pilots can react more quickly and effectively to problems than can ground controllers. The lifting bodies proved that a space plane can land on the glide and did not need power from an auxiliary source, such as jet engines, which increases a vehicle's weight and complexity.

The X-15, Dyna-Soar, and lifting bodies contributed fundamentally to technology and techniques for controlling the heating of a craft's skin upon reentry. The aerodynamics to reduce drag was studied, but it was learned that by positioning the body to reenter belly-first, the friction-induced heat could be spread over a larger area and that then only this area had to be provided with a heavy heat shield. Researchers also found that the delta-wing shape maintained lower average skin temperatures than a straight-wing configuration. Furthermore, the various metals in the skins of these experimental craft provided data on heat shedding and structural integrity at high temperatures and during vibration.

Finally, shuttle ancestors tested control and life support systems. The small attitude control rockets, such as those used under the X-15's wings and nose, gave pilots invaluable experience in handling a craft in vacuum and let engineers find the best placement and power for thrusters. Guidance systems, sensors, on-board instruments, escape systems, fuel mixtures and pressurization, use of devices such as fences and slats on wings to increase stability, and many other features received crucial trials. At the same time, medical researchers studied how the human body behaves under high acceleration and how to design suits and seats to protect the body.

Context

Released in 1968, the movie *2001, A Space Odyssey* depicts a sleek space plane streaking up from Earth to rendezvous with a huge wheel-shaped space station. Based on a novel by English writer Arthur C. Clarke, the movie indicates how deeply ingrained the space plane-space station concept had become among proponents for manned space flight, thanks primarily to von Braun. It also impressed audiences so deeply that the scene has since been an icon of a bold technological future. Such imaginative systems as that in the movie and the actual testing of rocket planes and lifting bodies thus figured centrally in both the professional and popular visions of space exploration and assumed a place in modern culture.

Space shuttle precursors also reveal much about the national and international politics behind large technological research programs. From the earliest German rocket planes through the testing of lifting bodies, involvement of military planners was typical and often crucial. In the United States, the Air Force initiated and conducted much of the research on the X-planes and lifting bodies. Even for the space shuttle, billed as a civilian program, the final design derived to a large extent from military requirements.

Although NACA, and later NASA, played key roles all along, these research programs needed the weight of the military for economic reasons. First of all, the extensive system of Air Force bases with generous resources and financing, Navy ships for recovery at sea, the many highly trained military pilots, and the numerous technical support personnel made programs cheaper to start up and run than if they had depended on building new facilities and assembling strictly civilian staffs. What is more important, the budget-conscious Congresses were much more likely to appropriate money for programs that had military participation and possible military applications. This militarism derived from competition with the Soviet Union, both military and civilian. That the Soviets might use space for manned weapons systems argued powerfully for a U.S. military presence in space as well. National prestige was also at stake: The Soviet Union had launched the first spacefarer in a capsule, and so the U.S. did not want the Soviets to get ahead in other manned ventures as well.

The shuttle precursors had more than a national effect, however. On the basis of rocket-plane and lifting-body research as well as that for the space shuttle, the Soviet Union, the European Space Agency, and Japan set up their own shuttle programs. Furthermore, more so than the American space shuttle, the X-planes are the direct antecedents of various proposals in the United States and Europe for a single-stage space plane. The Air Force-sponsored X-30 program, for example, has studied prototypes for a National Aerospace Plane, which will take off and land like a commercial jet liner and travel in space like the shuttle.

It is likely that twenty-first-century manned space vehicle research will aim at continuing the lineage of rocket planes, lifting bodies, and shuttles. A reusable, single-stage, Earth-to-orbit vehicle is necessary to minimize development and operation costs, construct and maintain a space station, provide inexpensive cargo service, and make space travel widely available.

Bibliography

Kleinknecht, Kenneth S. "The Rocket Research Airplanes." *In The History of Space Technology,* Eugene Emme, ed. Detroit, Mich.: Wayne State University Press, 1964: 189-211. Stiffly written and laced with aerospace jargon, only-occasionally defined, this article contains much specific information, especially about the technology developed for the X-1 and X-15 and the people involved.

Grey, Jerry. *Enterprise.* New York: William Morrow, 1979. An aerospace scientist, Grey participated in the development of U. S. space technology beginning in the 1950's. His enthusiasm for the space shuttle shines through clearly in his detailed history of the project, particularly the politics involved. Chapter 3 discusses the research with rocket planes and lifting bodies that prepared engineers for the space shuttle effort. With many photographs and drawings.

Ley, Willy. *Rockets, Missiles, and Men in Space.* New York: Viking, 1968. A thorough history of rocketry and spacecraft by a rocketry pioneer. The writing is quirky and leisurely but enjoyable as Ley discusses technology and scientists, many of whom he knew, in considerable detail. Appendix 1 is a substantial treatment of rocket planes up to the X-15. The text includes diagrams explaining technical and design matters and many rare photographs.

Miller, Jay. *The X-Planes, X-1 to X-29.* Marine on St. Croix, Minn.: Specialty Press, 1983. This well-illustrated book is the definitive study on the X-series experimental aircraft. It contains complete flight records and details each craft.

Stockton, William, and John Noble Wilford. *Spaceliner.* New York: Times Books, 1981. This somewhat sensationalized account by two *New York Times* science reporters mixes a lively description of the first shuttle flight with the political and technical history of the project. They mention shuttle precursors briefly. The book is most valuable as general background and is directed at a general audience. Some photographs.

Vogt, Gregory. *An Album of Modern Spaceships.* New York: Franklin Watts, 1987. Written for a young adult audience, this book has many illustrations of lifting bodies, early shuttle designs, and ideas for the next generation of space planes after the shuttle. The text provides a simple but informative historical review and then describes shuttle precursors non-technically.

Winter, Frank H. *Rockets into Space.* Cambridge, Mass.: Harvard University Press, 1990. A pleasantly written general history of the Space Age. The sixth chapter concerns the space shuttle, and Winter summarizes both the visionary and practical projects for a space plane. He pays particular attention to lifting bodies and alternative designs for the shuttle, although there is little technical detail. Illustrations of four early designs.

Roger Smith

THE SPACE SHUTTLE
APPROACH AND LANDING TEST FLIGHTS

Date: September 8, 1977, to December 6, 1985
Type of mission: Manned test flights

Enterprise, *the first space shuttle orbiter, tested the approach and landing techniques of the United States' Space Transportation System. Engineers also used it for ground handling, vibration, and landing brake net tests.*

PRINCIPAL PERSONAGES
  FRED W. HAISE,
  C. GORDON FULLERTON,
  JOE H. ENGLE, and
  RICHARD H. TRULY, NASA astronauts
  FITZHUGH L. FULTON, the shuttle carrier aircraft pilot
  THOMAS C. MCMURTRY, the shuttle carrier
    aircraft copilot

Summary of the Missions

In September, 1976, the National Aeronautics and Space Administration (NASA) unveiled the prototype for a new class of reusable manned spacecraft. The delta-wing vehicle, *Enterprise*, was the first space shuttle orbiter, the manned component of the space shuttle vehicle. The complete space shuttle comprised the orbiter attached to a large expendable fuel tank flanked by two Solid-fueled Rocket Boosters (SRBs). The space shuttle would be launched from a rocket, orbit like a spacecraft, then land like an airplane.

Before any of the reusable craft ventured into space, tests of the landing system had to be made; these tests were called Approach and Landing Tests (ALTs). The first orbiter, which bore the airframe designation OV-101, was designed to undergo such tests. Construction of OV-101 began on June 4, 1974. OV-101 was not intended for spaceflight, so it did not contain many of the systems needed for an orbital craft. For example, OV-101 did not have the three rocket engines necessary for boost to space, nor did it contain any of the orbital maneuvering engines. Such spaceflight components as the star trackers, unified S-band antennae, and rendezvous radar were also not part of the design. Since OV-101 would never venture beyond Earth's atmosphere, it was not covered with any of the thermal tiles (ceramic blocks of silicon material that can withstand extremely high temperatures and insulate the orbiter's aluminum structure from the heat of reentry) needed for protection during the fiery return from space. On OV-101, blocks of polyurethane foam simulated the ther-

mal protection tiles. Glass fiber panels on the nose cap and wing leading edges simulated the reinforced carbon-carbon structures planned for later orbiters.

On September 8, 1976, NASA Administrator James Fletcher met with President Gerald Ford to discuss the name of OV-101. For several months prior to the meeting, fans of the science-fiction television series *Star Trek* had conducted a letter-writing campaign suggesting the name *Enterprise*, the name of the spaceship in the series. After an estimated sixty thousand letters arrived at the White House and NASA headquarters, President Ford approved the name.

Slightly more than a week later, on September 17, 1976, Fletcher, U.S. senator Barry Goldwater, Rockwell International board chairman Willard Rockwell, and about five thousand others stood outside a hangar at Rockwell International's Palmdale, California, plant. At about 9:30 A.M., as the band played the theme music from *Star Trek*, for a cheering crowd, a red, white, and blue tractor towed the *Enterprise* from behind the hangar. *Enterprise* had a wingspan of 23.1 meters, a length of 37.1 meters, and weighed 65,000 kilograms.

On January 31, 1977, NASA and Rockwell engineers moved *Enterprise* out of Palmdale, 58 kilometers to the Dryden Flight Research Center at Edwards Air Force Base. *Enterprise* made the journey, which took nearly twelve hours, on a 90-wheel trailer. At Dryden, engineers had a modified commercial passenger jet waiting to be used with *Enterprise* for the landing tests. Since the space shuttle orbiter returned from space unpowered, getting the shuttle airborne to test the landing system posed a unique challenge. The solution was to mount the orbiter piggyback on top of a large aircraft which could carry it aloft and then release it. In addition, the aircraft could ferry space shuttle orbiters for cross-country travel.

On June 17, 1974, NASA purchased a Boeing 747 from American Airlines for use as the Shuttle Carrier Aircraft (SCA) and subsequently modified it to carry a 68,000-kilogram load on its back. All passenger accommodations — seats, galleys, and the like — were removed. Bulkheads and reinforcements had to be added to the fuselage's main deck. The 747's standard engines were replaced with higher-thrust engines. Three supports for the orbiter were added to the SCA's top, and tip fins were added to the horizontal stabilizer. Other changes to the aircraft included modifications to its trim system, air conditioning system, and electrical wiring, and the addition of an escape system for the flight crew.

On February 7, 1977, *Enterprise* was mounted atop the SCA for the first time. Eight days later, engineers began a series of taxi tests. During these, the 747 taxied along the runway at speeds just less than what was needed to become airborne. These tests showed no unusual problems, so preparations were made for the first flight of the tandem aircraft. The first five flights of the approach and landing test program were so-called captive inert flights to verify the handling of the SCA/space shuttle orbiter combination. For these, *Enterprise* did not carry a crew. The program then progressed to "captive active" flights, where *Enterprise* was "powered up" and occupied but remained attached to the SCA. Finally, there were the free flights, where *Enterprise* separated from the SCA and made an unpowered approach and landing.

For all but the last two free flights, which tried to duplicate the return from space of an orbiter as closely as possible, *Enterprise's* aft end was covered with an aerodynamic fairing. This fairing, or tailcone, smoothed the airflow off the orbiter's base, reducing drag and buffeting on the SCA's tail. For cross-country flights with operational orbiters, the fairing also protected the three space shuttle Main Engines.

The first captive inert flight occurred on February 18. For two hours and ten minutes, Fitzhugh L. Fulton and Thomas C. McMurtry flew the SCA/*Enterprise* combination over the California desert. This flight evaluated the airworthiness of the configuration and verified the preflight stability and control predictions. Four more captive inert flights followed during the next two weeks, each exploring a different portion of the flight envelope.

After the captive inert flights, the program moved into the captive active phase. The first captive active flight was on June 18, 1977. Fred W. Haise and C. Gordon Fullerton occupied the orbiter's flight deck. During a 56-minute flight, they tested various orbiter systems and gathered additional information about buffeting of *Enterprise's* vertical tail. Ten days later, astronauts Joe H. Engle and Richard H. Truly occupied Enterprise for the second captive active flight. Haise and Fullerton again occupied *Enterprise* for the final captive flight of the ALT program, a dress rehearsal for the free flights.

On August 12, 1977, *Enterprise* flew independently of the SCA for the first time. Haise and Fullerton piloted *Enterprise*, Fulton and McMurtry the SCA. About 48 minutes after takeoff, the piggyback aircraft reached an altitude of 8,650 meters. Fullerton and McMurty began a shallow dive, a maneuver called "pushover." At an altitude of 7,350 meters, Haise pushed the separation button in *Enterprise*, and seven explosive bolts fired; *Enterprise* was on its own. Fulton pitched the SCA down and rolled left to clear the free-flying space shuttle. Haise and Fullerton pitched the orbiter's nose up and gently rolled *Enterprise* to the right to clear the SCA. They straightened the craft, then eased the orbiter's nose up, practicing a landing flare maneuver. After this maneuver, the crew of the *Enterprise* executed a 180-degree turn to the left, align-

ing *Enterprise* with the Dryden runway. *Enterprise* touched down 5 minutes and 22 seconds after separating from the SCA.

There were some relatively minor problems encountered during the first free flight. For example, one of the General Purpose Computers (GPCs) stopped working at separation; however, the other on-board computers sensed the fault and automatically took the malfunctioning GPC off-line. Because Haise and Fullerton remained flexible, all mission objectives were achieved, and the handling, stability, and flight performance characteristics were as predicted.

Engle and Truly piloted *Enterprise* for its second free flight on September 13, 1977. During this flight, the pilots executed a 1.8-g turn to the left as they lined *Enterprise* with the runway, duplicating the maneuvers of an orbiter returning from space. The term "g" refers to the acceleration of gravity. Sitting still on Earth, one experiences an acceleration of 1 g, or a gravitational force of 1, the normal sensation of gravity. During periods of changing acceleration, such as a banking turn in an airplane, the so-called g-loading will change. As Engle and Truly banked the *Enterprise*, they felt as though they weighed twice as much as normal. The second flight lasted 5 minutes, 28 seconds.

Haise and Fullerton were again at *Enterprise's* controls on September 23, 1977, for the third free flight of the ALT program. During this flight, the pilots tested the orbiter's automatic landing system. After the flight, ALT program managers decided to proceed with the tailcone-off flights, eliminating two planned missions. For these flights, the tailcone was removed, duplicating the aerodynamic characteristics of an orbiter returning from space.

The fourth free flight, also the first tailcone-off flight, occurred on October 12, 1977. Engle and Truly flew OV-101. While the SCA climbed to separation altitude, the 747's tail was visibly buffeted by the turbulence coming off the base of the orbiter. After separation, Engle and Truly flew *Enterprise* on a straight course to Dryden. Without the tailcone, the drag on *Enterprise* was much higher, and the flight lasted 2 minutes, 35 seconds.

Haise and Fullerton again piloted *Enterprise* for the fifth and final free flight on October 26, 1977. This would be the first landing on a paved runway. (All previous landings had been on dry lakebed runways.) This free flight lasted 2 minutes, 5 seconds. During landing, after making a near touchdown, the orbiter's nose suddenly rose, then settled back down. The nose rose once more, made a second touchdown, skipped off the runway briefly, then finally settled down for a third, and final, touchdown. Despite the difficulties, no further free flights were deemed necessary, so the most visible part of the ALT program ended.

After the free flights, engineers evaluated one more aspect of the space shuttle's flight characteristics: how the SCA/orbiter combination handled on cross-country ferry flights. For the ferry flights, the orbiter's front support on the back of the 747 was lowered from 4 degrees to 2 degress to reduce drag and improve the aircraft's cruise characteristics. Following a series of tests, the SCA/*Enterprise* took off from Dryden on March 10, 1978, bound for the Marshall Space Flight Center (MSFC) in Huntsville, Alabama.

At MSFC, engineers joined *Enterprise* with an external propellant tank and two SRBs in a test stand. They then applied vibrations of varying frequencies and intensities to the vehicle to see how the space shuttle reacted. These tests simulated the vibrations expected during launch.

By early 1979, the so-called mated ground vibration tests were over, and on April 10, the SCA ferried the *Enterprise* from MSFC to the Kennedy Space Center in Cape Canaveral, Florida. At the Cape, *Enterprise* was assembled with the tank and boosters atop the mobile launch platform and taken to the launchpad. This was the first time a complete space shuttle was processed for launch, and technicians tested how well all the plumbing and electrical fittings at the launchpad interfaced with the vehicle. On May 1, 1979, a space shuttle vehicle stood on Launch Complex 39A for the first time.

Following the launchpad assembly and fit checks, *Enterprise* was used for several public exhibitions. In May, 1983, NASA took the *Enterprise* to the Paris Air Show. After the French exposition, *Enterprise* was also viewed by millions of people in West Germany, Italy, England, and Canada. The following year, from May through November, *Enterprise* was exhibited at the Louisiana World Exposition in New Orleans. For this trip, the orbiter was transported on a barge part of the way.

On November 18, 1985, *Enterprise* made its last flight, as NASA delivered it to the Smithsonian Institution at Dulles International Airport in Virginia. At a ceremony on December 6, NASA formally turned the *Enterprise* over to the Smithsonian. In June, 1987, however, NASA engineers found one more use for OV-101, as they tested an emergency net designed to keep returning orbiters from running off the runway.

Knowledge Gained

As the prototype space shuttle orbiter, *Enterprise* validated the design of a new family of spacecraft. Also, by testing the landing system, Enterprise proved the validity of using glide return for manned spacecraft. In the course of the flight test program, NASA engineers learned how to land a 90,000-kilogram, delta-winged spacecraft. The flights verified both pilot-guided and automatic approach and landing systems and showed that the orbiter could land on a paved runway. This had been particularly worrisome, because the orbiter lands at relatively high speeds (about 350 kilometers per hour) and tire wear is a problem. (In fact, subsequent landings by orbiters returning from space showed this to be a major problem.)

The ground vibration tests were a crucial step toward validating the entire space shuttle "stack" for launch. In the course of these tests, engineers at Marshall Space Flight Center ballasted the external tank with varying amounts of deionized water, simulating the consumption of propellants during flight. Also, the SRBs were ballasted with inert propellant for the tests. Thus, the shuttle's designers had actual measurements of the vehicle's response to dynamic flight conditions. These data were used to validate researchers' analytical math models and predictions of the space shuttle's in-flight behavior.

By taking the *Enterprise* to Kennedy Space Center for launch complex fit checks, engineers could verify that all the electrical, hydraulic, and fuel lines on the launchpad interfaced properly with the space shuttle. This was particularly important, because the shuttle used facilities originally built for the Apollo program in the 1960's. The *Enterprise* also provided a dress rehearsal for pad crews to handle and assemble a complete space shuttle before the first actual flight vehicles arrived.

Context

The idea of winged, reusable spacecraft had long appealed to engineers as the most economical means for opening the space frontier. Throughout the 1960's, NASA and military engineers discussed the possibilities of an aircraft-type space vehicle and advanced numerous designs. One of the best known was the United States Air Force Dyna-Soar, a manned glider planned for launch by a Titan 3 launch vehicle. Other designs, though never built, bore such exotic names as Triamese (a shuttle comprising three identical vehicles connected for launch, one of which would reach orbit), Meteor, and Astro Rocket. By the late 1960's, NASA planners were developing a program to succeed the Apollo manned lunar missions. They finally decided on a reusable space shuttle. In April, 1972, the United States Congress approved NASA's request to proceed with the space shuttle. On August 9, 1972, NASA managers authorized North American Rockwell (later, Rockwell International) to build the space shuttle orbiter.

Engineers proposed numerous designs for the space shuttle before choosing what became the *Enterprise*. The orbiter contained a cargo bay capable of carrying payloads up to 4.5 meters in diameter, 18.3 meters in length, and 29,500 kilograms in weight. The space shuttle would lift off like a rocket, orbit like a spacecraft, then glide back for an airplane-style landing.

Winged spacecraft had existed for nearly two decades, in the form of the North American X-15. Between 1959 and 1968, three of these remarkable aircraft made 199 flights. On several occasions, X-15's took their pilots above 80 kilometers, qualifying them as astronauts. Experiences with the rocket-powered X-15 aircraft, which made unpowered, high-speed landings, were directly applicable to the space shuttle program. Yet before the plans for the space shuttle could be put into practice, particularly the notion of landing a spacecraft like an airplane, a test vehicle was needed. The *Enterprise* served that function. Thus *Enterprise* bridged the gap between the X-15 research aircraft flights and the first space flight of the OV-102 orbiter *Columbia* in 1981.

Bibliography

Grey, Jerry. *Enterprise.* New York: William Morrow and Co., 1980. This popular book on the space shuttle program deals extensively with the politics of building the space shuttle and starting the program.

Hallion, Richard P. *On the Frontier: Flight Research at Dryden, 1946–1981.* NASA SP-4303. Washington, D.C.: Government Printing Office, 1984. This volume in the NASA History Series describes flight-testing activities at the Dryden Flight Research Center from 1946 to 1981. Provides an excellent perspective on flight research activities at Dryden up to the first space shuttle mission. Also included are chapters on the X-15 and "lifting body" programs, which were precursors of the space shuttle.

Joels, Kerry M., and Gregory P. Kennedy. *The Space Shuttle Operator's Manual.* Rev. ed. New York: Ballantine Books, 1987. This book provides a description of how the space shuttle flies and includes checklists and time lines of the space shuttle landing system. Descriptions of shuttle facilities and of accessories such as the shuttle carrier aircraft are also included.

National Aeronautics and Space Administration. *Space Shuttle.* NASA SP-407. Washington, D.C.: Government Printing Office, 1976. Written on a popular level, this source provides an early look at the benefits and capabilities of the space shuttle program. Although its projections on launch frequency and ground turn-around time proved too optimistic when space shuttles began flying, this document shows the expectations that NASA managers had for the vehicle at the time *Enterprise* was being tested.

Stockton, William, and John Noble Wilford. *Space-Liner: The New York Times Report on the Columbia's Voyage.* New York: Times Books, 1981. An excellent narrative of the space shuttle program up to the first flight of *Columbia* in April, 1981. Included in the narrative are chapters devoted to the design, development, and testing of the space shuttle system, including *Enterprise.*

Gregory P. Kennedy

SPACE SHUTTLE LIVING CONDITIONS

Date: Beginning April 12, 1981
Type of program: Manned spaceflight

Safe and comfortable living conditions in a microgravity environment must be provided for space shuttle astronauts in order to ensure the success of assigned scientific and military tasks.

PRINCIPAL PERSONAGES

MAXIME A. FAGET, Director of Engineering
and Development, Johnson Space Center
CHARLES A. BERRY, Director of Medical Research
and Operations, JSC
ROBERT E. SMILEY, head of the Crow Systems
Division, JSC
ROBERT F. THOMPSON, Program Manager of the
Space Shuttle Office, JSC
JAMES W. BILODEAU, the chief of crew procedures for
space shuttle missions
ALAN B. SHEPARD, head of the astronaut office,
Manned Spacecraft Center

Summary of the Program

Daily life aboard the U.S. space shuttle is much more luxurious than that on the earliest manned space missions, although interior decorating is not a high priority. "The decor could be called modern metal file cabinet," according to Rick Gore of *National Geographic* magazine (March, 1981). The nose section of the shuttle orbiter contains the bilevel, pressurized crew cabin. Approximately seventy-two cubic meters of space are available for astronaut activities. Commander and pilot observe and operate shuttle functions from the upper-level flight deck, and a work area for other crew members is located behind the command center. The lower level of the cabin, known as middeck, serves as living quarters for the entire crew; experiments requiring oxygen are also housed here during some missions.

Fans force the cabin air through an array of filters. Particulates, including dust, bacteria, dead skin cells, and hair, are filtered directly; lithium hydroxide canisters remove carbon dioxide, a waste product of respiration; and activated charcoal filters remove odors. Humidity is kept low to discourage bacterial growth. The astronauts' breathing mixture is 80 percent nitrogen and 20 percent oxygen. (Earth air is 78 percent nitrogen, 21 percent oxygen, and 1 percent other gases.) Internal cabin pressure is maintained at 14.7 pounds per square inch,

the pressure of Earth's atmosphere at sea level. Water — a by-product of the fuel cells that generate electricity — is provided to the crew cabin at a rate of 3 kilograms per hour. Cabin temperature is maintained by water-pipe heat exchangers which transfer heat to Freon-fueled cargo bay radiators.

Early spacefarers survived on unpalatable pastes squeezed out of devices resembling toothpaste tubes. Shuttle astronauts enjoy a wide variety of foods and even select their own menus. Each day's intake must provide 2,800 to 3,000 kilocalories and the recommended daily allowances of minerals and vitamins. Special attention is given to potassium, calcium, and nitrogen; in weightlessness, the human body loses these minerals first. Everything from scrambled Mexican eggs to candy-coated chocolates can be found among the thirty-seven beverages and ninety-two foods available.

The shuttle's galley, or kitchen, contains food lockers, which hold each day's meals in the order in which they will be consumed, a convection oven to heat precooked foods to 82 degrees Celsius, and a hot and cold water dispenser, which injects fluid into rehydratable food packages with a needle to prevent spillage. Velcro strips on lockers and walls allow food items to be fixed in place while meals are prepared. Fresh foods, including fruit, vegetables, and baked goods, are loaded shortly before launch. The galley also has a washing station for cleaning hands and utensils.

For certain special missions in which space is at a premium, the entire galley is removed and the astronauts make do with a cold water dispenser that can rehydrate a reduced menu of foods. A small, portable warmer substitutes for the convection oven. Emergency food, for unexpectedly lengthened missions, and additional snacks are stored in the pantry, which contains enough extra food to provide 2,100 kilocalories per day per astronaut for two additional mission days.

There is no refrigerator on board the shuttle because of weight considerations, so foods are prepared and packed in several ways for storage at room temperature. Freeze-dried, rehydratable food and drink come in small, sealed plastic bowls or pouches. Thermostabilized foods are heat sterilized, to discourage bacterial spoilage, and packaged in metal cans or flexible pouches. Meats are sterilized by exposure to ionizing radiation and packaged in flexible foil. Intermediate-moisture foods, such as dried apricots, dried peaches, and beef jerky, are packed in transparent plastic. Items such as nuts and cookies, designated "natural form foods," are packed in plastic containers, and condiments are provided in dropper bottles. Even salt and pepper are fluidized in water or oil to make them dispensable.

Mealtime in the shuttle may find the astronauts rooted to a surface by suction cups, sitting in a seat with meal trays strapped to their laps, upside down with trays attached to the ceiling, sitting at the small shuttle table with feet in loops, or even floating freely through the cabin. Knives, forks, and spoons three-quarters of the normal size are used for shuttle meals; their smallness helps minimize food spillage and cleanup problems for the vehicle's air recycling systems.

Keeping clean and disposing of waste is no easy task in space, but it is vital for the health of the space shuttle crew. The confined space and effects of microgravity could allow disease-inducing bacteria to multiply out of control if sanitation were not carefully maintained. The microgravity environment of low Earth orbit allows water droplets and crumbs of debris to float around the shuttle cabin, posing a threat to instrumentation and crew members.

Garbage from food preparation, used clothing, and waste from experimental activities are sealed in plastic containers for storage under the middeck floor until return to Earth. All areas of the spacecraft used by the crew are cleaned in flight with disposable wipes and disinfectant solution.

The space shuttle lavatory contains a small sink, a toilet, mirrors, and a light. Washing is performed with a minimum of water. Since fluid adheres to skin in a weightless environment, little soap or water is necessary. Crew members take short sponge baths rather than taking showers, as they did in Skylab. A fan sucks water into the basin drain, substituting for the ordinary effect of gravity on Earth. Hot water is dispensed from a gun set at temperatures from 18 to 35 degrees Celsius. Male astronauts wishing to shave must use ordinary shaving cream and safety razors, wiping the razors with disposable towels to keep whiskers from escaping into the cabin environment. Towels and personal necessities are clipped to the wall for easy access. The zero-gravity toilet drains by means of a fan, which transports the waste materials to a storage area. Originally, wastes were shredded; now, they are disinfected, deodorized, and dried for storage until return to Earth. In micro-gravity, to remain in place on the toilet, astronauts use a seat belt and place their feet in toeholds while holding onto handles. The toilet's cycle time is approximately fifteen minutes, so it can be used about four times in an hour. Spacesick astronauts use velcro-sealable pouches, similar to airsickness bags, which are disposed of in the commode. The same bags can be used as an alternative method of fecal collection if the zero-gravity toilet malfunctions.

Shuttle astronauts wear ordinary trousers, shirts, and jackets covered with Velcro-fastened pockets. Ordinary clothes contribute to a sense of normality and comfort as the astronauts go about their many tasks. For extravehicular activities (space-walks outside the vehicle or in the exposed cargo bay), protective spacesuits are required. Each suit is really a tiny

spaceship that provides a breathable atmosphere, comfortable temperatures, and adequate internal pressure. Propulsion is provided by separate manned maneuvering units.

Shuttle suits, called EMUs (Extravehicular Mobility Units), are much more sophisticated than those used on past space missions. In addition to offering greater flexibility and comfort, they can be put on relatively quickly without assistance. The modular suits have three primary and several secondary components, all of which come in several sizes so as to fit the exact dimensions of each astronaut. The main part of the suit consists of a hard covering for the upper torso and flexible pants, a detachable helmet with a visor, gloves, and a bag containing about twenty ounces of drinking water. The gloves are the only component that is fitted to each crew member. The torso is composed of many layers, each with an important job. A Teflon-coated outside layer resists tears and protects against damage from micrometeoroids, the next several layers of aluminized Mylar plastic reflect heat away from the suit, and the interior layers are made of insulating Dacron unwoven fabric. Inside the outer suit, a separate, water-cooled, liner garment is worn; it resembles long underwear. The small tubes sewn all over it conduct water to lower the astronaut's body temperature. Under the liner, there is a urine collection device that retains liquid waste for later transferral to the shuttle's waste management system. Physiological functions are controlled by the primary life-support system, which circulates and cools the water in the liner, absorbs carbon dioxide, supplies temperature-controlled oxygen for breathing, and maintains suit pressure. Filters remove unpleasant odors from the suit. All these complicated and vital functions are controlled by a tiny computer which displays information on the suit's chest section. Automatic checking programs test various suit functions, provide instructions for astronauts donning their suits, and sound an alarm if any suit functions begin to fail.

Sleeping accommodations vary with the demands of a particular mission. For uncrowded missions, one vertical and three horizontal bunks are available. Each sleeping pallet has storage areas, a light, pillows, a fan, a communications station, microgravity restraint sheets, and a noise-suppression blanket. For crowded missions, up to four sleeping bags attached to provision lockers are used, although some astronauts have preferred to catch naps while simply tethered to a cabin wall.

Communication among crew members in various parts of the shuttle is accomplished via simple headsets that connect to eight intercom terminals in the crew compartment. Books, tape recorders and tapes, playing cards, and other games are supplied, and a regular exercise regime prevents physiological deconditioning.

Knowledge Gained

Keeping humans alive and well in space is a tremendously complicated task. Space shuttle designers must plan each subsystem and then integrate it with the entire human systems design scheme. Shuttle missions carry up to eight astronauts

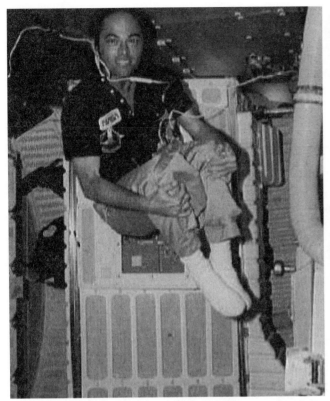

STS-1 Pilot Robert L. Crippen takes advantage of zero gravity to do some aerobics in the mid-deck area of Columbia *(left); Astronaut Rhea Seddon sits down to a meal in the mid-deck.* (NASA)

for periods of seven to fourteen days; shuttle designers therefore face challenges quite different from those presented by the three Skylab missions, with their smaller numbers of astronauts and longer-duration missions.

As the Space Transportation System (STS) became operational, much was learned about problems with the various living arrangements. Solutions to these problems were gradually refined as more missions were flown. Strong odors are particularly disturbing to persons working in the confined area of a spacecraft. It was found that the odor of certain fresh fruits, such as bananas, was terribly annoying, and such items had to be stored separately or eliminated from the cargo.

Spacesickness has been a problem for a number of astronauts on the shuttle, and scientists have made considerable efforts to understand and prevent the syndrome. Senator Jake Garn experienced significant spacesickness during his 1985 voyage on *Atlantis*. He graciously offered to serve as a "guinea pig" for important studies on the symptomatology and physiology of the complaint.

The zero-gravity toilet was notoriously unreliable during many of the shuttle missions. Problems with the system were particularly demoralizing to the crew, presented a potential health hazard, and contributed to cabin odors. After the first few missions, the shredder/slinger was removed, since it did

not function well in microgravity. The system was changed to allow wastes to be diverted to the side of the receiving vessel.

Debris in the cabin proved to be difficult to control, even with considerable prior planning. It is very important, for example, to have low-lint clothing and towels. During some experiments with small animals, most notably those on the April, 1985, *Challenger* mission, soiled bedding and unused food pellets escaped from the containment units and caused a major annoyance to the astronauts aboard. Such experiences have led to increasingly better designs for daily use items and experimental apparatus.

The maintenance of normal sleep cycles is very important to the crew's productivity and alertness. Schedules which keep astronauts on a regular twenty-four-hour cycle were found to be far superior to radically different schedules.

In the limited environment of a spacecraft, mealtimes become significant events. The space shuttle has the most sophisticated eating arrangements of any American spacecraft. It is possible that the preparation and consumption of food substitutes for the social and personal rituals that are lacking on board the shuttle.

Context

Just as astronauts' experiences on the three Skylab missions

Sullivan and Ride show sleep restraint equipment. (NASA)

affected the design of the space shuttle living quarters, crew members' experiences aboard the shuttle have helped engineers to design the U.S. Space Station and the spacecraft that may go to Mars and the Moon. Human performance and well-being in space is the single largest factor determining the success of piloted space missions. It is also the area in which the United States has less cumulative experience than the Soviet Union. Setbacks in the shuttle program have slowed U.S. progress in understanding human needs in space.

The Soviet Union's piloted missions have almost all been of long duration and involved only a few astronauts at a time. Psychological stress, depression, and conflict among Soviet cosmonauts seem usually to have emerged after thirty days. As of 1988, it was not clear whether any of these symptoms arise when astronauts spend cumulatively long periods in space over the course of many missions.

The demanding work of launching and repairing satellites or recovering damaged satellites for later repair on

Earth requires that the crew be in top-notch physical condition. Such projects have served as models for future situations in which spacefarers will be faced with demanding tasks under stressful conditions. The International Space Station will require high performance standards in the demanding work of constructing large space structures in orbit. What scientists learn about human performance from the shuttle will be valuable in planning the activities of space station workers.

The lengthy consideration of human daily life — from sleeping, to eating, to exercising — in the alien environment of space has helped to advance the sciences of ergonomics, human performance, and stress psychology and the technologies of air and water recycling, food sterilization, and food packaging. Systems on the shuttle provide rigorous, immediate feedback on theories about human behavior and human engineering. If the designs are faulty, that is instantly apparent in the demanding shuttle environment.

Keeping human beings alive and comfortable in an environment as hostile as space shows that mankind's domain can be extended far into the solar system. Although unpiloted missions offer new perspectives on the outer planets and the universe, manned missions are the key to moving human life into the realm of space.

Bibliography

Committee on Space Biology and Medicine. *A Strategy for Space Biology and Medical Science.* Washington, D.C.: National Academy Press, 1987. A description of the biological and medical areas that are important to space planning, this book covers topics ranging from human nutrition and reproduction to the behavior of plants in microgravity and developmental biology. Some sections are accessible to general readers; some are college-level material.

Connors, Mary M., et al. *Living Aloft: Human Requirements for Extended Space-flight.* NASA SP-483. Washington, D.C.: Government Printing Office, 1985. Discusses the physical and social stresses of living in space and explores medical considerations and possible ways to deal with those stresses. Includes sections on exercise, leisure time, the need for privacy, astronaut work schedules and other aspects of human performance, the selection of crews, communications, and responses to crises. For college-level readers.

Haynes, Robert. "Space Shuttle Food Systems." In *NASA Facts.* NF-150/1-86. Washington, D.C.: National Aeronautics and Space Administration, 1986. This edition of the *NASA Facts* Educational Publications Series deals with the foods provided for the space shuttle astronauts. It is well illustrated with pictures of the various food packs, lockers, ovens, and hand-washing facilities aboard the shuttle. Includes a picture of an astronaut eating from a food tray in space. Sample menus and actual foods carried on the first four space shuttles are listed. Accessible to a general readership.

Joels, Kerry M., and Gregory P. Kennedy. *The Space Shuttle Operator's Manual.* Rev. ed. New York: Ballantine Books, 1987. A delightful book that puts the reader in the driver's seat on board the space shuttle. The reader is treated to a step-by-step tour of all the shuttle's operations from launch to landing. Chapter 2 deals with daily life in space. Most of the material is understandable to high school students. A good series of more technical appendices on specific shuttle subsystems is included.

Lattimer, Dick. *Space Station Friendship.* Harrisburg, Pa.: Stackpole Books, 1988. An entertaining fictional account of a visit to a future space station. All the problems of living in space are well covered, from acquiring food, water, and breathable air to coping with confinement and isolation. The challenges of life in space are made clear to the reader by the various characters in the story. Suitable for teenagers and adults.

McElroy, Robert D., Norman V. Martello, and David T. Smernoff. *Controlled Ecological Life Support Systems: CELSS 1985 Workshop.* NASA TM-88215. Springfield, Va.: National Technical Information Service, 1986. A compendium of ideas for life-support methods on the space shuttle, the space station, and future space missions. Discusses systems for recycling water, disposing of waste, reconstituting gases for use as breathable air, and producing food from algae, bacteria, and higher plants. College-level material.

U.S. Air Force Academy. Department of Behavioral Sciences and Leadership. *Psychological, Sociological, and Habitability Issues of Long-Duration Space Missions.* NASA T-1082K. Houston: National Aeronautics and Space Administration, 1985. This book thoroughly assesses human engineering for current and future space missions. Topics covered include work-rest cycles; astronauts' sleep needs and performance levels; the design of astronauts' garments, food, and private accommodations; and the effects of all these factors on space personnel. Advanced college-level material.

Penelope J. Boston

SPACE SHUTTLE MISSION
STS-1

Date: April 12 to April 14, 1981
Type of mission: Manned Earth-orbiting spaceflight

Space Transportation System (STS) 1 was the first launch of the space shuttle. The primary objectives of this mission were to demonstrate a safe launch to orbit, test basic orbiter systems in space, achieve reentry, and land safely.

PRINCIPAL PERSONAGES

JOHN W. YOUNG, Commander
ROBERT L. CRIPPEN, Pilot
JOHN F. YARDLEY, Associate Administrator for Space
    Transportation Systems
DANIEL GERMANY, Director of Orbiter Programs
ROBERT F. THOMPSON, Manager of the shuttle pro-
    gram at Johnson Space Center
DONALD K. "DEKE" SLAYTON, Manager of the Orbital
    Flight Test Program
GEORGE ABBEY, Director of Flight Operations at
    Johnson Space Center
MAXIME A. FAGET, Director of Engineering and
    Development at Johnson Space Center
ROBERT H. GRAY, Manager of the Shuttle Projects
    Office at Kennedy Space Center
GEORGE F. PAGE, Director of Shuttle Operations at
    Kennedy Space Center
ROBERT E. LINDSTROM, Manager of the Shuttle
    Projects Office at Marshall Space Flight Center
MEL BURK, Shuttle Project Manager at Dryden Flight
    Research Center

Summary of the Mission

STS-1 was the first demonstration of a spacecraft that could take off like a rocket, fly into space, reenter the atmosphere, and land like an airplane — then be refurbished and launched again. The mission officially lasted 2 days, 6 hours, 20 minutes, and 52 seconds from launch to landing.

The first space shuttle mission was also the first of four demonstration flights, collectively known as the Orbital Flight Test Program, which were designed to test the shuttle before it would be declared officially operational. The program of more than eleven hundred carefully outlined tests and experiments was designed to verify the shuttle as a launch vehicle, living space, freight handler, instrument platform, and craft. The program was also intended to test ground operations and

personnel before, during, and after each launch. In the interest of safety, the trajectory, thrust, payload weight, and other stresses on the spacecraft were deliberately kept to a minimum. The orbiter's only two payloads were associated with the flight test program.

The development flight instrumentation package consisted of strain sensors and measuring devices to gauge spacecraft performance and stresses encountered during launch, flight, and landing. The aerodynamic coefficient identification package consisted of instruments to determine air velocity, temperature, pressure, and other aerodynamic characteristics during the flight. The third experiment, not carried on board, was the Infrared Imagery of Shuttle (IRIS). Its objective was to obtain a detailed infrared image of the orbiter's lower, windward, and side surfaces during reentry. These images were obtained using NASA's C-141 Kuiper Airborne Observatory aircraft positioned at an altitude of about 13,716 meters along the reentry path.

The first mission followed a series of taxi tests, unmanned and manned captive flight tests atop a modified Boeing 747 aircraft that concluded with free flights. Repeated problems with the *Columbia's* three liquid-fueled main engines and the thermal tiles that protected it from reentry heat had delayed the first launch by almost three years. On November 6, 1972, the NASA Associate Administrator told the House Committee on Science and Astronautics that the first manned orbital flight was slated for March, 1978. Technical problems caused NASA to reschedule the launch several times: to June, September, November, and December, 1979; again to March 30, 1980; and to March, 1981.

NASA selected John Young, the most experienced member of the astronaut corps, as commander. His pilot, Robert Crippen, had had no spaceflight experience but was very knowledgeable about the shuttle's complex computer system and was considered one of the best astronauts.

The first space-worthy orbiter, *Columbia*, was rolled out of its hangar at Rockwell International and towed to Edwards Air Force Base on March 8, 1979. The spacecraft was missing its main engines, its orbital maneuvering system (OMS) pods, and about 7,800 heat tiles. It would be mounted on its Boeing 747 carrier aircraft and flown to Kennedy Space Center, where the remaining work would be finished. Technicians added dummy tiles held on with tape to give a

Launch view of the Columbia *for the STS-1 mission, April 12, 1981.* (NASA)

clean, aerodynamic surface, but these parts fell off during a brief test flight along with one hundred real tiles. Nevertheless, the 747 took off for Florida on March 20, 1979, arriving there on March 24.

At Kennedy, the tiles were subjected to a "pull test." Many failed, and a total of 4,500 had to be removed and strength-

ened. Concerned that some might fall off in flight and have to be replaced, NASA sped work on its Manned Maneuvering Unit (MMU) jet backpack so it could be used for an in-orbit repair. The plan to prepare the MMU for STS-1, however, was dropped.

The main engines were installed by the summer of 1979,

while test engines at the National Space Technology Laboratories in Bay St. Louis, Mississippi, were still giving engineers problems. An aborted firing test on November 4, 1979, revealed a problem that could have affected Columbia's engines, so they were removed from the orbiter in January, 1980, for modifications. After further tests, the engines were reinstalled on the orbiter by August 3. They were removed again on October 9 and 10, then reinstalled early in November. Since the engines had never been fired as a cluster on the shuttle, NASA decided to have a 20-second test firing while the shuttle was sitting on Pad 39A. On November 3, 1980, the two solid-fueled rocket boosters were mated to the external fuel tank. The orbiter was attached on November 26, and the slow 5-kilometer trip to the pad began December 29. The engine test firing took place on February 20, 1981. Minor problems required that the target launch date be rescheduled for April 10.

With no further problems, the countdown began. About two hours before launch, Young and Crippen climbed into their seats on the flight deck. For this flight and the three to come, the Columbia was equipped with two rocket-propelled ejection seats. Between the time the shuttle cleared the tower and the point at which it reached an altitude of 30,480 meters, overhead panels could be blown off and the crew could be hurled to safety. Young and Crippen also wore U.S. Air Force high-altitude escape suits to provide oxygen and full-body protection at high altitudes. Ejection seats and pressure suits would be discontinued after the fourth mission.

During a built-in countdown hold at T minus 20 minutes (20 minutes prior to launch), the shuttle began to experience problems. One of the orbiter's five computers was not synchronized. Only three were needed, but NASA officials wanted all systems operational on the first launch, so the countdown was halted. It was recycled to T minus 23 minutes and held there for 28 minutes before officials restarted the clock. The timing problem happened again at T minus 16 minutes, so officials decided to postpone the launch. Correcting the problem took five hours, and the delay meant that the external fuel tank had to be emptied of its volatile propellants. A new launch was scheduled for two days later, on April 12. This time, the computers functioned normally. At 7:00 A.M. eastern standard time Columbia's three main engines fired, followed seconds later by the twin boosters. They generated 6.5 million pounds per square inch of thrust to lift the 2-million-kilogram shuttle from the launchpad.

The boosters burned out in 132 seconds and were jettisoned, parachuting to the Atlantic Ocean, where they would be towed back to the launch site by recovery ships. The shuttle continued its eastward climb. At an altitude of 50 kilometers, it was moving at a speed of 4,670 kilometers per hour. It passed Mach 4 in four minutes and at seven minutes was traveling at 28,161 kilometers per hour in an elliptical orbit. It dived slightly from 135 to 117 kilometers to pick up speed. It began climbing again at Mach 25 as the engines were throt-

tled back to provide 65 percent of the maximum thrust. The engines were shut down after 8 minutes, 34 seconds. Columbia was soon flying 28,161 kilometers per hour in an elliptical orbit. The tank was jettisoned to burn over the Indian Ocean. Over the next seven hours, the two OMS engines were fired four times to attain a final, nearly circular orbit of 277 kilometers.

A major milestone after the second burn was the opening of the orbiter's payload bay doors. Had the doors not opened, Columbia's final two OMS burns would have been used to position the orbiter to return to Earth after only five hours. On the fifth orbit, the doors were opened. The right door was opened and closed, followed by the left door. Then, both were opened. Young spotted dark patches on the OMS pods where thermal tiles had broken free. The loss of these tiles was not critical because they were not in an area of high reentry heat. Yet mission officials worried about whether any tiles were missing from the orbiter's underside, which the crew could not see. Some reports said that spy satellite cameras were turned toward Columbia to look at the bottom tiles, but neither NASA nor the U.S. Air Force would confirm those reports. At this point, the shuttle was orbiting with its open cargo hold pointing toward Earth.

The last OMS burn occurred at 7 hours, 5 minutes after launch and lasted 40 seconds. The crew conducted a long series of performance tests on the shuttle's smaller thrusters in the nose and tail of the orbiter. After their 13-hour day, the astronauts finally settled down for the night, remaining in their seats in case of an emergency. During the night, they complained of being cold, a problem traced to a malfunctioning heating valve.

The second day included more thruster tests and television transmissions, including a talk with Vice President George Bush. The attitude control system was given a workout, as the crew performed a series of yaws, rolls, and pitches. Young put the Columbia into a gravity-gradient attitude. Nose down and perpendicular to Earth, the ship became almost motionless, saving fuel that otherwise would have been required to maintain stability. The thrust monitoring instruments caused problems, failing to record the correct velocity. Crippen closed and reopened the doors again to determine if they had been warped during the alternately hot and cold temperatures of the 1.5-hour day/night orbits. He reported that an oxygen regulator valve was leaking slightly. Mission Control in Houston advised the astronauts that the problem was not serious as long as the valve was functioning.

During their twenty-first orbit, the astronauts donned their pressure suits and rehearsed for reentry the next day. Before going to sleep the second night, they performed a test of the inertial measurement unit in the nose. The device indicated Columbia's location in space and in what direction the spacecraft was pointing by keeping track of all maneuvers made previously.

Shortly after Mission Control had signed off for the night,

an alarm bell sounded on the flight deck. It had been triggered by a sudden drop in temperature in one of three Auxiliary Power Units (APUs). Two APUs were needed to provide power for the orbiter's aerodynamic control surfaces after reentry. A cold power unit might be difficult to activate. Houston asked the astronauts to recycle the heating system switch to engage the heater. They did that and then returned to sleep, but the problem persisted. They used heaters when they woke, but the temperature continued to drop. The crew knew that they would be in no danger, unless the temperature dropped below 26 degrees Celsius.

After the thirty-fourth orbit, more than 43 hours in orbit, the astronauts donned their pressure suits for the last time and closed the payload bay doors. Young used the thrusters to flip the *Columbia* so that the OMS engines at the rear faced in the correct direction for reentry. On the thirty-sixth orbit, the two 2,270-kilogram thrust engines fired for 160 seconds to slow the orbiter and bring its path inside the atmosphere. Its orbit was now an ellipse that intercepted the ground, and its speed was reduced by 320 kilometers per hour. Young turned the ship's nose forward to prepare for the 7,081-kilometer descent through the atmosphere, the longest glide of the space program.

Columbia entered the atmosphere over the eastern Pacific Ocean and headed north to begin the descent to California, entering at a nose-up angle of 40 degrees. The tiles began to heat, surrounding the cabin with a red glow. Communications were lost for 16 seconds when the temperature rose high enough to ionize the surrounding air and block radio signals. The craft emerged from the blackout at 57.3 kilometers at Mach 10.3. Young commanded the computers to put the orbiter into a series of S-shaped maneuvers to decrease speed. The ship crossed the California coastline at Mach 6.6 and flew over the dry lake bed at Edwards Air Force Base at a 16-kilometer altitude. The spacecraft caused a double sonic boom as it dropped below the speed of sound. The landing gear came down only seconds before contact. The rear wheels touched down first, then the nose gear. It was April 14, 1:21 P.M. eastern standard time. By April 29, *Columbia* was back in the processing facility at Kennedy Space Center to prepare for the second mission, scheduled for September.

Knowledge Gained

The primary objectives of STS-1 itself were simple: a safe ascent into orbit and return to Earth. The secondary objectives were to verify the combined performance of the entire vehicle through separation and retrieval of the boosters and to gather data on the orbiter's performance in space and during reentry and landing. Potential stresses on the spacecraft were kept to a minimum. The only payloads were those associated with verifying the shuttle system.

In addition to demonstrating a safe launch and landing, STS-1 demonstrated two vital systems, the payload bay doors with their attached heat radiators and the Reaction Control

System (RCS) thrusters used to maneuver the craft while in orbit. Opening and closing tests with the doors showed their movement to be jerkier than it had in tests on Earth, but the doors performed satisfactorily. The main engines, which had caused so many problems in development, met all requirements for start and cutoff timing, thrust direction, and flow of propellants.

The shuttle slightly overshot its path to orbit, about 3,000 meters at main engine cutoff. The inability of wind tunnel models to simulate the afterburning of hot rocket exhaust gases in the real atmosphere accounted for this miscalculation. The crew noted some momentary shaking of the ship when the main engines shut down, presumably caused by the dumping of excess fuel. That process was expected to add some velocity to the shuttle but did not. On-board cameras showed that the external tank separated from the orbiter as predicted after engine cutoff. Ascent temperatures proved low enough during development flights that NASA decided to delete the white reflective paint on the tank to save weight.

The launchpad's water deluge system, designed to lessen the powerful acoustic pressure waves seconds before and after lift-off, had to be modified before the next mission. Sensors and microphones on STS-1 showed the acoustic shock to be up to four times the predicted values in parts of the shuttle closest to the pad. The system dumps tens of thousands of liters of water onto the pad and into the flame trenches beneath the rockets to absorb sound energy that could damage the orbiter or its cargo. In the subsequent missions, rather than dumping water into the bottom of the flame trenches, water was redirected into the exhaust plumes of the boosters just below the nozzles. Energy-absorbing water troughs were placed over the exhaust openings. The changes reduced acoustic pressures 20 to 30 percent.

On missions such as STS-1 with moderate heating and cooling requirements, engineers found the total heat stress on the cooling system to be 15 percent lower than expected. Temperatures on the surface of the payload bay insulation ranged from minus 96 degrees Celsius to a peak of 127 degrees Celsius.

Fuel usage and thrusting power of the orbiter's reaction control system performed as expected in space, with the smaller maneuvering jets more fuel efficient than predicted. Two of the four vernier jets in the tail were observed to have problems when firing downward. The exhaust hit the aft body flap, or beaver tail, and eroded some protective tiles. It also reduced the jets' effectiveness. One solution considered later was to reorient the jets slightly. In the meantime, a protective coating was applied to the tiles on the body flap.

The crew found that the orbiter was nearly impossible to stop after maneuvers without overshooting its target and coming back to the required position. A group of microphones and other sensors in the payload bay showed that noise and stress were lower than expected. After STS 1, astronauts switched to wireless headphone radios because the trailing wire proved to be a nuisance for Young and Crippen.

Upon reentry, an unplanned correction was made in the angle of the body flap. Engineers had predicted an angle of 8 to 9 degrees for the flap at high altitudes. The actual angle required to maintain the proper reentry angle was 14 degrees. Onboard computers made the adjustment automatically. The reaction control thrusters showed a greater than expected influence on the orbiter's motion in the atmosphere. It indicated that reliance on the thrusters during reentry could be reduced or eliminated entirely.

The STS-1 approach and landing was fully manual, although computers could land it automatically if necessary. Stress gauges on the landing gear and crew reports both indicated that the shuttle landing was very gentle. The length required for the orbiter to roll to a stop was well within the 4,500-meter design limit, but the actual wheel touchdown point was considerably beyond the planned point because the shuttle demonstrated a greater tendency to glide near the ground than had been expected.

When it returned to be refurbished, Columbia showed that it had endured the rigors of spaceflight well. Workers had to replace some three hundred of the thirty-one thousand tiles that were either missing or damaged. Yet the vehicle showed no damage from reentry temperatures as high as 1,650 degrees Celsius. With the removal of ice-forming hardware on the external tank, deletion of certain insulation on the boosters, and a general cleanup of the pad area, fewer than forty tiles needed to be replaced after the fourth mission.

Context

STS-1, the first orbital flight of NASA's Space Transportation System, was the United States' first manned space mission in nearly six years. As the first flight of an entirely new type of spacecraft, the mission set several spaceflight records.

It was the first shuttle mission, the first NASA space mission to touch down on land, and the first use of solid-fueled rocket boosters for manned spaceflight. It was also the first recovery of a booster for reuse. Unlike previous spacecraft to that time, which had used ablative heatshields, the orbiter used ceramic tiles to protect it from reentry heat. STS-1 was the first maiden flight of a spacecraft carrying a human crew. Previous spacecraft had been tested unmanned first. The shuttle was the first vehicle to provide no emergency escape for the crew during launch as part of its basic design; after the first four STS orbital flight tests, ejection seats for the pilot and copilot were omitted from the design.

For the first time, hydrogen- and oxygen-fueled engines on a launch vehicle would be fired at ground level. Previously, such engines had been confined to upper stages. The main engines were the most advanced liquid-fueled rocket engines ever built, with the highest thrust for its weight of any engine yet developed. STS-1 marked the first time a winged craft encountered the forces of maximum

aerodynamic pressure in a launch through the atmosphere. Columbia's flight also marked the first time a winged spaceship reentered the atmosphere, passing through hypersonic and supersonic ranges to make a runway landing.

Even without its 29,484-kilogram advertised cargo capacity, the shuttle was the heaviest vehicle ever flown. The Apollo command and service modules and the lunar module placed into orbit had weighed 45,359 kilograms. Skylab had weighed 82,553 kilograms. The shuttle Columbia weighed 96,163 kilograms; it was the most massive vehicle flown to date. Its 37.18-meter length exceeded that of the Skylab orbital workshop by almost 11 meters. The shuttle orbiter's payload bay doors were the largest aerospace structures made of composite materials to that time.

Columbia was the first American manned spacecraft to provide a normal breathing atmosphere for its crew. It was an Earth-type mixture of 80 percent nitrogen and 20 percent oxygen at sea-level pressure, 14.7 pounds per square inch. The previous Mercury, Gemini, and Apollo spacecraft had provided pure oxygen at a pressure of about five pounds per square inch.

After numerous setbacks in developing the shuttle and a two-year delay in the first launch, STS-1 proved to be a morale booster for the American space program. It was the country's first manned space mission since the joint U.S.-Soviet mission, which ended July 24, 1975. Thousands of spectators viewed the launch and landing of the shuttle. The first launch renewed enthusiasm for the shuttle program and its goals. To the space agency and other shuttle supporters, the first mission represented the opening of a new era of space exploration. The shuttle was to be the United States' reusable space truck, providing routine access to low Earth orbit, not only for astronauts but also for scientists and other civilian passengers.

STS-1 raised hopes that the cost of space operations would be significantly reduced. Once normal operations began, according to the space agency, the shuttle would serve as a platform for deploying satellites, interplanetary probes, and scientific instruments such as the Hubble Space Telescope. From the shuttle, astronauts would repair ailing satellites. Research in low gravity would lead to new or reduced-cost drugs, metal alloys, and electronics materials. The shuttle was also NASA's long-awaited vehicle for shipping the parts of a permanently manned station into space.

Components of the shuttle had been tested separately, but until April, 1981, there was no proof that the shuttle system could reach orbit, perform a mission, and return for a landing. That was the goal of the orbital flight test program, which was to continue for three more missions. Design, development, test, and evaluation cost for the shuttle was $9.912 billion. Production of four orbiters, spares, ground support, and other costs would total an added $5.6 billion.

Proving the concept of a reusable winged spacecraft was more difficult than NASA had expected. The agency was at least partially vindicated by the first STS flight.

Bibliography

Allaway, Howard. *The Space Shuttle at Work*. NASA SP-432. Washington, D.C.: Government Printing Office, 1979. A look at the envisioned shuttle operations written after the initial glide tests but before first launch. Deals with an anticipated mission, uses of space, shuttle rationale, launch and landing profiles, working in space, and future uses. A seventy-six-page guide suitable for general audiences.

Chant, Christopher. *Space Shuttle*. New York: Exeter Books, 1984. Largely a picture guide. Covers initial concept through STS-14. Includes sections on the orbiter, propulsion, external tank and boosters, and ground and payload operations — as well as shuttle mission summaries. Also contains sections on future missions and principal contractors.

Kerrod, Robin. *Space Shuttle*. New York: Gallery Books, 1984. A full-color pictorial shuttle diary, covering training and ground and space operations. Also includes artist conceptions of future shuttle missions.

Lewis, Richard S. *The Voyages of Columbia: The First True Spaceship*. New York: Columbia University Press, 1984. Describes the shuttle program from concept and development to details of STS missions 1 through 9. Includes the budgetary, political, and technical debates that led to the shuttle's eventual configuration. Includes notes and index sections as well as a section with major mission statistics. Illustrated with numerous drawings and black-and-white photographs.

Reichhardt, Tony. *Proving the Space Transportation System: The Orbital Flight Test Program*. NASA NF-137-83. Washington, D.C.: Government Printing Office, 1983. A 20-page educational report summarizing the shuttle system and the objectives and results of the four-mission test program. Includes black and white photos and diagrams of shuttle components. For general audiences.

Trento, Joseph J. *Prescription for Disaster: From the Glory of Apollo to the Betrayal of the Shuttle*. New York: Crown Publishers, 1987. Written in the year after the 1986 *Challenger* accident. Offers a critical postaccident analysis of the shuttle program. Describes administrative and political decisions that led to the shuttle program and subsequent developmental problems that contributed to the accident. Based on interviews with all the NASA administrators and some other key officials. Includes an index to sources and interviews.

Wilson, Andrew. *Space Shuttle Story*. Twickenham, England: Hamlyn Publishing, 1986. Touches briefly on the early days of rocketry, experimental aircraft that led to the shuttle, and the development of the shuttle concept. Includes an index and a flight summary. Illustrated with color photographs.

Martin Burkey

SPACE SHUTTLE MISSION
STS-2

Date: November 12 to November 14, 1981
Type of mission: Manned Earth-orbiting spaceflight

The second flight of the shuttle Columbia *carried the first scientific experiments and tested the remote manipulator system for the first time. Although the flight was cut short because one fuel cell failed, more than 90 percent of the mission's goals were reached.*

PRINCIPAL PERSONAGES
JOSEPH HENRY ENGLE, Commander
RICHARD HARRISON TRULY, Pilot

Summary of the Mission

After the successful mission STS-1 in April, 1981, the National Aeronautics and Space Administration (NASA) was eager to prepare *Columbia* for another flight. NASA set a launch date in late September, 1981, for the shuttle's second trip to space. Yet there were several obstacles. First, many tiles from the heatshield had to be replaced because they had been damaged during the first flight. Then, as the rescheduled launch date of October 9 approached, nitrogen tetroxide was spilled on the tiles; some 370 of them needed to be replaced. Again, a new launch date was set, this time for November 4.

On November 4, the countdown proceeded smoothly until T minus 31 seconds, when it was stopped by a computer that detected a low pressure reading in a fuel cell oxygen tank. An attempt to override the computer manually failed. During this time, engineers who monitor the performance of the Auxiliary Power Units (APUs) noticed that the oil pressure was higher than expected. That caused concern over the ability of the APUs to restart at the end of the mission (the APUs supplied the rudder and elevons of the shuttle with power during reentry). In addition to the problems with the oxygen and oil pressures, the weather was worsening. Once more, NASA officials decided to postpone the launch. It was rescheduled for November 12.

The problems of November 4 were corrected. The difficulty with the oxygen pressure had been in the computer software, not in the shuttle itself. The oil pressure problem was corrected by replacing the oil in the gearboxes of the APUs and by replacing the oil filters.

The countdown toward the November 12 launch proceeded smoothly until November 11, when a multiplexer-demultiplexer, an instrument which mixes data for easier transmission, failed. A replacement had to be borrowed from

the shuttle *Challenger* and flown to Florida from California. The unit was installed by midnight, and the countdown proceeded again. *Columbia* was lifted into space at 10:10 A.M. on November 12 in a perfect launch.

As the second of four test flights scheduled for *Columbia*, STS-2 had the primary goal of collecting data on the performance of the shuttle. This flight was the first to carry a set of scientific experiments called OSTA 1, which had been prepared by NASA's Office of Space and Terrestrial Applications. OSTA 1 was mounted on a British-made Spacelab pallet. Spacelab was designed to function in much the same way as Skylab had but with less working space and with the ability to return to Earth.

During *Columbia's* second orbit, one of the fuel cells on board failed. The fuel cells serve a dual purpose. First, they generate the power needed to operate the electrical systems on the shuttle. Second, they provide a supply of clean drinking water for the astronauts. *Columbia* was equipped with three fuel cells. The orbiting shuttle could function well using two, and it could reenter the atmosphere with the power from only one — but the risks involved would be high.

There were some important data that the scientists at NASA needed to collect during reentry. Failure of the development flight instrumentation recorders on STS-1 had prevented the acquisition of important thermal and spacecraft maneuvering data. Failure of a second fuel cell on STS-2, or of its deterioration in power output, would cause these data to be lost for a second time. After careful evaluation of the situation, NASA officials decided to proceed with a "minimum mission" of fifty-four hours rather than the 124 hours originally scheduled for the flight. The crew was told of the decision on November 13.

Failure of the fuel cell was among the most surprising failures that could have occurred. Fuel cells have functioned well on many spaceflights, and a failure of the type that occurred on STS-2 had been seen only on test cells that had undergone hundreds of hours of operation.

Since spaceflight is such a complicated matter, NASA has attempted through the years to plan for the unexpected. The STS-2 flightplan had been written so that primary test objectives and experiments were concentrated in the first two and one-half days of the scheduled 124-hour mission. NASA had also developed a flightplan that would meet the majority of the goals in the event that the shuttle had to be brought home early. It was this plan that was used for the minimum mission of STS-2. The astronauts worked efficiently and accomplished 90 percent of the 258 priority flight objectives.

One of the major goals of the mission was the testing of the Remote Manipulator System (RMS). The robot arm on the RMS is also referred to by some as the Canada arm, since it was constructed in Toronto. The tests that astronaut Richard Truly performed on the arm went very well. The motions expected were observed. When Truly was using the backup system to put the RMS into its cradle for storage and landing, the shoulder of the arm would not respond, a malfunction later found to be the result of a broken wire. The arm was successfully operated in both the fully automatic and the manual modes. It was to be used on future flights to handle satellites and perform other grasping tasks in the cargo bay of the shuttle (there were no plans to use the arm to grasp a load on STS-2). A camera is mounted on the arm so that astronauts have a clear view of the area in which it is working. On this flight, the camera gave views of the cargo bay and of the astronauts displaying a sign in the window of the shuttle. The Reaction Control System (RCS) of the shuttle was also tested. The RCS changes the orientation of the shuttle relative to Earth. Extensive testing of the system was performed.

The scientific package OSTA 1 had several goals. One rather fundamental goal was to determine whether *Columbia* could provide a stable platform for conducting Earth surveys. Other goals included testing advanced technologies and instruments in space and gathering data about Earth's resources and environment.

The Shuttle Imaging Radar (SIR) used all of its 1,000 meters of photographic film taking pictures of the continents that passed under *Columbia*'s path. During one pass over the Sahara Desert, SIR penetrated the Selina Sand Sheet to reveal unsuspected river channels, significant geological structures, and possible Stone Age settlements. The shuttle multispectral infrared radiometer was designed to help identify rock types such as those found by the SIR.

The ocean color experiment was designed to improve Earth sensing over deep ocean areas. Areas of the East and West coasts of the United States, the Sea of Japan, and an area extending from southern Europe down the West African coast were all scanned by the instrument. Extensive cloudiness over much of the target areas resulted in data which were less useful than had been hoped.

The Measurement of Pollution from Satellites (MAPS) experiment charted concentrations of gases in the atmosphere. The instrument recording the data was on its first spaceflight, so all the data obtained were of a new class. Data on one pass revealed that the concentration of carbon monoxide in the middle troposphere (the layer of the atmosphere nearest Earth's surface) varied from 70 parts per billion over the Americas to 140 parts per billion over the Mediterranean.

The goal of the feature identification and location experiment was to classify Earth views into those of water, vegetation, bare ground, and snow or cloud cover automatically. Such a sensor would allow Earth resources satellites to operate with much less dependence on ground control targeting. Analysis showed that the system's data acquisition logic was effective.

The night/day optical survey of lightning collected some meteorological data, especially over Africa. Since this experiment required active participation from the crew, limited data were collected.

The Heflex bioengineering test was designed as a plant growth test. Its goal was to determine the amount of water in the soil (or other growth medium) that afforded the best plant growth. In a low-gravity environment, water gathers around a seed and can even drown it. Because of the abbreviated mission, however, only limited data could be extracted. The entire package would be relaunched aboard STS-3.

Overall, the scientific experiments worked well. The Spacelab pallet provided a stable structure on which both experiments and electrical power could be mounted. The pallet had been designed for experiments requiring a broad field of view or direct exposure to the space environment.

The shuttle's return to Earth went smoothly. *Columbia* entered the atmosphere about 40 kilometers south of the desired ground track. The position was easily corrected, although the incorrect entry position caused an experiment designed to capture an infrared (heat) image of STS-2 to fail. The data would have been used to improve thermal protection systems on future space vehicles.

A series of roll maneuvers were executed during *Columbia*'s reentry to gather data on the handling capabilities of the shuttle. Several of the maneuvers were performed manually while most of the descent was handled by the computers on board. Joseph Engle took manual control of the shuttle at an altitude of about 11 kilometers, a little higher than originally planned. (The computer had not taken into account the high winds at Edwards Air Force Base.) To compensate for the wind, Engle took the craft through a tight turn and then gradually brought *Columbia* back into alignment with the computer coordinates as the spacecraft faced into the wind at an altitude of 3.6 kilometers. The autoguidance system brought the shuttle down to 530 meters, where Engle again took control and landed *Columbia* manually. A planned crosswind landing was canceled because wind speeds were too high.

The main wheels of *Columbia* touched the runway at 4:23 P.M., November 14. Forty-two minutes later, the smiling astronauts came out to greet the ground crew and inspect their spacecraft from the outside. Their pride in a job well done was evident as they addressed a crowd of reporters and space program employees several hours later. Both men were extremely pleased with *Columbia* and the way it had flown.

Truly said *Columbia* was "the real hero of the mission, really a fine machine." Engle added, "You can spread the word around: We got us a good one."

As evening came, the shuttle was turned over to the ground teams from Kennedy Space Center as they began the turnaround for the next mission, which was scheduled for March 12, 1982.

Knowledge Gained

At lift-off for STS-1, shock waves from the blast of the rocket motors had bent several of *Columbia's* fuel tank supports. A newly developed water-deluge system installed at the base of the launchpad sprayed 1.25 million liters of water into the rocket exhaust for the STS-2 launch and damped the shock wave by more than 75 percent.

Columbia handled very well upon entry into the atmosphere. A number of pitch and roll maneuvers were needed to add to the data on the handling characteristics of the shuttle. Instrument failure on STS-1 had caused all such data on that flight to be lost. Although weather conditions were less than optimal for landing, with strong winds and low visibility, the pilot experienced no handling problems. The crew's ability to stop the shuttle after 2.1 kilometers of ground roll was encouraging to engineers at NASA as they looked foward to the use of the 4.5-kilometer runway at the Kennedy Space Center.

The robot arm of the RMS performed the movements that were expected of it. Yet the failure of one movement using the backup system for putting the arm back into its cradle was a concern to Truly. This failure was found to be the result of a broken wire in the system.

The heatshield tiles sustained much less damage than they had sustained on STS-1. This improved performance was largely the result of the water which had been sprayed into the rocket blast at lift-off. In addition, some of the tiles in higher risk areas had been strengthened.

The scientific experiments gathered much useful data. The shuttle imaging radar detected several underground features in North Africa. MAPS provided new data on gaseous pollutants in the atmosphere. The Spacelab pallet proved an excellent support for the instruments and their electrical supply. *Columbia's* maneuverability made collection of some data from the feature identification experiment possible.

Context

STS-2 came at a time when support for space exploration and experimentation was not strong. NASA's budget was tight, and other space programs were in danger of being terminated. STS-1 had been a tremendous success, but that success had not carried into the political arena, where funding originates. The shuttle was being promoted as a reusable spacecraft that could fly on a routine basis. Not everyone was convinced that it could, or should, do that. It was important that STS-2 be a successful mission.

The scientific payload performed well, and many useful data were returned. The shuttle vehicle itself also performed well, and the crew completed 90 percent of the primary goals of the mission while flying only half the time scheduled, a sure indication that STS-2 was indeed a success. The only significant failure occurred in a component (the fuel cell) that had an excellent record in space as well as in the test laboratory. Clearly, the second test flight of the space shuttle *Columbia* contributed significantly to the success of subsequent flights of the shuttles.

Bibliography

Allen, Joseph P., and Russell Martin. *Entering Space: An Astronaut's Odyssey*. New York: Stewart, Tabori and Chang, 1984. A description of a typical flight of the space shuttle, although other aspects of the exploration of space are included. Illustrated with excellent photographs.

Bond, Peter. *Heroes in Space: From Gagarin to Challenger*. Oxford: Basil Blackwell, 1987. A history of spaceflight that focuses on the astronauts and cosmonauts. Its last two chapters give details of the shuttle flights. Well written.

Cassutt, Michael. *Who's Who in Space: The First Twenty-five Years*. Boston: G. K. Hall and Co., 1987. A comprehensive biographical work covering manned space-flight. Includes essays on programs, biographical profiles organized by nationality, a summary of all manned flights from April, 1961, to April, 1986, and a chronological log of flights. Suitable for the lay reader.

Collins, Michael. *Liftoff: The Story of America's Adventure in Space*. New York: Grove Press, 1988. A survey of the U.S. space program that includes a chapter on the space shuttle. Speculation on future flights is included. Contains many line drawings but few photographs.

Cross, Wilbur, and Susanna Cross. *Space Shuttle*. Chicago: Children's Press, 1985. A summary of the shuttle and its intended use, Mission Control, the astronauts, and the future of the shuttle program. Contains many color photographs. Written for children but suitable for adults.

Furniss, Tim. *Manned Spaceflight Log*. Rev. ed. London: Jane's Publishing Co., 1986. Contains a summary of each space-flight from Vostok 1 in April, 1961, to Soyuz T-15 in March, 1986. Data include flight name, crew, launch site, and significant accomplishments of the mission.

National Aeronautics and Space Administration. *NASA: The First Twenty-five Years, 1958 – 1983*. NASA EP-182. Washington, D.C.: Government Printing Office, 1983. Highly condensed information about the space missions of the United States, including the first ten STS flights. Written for a general audience.

Time-Life Book Editors. *Life in Space*. Boston: Little, Brown and Co., 1984. Contains a pictorial as well as a written history of the first six shuttle flights. Also contains histories of the other major space programs of the United States. Includes color photographs.

Vogt, Gregory. *The Space Shuttle: Projects for Young Scientists*. New York: Franklin Watts, 1983. The goal of the author is to encourage young scientists (primarily high school students) to consider becoming involved in space research by designing a project to fly in space. Examples of successful projects are given, and steps to follow in developing the project are discussed.

Yenne, Bill. *The Encyclopedia of U.S. Spacecraft*. New York: Exeter Books, 1985. Reprint, 1988. Contains a section covering the first seventeen shuttle flights as well as a general description of the vehicle and its function.

Dennis R. Flentge

SPACE SHUTTLE FLIGHTS, 1982

Date: March 22 to November 16, 1982
Type of mission: Manned Earth-orbiting spaceflight

The National Aeronautics and Space Administration (NASA) followed the success of the first two Space Shuttle flights in 1981 with two additional test flights to determine the flightworthiness of the vehicle. It then declared the Space Transportation System operational and flew the first commercial payloads.

PRINCIPAL PERSONAGES

JACK R. LOUSMA, STS-3 Commander
C. GORDON FULLERTON, STS-3 Pilot
THOMAS K. "KEN" MATTINGLY, STS-4 Commander
HENRY W. HARTSFIELD, STS-4 Pilot
VANCE D. BRAND, STS-5 Commander
ROBERT F. OVERMYER, STS-5 Pilot
JOSEPH P. ALLEN, STS-5 Mission Specialist-1
WILLIAM B. LENOIR, STS-5 Mission Specialist-2

Summary of the Missions

At 11:00 A.M. eastern standard time, March 22, 1982, *Columbia* began its third trip to space only one hour after its scheduled launch time. The launch followed the smoothest countdown in the brief history of the space shuttle program. About four minutes into the flight, Charles Gordon Fullerton noticed that the temperature of the oil in the third Auxiliary Power Unit (APU) was rising. After about seven minutes the temperature alarm from the APU was acknowledged, and thirty seconds later the APU was deactivated. (The auxiliary power units swivel the rocket engines during launch and operate the rudder and elevons during the return through the atmosphere, when the shuttle flies like an aircraft.) The shuttle can operate safely with only two of the APUs functioning properly. All three APUs worked well when *Columbia* returned to Earth.

As the third of four test flights of the shuttle, STS-3 was to collect much information on the shuttle itself. The development flight instrumentation, designed to inform flight engineers on exactly how *Columbia* operates under various flight conditions, was carried on this flight just as it had been on STS-1 and STS-2. The data from this equipment were collected as *Columbia* entered the atmosphere and glided to a landing in White Sands, New Mexico.

Columbia underwent passive thermal tests to determine its response to the varying temperatures found in space. The tail

was aimed at the Sun for thirty hours, the nose for eighty hours, and the cargo bay for twenty-six hours. The entire shuttle was rotated like a rotisserie (a maneuver known as the barbecue mode) for ten hours. *Columbia* demonstrated that it could tolerate the extremes of space well. The only difficulty came after the cargo bay doors had been on the dark, cold side of the shuttle. The doors could not be closed because of a frozen latch. When the shuttle was rolled over and the doors were heated by the Sun, the latch worked well and the doors closed securely.

The Remote Manipulator System (RMS) was first scheduled for several hours of testing without a load. During this session, six data acquisition cameras, the arm's wrist camera, and the aft starboard payload bay camera failed. Without the wrist camera and the payload bay camera, the larger of the two payload packages to be handled could not be grasped. Although manipulating the 360-kilogram contamination monitor would have been a better test of the arm's capabilities, the 160-kilogram Plasma Diagnostics Package (PDP) provided a good load for the arm to handle.

A Get-Away Special payload canister made its first trip into space on STS-3. Experiments sent to space in these canisters must be entirely self-contained, except for a switch that activates them. The first experiment measured temperature, vibration, noise, and pressure continuously through launch and at ten-minute intervals during the entire flight.

The OSS-1 scientific payload prepared by NASA's Office of Space Science and Applications (for which it is named) contained nine experiments. Some of the experiments provided data for immediate use while some were so-called pathfinder experiments for more detailed studies to be conducted on future flights. Four of the experiments dealt with the environment surrounding the orbiter. The induced environment contamination package was a set of ten instruments that sampled whatever pollution *Columbia* generated. The shuttle induced atmosphere package studied the effects of gases and vapors on the optical quality of the space near the orbiter.

The plasma diagnostics package measured the characteristics of the rarefied gases in space. It measured the electric and magnetic fields within 13.7 meters of the orbiter. The vehicle charging and potential experiment was designed to indicate whether the shuttle builds static charges as it sweeps through space.

Two experiments studied the Sun. The X-ray polarimeter studied X-rays emitted during solar flare activity on the Sun. These data aid astronomers in determining how flares develop. The second experiment studied the ultraviolet radiation emitted from the Sun. A correlation of ultraviolet radiation intensity and long-term climate changes on Earth is of great interest to scientists.

The last of the OSS-1 experiments studied the effects of low gravity on the production of lignin in plants. Lignin is a polymer that provides strength for plants to grow upward against the pull of gravity. While it is essential for plant growth, it can interfere with the extraction of wood fibers that make paper and chemical cellulose. Analysis of the plants for lignin was completed soon after STS-3 returned to Earth.

Electrophoresis is a process that uses an electrical charge to separate cells. On Earth, gravity causes the cells to settle out of the mixture, and separation is not easily accomplished. The cells separated quickly in the low-gravity environment of the shuttle, but the samples were ruined when a storage freezer failed.

The Heflex Bioengineering Test (HBT) was designed to determine the amount of soil moisture that promotes the best plant growth. These data would be used in later Heflex experiments carried on Spacelab 1. Dwarf sunflower seeds were planted in pots with varying amounts of moisture in the growth medium. Analysis of the experiment was based on the amount of growth measured in the plants compared with the amount of growth measured in plants grown on Earth using the same moisture conditions. In low gravity, moisture gathers around the seeds and can drown them.

Along with the varied experiments developed by professional scientists for STS-3 was an experiment developed by Todd Nelson, an eighteen-year-old senior from Southland Public School in Adams, Minnesota. Nelson was one of ten finalists in the first national Shuttle Student Involvement Project (SSIP). SSIP was cosponsored by NASA and the National Science Teachers Association (NSTA) and was intended to stimulate the study of science and technology in the nation's secondary schools by providing access to research on the space shuttle. Nelson's experiment, the insects in flight motion study, was to examine the effects of a low-gravity environment on the flying activities of the velvet bean caterpillar moth, the housefly larva, and adult honeybee workers. After several days in space, most of the insects walked on the chamber walls instead of flying (or floating).

After seven days in space, the crew was ready to come home, but water-logged runways on the "dry" lakebed at Edwards Air Force Base forced NASA to delay the landing one day and relocate the site to the White Sands Missile Range in New Mexico. On March 30, Lousma and Fullerton brought *Columbia* safely home. During the descent, the shuttle was taken through a set of maneuvers to provide the engineers with more data to use in analyzing the flight capabilities of the shuttle.

Touchdown occurred at 11:04 A.M., 8 days and 5 minutes after lift-off and after a journey of almost 5,400,000 kilometers. The mission had accomplished its goals, and, apart from a few minor problems, the shuttle had performed beautifully. NASA had taken another step toward bringing the space shuttle into routine operation.

The space shuttle Orbital Flight Test (OFT) program initially comprised about six flights, but shortly before the program started two of those flights were eliminated from the plan. The primary objective of the fourth OFT (which, like the previous three, used the spacecraft *Columbia*) was to put the shuttle through the harshest conditions it could endure in routine operations. The flight would test the Remote Manipulator System (RMS), fly with portions of the spacecraft aimed at space to test the "cold-case" attitude, measure contaminants around the spacecraft with a monitor, measure spacecraft reactions through a package of Development Flight Instrumentation (DFI) designed to measure vehicle responses to flight, and, by way of an Induced Environmental Contamination Monitor (IECM), determine whether the shuttle polluted its own environment.

The payload for STS-4 was not selected until after the OFT program started and was ultimately given to the United States Air Force, a major user of the shuttle. Officially, the payload was known only as DOD 82-1 (meaning the first Department of Defense payload in fiscal year 1982) and the Air Force sought to keep all details classified. In addition, the shuttle carried the first Get-Away Special payload and the first major materials science experiments for the shuttle program. Operation of these payloads was considered secondary, although DOD 82-1 was designed to quantify the shuttle environment for that agency's prospective users and to provide its managers with actual mission experience.

Launch originally was scheduled for early 1982 but was postponed several times because of difficulties in the preceding shuttle missions. It was finally scheduled for June 27, 1982, with the landing to take place on July 4. Launch was almost postponed one more time when a freak storm pelted Kennedy Space Center with rain and hail, but a quick survey of the shuttle revealed no major damage. The heatshield tiles, however, were found to have four hundred small dents and were calculated to have absorbed some 200 kilograms of water. The heatshield tiles are coated with a commercial waterproofing agent to prevent water absorption, but the pits created new openings. Although the water would not impair the mission, engineers were concerned that the water would boil out during reentry and crack tile surfaces open, requiring extensive replacement after the flight. It was decided to proceed with the mission and to dissipate the water by pointing the belly of *Columbia* at the sun for an extended period of time. The small pits in the tiles were repaired by applying a slurry of ground tile by hand. This was completed an hour before propellant loading started.

Columbia was launched at 11:00 A.M. eastern daylight

time, June 27, 1982. The ascent trajectory was some 2,500 meters too low one minute into the flight because of a slightly lower-than-normal impulse from the twin solid-fueled rocket boosters. This depression delayed milestones in the remainder of the ascent by as much as 15 seconds, and the main engines had to burn 2 seconds longer than planned to compensate. (This would have cost 900 kilograms of payload in a fully loaded flight.) In addition, the two solid rocket boosters were lost during recovery. Their parachutes deployed late and were still partially reefed when they struck the water at 549 kilometers per hour, causing their casings to spring and allowing them to leak and sink.

Columbia was placed into a circular 241-kilometer-high orbit. The craft initially went into the "gravity-gradient" attitude, with its tail pointed to Earth, in order to allow the DOD 82-1 payload to scan the horizon, but *Columbia* developed a tendency to drift out of position. This behavior was later attributed to the slight motion of the shuttle as water evaporated from its tiles.

During the flight, the Mission Commander, Thomas K. Mattingly, and the Pilot, Henry W. Hartsfield, continued experiments started on the STS-1, 2, and 3 missions. Each flight carried the shuttle through a series of more rigorous environmental tests. Where STS-3 had conducted "hot-case" tests, STS-4 conducted "cold-case" experiments, exposing selected portions of the shuttle to deep space for extended periods to determine the effects of such exposure. These were accomplished by flying in various positions such that the nose and payload bay would be shaded from the Sun. (The direction a spacecraft points is immaterial, except when a certain orientation is necessary to expose solar cells to the Sun and direct telescopes at targets, among other things.) The bottom of the spacecraft was exposed to the Sun for 33 hours in order to evaporate water that had been absorbed in the preflight hailstorm.

One result of extended cold-case testing was that the payload bay doors became misaligned and would not close. The orbiter apparently had warped slightly from uneven heating, an expected effect. Rotating the shuttle for several minutes, like a barbecue, evened the heat distribution and allowed the doors to close. Another test measured how long hydraulic fluid could be left static in its lines without periodic warming for recirculation; that would affect how much electrical power had to be reserved for spacecraft housekeeping.

Payload operations probably were similar to those of STS-3 in many respects, since the crew was operating a similar suite of plasma physics instruments. Payload control, in this case, was the U.S. Air Force Satellite Control Facility in Sunnyvale, California. The Big Blue Cube, as it is known, served as mission control for several classified Air Force satellite projects. The 82-1 payload was comparable in size and capability to the single-pallet Spacelab payloads carried on STS-2 and STS-3, although a different Experiment Support Structure (ESS) was used. This one was made up mainly of large tubing that held

a boxlike structure with standard mounting points for instruments. These instruments included Cryogenic Infrared Radiance Instrumentation for the Shuttle (CIRRIS), the Horizon Ultraviolet Program (HUP), the Space Experiments with Plasmas in Space (SEPS), the Sheath and Wake Charging (SWC) experiments, a solar coronagraph, and an autonomous navigational aid called a space sextant. The payload's mass was 11,021 kilograms.

CIRRIS 1 was a liquid-helium-cooled, 15.2-centimeter telescope sensitive to the 2.5-to-25-micron spectral range of the infrared spectrum. Its objective was to provide fine-resolution spectral data on the constituents of the upper atmosphere by scanning Earth's limb (horizon) from a 30-to-300-kilometer altitude. HUP had a similar objective but was designed as a smaller instrument since it did not have to be cooled to cover the ultraviolet spectrum from 110 nanometers (short wavelength) to 400 nanometers (the edge of visible, violet light). Together, these two instruments were part of a larger military program conceived to understand the space environment as it affects sensors used in a variety of spacecraft and what could be expected when observing targets near the horizon and in deep space.

The plasma instruments in the SEPS/SWC configuration, and the solar coronagraph, were designed to study the space environment. SEPS/SWC included ion and neutral mass spectrometers to analyze the environment directly and plasma probes of various types to measure electrical charging effects that might hamper experiments. The coronagraph was used as a means of photographing particles near the shuttle by "back-lighting" with sunlight. This would provide a measure of particulate contamination around the craft.

The space sextant received its field test on *Columbia*. With two telescopes operating like a mariner's handheld sextant, it would fix a spacecraft's position to within 250 meters and its attitude to within 0.5 degree, without the need for contact with the ground.

Orientation toward Earth was also used for the Night/day Observations from Space of Lightning (NOSL) experiment, which used a modified Bolex motion picture camera with a special light sensor to observe lightning from above the cloud tops. The crew also took a number of photographs from the aft payload bay windows using a 35-millimeter film camera equipped with an image intensifier in an attempt to collect data that would aid in an understanding of the shuttle glow phenomenon observed on STS-3.

Also, during the flight the crew performed the first operations on two promising materials processing facilities, the Continuous Flow Electrophoresis System (CFES) and the Monodisperse Latex Reactor (MLR), both located in the middeck for crew operation.

CFES was designed to separate biological materials by setting up an electrical field at right angles to the flow of a buffer solution containing cells or proteins and enzymes. As the buffer flows through the chamber, the field pulls with different strengths on the components, and they separate according to mass and charge. The technique is limited on Earth because the electric field also heats the fluid and causes convection currents, which remix the separated components.

The MLR carried chemical reactors used to heat and mix styrene solutions that would form microspheres of one size, depending on ingredients and conditions. The microspheres are used for calibrating electron microscopes, filters, and pores in cells, among other things. Their size is limited, as larger microspheres become buoyant and will form a sort of cream on the surface.

The Get-Away Special experiment was the first of its kind to be carried by a shuttle. It had been purchased for use by students at Utah State University, the University of Utah, and other colleges in the state. GAS 1, as this one was known, included experiments on seed germination, brine shrimp growth, composite curing, oil-water mixing, soldering, fruit fly growth, and metals alloying. Unfortunately, GAS 1 could not be activated after repeated attempts. Although it would be flown again, its failure on STS-4 frustrated the students who had built it. After analyzing possible problems, NASA finally gave the crew permission to hot-wire a section of shielded cabling going into the payload bay, and an "on" signal was received on the flight deck. To protect the biology experiments, power was left on until after landing rather than being shut down in flight.

The Remote Manipulator System (RMS) was used again on STS-4. This time it picked up the 394-kilogram induced-environment contamination monitor and held it at 92 programmed "pause and flyby points," where it collected samples and analyzed the environment for postflight study. Stereo cameras photographed dust floating in the region. The handling of the monitor also provided data on how the RMS behaves when loaded.

Reentry was scheduled for July 4, with the shuttle landing at Edwards Air Force Base, where President Ronald Reagan and his wife, Nancy, would be watching. The angle of attack at entry was 40 degrees, a few degrees higher than most missions would use, in order to generate new data. *Columbia* also flew one of the greatest cross-range entries, 930 kilometers. Touchdown occurred at 12:09 P.M., Pacific daylight time, on conrete Runway 22, the first hard-surface landing for the shuttle.

After four very successful test flights of the space shuttle, NASA had high hopes for the first commercial flight of *Columbia*. The STS-5 mission carried the first two satellites to be launched into orbit from the space shuttle (rather than from Earth using an expendable booster). Also aboard *Columbia* were three experiments designed by high school students, a Get-Away Special sponsored by West Germany, and more engineering tests on the shuttle's performance.

This flight also marked the first time that four astronauts were launched into space simultaneously. It was also the first flight for a new type of astronaut— the mission specialist. Mission specialists are trained in satellite deployment, payload support, Extravehicular Activity (EVA, or walking in space), and operation of the remote manipulator system.

The countdown for the November 11 launch proceeded exceptionally well despite a few setbacks, none of which affected the launch time, and STS-5 began at 7:19 A.M., November 11, 1982. Ascent to orbit went smoothly. The solid-fueled rocket boosters did not perform up to expectations, and the main engines had to carry the extra load necessary for boosting *Columbia* into a 296-kilometer circular orbit.

One of the most important tasks was the deployment of the two satellites. The SBS-3 communications spacecraft owned by Satellite Business Systems (SBS) was scheduled to be deployed on the sixth orbit, approximately eight hours after lift-off. At 3:17 P.M. explosive bolts that were holding down a powerful spring were fired, and SBS-3 was launched from the shuttle at a speed of about one meter per second. The shuttle then turned its well-protected underside toward the satellite and changed its orbit slightly to protect its windows from exhaust debris from the satellite's rocket motor. The satellite was now under the control of SBS, not the crew of the *Columbia*.

Forty-five minutes later, the Payload Assist Module's solid rocket motor was fired, and SBS-3 was carried into a highly elliptical orbit. On November 13, the satellite's apogee kick motor was fired, putting SBS-3 into a circular geosynchronous orbit. In such an orbit, the satellite appears stationary relative to Earth's surface.

The second satellite, Canada's Anik/Telesat, was successfully deployed using the same technique on November 12. Astronaut Allen reported to Earth, "Two for two. We deliver." The crew posed for a picture during a public telecast and displayed a sign: "Ace Moving Co. We deliver. Fast and courteous service."

There were three Shuttle Student Involvement Project (SSIP) experiments conducted on STS-5. D. Scott Thomas of Johnstown, Pennsylvania, designed an experiment to study surface tension convection. Surface tension is often described in terms of the "skin" on the surface of a liquid. Convection is a circulation pattern in a heated fluid. In the low-gravity environment of space, these surface effects can be more easily studied than on Earth.

Aaron K. Gilette of Winterhaven, Florida, proposed a study of the reaggregation of cells of particular sponges. In seawater on Earth, these sponge cells reassemble themselves into perfect sponges. Gilette wanted to determine if the absence of gravity would change the way in which the sponge cells performed.

Michele A. Issel of Wallingford, Connecticut, designed an experiment to determine if geometrically perfect crystals of triglycine sulfate can be grown in micro-gravity. Gravity is thought to cause defects in crystals grown on Earth.

The Ministry of Research and Technology of West Germany purchased a Get-Away Special that studied the behavior of a molten mixture of mercury and gallium. On

Earth, the mixture separates into liquid mercury and liquid gallium. The experiment used X rays to study the effects of low gravity on dispersion of mercury droplets into gallium and particle movement resulting from convection, as well as other properties of the mixture.

A study of spacesickness was conducted during STS-5. Since half the crew of the first four flights suffered some degree of spacesickness, its effect on the productivity of a mission could be significant. Astronaut Allen wore electrodes that recorded his eye motion during launch, both Allen and Lenoir attached the electrodes during the time of adjustment to low gravity, and Lenoir wore them during reentry. Eye movement data provide some indication of the nervous system's adaptive reactions to a lack of gravity. Astronauts Lenoir and Robert Overmyer suffered from spacesickness, and the Commander, Vance Brand, felt some discomfort. Allen was unaffected. The motion sickness caused some rearrangement of schedules for the astronauts who were performing at less than maximum capacity.

Allen and Lenoir were to participate in the first EVA of the space shuttle program. The activity was postponed from Sunday to Monday because of Lenoir's spacesickness early in the flight. Problems developed soon after the astronauts put on their spacesuits. A fan in Allen's life-support pack started to cause undue amounts of noise, and Lenoir's pack held a pressure that was more than 10 percent too low. Both suits could have been used if an EVA had been necessary for successful completion of the flight. Yet, since a spacewalk was not essential for completion of STS-5, mission managers decided to cancel the EVA.

Photographs taken on STS-3 had shown that the orbiter glowed in the dark. STS-5 studied the glow that covered the tail section, engine pods, and other shuttle features during the flight. Scientists were concerned about the effect this glow could have on faint effects measured by optical instruments mounted in the payload bay.

During reentry, a set of instruments collected more data on the handling characteristics of the shuttle and on the autoguidance system. The reentry and landing on the hard-surface runway at Edwards Air Force Base went very well. In fact, the landing, which took place on November 16, was so smooth that Brand asked, "Are we on the ground yet?"

Knowledge Gained

Thermal testing of the shuttle showed that the engineering design and materials withstood the temperature extremes of space very well. The failure of the cargo bays to latch properly after a cold soak during STS-3 was easily corrected by exposure of the doors to the Sun. Engineers did, however, check the system when it returned so that improvements could be made that would eliminate the problem. The shuttle's Orbital Maneuvering System (OMS) engines worked well after extended cold soaks.

The OSS-1 scientific package brought back information

that sent the scientists to their laboratories to prepare for future experiments in space. Although the samples from the electrophoresis experiment were ruined, the performance of the instruments in space had been successful. Eventually, the electrophoresis work would develop sufficiently to have a mission specialist dedicated to the electrophoretic work as part of the crew on the twelfth space shuttle mission. Manufacture of monodisperse polystyrene latex microspheres on STS-3 also demonstrated the viability of materials processing in space.

An unexpected discovery came when the astronauts photographed the shuttle glowing in the dark. The glow could present problems for future experiments that require detection of very low intensity radiation. The RMS worked especially well. It handled the plasma diagnostics package effectively and performed much as it had in its tests on Earth.

Much of what was learned from the STS-4 mission was in the form of engineering data, since this was the last of four missions intended to clear the shuttle for routine operations. The mission established a modest set of boundaries within which the vehicle could safely operate. All systems were validated and shown to operate as designed despite a number of minor problems. The remote manipulator arm functioned with precision when commanded to carry a payload (the contamination monitor) to a number of programmed points.

The contamination monitor recorded no exhaust from the solid-fueled rocket boosters (which spew aluminum oxide and hydrochloric acid) but did record water vapor as the main contaminant in orbit. The vapor largely came from the heatshield as it was baked out and from thruster firings. One set of measurements recorded pollutant patterns around the payload bay, but a second set, showing thruster exhaust pressures, was lost as a result of electrical problems.

For the most part, the materials science experiments proceeded as planned. The CFES separated human liver and kidney cells in the first run and rat and egg albumins in the second; each lasted about seven hours. McDonnell Douglas, the developer of the CFES, claimed that the system produced high-purity samples with a density 463 times greater than what could be produced by similar systems on Earth.

The MLR produced quantities of latex microspheres of great uniformity in two of four reactors. About 55 percent of the chemicals reacted as planned. These were used as seeds for experiments on growing even larger microspheres on later flights.

The Get-Away Special experiment did not function. Although it was activated and the computer ran, no data were recorded because the tape recorder failed. Most of the other experiments failed as well. Some were found to have blown fuses or lost batteries before launch, and some were frozen by the extended bottom-sun attitude, which caused payload temperatures to drop lower than anticipated.

The NOSL experiment produced several minutes of film showing lightning from above the cloud tops. Results from the Air Force instruments were largely classified. The handheld photography of the shuttle glow effect returned a number of photographs of the tail surfaces of the shuttle glowing at night. The photographs were partially censored to prevent the Air Force payload from being seen.

An important goal of the STS-5 mission was the successful deployment of the two satellites. The first use of the deployment equipment provided valuable information on the performance characteristics of that equipment and of the shuttle. SBS-3 was released from the shuttle within 1,500 meters of its target, and Anik/Telesat within 150 meters.

Data on motion sickness were collected throughout the mission. The two astronauts studied provided the extremes of no effect (Allen) and vomiting (Lenoir). In addition, initial testing of the new spacesuits gathered valuable data for future flights. Although the spacewalk was canceled, the suits were tested in low-pressure situations and worked well.

Experiments from SSIP provided useful data about convection within the surface of a heated liquid, about the cellular communication process that causes separated sponge cells to group together to form sponges, and about crystal formation in space. The last experiment was expected to provide information useful in commercial processing of materials in space.

Data were collected that would improve the performance of the shuttle on future flights. The braking system was deliberately given a severe test after the shuttle had landed.

Context

The third test flight of *Columbia* was successful from almost all perspectives. For the first time, the shuttle launch occurred on the originally scheduled date. Its experimental payload collected large quantities of useful data. Initial steps toward materials processing succeeded, and the electrophoresis experiment laid the groundwork for experiments that would repeatedly ride the shuttle. The first experiment designed by a high school student was used on the flight and generated good data. The problems that arose during the mission were minor and had no effect on the accomplishments of the crew and the shuttle.

STS-4 confirmed the space shuttle's capability to operate as a reusable spacecraft. The long bottom-sun attitude confirmed that the shuttle indeed warps slightly with uneven heating. Although the effect was observed on STS-3 during top-sun studies, some engineers had believed that foreign matter had been jamming the doors. With the flight of STS-4, the heat-warping theory was proved. The solution, in any event, was the same: The orbiter is slowly rolled for 15 minutes to even out the heat load. This procedure is now standard in pre-entry preparation.

The failure of the Get-Away Special experiment demonstrated that even though the program may seem amateur to NASA, it still required as much careful handling and testing as any "professional" experiment in order to work properly in space. More important, the experience gained by the students showed it to be a valuable tool for educating students about careers in space. GAS 1 also set the pattern for many of these payloads to be purchased by civic and professional groups and given to students.

The CFES purification runs confirmed that the absence of gravity's effects would eliminate convective flow caused by electrical currents. In like manner, the MLR showed that a commercially salable material could be manufactured in space, although the value had to be high to justify the small quantities that could be accommodated.

Data from the shuttle glow photography were somewhat inconclusive but showed the need for further investigation. The spectral grating over the camera lens showed that the shuttle glow phenomenon grows stronger toward the infrared end of the spectrum. The shuttle windows, however, are designed with a strong infrared cutoff to prevent sunlight entering the windows from overloading the environmental system. Thus, the photographs were cut off just at the point where they may have supplied useful data. They did, however, reveal that the glow was brightest on the "ram" side, where the exposed surfaces of the shuttle struck the thin upper atmosphere.

As the flights of the shuttle *Columbia* began in April, 1981, there had been many who were skeptical about the value of a reusable space vehicle and many who believed that the goals of the shuttle program could not be achieved. STS-1 had been extremely successful. Performance of crews preparing the shuttle for its next flight improved with each mission. Launch on the scheduled day at the scheduled time was achieved. Yet, even after the four successful test flights of *Columbia*, many people were not committed to the shuttle program.

STS-5 was probably the most significant manned spaceflight for the United States (and for NASA) since the landing of men on the Moon in 1969. The successful deployment of two satellites was a forceful rebuttal to the critics who had once said that no meaningful cargo would be hauled by the shuttle, that no one would buy space on it, and that it would not perform as predicted.

In spite of the importance of the mission, STS-5 passed relatively quietly. There were several reasons for its modest reception. First, there were no significant problems in the countdown, and the launch proceeded uneventfully. Second, neither the shuttle nor the crew had major difficulties during the flight. Third, shuttle flights were becoming more common, and general interest of the news media and the public was not as intense as it had been for STS-1.

Each of these factors was considered positive by NASA. The long-range goal of the shuttle program was to have a fleet of shuttles fly every two weeks. Routine, uneventful shuttle missions had to become a reality before the fleet could be built to its desired size. Clearly, STS-5 was a major step in the direction of routine shuttle flight.

Bibliography

Allen, Joseph P., and Russell Martin. *Entering Space: An Astronaut's Odyssey.* New York: Stewart, Tabori and Chang, 1984. This book describes a typical space shuttle flight in addition to other aspects of space exploration. Includes excellent illustrations. Intended for a general, nontechnical audience.

Baker, David. *The History of Manned Space Flight.* New York: Crown Publishers, 198. Rev. ed., 1985. A comprehensive history of manned missions, this large volume is generously illustrated. It offers the beginner an overview of human spaceflight and includes a helpful index.

Bond, Peter. *Heroes in Space: From Gagarin to Challenger.* Oxford, England: Basil Blackwell, 1987. A refreshing change from more technical overviews of the space program, this book focuses on the personal aspects of spaceflight: the men and women who risk their lives in space. Biographical profiles of astronauts and cosmonauts, as well as essays on space programs and a summary of all manned spaceflight from 1961 to the spring of 1986.

Cassutt, Michael. *Who's Who in Space: The First Twenty-five Years.* Boston: G. K. Hall and Co., 1987. A comprehensive work covering manned spaceflight, this book includes essays on various programs, biographical sketches organized by nation, and capsule summaries of all manned flights from April, 1961, to April, 1986.

Covault, Craig. "Shuttle Gears Towards Operational Era." *Aviation Week and Space Technology* 117 (July 5, 1982): 18-19. Detailed article highlighting operational aspects of the STS-4 mission. Written for those familiar with spaceflight.

Dooling, Dave. "Space Shuttle Columbia Passes Its Final Exam." *Space World.* 5-10-226 (October, 1982): 10-13. General article on the STS-4 mission with day-by-day summaries of activities. Written for the educated reader with an interest in space.

————. "USAF Cargo for Space Shuttle." *Space World.* 5-6-222 (June/July, 1982): 9-11. Provides overview of U.S. Air Force plans for payloads to fly aboard the space shuttle.

Furniss, Tim. *Manned Spaceflight Log.* Rev. ed. London: Jane's Publishing Co., 1986. Informative summaries of each manned spaceflight from Vostok 1 to Soyuz T-15. The information includes crew members, launch sites, and the significant accomplishments of each mission.

Time-Life Book Editors. *Life in Space.* Boston: Little, Brown and Co., 1984. In addition to general coverage of other major U.S. spaceflights, this book details the first six space shuttle missions. Illustrated.

Vogt, Gregory. *The Space Shuttle: Projects for Young Scientists.* New York: Franklin Watts, 1983. In keeping with the goal of encouraging high school students to become involved in the exploration of space, this book provides examples of successful experiments designed by students. Step-by-step project development is discussed.

Yenne, Bill. *The Encyclopedia of U.S. Spacecraft.* New York: Exeter Books, 1985. Reprint, 1988. The section on the space shuttle covers the first seventeen flights. Includes a general description of the vehicle and its many functions. Suitable for the lay reader.

Dennis R. Flentge (STS-3)
Dave Dooling (STS-4)
Dennis R. Flentge (STS-5)

SPACE SHUTTLE FLIGHTS, 1983

Date: April 4 to December 8, 1983
Type of mission: Manned Earth-orbiting spaceflight

A busy year, 1983 was filled with "firsts" for the fledgling space shuttle fleet. The second orbiter, Challenger, *would begin its all-too-short career. America would see its first woman astronaut, African American astronaut, non-American astronaut, dual spacewalk, and fully equipped, spaceborne science laboratory.*

PRINCIPAL PERSONAGES

PAUL J. WEITZ, STS-6 Commander
KAROL J. BOBKO, STS-6 Pilot
DONALD A. PETERSON, STS-6 Mission Specialist-1
F. STORY MUSGRAVE, STS-6 Mission Specialist-2
ROBERT L. CRIPPEN, STS-7 Commander
FREDERICK H. "RICK" HAUCK, STS-7 Pilot
JOHN M. FABIAN, STS-7 Mission Specialist-1
SALLY K. RIDE, STS-7 Mission Specialist-2
NORMAN E. THAGARD, STS-7 Mission Specialist-3
RICHARD H. TRULY, STS-8 Commander
DANIEL C. BRANDENSTEIN, STS-8 Pilot
DALE A. GARDNER, STS-8 Mission Specialist-1
GUION S. BLUFORD, STS-8 Mission Specialist-2
WILLIAM E. THORNTON, STS-8 Mission Specialist-3
JOHN W. YOUNG, STS-9 Commander
BREWSTER H. SHAW, STS-9 Pilot
OWEN K. GARRIOTT, STS-9 Mission Specialist-1
ROBERT A. PARKER, STS-9 Mission Specialist-2
ULF D. MERBOLD, STS-9 Payload Specialist-1
BYRON K. LICHTENBERG, STS-9 Payload Specialist-2

Summary of the Missions

The third year of space flight activity for the shuttle program introduced the second orbiting member of the fleet. *Challenger* was not a new space vehicle, having seen life as a so-called static test article during 1978. It was later retrofitted with all of the necessary equipment to make it space worthy and rolled out of the assembly plant in 1982.

Challenger's maiden flight began on April 4. Its mission was to deploy the first Tracking and Data Relay Satellite (TDRS) — the first link in the National Aeronautics and Space Administration's (NASA) spaceflight tracking and communications network. This system would replace NASA's aging ground tracking stations.

In addition to the deployment of TDRS-1, two of *Challenger's* astronauts performed the first Extravehicular Activity (EVA or "spacewalk") without the aid (or encumbrance) of umbilical cords attached to their vehicle. *Challenger* landed at Edwards Air Force Base on April 9, after a successful mission lasting 5 days, 23 minutes, and 42 seconds.

Much attention was paid to Sally Ride when she was assigned to be the first woman to ride the shuttle. It had been twenty years since the first woman, Soviet cosmonaut Valentina Tereshkova, had flown into space aboard Vostok 6. Although she flew as part of a double flight, with Valeri Bykovsky being launched first in Vostok 5, more emphasis was placed on her presence in space. The time lapse between Tereshkova's flight and that of the second Soviet spacewoman, Svetlana Savitskaya in Soyuz T-7 in 1982, seems to add credence to the common belief in 1963 that Tereshkova's flight was another of Nikita Khrushchev's attempts to outshine the United States. Bykovsky had set a record during his flight, spending nearly five days in orbit alone, a fact lost in history. Ride, an astronaut since 1978, had been Capsule Communicator (CapCom) on STS-2 and STS-3. In the eyes of the news media, Ride's presence on STS-7 overshadowed the rest of the crew, as well as the goals of the mission.

The Commander of the flight was U.S. Navy Captain Robert L. Crippen, the only veteran astronaut of the crew, having been the Pilot on the first space shuttle mission in April, 1981. He became a NASA astronaut in 1966, transferring from the canceled United States Air Force Manned Orbiting Laboratory program. His pilot was U.S. Navy Captain Frederick H. Hauck. He and the rest of the crew were selected as astronauts in 1978. In addition to Ride, two other mission specialists were aboard, U.S. Air Force colonel John Fabian and a medical doctor, Norman Thagard.

Challenger blasted off from Launch Complex 39A at 7:33 A.M., eastern daylight time, on June 18, 1983, and traveled flawlessly into space. During the spacecraft's first orbit of Earth, its Orbital Maneuvering System (OMS) engines placed the orbiter in a nearly circular 297-kilometer orbit.

The main activity for the first day was the launching of the Canadian Anik/Telesat, a telecommunications satellite tucked inside *Challenger's* payload bay. When *Challenger* was at the proper position in space, Anik was sent on its way at 5:02 P.M. *Challenger* maneuvered away from the satellite and forty-five minutes after deployment, the first stage of the

McDonnell Douglas Payload Assist Module, Delta class (PAM-D) engine fired. Three days later, the second-stage engine placed the Anik into geosynchronous orbit (that is, an orbit where the satellite's velocity and altitude make it appear to hover over one spot on Earth's surface).

On the second day of the mission, the Indonesian Palapa-B communications satellite was successfully deployed at 9:33 A.M. and placed into geosynchronous orbit. Palapa, like Anik, was deployed from the orbiter within 0.15 degree of its planned location, well within the orbiter specification of 2.0 degrees. Once Palapa was well on its way, the Remote Manipulator System (RMS) arm and the Shuttle Pallet Satellite (SPAS-01) were prepared for their upcoming activities.

On the third day, experiments with SPAS attached to the RMS were conducted, as were other experiments carried in the orbiter's middeck and in the payload bay. SPAS was the first satellite tailored specifically to the reusability of the shuttle. It was developed by a West German firm and carried ten experiments furnished by West Germany, the European Space Agency (ESA), and NASA.

In *Challenger*'s cargo bay was the OSTA-2 (sponsored by NASA's Office of Space Science and Applications and named for the acronym of that office's predecessor organization, the Office of Space and Terrestrial Applications). OSTA-2 was the first NASA materials processing payload to fly in the shuttle's payload bay and was a cooperative project with the West German Ministry of Research and Technology. It was completely automated and comprised four instrument packages containing six experiments designed to study fluid dynamics, transport phenomena, and metallurgy. The instrument packages were attached to a frame-like structure called a mission-peculiar equipment support structure (MPESS).

Three German materials processing payloads included in OSTA 2 were in Get-Away Special (GAS) canisters. Two of the canisters contained "stability of metallic dispersions" experiments, while the third held the Particles at a Solid-Liquid Interface (PSLI) experiment. The NASA materials processing payload element, the Materials Experiment Assembly (MEA), contained two general-purpose rocket furnaces and the Single-Axis Acoustic Levitator (SAAL).

The next day, the Continuous Flow Electrophoresis System (CFES) completed three runs to separate materials to purity levels four times higher than those possible on Earth. The CFES functions on the principle that molecules of substances with different structures and sizes contain different electrical charges and, therefore, can be separated within an electrical field. While electrophoresis can be performed under normal Earth gravity, heavier molecules tend to settle to the bottom of the device. There is none of this settling in the microgravity of orbital flight.

Activities on the fifth flight day began when SPAS-01 was gently lifted from its berth in the payload bay and deployed. Once it had deployed the satellite, *Challenger* moved away to fly in formation with it. For the next 4.5 hours tests were conducted, first with SPAS-01 about 300 meters away and then at distances of about 60 meters. During these activities, cameras mounted on SPAS were used to photograph *Challenger*, providing the first pictures of an entire orbiter in space.

Challenger was scheduled to make the first landing at Kennedy Space Center, but poor weather conditions there resulted in a decision to delay the landing for two orbits. The landing took place at 6:57 A.M., Pacific daylight time, on the runway at Edwards Air Force Base in California. *Challenger* and its crew had spent 6 days, 2 hours, 23 minutes, and 59 seconds in flight and traveled more than 4 million kilometers.

STS-8 would mark the first night launch and landing of a shuttle. Mission commander for STS-8 was U.S. Navy captain Richard H. Truly, making his second spaceflight. Truly had been an astronaut with the National Aeronautics and Space Administration (NASA) since 1969 and was the Pilot on STS-2 in November, 1981. U.S. Navy commander Daniel C. Brandenstein, the Pilot, was making his first spaceflight. Brandenstein was selected as a NASA astronaut in 1978 and had served as Capsule Communicator (CapCom) between the shuttle orbiter and Mission Control in Houston on STS-1 and STS-2.

The three Mission Specialists on the flight were also space rookies. U.S. Navy Lieutenant Commander Dale A. Gardner was selected as a NASA astronaut in 1978 and had served on the support crew for STS-4. U.S. Air Force lieutenant colonel Guion S. Bluford, who became an astronaut in 1978, would make history by becoming the first African American to fly in space. William E. Thornton, a medical doctor, was selected as a NASA scientist-astronaut in 1967. At fifty-four, Thornton would become the oldest person to fly in space up to that time. He was a support crew member for the three Skylab missions and was a crew member on the Skylab Medical Experiments Altitude Test (SMEAT), a 56-day Skylab simulation mission made in 1972 to collect medical data and evaluate equipment.

STS-8 lifted-off at 2:32 A.M., eastern daylight time, August 30, 1983. It was only the second night launch in the history of the United States' manned space program. Apollo 17 had been launched at night in 1972.

Thunderstorms menaced Launch Complex 39A, where *Challenger* stood, and delayed the launch 17 minutes into the launch window, at 2:32 A.M. Night turned to day for a short time as *Challenger* lifted from its launchpad. A beautiful sight in daylight, this nighttime shuttle launch was simply magnificent. The two Solid-fueled Rocket Boosters (SRBs) and three main engines lit the area enough for viewers to read a book at the press site, some 5 kilometers away.

The ascent itself was perfect. *Challenger* entered a 277-kilometer circular orbit with an inclination to the equator of 28.45 degrees. Only after the flight was it discovered that the carbon lining in one of the two SRB nozzles had come dangerously close to burning completely through. Had this occurred, hot gases would have escaped through the hole in the nozzle, which could have sent *Challenger* veering off course.

During the first orbit, the crew communicated with Mission Control through the Tracking and Data-Relay Satellite (TDRS) for the first time ever. The quality of communications through TDRS was reported to be as good as any in the shuttle program. *Challenger* would test TDRS throughout the mission. TDRS would be vital for transmitting the large volumes of data anticipated during the first Spacelab flight late in 1983. The completed TDRS system would consist of three nearly identical satellites and would provide the shuttle with communications, both audio and video, over approximately 80 percent of its ground track.

The only deployable payload of the mission was the Indian National Satellite (INSAT-1B), which provides radio and television communications and carries data collection equipment for weather surveillance and forecasting. INSAT-1B was deployed at 3:49 A.M., August 31, and was placed into a geosynchronous orbit at 74 degrees east longitude.

During the first two days in orbit, Bluford and Gardner operated the Continuous Flow Electrophoresis System (CFES), which was being flown on its fourth shuttle mission. CFES is a commercial project sponsored by McDonnell Douglas, in conjunction with Johnson & Johnson. Electrophoresis is a process of separating materials by passing them through an electrical field. In zero gravity, electrophoresis results in much purer materials. For the first time, on STS-8, live cells were used as samples. Six samples were run through CFES — two each of kidney, pituitary, and pancreas cells. The samples were removed from the shuttle immediately after landing for analysis.

During the third and fourth days in flight, the crew maneuvered the Payload Flight Test Article (PFTA) at the end of the Remote Manipulator System (RMS). The PFTA, which resembled a large dumbbell, was 6 meters long and 4.7 meters wide. Weighing 3,855 kilograms, it was the heaviest object to be lifted thus far by the RMS. The tests went well, with the crew observing how the elbow, wrist, and shoulder joints of the robot arm functioned with a large mass at the end of it. As had been the case on earlier shuttle missions, the arm worked perfectly. The PFTA was not released from the RMS, however, as it was a passive object with no propulsion system to control it.

A student experiment involving the ability to use biofeedback techniques in zero gravity was conducted by Thornton. During four tests, his skin-surface temperature, heart rate, skin response, and muscle activity were measured. Six rats were carried into space in the Animal Enclosure Module (AEM), which was being tested for the first time. The purpose of the test was to demonstrate the ability of the AEM to support live animals without compromising the health or comfort of the crew. The rats behaved well and remained healthy throughout the flight. Four small, self-contained Get-Away Special payloads flew in the payload bay. These included a cosmic-ray experiment designed to help determine the effect of these rays when they pass through memory cells. A photographic

emulsion experiment evaluated the effect of the orbiter's gaseous environment on film emulsion. This would be important for future missions requiring highly sensitive cameras and films. A Japanese snow crystal experiment that did not function properly on STS-6 flew on this mission; the experiment was designed to determine whether snow can be made in space. A contamination monitor was flown to test the reaction of different materials to atomic oxygen molecules, which exist just outside Earth's atmosphere.

Eight Get-Away Special canisters in the payload bay carried approximately 260,000 United States Postal Service philatelic covers commemorating the twenty-fifth anniversary of NASA. These covers were sold to collectors by mail order after the flight. The money from the sale of the covers was split evenly between NASA and the U.S. Postal Service.

Thornton conducted tests throughout the flight concerning space adaptation syndrome, or space sickness. These tests involved monitoring brain waves and observing the nervous system, the eyes, and fluid shifts within the body. NASA was trying to determine why so many astronauts suffer from space sickness and why ailments vary from person to person.

After six days in orbit, *Challenger* landed at Edwards Air Force Base in California; it was the first night landing of the shuttle program. Xenon floodlights illuminated runway 22 at Edwards as Truly and Brandenstein brought *Challenger* to a pinpoint landing at 12:40 A.M., Pacific daylight time, on September 5, 1983.

The ninth space shuttle mission, STS-9 saw the first flight of Spacelab, built by the European Space Agency (ESA). It also marked the return of *Columbia* to space after a year of modifications following the STS-5 mission. Spacelab was housed in the payload bay of the orbiter. The module was 4 meters in diameter and 7.5 meters in length. The crew entered through a connecting tunnel 1.1 meters in diameter that led from the air-lock module of the orbiter to the Spacelab, where crewmen could work in their shirt-sleeves because of the module's pressurized environment.

Columbia was originally scheduled to launch Spacelab 1 on October 28, 1983. When the problem with insulation in one of the recovered STS-8 Solid-fueled Rocket Booster (SRB) nozzles was discovered, STS-9 was already on the launch pad. The SRBs were closely checked, and a similar problem was discovered on one of the STS-9 nozzles. The decision was made to roll *Columbia* back to the vehicle assembly building to replace the suspect nozzle assembly, as a precaution. The rollback caused a two-month delay of the launch.

Spacelab 1 marked the first American spaceflight to conduct twenty-four-hour around-the-clock operations. The six-man crew was split into the red shift and the blue shift, each working twelve-hour days. The red shift was led by John W. Young, making a record sixth spaceflight. He had flown on Gemini 3, Gemini X, Apollo 10, Apollo 16 and STS-1. He was also chief of the astronaut office. Joining Young on the red shift was Mission Specialist Robert A. Parker, making his first spaceflight. Selected as an astronaut in 1967, Parker had been a support-crew member for Apollo 15 and Apollo 17 and had served as a Program Scientist for the Skylab missions. Payload Specialist Ulf Merbold from West Germany was the third member of the red team. He became the first non-American to fly aboard an American spacecraft. Merbold was a materials engineer in Stuttgart, West Germany. He was selected as a Payload Specialist by the ESA in 1978.

The blue shift was led by U.S. Air Force major Brewster H. Shaw, the mission's pilot, who was making his first space-flight. He had become an astronaut in 1978 and had served as a capsule communicator on STS-3 and STS-4. Mission Specialist Owen K. Garriott joined Shaw on the blue team. Garriott was making his second spaceflight, having spent 59 days in space aboard the second Skylab mission in 1973. The third blue team member was Payload Specialist Byron K. Lichtenberg of the Massachusetts Institute of Technology. He had been selected as a Payload Specialist in 1978 and was making his first spaceflight.

STS-9 lifted off on schedule from its launch pad at 11:00 A.M., eastern standard time, November 28, 1983. The ascent was perfect, and *Columbia*, with Spacelab aboard, was placed into a 250-kilometer circular orbit inclined to the equator at 57 degrees. Less than two hours after launch, Garriott, Merbold, and Lichtenberg floated through the tunnel to begin activating Spacelab and its experiments. Once Spacelab was activated, control of the mission switched from the Johnson Space Center in Houston, Texas, to the Marshall Space Flight Center in Huntsville, Alabama.

This first Spacelab mission tested the module in five major science areas: Earth studies, space physics, astronomy, life science, and materials science. Tests on human physiology and biology occupied most of the first two days, with much time spent studying the human body and its ability to adapt to weightlessness. The main experiments in this area involved a crew member placing his head in a dome with a pattern of rotating dots giving him the sensation of spinning. (It is believed that vestibular responses to body movement and acceleration are associated with rotational eye movements.) The crew member's eye rotation was measured and recorded in the dome. Eye rotation was also measured while a subject was spinning in a device called a body restraint system.

Another experiment, known as the "hop and drop," tested the influence of weightlessness on basic postural reflexes nor-mally experienced on Earth. This experiment was done by placing elastic cords connected to the floor around a crew member's shoulders and having him hop up and down while blindfolded, in an effort to determine changes in the inner ear. The "drop" portion of the experiment involved falling with the aid of elastic cords pulling the subject toward the floor, while nerves in his lower leg were electrically stimulat-ed. The subject's physiological responses to this stimulation were monitored and recorded, providing data on responses to brief periods of acceleration during weightlessness.

Materials processing began on the third flight day, with Merbold switching on three high-temperature furnaces. A materials sciences facility contained equipment shared by investigators from ten European countries. Two of the fur-naces had to be repaired by the crew, but all three brought back excellent data on the study of crystal growth, fluid physics, and the processing of metals and glass in space.

The far ultraviolet space telescope did not work as expect-ed because of fogging on its film. It was at first believed that the fogging was caused by the shuttle glow phenomenon — which occurs when the shuttle itself gives off light interfer-ence — but it was later determined that two bands of ionized oxygen circling the globe caused the problem. These bands were first photographed from the Moon by an ultraviolet camera aboard Apollo 16 in 1972. A wide-angle camera pho-tographed deep-space targets, including a thin band of stars and gas between the Magellanic Clouds. An X-ray spectrome-ter observed the Perseus cluster of galaxies and measured X-ray emissions from the Cassiopeia A supernova remnant.

The flight was proceeding so well, and the data return was so massive, that NASA officials extended the mission by one day. The crew spent part of the extra day conducting experi-ments, and then they packed and stowed equipment in prepa-ration for return to Earth.

Four hours before the de-orbit burn to bring *Columbia* home, two of the orbiter's five on-board computers shut down. NASA wanted time to study the problem, so the land-ing was delayed. Because of the delay, *Columbia* made its first-ever descent to Edwards from the north. The reentry took *Columbia* northward over eastern China and the Soviet Union, then over the Aleutian Islands at Mach 23, or 23 times the speed of sound. Young and Shaw piloted the heaviest landing payload of the shuttle program to a perfect touch-down on runway 17 at Edwards Air Force Base at 3:47 P.M., Pacific standard time, December 8, 1983.

Knowledge Gained

Challenger's maiden flight increased NASA's capabilities to deliver payloads to space — especially the large ones it would need to continue its exploration of the solar system and to study the Earth. Many of *Challenger's* systems were improved versions of what had gone into its predecessor. *Columbia* was "overbuilt" to withstand the stresses of flight determined by calculation and wind tunnel testing. The overestimation of some of these stresses was demonstrated in the Approach and Landing Tests and the early flights of *Columbia*. By lightening the structure of *Challenger* (and future orbiters), larger and heav-ier payloads could be carried without sacrificing performance.

The first spacewalks of the Shuttle era demonstrated that astronauts could become self-contained spacecraft, working free of their vehicle. This capability would lead to work on satellites and the Space Station without astronauts getting tangled by air and communications lines. These first spacesuits with interchangeable parts permitted EVAs by any astronaut, since each suit did not have to be customized for an individual. This would also cut some of the expense of these extravehicular activities.

For more than twenty years, NASA's ground tracking system relied on the kindness of foreign heads of state to permit the operation of the giant antennas. The system used antiquated equipment and did not cover every part of a spacecraft's orbital track. The TDRS system would utilize only two active satellites, yet would provide greater coverage of a mission. More important, it permitted communications with manned spacecraft reentering the atmosphere. A returning spacecraft creates a layer of ionized air molecules as it plunges back into the atmosphere. This caused the familiar "blackout" of communications during one of the most critical phases of the flight. A TDRS orbiting 35,800 kilometers above Earth is able to communicate with a reentering spacecraft because the ionization layer forms on the leading (or downward) edge of the craft.

Norm Thagard's main objectives for the STS-7 mission were to gather information on motion sickness and cardiovascular deconditioning countermeasures. Many of the astronauts and cosmonauts who have flown in space suffer from a malady similar to seasickness. After a few days, they become accustomed to the weightless conditions and can function normally. Experiments on space sickness have been flown on subsequent shuttle flights and, although a "cure" has not been found, methods of lessening its effect on crewmen have worked. One such remedy is to have the crewman move slowly and deliberately, while using visual cues for establishing "up" and "down."

When a person has been in space for a while, the cardiovascular system (heart, veins, and arteries) becomes "lazy" since it does not have to fight gravity as it does on Earth. Thagard conducted experiments to find ways of keeping the system in shape. These experiments, along with those conducted on subsequent shuttle flights, have shown that exercise is necessary even in space. By walking on a treadmill, an astronaut can give his cardiovascular system a workout and resume his duties with renewed energy.

Evaluation of the OSTA 2 flight hardware and experiments showed that the payload's objectives were met. The MPESS was demonstrated to be a very effective carrier for both shuttle payload integration and flight operations. The MEA and experimental furnaces functioned as expected and provided good specimens in the NASA materials processing investigations. The SAAL malfunctioned and was only a partial success. It was designed to process up to eight samples that are injected sequentially by a carousel mechanism into a furnace chamber and levitated by an acoustic energy field to hold the sample free of the container. The German materials processing payloads were evaluated and considered to have performed satisfactorily. These experiments help scientists better understand how a variety of metals and liquids react to the environment of space, an understanding that will aid in the development of structures and propellants for future

spaceflight, as well as newer and better medicines, materials, and energy sources to be used on Earth.

As on STS-6, the Monodisperse Latex Reactor (MLR) experiment was carried in the middeck area of *Challenger* for STS-7. It used the weightless conditions of spaceflight to study the feasibility of making monodisperse (identical size) spheres of a plastic-like substance called latex. These spheres may have major medical and industrial research applications. They can also be manufactured larger and more economically in space once an MLR is placed permanently in orbit aboard the international space station.

Above all, STS-8 demonstrated the capability of launching and landing the space shuttle at night. Launch windows, the period of time during which a mission may be launched, are determined by many factors, including payload requirements and lighting conditions at emergency landing sites. The capability to launch whenever desired and to land in the dark gave the shuttle greater flexibility.

The continuous flow electrophoresis system separated live cells (from the pancreas, pituitary glands, and kidneys) for the first time. The results of cell purification have become essential to medicine. Insulin from the pancreas is used to aid diabetes, growth hormones from the pituitary gland can be used to treat burns, and an enzyme from the kidney can dissolve blood clots.

India's INSAT-IB was properly deployed in geostationary orbit despite problems after leaving *Challenger*'s payload bay. The folded solar array would not deploy as planned after the satellite reached its orbit. After its initial difficulties, however, the solar array was properly unfolded.

The Tracking and Data-Relay Satellite System (TDRSS) tested well for the first time, although there were problems with ground equipment in White Sands, New Mexico. The S-band link was used for voice, commands, and telemetry data. The Ku-band link provided high-quality television pictures.

The animal enclosure module was tested successfully with six rats carried aboard. The rats seemed to adjust well to their new environment, although no tests were run on the animals. Their food consisted of nutrient bars developed at Ames Research Center, and their water was obtained through raw potatoes.

Thornton's medical tests designed to collect data on physiological changes associated with space adaptation syndrome recorded symptoms ranging from dizziness and mild nausea to vomiting. The space sickness usually subsided in a day or two, but sick time in flight is extremely expensive. The hope was that enough would be learned about space sickness to find ways to prevent it in the future.

Columbia and its Spacelab 1 cargo gathered two trillion bits of digital scientific data, more than twenty million frames of television data, and several thousand photographs to document the flight's science output. The Spacelab module itself proved to be a flexible tool for a multitude of science studies. STS-9 was really an experimental flight to determine how useful Spacelab could be.

The TDRSS was instrumental in allowing the crew to

send long transmissions of real-time data to ground scientists. With communication continuous for up to 40 minutes at a time, the crew and ground controllers were able to converse at length concerning the experiments on board. This resulted in more data and allowed the crew to receive instructions on the maintenance and repair of some equipment that had malfunctioned.

A grille spectrometer, used for the first time in space, detected gases thought not to be present at certain upper levels of the atmosphere. Carbon dioxide in the thermosphere and water vapor in the mesosphere were discovered. The materials processing experiments produced protein crystals several hundred times larger than those possible to manufacture on Earth. In the area of vestibular experiments, tests using the rotating dome revealed that humans rely heavily on vision for orientation in space. As the body adjusted to space, the motion in the dome gave the person the impression he was moving in the opposite direction.

The most startling data of the mission involved tests concerning the inner ear. Researchers varied the temperatures of a subject's ears and discovered that, with his ear temperatures varied, a subject experiences the sensation of rotating when he is actually sitting still.

In addition to the scientific experiments, engineering tests were run on the Spacelab itself. The orbiter was turned at different angles toward the Sun to bake and cool the Spacelab in an effort to see how the module reacted to temperature variation. The crew also participated in a fire drill, scrambling quickly out of Spacelab in a test to determine how much time would be needed to evacuate the module in case of emergency. Both crew and ground personnel learned much about around-the-clock operations — including ways to accommodate sleeping crew members while still accomplishing work.

Context

Challenger began its life as a static test article, destined to be a landlocked spacecraft sitting in some museum. Fate, and Congressional cuts, provided it with a ticket to space and, eventually, an unwanted place in history. During its maiden flight on STS-6, *Challenger* became the first operational orbiter. *Columbia*, it predecessor, was built to be a test vehicle. *Challenger* weighed nearly 1,200 kilograms less than *Columbia* and had improved systems. It would complete eight more missions before meeting its end in January, 1986, as the orbiter on the STS 51-L mission.

Materials processing experiments carried aboard *Challenger* in future Spacelab missions would begin making some of the spage-age products only thought possible before STS-7. Spacelab is built by the European Space Agency and consists of interchangeable modular units. By combining these units, different experiments can be performed in the shuttle's payload bay. Most of these experiments are too large to be run inside the crew compartment.

Crippen became the first astronaut to fly the shuttle into space four times. In addition to STS-1 and STS-7, he commanded STS 41-C in April, 1984 (which featured the first in-orbit repair of an ailing satellite), and STS 41-G. The latter marked the first flight of a seven-person crew, including two female astronauts, Ride and Kathryn Sullivan (who performed the first spacewalk by an American woman).

By the time STS-8 flew in August, 1983, the Space Transportation System had already proved valuable. NASA had scheduled many space shuttle missions requiring precise launch times, not all of which would occur during daylight. With a flawless night launch and landing, STS-8 paved the way for future missions and gave NASA a boost of confidence that the shuttle was, indeed, a flexible vehicle. The ability to schedule launches regardless of lighting constraints was a major step forward in the shuttle program.

Furthermore, the testing of the TDRS was the first step toward elimination of the costly ground stations located around the world along the orbital path of the shuttle. With one TDRS, *Challenger* was able to receive and transmit high-quality data continuously for more than 40 minutes at a time. With a second TDRS placed in orbit, shuttle communications would be continuous for more than 80 percent of each orbit. An improved version of the TDRS would be vital for transmitting and receiving an even greater amount of data.

Guy Bluford became the first African American to travel into space, but he did not garner as much attention as had Sally Ride two months earlier as the first American woman in space (on STS-7).

William Thornton, by becoming the oldest person to fly at age fifty-four, broke another barrier. Age apparently is not a factor in spaceflight, provided the person is in good health. Thornton was not affected by spaceflight any more or less than his younger colleagues. He downplayed any rigors associated with flying in space. Again, space was being opened up to many people regardless of age.

Prior to STS-9, no manned spaceflight had ever been attempted that required so much interaction among different countries. During STS-9, a laboratory built by the ESA was flown inside a spacecraft built by the United States. Although the flight was not controlled in Europe after the science operations began, as would be the case with later Spacelab flights, the amount of planning and execution for the multitude of experiments on Spacelab ushered in a new era of international cooperation in space.

Spacelab 1 was an introduction to permanent scientific laboratories in space. Future space stations will undoubtedly use the lessons learned in Spacelab concerning everything from the types of experimental payloads to the coordination of twenty-four-hour work schedules. The space shuttle was stretched to its endurance limits — especially since its mission was extended by one day. The ten-day STS-9 mission was the longest shuttle mission up to that time and came close to the maximum duration possible for a shuttle flight without extensive modifications to the orbiter. Despite computer problems prior to the de-orbit burn, *Columbia* demonstrated the shuttle's flexibility by flying five extra orbits and still executing a perfect landing.

The value of having a crew on board was demonstrated on STS-9. The crew was called upon repeatedly to repair equipment and alter procedures to ensure proper data collection. The fluid physics module, high rate data recorder, materials science rack, and metric camera were all repaired by the crew— in addition to the computers. The interaction between crew and ground in devising methods of fixing

equipment was essential to the success of the mission.

The Spacelab program was created in December, 1972, by the European Space Research Organization (ESRO), predecessor to the European Space Agency. In August, 1973, Spacelab became a confirmed ESRO program and an integral part of the Space Transportation System. Ten years later, the planners saw their dreams come to fruition as the concept of a space laboratory worked in a manner never before demonstrated. Spacelab, a prelude to future space stations, was now a viable concept for space science, representing a first step toward a permanent human presence in space.

Bibliography

Bond, Peter. *Heroes in Space.* New York: Basil Blackwell, 1987. A fast-paced, readable account of every manned space mission up through the *Challenger* explosion. Contains few illustrations, but the appendix lists all American and Soviet manned spaceflights through May, 1987. A good primer for anyone first learning about manned spaceflight.

Furniss, Tim. *Manned Spaceflight Log.* Rev. ed. London: Jane's Publishing Co., 1986. This comprehensive, nontechnical volume covers spaceflight since Yuri Gagarin's mission in April, 1961. Presents a concise profile of flights by astronauts, cosmonauts, and space travelers from the other nations. Contains black-and-white photographs.

————. *Space Flight: The Records.* London: Guinness Superlatives, 1985. A listing of every Soviet and American manned spaceflight in chronological order, with a paragraph synopsis of each flight. Also contains Guinness records on spaceflight and a detailed account of every person who had flown in space, through May, 1985.

————. *Space Shuttle Log.* London: Jane's Publishing Co., 1986. An in-depth look at the space shuttle and its first twenty-two flights. Design concepts for the Space Transportation System, as well as an overview of the space shuttle systems, are covered in the first sections. Also included is a concise accounting of each shuttle flight, along with photographs from the missions.

Kerrod, Robin. *Space Shuttle.* New York: Gallery Books, 1984. Written for general audiences. This book is replete with color photographs of the space shuttle that convey well the essence of this marvelous vehicle and the people who fly it. The components of the Space Transportation System, and highlights of the first dozen missions, are presented, as well as a fanciful look toward the future of space exploration.

Otto, Dixon P. *On Orbit: Bringing on the Space Shuttle.* Athens, Ohio: Main Stage Publications, 1986. Dixon examines how the early designs of the space shuttle developed into the actual flying machine and gives an account of each of the first twenty-five flights. Included are the names of the crew members, the principal payloads, and the objectives of each flight. The book contains a large number of black-and-white illustrations to highlight each mission.

————. *On Orbit: The First Space Shuttle Era, 1969 – 1986.* Athens, Ohio: Main Stage Publications, 1986. Beginning with approval of the shuttle design in 1972, this book covers every shuttle flight through the *Challenger* explosion. Includes a few color and many black-and-white photographs. Mission objectives — along with crew members, launch date, landing date, duration, orbits, and cargo weight — are listed for each flight. Suitable for general audiences.

Ride, Sally, with Susan Okie. *To Space and Back.* New York: Lothrop, Lee and Shepard Books, 1986. The personal account of life aboard the space shuttle during STS-7, this book is directed to a younger audience. Readers of all ages will find the color photographs breathtaking. The adventure begins on the morning of launch and uses photographs from many shuttle flights for illustration. A glossary of terms is provided to make the reading easier.

Smith, Melvyn. *An Illustrated History of Space Shuttle: X-15 to Orbiter.* Newbury Park, Calif.: Haynes Publications, 1986. A concise, highly-detailed overview of the space shuttle and the experimental aircraft that led to its design, the book spans the period from 1959 through 1985. Contains many rare photographs of the early shuttle lifting bodies, which help to show how the shuttle orbiter came to look as it does. Information is presented chronologically.

Wilson, Andrew. *Space Shuttle Story.* New York: Crescent Books, 1986. Through more than one hundred color photographs, the space shuttle's history from the early days of rocketry to the destruction of *Challenger* is told. Aimed at a general audience, this volume emphasizes the men and women who fly the spacecraft.

Yenne, Bill. *The Astronauts.* New York: Exeter Books, 1986. Directed toward the general audience, this volume details the manned space programs of the United States and the Soviet Union. It is illustrated with several hundred color and black-and-white photographs taken aboard the spacecraft. Although the book does not give the complete personal backgrounds of every space traveler, it does provide insight into human travel in space.

————. *The Encyclopedia of U.S. Spacecraft.* New York: Exeter Books, 1985. Reprint, 1988. A complete, nontechnical reference of spacecraft built and launched by the United States. Contains photographs of each of the spacecraft, most of them in color. The book gives details of the spacecraft, as well as launch dates and the disposition of the craft. Also covers the American launch vehicles and contains a glossary of acronyms and abbreviations.

Russell R. Tobias (STS-6)
Russell R. Tobias (STS-7)
Christopher F. Dickens (STS-8)
Christopher F. Dickens (STS-9)

SPACE SHUTTLE
MISSION STS-6

Date: April 4 to April 9, 1983
Type of mission: Manned Earth-orbiting spaceflight

Space Transportation System mission 6, the maiden voyage of the space shuttle Challenger, *deployed the first Tracking and Data-Relay Satellite. The extravehicular mobility unit, used for the first time, permitted two astronauts to work in space unencumbered by oxygen hoses and communication lines attached to their spacecraft.*

PRINCIPAL PERSONAGES
    PAUL J. WEITZ, Mission Commander
    KAROL J. BOBKO, Mission Pilot
    F. STORY MUSGRAVE and
    DONALD H. PETERSON, Mission Specialists

Summary of the Mission

The "shakedown cruise" of the space shuttle *Challenger*, Space Transportation System (STS) 6 was designed to test the second orbiter in the National Aeronautics and Space Administration (NASA) fleet and to deploy a satellite that would be the first link in a continuous communications system for future missions. The flight marked the first time a manned mission was flown without a flight-tested vehicle and with no means of escape if a problem arose during launch.

The Commander of STS-6 was Paul Weitz, a retired Navy captain who was selected with the fifth group of astronauts in April, 1966. The only experienced member of the crew to have flown in space, he had been the pilot of the 28-day Skylab 2 mission, which was launched May 25, 1973. The STS-6 Pilot was United States Air Force colonel Karol J. Bobko. He was transferred from the Air Force's canceled Manned Orbiting Laboratory program in September, 1969. He served in a support capacity for the Skylab and Apollo-Soyuz programs, as well as for the shuttle approach and landing tests.

Two Mission Specialists were on the flight, Dr. Story Musgrave and retired U.S. Air Force colonel Donald Peterson. Musgrave became an astronaut in 1967, served as backup Science Pilot for Skylab 2, and was the Capsule Communicator for the second and third Skylab missions. He participated in the design and development of all space shuttle Extravehicular Activity (EVA) equipment, including life-support systems, spacesuits, air locks, and manned maneuvering units. Peterson became an astronaut in September, 1969, and served on the astronaut support crew for Apollo 16.

The major payload for the flight was the Tracking and Data-Relay Satellite (TDRS), the largest privately owned telecommunications satellite ever built. TDRS-A was the first of three identical spacecraft planned for the Tracking and Data-Relay Satellite System (TDRSS) to be placed into geosynchronous orbit (that is, an orbit where the satellite's velocity and altitude make it appear to hover over one spot on Earth's surface). TDRS-B would be placed into orbit by STS-8, while TDRS-C would be deployed by STS-12 and would be an in-orbit spare that could be moved to replace either of the other satellites.

The two operational satellites would be in orbit 130 degrees apart to permit the use of a single ground station (located in New Mexico), instead of the two needed if they were 180 degrees apart. Initially, the TDRSS was used to support space shuttle flights, providing a communications link for about 98 percent of each orbit. Without the TDRSS, only about 30 percent of an average orbit is covered, because of the scarcity of ground tracking stations. Later, TDRSS would support other spacecraft, including Landsat and the space station *Freedom*.

Challenger, designated STA-099, was built as a static test article in 1976. It was named for the American research vessel that made extensive oceanographic cruises of the Atlantic and Pacific oceans between 1872 and 1876. Being ships of the new ocean of space, each of the orbiters was named for a famous oceangoing vessel. In addition, *Challenger* was the name of the last lunar module to land on the Moon during the Apollo 17 mission. The spacecraft are designated as Orbiter Vehicles, or OVs. Thus *Enterprise*, the first Orbiter Vehicle to be built, is designated OV-101, *Columbia* is OV-102, *Discovery* is OV-103, and *Atlantis* is OV-104. Although *Enterprise* was designated as an Orbiter Vehicle, it was never designed to go into space. Instead, it was used to test the gliding and landing characteristics of the orbiter's design.

Early in 1978, *Challenger* was completed and rolled out from Rockwell International's main orbiter assembly plant in Palmdale, California, to the Lockheed-California test building across the road. There, engineers bonded approximately 2,700 pads to *Challenger's* aluminum structure so that load jacks could apply stresses which simulate the complex dynamic environment that the orbiter would experience during the different phases of flight. Stresses up to 140 percent of the design limits could be applied, but researchers limited the stress they applied to 120 percent. They wanted to test the structure and still ensure a safe margin so that *Challenger* could be used for actual space missions. In order to save the expense of building a fifth orbiter, something not provided for in NASA's budget, NASA and Rockwell agreed to move STA-099 back to its assembly bay. During 1979, modifications were made to equip it for spaceflight. On June 30, 1982, *Challenger* (redesignated OV-099) was rolled out of the Palmdale plant, transported overland to the nearby NASA Dryden Flight Research Facility at Edwards Air Force Base, loaded atop the Boeing 747 shuttle carrier aircraft, and flown to the Kennedy Space Center in Florida. STS-6 was scheduled for launch on January 27, 1983, having been stacked and rolled out to Launch Complex 39A at Kennedy Space Center in late November. Serious problems with the Space Shuttle Main Engines (SSMEs), however, delayed the launch until April, 1983.

Prior to the first flight of any orbiter, the SSMEs are qualified by means of a Flight Readiness Firing (FRF). The three SSMEs, each providing 1,670 kilonewtons of thrust at lift-off, burn for approximately 8 minutes during launch. The FRF is a 20-second test, performed with the shuttle on the launch-pad, which ensures that the SSMEs are functioning properly. The FRF also allows for a complete rehearsal of the activities leading to a launch, without involving the crew or igniting the Solid-fueled Rocket Boosters (SRBs).

After the FRF, which took place on December 18, 1982, engineers found a high concentration of hydrogen gas near the aft end of the orbiter, where the SSMEs are located. Liquid hydrogen is the fuel for the SSMEs and the presence of the gas after the firing indicated a possible leak in one or more of the engines. A second FRF was ordered for January 26, 1983.

After this second firing, a crack was found in the hydrogen line of one of the engines. In order to fix the problem, the engine had to be replaced, causing a one-month delay. A liquid oxygen leak was discovered in the replacement engine. This delayed the launch until mid-March. In early March, cracks were found in the hydrogen lines of *Challenger's* other two main engines. At the same time, several strong storms blew into the area, causing the TDRS to become contaminated with dust and sand. Cleaning of the satellite and repair of the engines pushed the launch date back to April 4.

On April 4, 1983, at 1:30 P.M. eastern standard time, *Challenger* was launched. The launch went flawlessly and, after two short burns of the orbital maneuvering system engines, STS 6 was safely in a circular orbit at an altitude of 265 kilometers. Preparations for the deployment of TDRS and its Inertial Upper Stage (IUS) propulsion unit were immediately begun. Ten hours after lift-off, the 22-meter-long, 17,000-kilogram combination was raised 59 degrees. Small springs gave the TDRS/IUS a gentle push and sent it on its way as *Challenger* was 2,300 kilometers east of a point above Rio de Janeiro.

The TDRS deployment plan called for the first stage of the two-stage IUS to burn for 2 minutes and 31 seconds, placing the high point of the orbit (the apogee) at 35,888 kilometers. This stage would be dropped and the second stage would burn for 1 minute and 43 seconds, circularizing the orbit at the apogee. Once in its geosynchronous orbit, *Challenger* would jettison its second stage, and TDRS would open up like an awakening butterfly.

Approximately 80 seconds into the second-stage burn, an oil-filled seal on the steering mechanism of the stage's engine deflated. This caused the engine to remain at an extreme angle and sent the spacecraft tumbling. Quick action by the controllers allowed TDRS to be separated from the IUS before the gyrations tore the spacecraft apart. TDRS was left nearly powerless, in an orbit 35,021 kilometers by 22,013 kilometers.

In the meantime, *Challenger* continued on its successful flight. On April 7, the first shuttle Extravehicular Activity (EVA), or spacewalk, began with Musgrave exiting the orbiter through the air lock on the lower deck at 4:21 P.M. He and Peterson had donned their Extravehicular Mobility Units (EMUs), or spacesuits, and spent 3.5 hours in the air lock breathing oxygen to purge nitrogen from their blood. After floating out into *Challenger's* payload bay, they attached tethers to wires running the length of the bay on either side. These would prevent the pair from accidentally drifting off into space. Handrails along the perimeter of the bay were used to travel to various workstations. All planned tasks for the EVA were successfully performed, and the EVA was completed at 8:15 P.M.

The final full day in orbit was spent stowing equipment and experiments for reentry and landing. Of the thirty test objectives scheduled for the flight, only one could not be accomplished. The de-orbit burn to slow *Challenger* down for reentry was performed on schedule, and the spacecraft glided in for a landing on the dry lake bed at Edwards Air Force Base in California.

At 11:54 A.M. Pacific standard time on April 9, 1983, *Challenger* rolled to a stop some 12 kilometers beyond where its main wheels had touched down. STS-6 had come to an end. The flight had lasted 5 days, 23 minutes, and 42 seconds and covered some 3.4 million kilometers in orbit.

Knowledge Gained

The successful deployment of the TDRS showed that large payloads could be transported to low-Earth orbit by the shuttle. This would prove valuable for satellites too big for conventional launch vehicles. Despite the problems TDRS had after deployment, the crew of STS-6 had released it at the proper time and place.

About six hundred of the bulky heat-resistant tiles that cover the upper surfaces of *Columbia* had been replaced by Advanced Flexible Reusable Surface Insulation (AFRSI) blankets on *Challenger*. Each blanket is made of a sewn com-

Views of EVA performed during STS-6. (NASA)

posite quilted fabric with the same silica material as the tiles sandwiched between the inner and outer layers. By replacing the bulky tiles, technicians could reduce the overall weight of the spacecraft by approximately 900 kilograms. This decrease in orbiter weight allowed for heavier payloads to be taken up to orbit and returned.

The cockpit featured new "heads up" visual displays, which project essential flight control information for landing onto a clear screen between the pilots and the front windows. This setup permitted them to monitor the information while continuing to watch where they were going.

The 4 hours and 10 minutes that the two mission specialists spent in *Challenger's* cargo bay gave the new spacesuits an excellent test. The suits, unlike the ones used in prior space programs, are not custom-made for each astronaut. Instead, they are "off the rack." Their main components — the hard-shell torso, helmet, flexible arms and gloves, pants, and boots — come in various sizes. An individual's suit is fitted to him

or her, allowing hundreds of different suit combinations and saving millions of dollars in building customized suits for each astronaut; this approach also permits every astronaut the opportunity to perform a spacewalk.

The two spacesuits performed perfectly, adjusting for increased activity and, consequently, increased oxygen consumption. Even though the suits were pressurized, they were still flexible enough to permit freedom of movement without overexertion. Tools could be used easily, and hardware such as screws, nuts, and bolts could be removed and replaced. These spacesuits would make possible the in-orbit repair of satellites and, if necessary, the repair of an orbiter itself.

One of the experiments carried on board *Challenger*, the Continuous Flow Electrophoresis System (CFES), verified that the device would separate materials to purity levels four times higher than those possible on Earth. The CFES works on the principle that molecules of substances with different structures and sizes contain different electrical charges and, therefore, can be separated within an electrical field. Although electrophoresis can be performed under normal Earth gravity, heavier molecules tend to settle to the bottom of the device. In the microgravity of orbital flight, there is none of this settling.

Another experiment, the Monodisperse Latex Reactor (MLR) experiment, used near weightlessness of spaceflight to study the feasibility of making monodisperse (identically sized) spheres of a plasticlike substance called latex. These spheres may have major medical and industrial research applications.

Context

STS-6 was the sixth flight of the Space Transportation System, the second operational flight and the first flight of an operational orbiter. For its first five missions, *Columbia* was an experimental spacecraft, requiring heavy ejection seats (available for use only on the first four flights) and structural supports for such things as the landing gear doors, main engine thrust frames, and propellant tanks. These heavier items were necessary because of the uncertainty of the stresses to which the orbiter would be subjected. As an operational spacecraft, *Challenger* could do away with these heavy structures, thereby trimming 1,128 kilograms.

Ground controllers were able to save TDRS through a series of maneuvers over a 58-day period. Small thrusters on the satellite, designed only for small orbital corrections, were fired thirty-nine times for a total of 40 hours of burn time. By performing the maneuvers at precisely the right point in the orbit, the ground controllers could raise the lowest point of its orbit (perigee) to equal the apogee. TDRS was in place and was used during STS-9 to relay data and voice communications to and from ground stations. The IUS problem was resolved — but not in time to permit the launch of the second TDRS on STS-8. The IUS problem also caused the postponement of the STS-10 flight, a Department of Defense mission that would use the IUS to deploy a military satellite. Eventually, the STS-10 mission was flown (as STS 51-C), and the IUS performed perfectly. TDRS-B was scheduled for the

STS 51-E mission, but the flight had to be canceled because of internal problems with TDRS; the TDRS-B was in the payload bay of *Challenger* for the ill-fated STS 51-L flight.

In thirty-three months of service, *Challenger* flew ten missions, traveling more than 41 million kilometers in orbit. In addition to the work done on STS-6, *Challenger* accomplished many impressive feats. STS-7 saw the first flight of an American woman in space in June, 1983. In August, STS-8 featured the first black astronaut and the first night launching and landing of the shuttle (which was also the first night landing of any American manned spacecraft).

Challenger's fourth mission, STS 41-B, in February, 1984, included the first use of the Manned Maneuvering Unit (MMU). The MMU is the Buck Rogers-style back-pack that astronauts use during an EVA to travel to and from the orbiter and other spacecraft. At the conclusion of this mission, *Challenger* made the first landing at Kennedy Space Center. It was the first time a spacecraft landed at the same location from which it had been launched.

On its next flight, STS 41-C, *Challenger's* crew used the MMU to fix the Solar Maximum Mission satellite (better known as Solar Max). This was the first time an ailing satellite had been retrieved, repaired, and replaced into orbit. STS 41-G marked the first flight of a seven-person crew, including two women astronauts, Sally Ride (becoming the first American woman to fly twice into space) and Kathryn Sullivan (who performed the first EVA by an American woman).

Three Spacelab missions were flown on *Challenger*: in April, 1985, in July, 1985, and the first German Spacelab mission, in October, 1985. Spacelab was built by the European Space Agency and consists of interchangeable modular units. By combining these units, different experiments can be performed in the shuttle's payload bay. Most of these experiments are too large to be run inside the crew compartment.

Challenger's last and, unfortunately, its most memorable flight was STS 51-L. On January 28, 1986, the spacecraft and crew of seven were lost when a leak in one of the joints of its right-hand SRB caused the vehicle to break up during launch.

Bibliography

Allen, Joseph P., and Russell Martin. *Entering Space: An Astronaut's Odyssey*. New York: Stewart, Tabori and Chang, 1984. This personal account of a space shuttle flight describes the tension of the countdown and launch, the fun and difficulties of living and working in space, the awe-inspiring views of Earth from above, and the fiery return to the ground. Included are more than two hundred color photographs taken during many shuttle and pre-shuttle spaceflights. The experiences of other astronauts are presented in an effort to present a balanced view of space travel. Written for general audiences.

Furniss, Tim. *Manned Spaceflight Log*. Rev. ed. London: Jane's Publishing Co., 1986. A complete, nontechnical look into each manned spaceflight since Yuri Gagarin was launched into space in April, 1961. The book presents a concise look at flights by American astronauts, Soviet cosmonauts, and space travelers from the other nations who have flown with them. Each flight is listed with specific data, black-and-white photographs, and an unbiased account of the major events of the mission.

————. *Space Shuttle Log*. London: Jane's Publishing Co., 1986. An in-depth look at the space shuttle and its first twenty-two flights, from STS-1 through STS 61-A. The first part of the book examines the design concepts for the Space Transportation System and provides an overview of the space shuttle systems. The remainder gives a concise account of each shuttle flight, along with black-and-white photographs from the missions.

Hallion, Richard. *On the Frontier*. NASA SP-4303. Washington, D.C.: Government Printing Office, 1984. Part of the NASA History Series, the book explores the experimental aircraft that provided the necessary research for the development of the space shuttle. It is very readable and contains many photographs of these pioneer aircraft. The author chronicles the evolution of these craft from the first plane to break the sound barrier through the space shuttle orbiter. A complete annotated listing of the flights of each of the vehicles, as well as an extensive bibliography, completes this authoritative work.

Kerrod, Robin. *Space Shuttle*. New York: Gallery Books, 1984. Written for general audiences, this book is full of color photographs of the space shuttle. The author includes very little text because the illustrations convey so well the essence of this marvelous vehicle and the people who fly it. The components of the Space Transportation System are detailed, and highlights of the first dozen missions are presented. Also included is some speculation on the future of space exploration.

Otto, Dixon P. *On Orbit: Bringing on the Space Shuttle*. Athens, Ohio: Main Stage Publications, 1986. The author takes a look at the early designs of the space shuttle and how they developed into the actual flying machine. Then he gives an account of each of the first twenty-five flights, including the names of the crew members, the principal payloads, and the objectives of the flight. The book contains many black-and-white illustrations to highlight each mission.

Smith, Melvyn. *An Illustrated History of Space Shuttle: X-15 to Orbiter*. Newbury Park, Calif.: Haynes Publications, 1986. A concise, highly detailed look at the space shuttle and the experimental aircraft that led to its design, the book spans the period from 1959 through 1985. Written for the general audience, it contains many rare photographs of the early lifting bodies, which help to show how the shuttle orbiter came to look as it does today. Organized chronologically.

Wilson, Andrew. *Space Shuttle Story*. New York: Crescent Books, 1986. More than one hundred color photographs are used to trace the history of the space shuttle from the early days of rocketry to the destruction of *Challenger*. This clearly written book is aimed at a general audience. Little detail is given on the technical aspects of the shuttle; the emphasis is on the men and women who fly the spacecraft.

Yenne, Bill. *The Astronauts*. New York: Exeter Books, 1986. This book, directed toward a general audience, presents an account of the manned space programs of the United States and the Soviet Union. It is illustrated with several hundred color and black-and-white photographs taken aboard the two countries' spacecraft. It also tells about the non-American and non-Soviet passengers who flew as guests on some of these flights. The book does not give the complete personal backgrounds of each astronaut and cosmonaut, but it does provide an intriguing insight into the human aspect of space travel.

————. *The Encyclopedia of U.S. Spacecraft*. New York: Exeter Books, 1985. A complete, nontechnical reference on the manned and unmanned spacecraft built and launched by the United States. Photographs of each of the spacecraft, most of them in color, accompany the text. The book gives details of the spacecraft, as well as launch dates. It also provides details of the American launch vehicles and a glossary of acronyms and abbreviations.

Russell R. Tobias

SPACE SHUTTLE FLIGHTS, 1984

Date: February 3 to November 16, 1984
Type of mission: Manned Earth-orbiting spaceflight

The National Aeronautics and Space Administration (NASA) had a very productive year in 1984. It achieved a number of firsts, overcame a few nagging problems, and displayed the ability of humans to play an important role in space flight with the capture and return of two failed satellites.

PRINCIPAL PERSONAGES

VANCE D. BRAND, STS 41-B Commander
ROBERT L. GIBSON, STS 41-B Pilot
BRUCE MCCANDLESS, STS 41-B
    Mission Specialist-1
ROBERT L. STEWART, USA, STS 41-B Mission
    Specialist-2
RONALD E. MCNAIR, STS 41-B Mission Specialist-3
ROBERT L. CRIPPEN, STS 41-C Commander and
    STS 41-G, Commander
FRANCIS R. "DICK" SCOBEE, STS 41-C Pilot
GEORGE D. NELSON, STS 41-C Mission Specialist-1
TERRY J. HART, STS 41-C Mission Specialist-2
JAMES D. VAN HOFTEN, STS 41-C Mission Specialist-3
HENRY W. HARTSFIELD, STS 41-D Commander
MICHAEL L. COATS, STS 41-D Pilot
RICHARD A. MULLANE, STS 41-D Mission
    Specialist-1
STEVEN A. HAWLEY, STS 41-D Mission Specialist-2
JUDITH A. RESNIK, STS 41-D Mission Specialist-3
CHARLES D. WALKER, STS 41-D Payload Specialist-1
JON A. MCBRIDE, STS 41-G Pilot
KATHRYN D. SULLIVAN, STS 41-G Mission
    Specialist-1
SALLY K. RIDE, STS 41-G Mission Specialist-2
DAVID C. LEETSMA, STS 41-G Mission Specialist-3
PAUL D. SCULLY-POWER, STS 41-G Payload
    Specialist-1
MARC GARNEAU, STS 41-G Payload Specialist-2
FREDERICK H. "RICK" HAUCK, STS 51-A,
    Commander
DAVID M. WALKER, STS 51-A, Pilot
JOSEPH A. ALLEN, STS 51-A, Mission Specialist-1
ANNA L. FISHER, STS 51-A, Mission Specialist-2
DALE A. GARDNER, STS 51-A, Mission Specialist-3

Summary of the Missions

By the end of 1983, NASA determined that it had become necessary to find a new "numbering" system for the space shuttle flights. While the first nine went aloft in sequence, STS-10 was delayed due to the problems with the boosters of satellites similar to the one scheduled to be deployed during the mission. This moved flight eleven into the tenth spot and shifted the others up one place. Foreseeing the possibility of other such shifts, NASA developed a system to identify a mission by its payload. Each mission would carry a two-number and one-letter designation, based upon the originally scheduled launch date. The first number would represent the last digit in the fiscal year of the launch. NASA's fiscal year runs from October 1 through September 30. The second digit would designate the launch site: "1" for the Kennedy Space Center and "2" for Vandenberg Air Force Base. The letter would indicate the chronological order of the flight during that fiscal year. Thus, the first mission of Fiscal Year 1984 (which was STS-9) would be STS 41-A.

Challenger's STS 41-B mission began at 8:00 A.M. on February 3, 1984, from the Kennedy Space Center, Florida. On board was a crew of five, led by Mission Commander Vance Brand, a veteran of STS-5. He was accompanied by rookies Robert L. "Hoot" Gibson, the pilot, and Bruce McCandless, Ronald McNair, and Robert Stewart, the mission specialists.

They carried with them into orbit two communication-satellites for deployment, which were released without mishap. Unfortunately, the attached Payload Assist Module, Delta class (PAM-D), rocket motor on each malfunctioned and left the satellites in useless orbits well below their intended geosynchronous places above the equator. The West German Shuttle Pallet Satellite (SPAS-01A) was on its second mission, having been deployed and retrieved during STS-7. This time it was not deployed because of an electrical problem with the Remote Manipulation System (RMS) arm.

The highlight of the mission was the first test of the Manned Maneuvering Unit (MMU), a jet-propelled back-pack. Bruce McCandless became the first human Earth-orbiting satellite when he ventured 15 meters from the orbiter using the MMU. He and Robert Stewart later practiced the maneuvers necessary for the retrieval and repair operation of the Solar Maximum Satellite planned for the next flight.

STS 41-B landed on schedule at Kennedy Space Center on February 11, after a flight lasting 7 days, 23 hours, 15 minutes, and 55 seconds. *Challenger* had traveled more than 5.3 million kilometers.

NASA's most daring shuttle mission to date began on the morning of April 6, as *Challenger* was launched for the STS 41-C mission and placed into a 491 kilometer orbit. The next day, the remote manipulator system released the Long-Duration Exposure Facility (LDEF). LDEF, a large test platform covered with 57 experiments, mounted in 86 removeable trays, was scheduled to be retrieved in September, 1986. Launch delays and the STS 51-L accident caused LDEF to remain is orbit for more than six years.

On the third day of the flight, *Challenger's* orbit was raised to 502 kilometers for the rendezvous with the ailing Solar Maximum Mission (Solar Max) observatory. Decked-out in his MMU, George Nelson attempted to grapple the satellite without success. After rethinking their plans and restabilizing the satellite overnight, NASA gave the okay to try the capture again. This time Nelson and van Hoften succeeded. Solar Max was then berthed in its special cradle in *Challenger's* payload bay for repairs. After several of its systems were replaced, Solar Max was redeployed on April 12, the third anniversary of the first shuttle mission. *Challenger* glided to a landing at Edwards Air Force Base on April 13, having completed a mission lasting 6 days, 23 hours, 40 minutes, and 7 seconds.

STS 41-D saw the maiden voyage of the orbiter *Discovery*, named for Captain James Cook's British ship that made voyages of discovery in the Pacific Ocean in the 1770's. The crew for the mission was commanded by Henry Hartsfield, a veteran astronaut who had served in a support capacity for the Apollo 16 lunar landing mission and for the three Skylab orbital flights. For 41-D, the pilot was U.S. Navy Commander Michael Coats, who had become an astronaut in 1978. The three mission specialists for the flight were from the same astronaut class, U.S. Air Force lieutenant colonel Richard Mullane, Steven Hawley, and Judith Resnik.

Joining them on the flight was Payload Specialist Charles Walker, chief test engineer for the McDonnell Douglas project on Electrophoresis Operations in Space (EOS). He was the first nonprofessional astronaut to fly into space under a NASA policy that allows major space shuttle customers to have one of their own people on board to operate their payloads. His major role on the flight was to operate one of the experiments carried onboard, the Continuous Flow Electrophoresis System (CFES). The CFES works on the principle that molecules of substances with different structures and sizes contain different electrical charges and, therefore, can be separated within an electrical field. Although electrophoresis can be performed under normal Earth gravity, in that environment heavier molecules tend to settle to the bottom of the device.

Discovery was to carry Leasat-1, a communications satellite, which measures 4.26 meters in diameter and 6 meters in length when deployed. The Leasat, designed by Hughes Aircraft Company, is unique in that it does not require a separately purchased upper stage for deployment.

A major experiments package on board *Discovery* was the OAST-1, sponsored by NASA's Office of Aeronautics and Space Technology (OAST). OAST-1 consisted of three major experiments: the solar array experiment, the dynamic augmentation experiment, and the solar cell calibration facility. The experiments were carried aboard a frame-like structure called the mission-peculiar equipment support structure (MPESS). Attached to a second MPESS was the large format camera, a high-altitude aerial metric stereographic mapping camera. The 408-kilogram camera can produce 2,400 negatives from 31.7 kilograms of film, including two types each of black-and-white and color.

A problem with one of *Discovery's* general purpose computers caused the June 25 launch to be canceled nine minutes before lift-off was to occur. Everything proceeded normally for launch the next day and the space shuttle main engines (SSMEs) were given the firing command. The engines are ignited in sequence, SSME 3 (engine number 3) first, followed by SSME 2 and SSME 1, at 120-millisecond intervals. SSME 3 ignited and so did SSME 2, but an indication of irregular operation of the main fuel valve of SSME 3 during the start transient (from ignition to full thrust) caused the engine to be shut down by the on-board computers. SSME 2 was commanded off and SSME 1 was never given the command to start. The launch vehicle systems were made safe, and the crew exited the vehicle.

As a result of the aborted launch, the vehicle was moved back to the vehicle assembly building, and SSME 3 was replaced. A decision was made to remanifest the mission, using the same flight crew but combining the payloads from STS 41-F with STS 41-D. Launch of STS 41-D was set for August 29. Leasat-1 was replaced by Leasat-2 at the request of Hughes and scheduled to be launched by STS 51-A. The large format camera was replaced by the two STS 41-F communications satellites, Telstar-3 and SBS-4 (Satellite Business System-4).

After a computer-related one-day delay, STS 41-D was launched on August 30, 1984, at 8:41:50 A.M. and *Discovery* was placed in an early circular orbit at 297 kilometers. The flight proceeded well, with each of the three communications satellites being deployed on time and in the correct location. Subsequently, all three attained geosynchronous orbit. The solar array experiment was extended and tested on two separate days. The flight of STS 41-D ended after 6 days, 56 minutes, and 4 seconds with touchdown of the main landing gear on the dry lakebed at Edwards Air Force Base in California. The spacecraft rolled for approximately 3,130 meters and came to a complete rest at 6:38:54 A.M., Pacific daylight time.

The STS 41-G mission was the thirteenth U.S. space shuttle flight and the sixth for the orbiter *Challenger*. The mission marked the first time seven persons were launched on one spacecraft, the crew consisting of five men and two women, another new feature. Commanding the eight-day flight in Earth orbit was shuttle veteran Robert L. Crippen, on his fourth spaceflight. The pilot was Jon A. McBride, making his first flight into space. The three mission specialists were Sally K. Ride, America's first woman in space, flying her second mission; Kathryn D. Sullivan, on her first flight; and David C. Leestma, also on his first mission. Sullivan and Leestma were to perform a spacewalk during the mission. Two payload specialists were also assigned to this flight: Marc Garneau of the National Research Council, the first Canadian in space, and Paul Scully-Power, an Australian by birth but now a U.S. citizen, a civilian oceanographer employed by the U.S. Navy.

The payload for 41-G consisted of a NASA Office of Space and Terrestrial Applications package (OSTA-3), which included the Shuttle Imaging Radar (SIR-B); the large format camera; the Measurement of Air Pollution from Satellites (MAPS) payload; the Feature Identification and Location Experiment (FILE); the Earth Radiation Budget Experiment (ERBE), which included the ERBE Satellite, the ERBE non-scanner and scanner, and the Stratospheric Aerosol and Gas Experiment (SAGE) 2 ; eight Get-Away Special (GAS) canisters; and the Orbital Refueling System (ORS) experiment.

After one of the smoothest countdowns on record, the shuttle lifted off at 7:03 A.M., eastern standard time, October 5, 1984. The vehicle performed a nominal ascent with a 57-degree heading alignment over the Atlantic. All stages of the ascent went according to plan, placing *Challenger* and its crew into orbit.

Initial activities included the activation of the scientific-payload in the payload bay and the checkout of the robot arm. The ERBE satellite was deployed by the RMS 11 hours and 15 minutes after launch. During the second flight day, Garneau began his series of experiments, and oceanographer Scully-Power made visual observations of the world's oceans. The crew experienced difficulty in directing the on-board Ku-band antenna drive mechanism to lock on to the Tracking and Data-Relay Satellite (TDRS) in order to relay important scientific data obtained by the Earth-pointing instruments in the payload bay back to Earth. Faced with a serious problem that threatened to erase most of the data that had been collected, the crew finally succeeded in working around the problem by maneuvering the whole orbiter to allow the antenna to lock on to the TDRS, thereby saving the all-important data.

Several minor problems plagued the crew during their flight, but none was serious enough to hamper their performance. Both Garneau and Scully-Power and the NASA members of the flight crew conducted extensive visual observations of geological and ocean features during the mission, in particular Sullivan, a qualified geologist, who complemented the observations of oceanographer Scully-Power and the

on-board instruments.

On flight day 7, the crew completed their planned EVA, delayed from flight day 5. Leestma was first out, followed by Sullivan, who became the first American woman to conduct a spacewalk. Her EVA came only three months after Soviet cosmonaut Svetlana Savitskaya became the first female to conduct an EVA, nineteen years after the first spacewalk by Soviet cosmonaut Alexei Leonov. The two astronauts spent the next three and one-half hours working in the open payload bay of *Challenger*, mostly at the aft bulkhead with the equipment that was being developed for future satellite refueling missions. Leestma performed the majority of this work with Sullivan assisting him and taking photographs to record his activities.

Leestma and Sullivan also successfully reconfigured the troublesome Ku-band antenna and inspected the SIR-B antenna. After Leestma performed a somersault from the middle of the payload bay to retrieve a dislodged air lock safety valve cover, the two astronauts stowed equipment and moved back into the airlock, completing their EVA of 3 hours, 27 minutes. Once post-EVA operations had been completed, the crew conducted further simulated fuel transfer experiments.

The crew spent the next day resting and stowing equipment for their return to Earth. The two payload specialists also took advantage of the final full day in space to complete their individual research programs. *Challenger* touched down on runway 13, at the Kennedy Space Center, at 12:26 P.M., eastern daylight time. The flight lasted 8 days, 5 hours, 23 minutes, and 33 seconds. *Challenger* had completed 132 orbits of Earth and traveled nearly 7 million kilometers.

The successful completion of the Solar Max repair during STS 41-C gave NASA great hope of retrieving the two satellites stranded after their deployment on STS 41-B. The flight crew assigned to the STS 51-A mission would also deliver two satellites of their own. One was Leasat-1, similar to the one deployed on STS 41-D, and the other Telesat-H, an HS-376 satellite like the two they were to retrieve. This would be the most ambitious flight to date.

Discovery was launched on November 8, 1984, at 7:15 A.M., eastern standard time. The crew was headed by U.S. Navy Captain Frederick Hauck, who occupied the left-hand seat in the cockpit. In the right-hand seat was Navy Commander David Walker, who served as pilot. Rounding out the crew were the Mission Specialists Joseph Allen, Ph.D., Anna Fisher, M.D., and Navy Lieutenant Commander Dale Gardner. Following a normal ascent to orbit, *Discovery* entered its planned orbit 302 kilometers above Earth.

Before they could attempt the satellite rescues, the crew had to deploy their own satellites. Telesat-H, also known to its Canadian users as Anik-D2, was released on the second day of the flight. Forty-five minutes later, the PAM-D motor fired and sent Anik on its way toward geosynchronous orbit. Leasat-1 rolled out of the payload bay the next day and it, too, achieved its desired orbit. Shortly after that, *Discovery* began a series of maneuvers to catch the errant Palapa B-2 satellite.

On day five of the mission, the crew rendezvoused with Palapa, and Mission Specialists Allen and Gardner performed a spacewalk to capture it and place it in *Discovery*'s payload bay. Difficulties with the RMS during the rescue forced the astronauts to change their procedures for the next retrieval. The next day, they successfully wrestled Westar 6 into the payload bay. Later, the crew posed for pictures with signs reading "2 up, 2 down" and "Ace Repo Co." With their primary objectives behind them, the crew of *Discovery* settled down to complete the remainder of their experiments and "watch the world go by the window."

On November 16, at 6:59 A.M., *Discovery* touched down at Kennedy Space Center. They had accomplished all of their goals during their 7 day, 23 hour, 44 minute, 56 second mission.

NASA ended 1984 with eyes toward the future. A busy year lay ahead with ten flights scheduled, including three with the Spacelab scientific laboratory.

Knowledge Gained

STS 41-B began the year with a flight of fantasy, where science fiction met reality. Two astronauts flew free of their spacecraft in much the way their childhood hero, Buck Rogers, did. The future of space travel, with its space stations and planetary journeys, was closer. The mission, too, would illustrate that getting there would not be an easy thing.

Two nearly-identical satellites were successfully deployed from *Challenger*, but failed to reach their desired orbits due to identical problems with their rocket motors. The problems were easily identified and easily corrected. The temporary blemish on NASA's armour was rapidly removed just a few months later when the satellites were plucked out of orbit and returned to earth on STS 51-A.

STS 41-B carried rats on board to study whether weightlessness relieves the symptoms of arthritis. Six experiments on SPAS-01A investigated materials processing in microgravity. Two experiments on SPAS-01A were designed to measure the drift of a spacecraft and the other to set calibration standards for solar power cells in direct sunlight systems on Earth.

If NASA was to maintain its ambitious flight schedule, it would be necessary to cut down on the turnaround time between missions using the same orbiter. By the time the Space Station would be assembled, NASA had hopes of flying a mission every two weeks — even though it had only four orbiters in the fleet. Part of the delay between flights was created by preparations for and the return of the orbiter to the Kennedy Space Center. By landing at the Kennedy Space Center Shuttle Landing Facility, the delay would be virtually eliminated. STS 41-B proved this was not only possible, but a relatively easy feat to accomplish.

The cost per kilogram for launching the space shuttle is enormous. NASA wanted to make each flight more efficient by packing in as much "paid" cargo as possible. On STS 41-C, *Challenger* carried a large free-flying satellite up to orbit and threw in the rescue and repair of an ailing solar observatory. Lessons learned during the STS 41-B spacewalks and tests of the manned maneuvering unit, led to the ability of an astronaut on STS 41-C to fly out to a satellite, grab onto it, and haul it back to the payload bay.

The Long Duration Exposure Facility (LDEF) was designed to be a cost-effective way of getting a large number of passive experiments into orbit. Once the satellite was deposited into the proper orbit by the shuttle's manipulator arm, it could remain in space for extended periods of time. A later flight would then take up a payload or two and return with LDEF.

The construction of a space station would require many trips by astronauts to the assembly site, leaving the protective cocoon of the orbiter behind. The astronauts would have to use special tools designed to work in the near weightless conditions of low-Earth orbit. The retrieval, repair and redeployment of Solar Max demonstrated that these skills and tools were available. With practice, these techniques could be taught to future generations of astronauts.

All fifteen of STS 41-D's test objectives were met, yet only 85 percent of the CFES samples were processed. Walker proved his value on the flight when he performed in-flight maintenance on the unit and operated the control system manually when difficulties developed in the automated system. Walker returned from orbit with quantities of an unidentified processed hormone to be used in human clinical tests relating to the treatment of diabetes. Unfortunately, the samples later proved to have been contaminated.

The solar array experiments carried active solar cell modules located on panels near the top. All other solar array panels carried dummy cells. It was neither necessary nor cost effective to cover the entire array with active cells. The array extensions and retractions to both the 70 percent and 100 percent levels were successful. The dynamic augmentation experiment showed that the array was able to dampen deflections introduced, well within desired ranges. Calibration of the solar cells by the solar cell calibration facility was satisfactorily completed.

The successful deployment of the Telstar and SBS 4 satellites once again displayed the ability of the Space Transportation System to deliver its payloads on time and with pinpoint accuracy. The problems with the Payload Assist Modules which had prevented previous satellites from reaching geosynchronous orbit had been fixed.

The capability of carrying a large satellite within the payload bay and then gently releasing it was demonstrated. The deployment method was unique, too, for the satellite was mounted horizontally in the orbiter's payload bay. Ejection of the satellite was initiated when locking pins at four contact points were retracted. An explosive device then released a spring that ejected the spacecraft in a motion not unlike the throwing of a Frisbee. This motion gave the satellite the momentum to separate from the shuttle and gyroscopic stability during the forty-five minutes between deployment and ignition of its first-stage motor. The HS-376 satellites were stabilized by spinning them on a platform before ejecting them.

STS 41-G was a science-oriented flight, with a broad range of observations and experiments. Despite the cramped confines of the flight deck and middeck, the crew was able to obtain significant results. As a result of reduced operating time, the SIR-B experiment was able to provide only 40 percent of the images planned. Nevertheless, the black-and-white images produced from millions of microwave radar pulses recorded were far superior to the images produced by SIR-A on STS-2 in 1981.

The large format camera recorded 2,300 photographs of every continent on Earth and provided 23-by-46-centimeter photographs which assisted environmentalists and geologists looking for pollution and new sources of minerals around the globe. In addition, several important photographs were taken of the town of Kyshtym in the Soviet Union, where, during the winter of 1957-1958, a nuclear accident had occurred.

The MAPS experiment provided important information on what effects industrial wastes have on the air after they are released into the atmosphere, taking measurements of the ozone layer from orbit and measuring levels of carbon monoxide in the troposphere. The FILE project continued experiments designed to assist in the production of more reliable and efficient remote-sensing instruments on later satellites and spacecraft.

The ERBE satellite provided data on the levels of solar energy absorbed over the different regions of Earth, as well as levels of thermal energy which the Earth returned to space. This information will provide an understanding of how the natural balance of thermal energy is affected by volcanic dust and industrial and aerosol pollution.

STS 51-A finished the year with a flurry of activity. The well-orchestrated deployment of two different satellites, coupled with the retrieval of the two STS 41-B satellites, proved that the shuttle crew was well-trained and capable of adjusting to problems as they developed. Numerous orbital changes required to accomplish these tasks displayed the astronauts' ability to rendezvous with unmanned spacecraft and operate efficiently within close proximity of their target. This would be necessary if a permanently manned space station was to be built.

Another requirement for building a space station would be the ability of astronauts to maneuver massive objects in space, bring them together with great precision, and fashion them into structures. Although objects in orbit have very little weight — that is, the force of gravity pulling them down is nearly canceled by their forward motion — they retain their original mass and respond readily to the laws of motion. It takes a great deal of effort to move massive objects short dis-

tances. An accelerating object is difficult to stop. Untethered tools and small parts drift away without notice.

The 51-A crew, along with the 41-C crew, demonstrated they could grapple satellites and move them where they wanted them to be. They were able to open doors and panels and replace components with ease. Their experiences would lead to the development of tools and work platforms necessary to complete more difficult tasks to come.

Context

The STS 41-B mission led many outside of NASA to criticize the space agency's ability to operate the space shuttle with any reliability. Much of the blame for the failure of the two satellites to achieve orbit was placed on *Challenger* and its crew. Subsequent investigation into the incidents, showed that the failures were caused by hardware failure on the satellites' solid rocket motors. Each satellite was insured, but it would be necessary for NASA to pull off a spectacular "rescue" of these satellites to regain public confidence in the shuttle program.

Much of the disappointment on STS 41-B was overshadowed by the first untethered spacewalk. The Manned Maneuvering Unit demonstrated the capability to propel an astronaut in a controlled manner to an object several meters away from his craft. This capability would be put to use, not only on the next flight, but also on the mission to retrieve the errant satellites. The original concept behind the space shuttle was the use of it in the construction of a space station and in the transportation of materials and personal to and from the station. The deployment of satellites from the shuttle was conceived as a means of providing a purpose for the shuttle until a space station was approved by Congress. The capability to service the space station also meant that malfunctioning satellites could be retrieved or repaired in place. This capability was demonstrated on the STS 41-C mission.

The Solar Maximum Mission spacecraft was one of the first designed with replaceable modules, which provided a means for in-flight repairs. This concept would be used in other spacecraft, including the Hubble Space Telescope. When optical problems were discovered in the Space Telescope, a repair mission was planned to replace some of the malfunctioning components and to fit the telescope with optical corrections. Without this, the science gained from Hubble would have been a fraction of its actual return.

After the spectacular rendezvous, retrieval and repair of Solar Max, the solar observatory was placed back into orbit. It continued its mission of collecting data on solar flares and other solar activity until November, 1989, when it reentered the atmosphere and was destroyed. A second rescue mission to raise its orbit had been proposed, but it never came to pass in the wake of the *Challenger* accident in January, 1986.

The techniques for rendezvousing with an orbiting spacecraft were refined during the STS 41-C mission. In addition, tools and equipment for the repair led to the development of hardware to be used in the construction of a manned space station. Some of the tools would be used in the rescue, repair and retrieval of several satellites, including the Hubble Space Telescope.

The Long Duration Exposure Facility was designed to remain in space for a period up to two years. It had been scheduled to be plucked from orbit during the STS 61-I mission in September, 1986. The *Challenger* accident cancelled all flights until 1988. Other payload priorities pushed the retrieval back to January, 1990, when it was captured by the crew of *Columbia* during STS-32.

With the completion of mission STS 41-D, NASA's Space Transportation System fleet of orbiters numbered three. Eventually, *Discovery* was assigned to flight operations at Vandenberg's shuttle launch facility. There, it would be used to launch Department of Defense payloads into polar orbits. The loss of *Challenger* and the subsequent cancellation of shuttle launch activities at Vandenberg freed *Discovery* to return to duty at Kennedy Space Center. *Discovery* was chosen to make the first shuttle flight after the *Challenger* accident.

Inspection of the main engine after the June 26 abort revealed slightly damaged insulation on one of the engine fuel ducts directly above the main fuel valve actuator, leading to speculation that liquid nitrogen leaked on the valve actuator. Single-engine tests at the National Space Technology Laboratories (renamed the Stennis Space Center in 1988) in Bay St. Louis, Mississippi, could not confirm that the insulation damage caused the failure. No exact cause of the failure was revealed, leading NASA officials to believe that it was a momentary clog that led to the shutdown. A year later, during the attempted launch of the STS 51-F/Spacelab 2 mission, another "stuck" valve was found, this time in one of *Challenger*'s main engines; the problem was blamed on an overly sensitive sensor which detected the sluggishly opening valve and sent an abort signal. Judith Resnik, who spent a few terrifying moments during the abort, would be aboard the ill-fated STS 51-L.

The observations from the STS 41-G mission proved that the shuttle was a valuable and reliable platform from which to conduct Earth-observation experiments for up to ten days in duration (at that time, the limit of on-board consumables in the orbiter). The crew's ability to overcome hardware failures

also ensured the success of the flight and once again demonstrated the value of having a human crew on hand to repair and operate equipment. Earlier operations from STS-2 and STS-3 established guidelines for this type of mission, but it was 41-G that finally secured the plans for other Earth resources data-gathering flights later in the program.

Information from the instruments flying on the vehicle was also supplemented by the trained observations of the NASA crew and payload specialists such as oceanographer Scully-Power. Flying payload specialists on the mission made it possible to make expert observations and gather specific data.

The orbital refueling system experiment was also an important step in future refueling operations in space using the shuttle as a space tanker to extend the operational usefulness of satellites in Earth orbit. The Landsat 4 satellite was the first satellite targeted for such a fuel transfer, and the 41-G experiments proved in practice what had been in the planning stages for several years. In addition, the flying of a mixed and international crew in the cramped confines of the orbiter provided valuable information to the designers of the U.S. Space Station and to planners of later Spacelab missions.

Crippen's return to flight after the 41-C mission was useful in determining the optimum level of training needed by a crew commander between flights, as a step toward more frequent flying of the shuttle's crew. Determining the desirability of exchanging crews as opposed to flying a core crew who would remain together as a team on different missions (as the Soviet space station crews did) was another purpose of the experiment.

STS 51-A concluded the year on a high note. The inclusion of astronauts on space missions proved to be valuable during the retrieval of the two HS-376 satellites. Their ability to adjust to problems with equipment and to manually grab the satellites would encourage NASA to attempt other rescues and repairs. This level of confidence would result in successful satellite repairs during STS 51-I, STS-49, and STS-61.

The satellites returned to earth by the crew of STS 51-A would be refurbished and launched into orbit. Westar 6 was sold to the Asia Satellite Telecommunications Co., Ltd. and became AsiaSat 1 following its successful launch atop a Republic of China Long March 3 vehicle on April 7, 1990. Palapa B2, launched six days later on a Delta vehicle, was redesignated Palapa

Bibliography

Allen, Joseph P. and Russell Martin. *Entering Space: An Astronaut's Odyssey.* New York: Stewart, Tabori, and Chang, 1984. This exquisite book was written by one of the STS 51-A mission specialists and includes his firsthand account of the flight. Suitable for all audiences, it describes the experiences of spaceflight, both the routine chores of living and working and the excitement and drama of being in space. It presents over 200 color photographs displaying the beautiful views available in space as well as the activities performed by astronauts.

Chaikin, Andrew. "Solar Max: Back from the Edge." *Sky and Telescope* 67 (June, 1984): 494-497. This article provides an overview of the Solar Maximum Repair Mission, recounting its successes and failures for the general reader.

Cooper, Henry S. F., Jr. *Before Lift-Off: The Making of a Space Shuttle Crew.* Baltimore: The Johns Hopkins University Press, 1987. An account of the preparations undertaken by a group of astronauts and training officers, which result in a crew for a space shuttle mission. The author obtained special permission to document the training program of the 41-G crew, from their selection to just after the mission, and a vivid text which records in detail the arduous training program was the result. The text identifies the problems and successes of the training program and follows the crew, their training team, and development of the mission itself from the autumn of 1983 through to the fall of 1984. Suitable for a general audience.

Furniss, Tim. *Manned Spaceflight Log.* Rev. ed. London: Jane's Publishing Co., 1986. An overview of spaceflight since Yuri Gagarin's pioneering mission in April, 1961. The book covers the flights by American astronauts, Soviet cosmonauts, and space travelers from the other nations who have flown with them. Includes an unbiased account of the major events of each mission.

————. *Space Shuttle Log.* London: Jane's Publishing Co.,1986. An in-depth history of the space shuttle and its flights, from STS-1 through STS 61-A. Features details about the design of the space shuttle vehicle and an overview of the shuttle systems. A concise account of the shuttle flights is provided, along with black-and-white photographs from each mission covered.

MacKnight, Nigel. *Shuttle.* Nottingham, England: MacKnight International, 1985. A history of the space shuttle, this copiously illustrated book traces the design of the vehicle and the various aspects of a typical mission. The author, a British journalist, takes the reader on a tour of the Kennedy Space Center in Florida and shows a shuttle vehicle being prepared for flight, from the stacking of the solid-fueled rocket boosters to the vehicle rollout to the launchpad. Articles about and interviews with astronauts help to emphasize the personal side of each flight.

McMahan, Tracy, and Valerie Neal. *Repairing Solar Max: The Solar Maximum Repair Mission.* Washington, D.C.: Government Printing Office, 1984. Describes the preparations for the Solar Max repair and the people who worked on STS 41-C. Tells of the successful operations and the problems encountered. This reference also discusses the new era of orbital repairs of spacecraft.

Otto, Dixon P. *On Orbit: Bringing on the Space Shuttle.* Athens, Ohio: Main Stage Publications, 1986. Covers the early designs of the space shuttle. An account of each of the first twenty-five flights is given, including the names of the crew members, the principal payloads, and the objectives of each flight. Illustrated.

Powers, Robert M. *Shuttle: The World's First Spaceship.* Harrisburg, Pa.: Stackpole Books, 1979. Despite having been written before the first flight of the Space Transportation System, this book contains excellent descriptions of the space shuttle systems and the types of missions conducted. There are very good explanations of why and how the environment of space is used for a variety of scientific and technological applications. The book includes many paintings and drawings, a glossary, and an index.

Smith, Melvyn. *An Illustrated History of Space Shuttle: X-15 to Orbiter.* Newbury Park, Calif.: Haynes Publications Inc., 1986. A highly detailed overview of the space shuttle and the experimental aircraft which led to its design, the book spans the period from 1959 through 1985. Written for the general audience, it contains many rare photographs. Information is presented chronologically.

Wilson, Andrew. *Space Shuttle Story.* New York: Crescent Books, 1986. Traces the history of the space shuttle from the early days of rocketry to the *Challenger* accident. Aimed at a general audience, Wilson's book contains little detail; the emphasis is on the men and women who fly the shuttle. Contains more than one hundred color photographs.

Yenne, Bill. *The Astronauts.* New York: Exeter Books, 1986. This book presents an account of the manned space programs of the United States and the Soviet Union. Illustrated with several hundred photographs taken aboard the two countries' spacecraft, it also tells about the non-American and non-Soviet passengers who flew as guests on some of these flights. While Yenne does not give the complete personal background of each astronaut and cosmonaut, he does provide an intriguing insight into the human aspect of space travel.

Beth Dickey (STS 41-B)
Robert E. Davis (STS 41-C)
Russell R. Tobias (STS 41-D)
David J. Shayler (STS 41-G)
Marc D. Rayman (STS 51-A)

USA IN SPACE

ALPHABETICAL INDEX

CATEGORY INDEX